Benchmark Papers
in Microbiology

Series Editor: Wayne W. Umbreit
Rutgers—The State University

Benchmark Papers
in Microbiology / 15

A BENCHMARK ® Books Series

RECOMBINANT DNA

Edited By

K. J. DENNISTON

**National Cancer Institute,
National Institutes of Health**

and

L. W. ENQUIST

**National Cancer Institute,
National Institutes of Health**

Dowden, Hutchinson
& Ross, Inc.
STROUDSBURG, PENNSYLVANIA

12/1983
Biol.

Copyright © 1981 by **Dowden, Hutchinson & Ross, Inc.**
Benchmark Papers in Microbiology, Volume 15
Library of Congress Catalog Card Number: 80-14100
ISBN: 0-87933-378-2

83 82 2 3 4 5
Manufactured in the United States of America.

LIBRARY OF CONGRESS CATALOGING IN PUBLICATION DATA
Main entry under title:
Recombinant DNA.
   (Benchmark papers in microbiology; v. 15)
   Includes bibliographical references and index.
   1. Recombinant DNA—Addresses, essays, lectures.   I. Denniston, K. J.
   II. Enquist, Lynn W.
[DNLM: 1. DNA, Recombinant.   W1 BE517M   v. 15 / QU58
R311]
QH442. R37   574.87'3282   80-14100
ISBN 0-87933-378-2

Distributed world wide by Academic Press,
a subsidiary of Harcourt Brace Jovanovich,
Publishers.

CONTENTS

Contents

PART III: BASIC TECHNIQUES

Contents

SERIES EDITOR'S FOREWORD

The decade of the seventies witnessed such remarkable advances not only in our understanding of at least the prokaryotic gene but also, and perhaps more important, in our ability to manipulate and characterize such genetic elements. A major part of these new developments is the concept and technology of "recombinant DNA." Even individuals not working in the field (which constitutes the vast majority of biological scientists) have been markedly influenced by recombinant DNA developments, and new horizons have opened before us. Drs. K. J. Denniston and L. W. Enquist, as active contributors to this field, have collected for us the papers that describe the concepts, the methods, and the opportunities of this new area. One gets a sense of the excitement of the new developments, and their discussions generate a desire to understand these achievements in which we ourselves are not engaged and in which the terminology and even the methodology seem incomprehensible. How, indeed, is one to join DNA molecules, and then prove that one has joined them? How is one to recognize a plasmid? How can a phage carry a "foreign" DNA? How can one identify specific DNA fragments? How can one separate a specific gene from the thousands of others with which it occurs? How can one control what is obtained, lest one make an organism carrying genes that might destroy us? How can one insert an isolated or reconstructed gene into a living organism and have it develop therein? All these questions (and many more) were essentially unanswerable in 1970; but by 1980 there were clear, definitive, positive answers, surely progress that should not go unnoted.

The developments under the general title "recombinant DNA" are known to a relatively few workers in this area of knowledge, but increasingly to a wider group not specifically concerned with microbial and eukaryotic genetics. It is to this group (with a minimum background in genetics) that this volume will be especially useful. These individuals do not have either the time or the knowledge to search the scattered literature and are not likely to respond to "Nonchromosomal Antibiotic Resistance in Bacteria: Genetic Transformation of *Escherichia coli* by R-Factor DNA" (Paper 7), for example. Yet when selected by people who do know its significance and who explain just what its significance was (and is), such a paper is not only read but becomes an important living part of the basic structure of a science. Of course, in almost any vital development, there are many papers on a given facet of the problem. One relies on the editor to select from among them, and, of course, the selections reflect the sometimes subjective judgments of the editors. Needless to say, other papers could have been selected, and an essentially similar pattern would have emerged. There is, however, a new difficulty that has arisen in that certain journal editors will not allow the reproduction of such material (even for a fee), evidently regarding it as a personal possession rather than as a part of science. Authors who publish in such journals face the restriction that their work may never be known to a larger public because other papers will be selected for such reproduction simply because of editorial policy. Editors of Benchmark and similar series are thus not necessarily responsible for the precise papers selected and might have, on occasions, preferred papers whose inclusion was not possible because of the inability to obtain permission for reprinting. This is clearly a restriction on the free flow of scientific information.

W. W. UMBREIT

CONTENTS BY AUTHOR

Contents by Author

RECOMBINANT DNA

INTRODUCTION

A fusion of ideas and techniques gave rise in the early seventies rather explosively to a new era in molecular biology. Daniel Nathans (1979), in his Nobel address, called this era "the new genetics." It is now possible for scientists to manipulate DNA molecules, reintroduce them into cells, and occasionally cross major species barriers. Genes and control elements of higher organisms, once seemingly intractable to molecular methods, can now be analyzed in detail, often at the level of DNA sequence. An integral part of the new genetics is the technology of "recombinant DNA": the joining together of diverse DNA molecules. This terminology, that prior to 1972 would have meant almost nothing, is now a part of the lexicon of the scientific and popular press. Several other terms are often interchanged when describing this technology, including "cloning" and "genetic engineering." Molecular cloning is a fairly accurate description of what happens in recombinant DNA technology in that a single DNA fragment is joined to a self-replicating unit (a "vector," see Part II), and many copies of the original molecule are produced. Unfortunately, the conceptual leap from cloning molecules to cloning individuals is made too easily, and the ethical problems of this latter technology are transferred to the former. At the moment, there is a vast expanse between the two concepts of cloning. Similarly, "genetic engineering" is an accurate, if somewhat imaginative, description of the art of joining DNA molecules together. One need only listen to a discussion of recombinant DNA experiments among a group of scientists using phrases like "blunt-end joining," "trimming back," "filling in," "hooking up," or "flipping around" to visualize some kind of microscopic construction work.

Recombinant DNA experiments gave rise to many other new, nonscientific entities, including: abbreviations, bureaucracy and multiauthored papers. RAC (Recombinant DNA Advisory Committee), ORDA (Office of Recombinant DNA Activities), MUA

1

(memorandum of understanding and agreement), EK1–EK2 (levels of biological containment), P1-P4 (levels of physical containment) and so forth, all are entrenched in the rules and regulations engendered by studies on recombinant DNA.

Recombinant DNA research is not one study or discipline but rather a variety of techniques that can be used for many kinds of experiments in molecular biology. This collection of papers is our attempt to define some of the critical work that created this remarkable technology. Some of the discoveries that form the foundation of this field have already been recognized: witness the 1978 Nobel prizes in Physiology and Medicine to Drs. Werner Arber, Hamilton Smith, and Daniel Nathans and the 1980 Nobel prizes in Chemistry to Drs. Paul Berg, Walter Gilbert, and Frederick Sanger. The field continues to expand rapidly, but the basic concepts set down by the papers contained here remain the backbone of recombinant DNA experiments. Some of the papers are reports of original work that opened up new areas; others are excellent reviews or elegant utilization of the technology. We sought to include those papers that are cited often, even though their methodology may be outdated, simply because they establish valuable technical ideas.

The papers included are for the most part devoted to work done using *Escherichia coli*, its phages and plasmids. Two papers are included describing the development of a recombinant DNA system in yeast. Certainly other systems are available including *Bacillus* systems (Young et al. 1977) and animal virus systems like SV40 (Ganem et al. 1976; Goff and Berg 1976, 1979; Hamer 1977). We chose to emphasize *E. coli* simply because of its central role in the development of recombinant DNA technology. We apologize for the omission of many papers obviously significant or incisive, but the final choice was limited primarily by space and inability to obtain permission for reprinting of some papers. Therefore, the final selection was influenced by our own prejudices.

The commentaries preceding each section are intended to provide continuity in the ideas and to give some historical perspective. Where possible, references to current reviews or recent papers are included, but the literature review is not intended to be exhaustive.

The papers are arranged in five sections. There is considerable overlap between sections, but the organization enables the reader to examine first those investigations that formed the foundations of recombinant DNA technology, to read those papers that detail the concepts and construction of vectors, to study the variety of methods evolving for use of such vectors in recombinant DNA studies, to see how these methods were used to analyze specific genes of complex organisms and, in the last section, to see one of the applications of this technology, namely, the expression of eukaryotic genes in *E. coli.*

REFERENCES

Ganem, D., A. L. Nussbaum, and G. C. Fareed. 1976. Propagation of a Segment of Bacteriophage λ DNA in Monkey Cells after Covalent Linkage to a Defective Simian Virus 40 Genome. *Cell* **7**:349–359.

Goff, S. P., and P. Berg. 1976. Construction of Hybrid Viruses Containing SV40 and λ Phage DNA Segments and Their Propagation in Cultured Monkey Cells. *Cell* **9**:695–704.

Goff, S. P., and P. Berg. 1979. Construction, Propagation and Expression of Simian Virus 40 Recombinant Genomes Containing the *Escherichia coli* Gene for Thymidine Kinase and a *Saccharomyces cerevisae* Gene for Tyrosine Transfer RNA. *J. Mol. Biol.* **133**:359–383.

Hamer, D. 1977. SV40 Carrying an *Escherichia coli* Suppressor Gene. In *Recombinant Molecules: Impact on Science and Society*, ed. R. F. Beers, Jr. and E. G. Bassett, pp. 317–335. New York: Raven Press.

Nathans, D. O. 1979. Restriction Endonucleases, Simian Virus 40, and the New Genetics. *Science* **206**:903–909.

Young, F. E., C. Duncan, and G. A. Wilson. 1977. Development of the *Bacillus subtilis* Model System for Recombinant Molecule Technology. In *Recombinant Molecules: Impact on Science and Society*, ed. R. F. Beers, Jr. and E. G. Bassett, pp. 33–44. New York: Raven Press.

Part I

FOUNDATIONS

Editors' Comments
on Papers 1 Through 9

One far-reaching, but seemingly trivial, fact established prior to the investigations described in the papers reprinted in Part I was that the general structure of DNA from all organisms was more or less the same. It was the arrangement of the four nucleotides that ultimately specified differences between organisms. It is not entirely clear just when someone envisioned joining two unrelated DNA molecules together, but certainly the idea followed from these first principles. However, it was not so obvious how to bring about such joining in a controlled manner. In 1967, the first breakthrough in solving this problem occurred with the discovery of DNA ligases, enzymes that repair nicks in DNA and join DNA molecules annealed by complementary single-stranded ends (Gefter et al. 1967; Olivera and Lehman 1967; Weiss and Richardson 1967; Zimmerman et al. 1967). It was clear that one of the keys to joining unrelated DNA molecules by these new ligase enzymes would be to establish general methods to generate single-stranded, complementary ends in DNA molecules. During this period, the activity of another enzyme, calf thymus terminal deoxynucleotidyl transferase (terminal transferase) was described by Bollum and colleagues (Kato et al. 1967). This provided the second break-though. The enzyme was first characterized by its ability to catalyze stepwise addition of nucleotide residues to the 3' termini of DNA molecules. Papers 1 and 2 describe a general method using terminal transferase and ligase to join DNA molecules together. These seminal papers provided technological and conceptual insights that set the framework for future recombinant DNA work.

Although Paper 1 by Lobban and Kaiser appeared in print in 1973, reports of their experiments were widespread more than a year before. Lobban and Kaiser began with DNA from bacteriophage P22, a molecule that was known to have no single-stranded ends. They used terminal transferase to add stretches of oligo (dA) to the ends of one P22 DNA population and stretches of oligo (dT) to the ends of another P22 DNA population. They then mixed these two populations and showed that the two DNAs could anneal to one another via these synthetic complementary "tails." This technique has been called terminal transferase "tailing." The meticulous attention to detail and characterization of products and by-products in this paper is noteworthy.

Concurrently, in the same department at Stanford University, Jackson, Symons, and Berg (Paper 2) developed a similar system, but their paper contained the seed of more startling experiments, notably the joining of DNA from a eukaryotic DNA tumor virus (SV40) to a DNA segment containing the bacteriophage lambda

DNA replication functions and the entire *E. coli* galactose operon. As these authors noted, this particular recombinant DNA molecule was, in effect, a trivalent biological agent, potentially capable of replication and expression in *E. coli* as well as in eukaryotic cells. While neither group tested the biological activity of these novel recombinants, it is remarkable how clearly each perceived the far-reaching utility of their discoveries. For example, from the discussion of Lobban and Kaiser's paper:

> The apparent generality of the method described here for the joining of any pair of double-stranded DNA molecules to each other may provide a new way to study eukaryotic genes and their controlling elements. Any block of genes from any organism could be inserted into the genome of a temperate bacteriophage to generate a specialized transducing phage. The genes borne by the phage could then perhaps be studied in bacteria, possibly even at the level of fine structure genetic analysis. Alternatively, the phage could be used to produce many copies of the genes to be used to detect specific DNA-binding macromolecules in the donor organism. In short, it may become possible to apply the techniques of bacterial genetics and biochemistry to genes from eukaryotic cells.

Paper 3 also came from the department of biochemistry at Stanford. It describes an alternative approach to that presented in the first two selections. This work by Mertz and Davis involved the restriction enzyme Eco R1, a novel endonuclease produced by certain plasmids carried in *E. coli*. [For a review on restriction enzymes, see Nathans and Smith (1975) and also Paper 28 in this volume.] Pioneering work by Smith and Wilcox (1970) had established the novel site specificity of another restriction enzyme (Hind II) from *Haemophilus influenzae*. The investigation by Mertz and Davis revealed that the enzyme Eco R1, like Hind II, cleaved DNA at specific sites. Unlike the enzyme described by Smith and Wilcox, Eco R1 made "staggered" breaks in the DNA generating "cohesive" ends. These authors provided evidence that such Eco R1 cohesive ends were identical for all DNA substrates. It was this work that paved the way for recombinant DNA experiments using restriction enzymes to generate the DNA fragments. This concept was stated explicitly by Mertz and Davis:

> R1 endonuclease, in conjunction with DNA ligase, provides a means for *in vitro*, site-specific recombination: any two DNAs with Eco R1 cleavage sites can be "recombined" at their restriction sites by the sequential action of R1 endonuclease and DNA ligase.

Paper 4 in this section by Sgaramella established a method

other than generation of cohesive ends to join diverse DNA molecules. Sgaramella found that the DNA ligase isolated from phage T4-infected *E. coli* could "butt-join" double-stranded DNA molecules in the absence of any cohesive ends. Such "blunt-end" ligation, as it is now termed, is used routinely in current recombinant DNA experiments. It is interesting to note the unsuccessful attempt by Sgaramella to join phage P22 DNA to Eco R1 cleaved SV40 DNA by this method. The experiment could not have succeeded because, unknown to Sgaramella at that time, the Eco R1 ends were single-stranded. What was significant was the idea of joining tumor virus DNA to bacterial virus DNA, a concept brought to fruition by Jackson, Symons, and Berg (Paper 2).

By 1973 it was clear that any DNA molecule could be joined to another by several methods. However, the construction of tumor virus-bacteriophage recombinant DNA set the stage for more than the development of a powerful new tool for the study of complex organisms. It also generated a wave of concern about the potential hazards and misuse of recombinant DNA research (Curtiss 1976; Sinsheimer 1977). This concern led to much popular discussion (see cover story of *Time*, April 18, 1977, for example) and, ultimately, gave rise to the unprecedented moratorium on recombinant DNA experiments (see Part II) and the establishment of a new system of "guidelines" for recombinant DNA research (Recombinant DNA Research Guidelines 1976).

Until the early 1970s, the analysis of DNA molecules and fragments derived from them could be measured only by rather insensitive methods including changes in viscosity and differential sedimentation through sucrose gradients. An important advance was the establishment of agarose gel electrophoresis; a technique that separated DNA fragments by size (Helling et al. 1974; Sharp et al. 1973). The use of ethidium bromide to stain DNA in the agarose gel permitted rapid, direct examination of DNA fragments for estimation of size, quanitity and, to some extent, conformation. These methods were soon to become a routine aspect of almost any recombinant DNA experiment.

The first recombinant DNA work involved simple joining of diverse DNA segments in the test tube. The next step was to introduce these molecules back into living cells. The basic concepts were first established in *E. coli* and involved solving at least three different problems. The first was to find a way to get DNA into *E. coli*. Another was to get this newly introduced DNA to replicate. The third problem was to develop a sensitive selection method for those cells that had received a new DNA fragment.

9

Introducing DNA into bacterial cells was not a novel concept; indeed, such work goes back to the discovery of transformation by Griffith in 1928. (The term "transformation" is used to indicate uptake of DNA to produce a cell with a new trait. The term "transfection" is used to indicate uptake of viral DNA that results in a lytic infection or plaque formation.) Surprisingly, it was not until 1960 that such manipulation of *E. coli* was established firmly by Kaiser and Hogness (1960), and even then, this particular method was specific, based on coliphage lambda DNA. About the same time, several groups reported transfection of *E. coli* by a variety of phage DNAs (see, for example, Benzinger et al. 1967). Benzinger's group subsequently established the use of spheroplasts (cells with the cell wall partially removed) for phage DNA work. Paper 5 describes the spheroplast methodology that was subsequently adopted for some recombinant DNA experiments. An alternative method was reported by Mandel and Higa in 1970 (Paper 6) and in fact became the method of choice for most recombinant DNA work. They discovered that if *E. coli* were treated with calcium ions, phage DNA could be taken up, giving rise to infectious particles.

The calcium treatment was subsequently modified by Cohen and colleagues (Paper 7) who adapted it for use with plasmid DNA. By demonstrating that autonomously replicating plasmids could be introduced as pure DNA into *E. coli* and subsequently propagated, Cohen, Chang, and Hsu provided the foundation for future research in recombinant DNA using plasmid vectors. This methodology, which established lines of cells containing the progeny of a single plasmid DNA molecule, provided the means to do experiments that previously were only possible with phage particles. The plasmids Cohen and coworkers were studying were self-replicating circular DNA molecules that grew in *E. coli* but were not essential for normal cell development. These plasmids encoded antibiotic resistance functions and were called R-factors. In fact, the study of antibiotic resistance genes suggested that they were rather mobile, moving from plasmid to plasmid in what seemed then to be mysterious ways. Perhaps the facility with which plasmids could carry extra genes stimulated the pioneering work by Cohen and his colleagues. Using their modification of Mandel and Higa's technique, Cohen and Chang began introducing R-factor DNA into *E. coli* to determine precisely the size and location of the antibiotic resistance genes. Similarily, they set out to locate the replication functions on the circular R-factor DNA. The work described by Cohen and Chang (1973) solved the three problems outlined above. They use controlled shearing followed by calcium-promoted transformation of *E. coli* to isolate a small, self-replicating plasmid that expressed resistance to tetracycline. The plasmid was named subsequently pSC101 (see Paper 8). It was ideally suited for recombinant DNA experiments: it was self-replicating; it was subsequently shown to have a single Eco R1 site in a nonessential region so the Mertz and Davis joining technique could be used; it had a strong selectable marker (tetracycline resistance) so that cells carrying pSC101 containing new DNA fragments could be recognized. This plasmid was used to do the first complete recombinant DNA experiments in which foreign DNA was joined to pSC101 DNA in the test tube, introduced into *E. coli*, replicated many times, the DNA reisolated and studied, and finally, even the expression of functions by the recombinant DNA was studied in *E. coli*. Two of these prophetic papers by Cohen and colleagues are presented here (Papers 8 and 9) and the third by Morrow and co-workers is reprinted in Part 5 (Paper 34; and also Wensink et al. 1974).

An interesting side issue came about later during analysis of pSC101 and its parent R-factor R6-5. Cohen and Chang (1977) established that pSC101 did not arise from R6-5 in the simple manner proposed initially (see Cohen and Chang 1973). In any case, pSC101 was the first potent "vector" for carrying diverse DNA molecules and replicating them in *E. coli*. Its significance cannot be overstated.

REFERENCES

Benzinger, R., J. Delius, R. Jaenisch, and P. H. Hofschneider. 1967. Infectious Nucleic Acids of *Escherichia coli* Bacteriophage. 10. Preparations and Properties of *Escherichia coli* Competent for Infectious DNA from Bacteriophages ΦX174 and M13 and RNA from Bacteriophage m12. *Eur. J. Biochem.* **2**:414–428.

Cohen, S. N., and A. C. Y. Chang. 1973. Recircularization and Autonomous Replication of a Sheared R-factor DNA Segment in *Escherichia coli* Transformants. *Natl. Acad. Sci. (USA) Proc.* **70**:1293–1297.

Cohen, S. N., and A. C. Y. Chang. 1977. Revised Interpretation of the Origin of the pSC101 Plasmid. *J. Bacteriol.* **132**:734–737.

Curtiss, R., III. 1976. Genetic Manipulation of Microorganisms: Potential Benefits and Biohazards. *Ann. Rev. Microbiol.* **30**:507–533.

Gefter, M. L., A. Becker, and J. Hurwitz. 1967. The Enzymatic Repair of DNA. I. Formation of Circular λ DNA, *Natl. Acad. Sci. (USA) Proc.* **58**:240–247.

Helling, R. B., H. M. Goodman, and H. W. Boyer. 1974. Analysis of Endonuclease R. *Eco*R1 Fragments of DNA from Lambdoid Bacteriophages and Other Viruses by Agarose-gel Electrophoresis. *J. Virol.* **14**:1235-1244.

Kaiser, A. D., and D. S. Hogness. 1960. The Transformation of *Escherichia coli* with Deoxyribonucleic Acid Isolated from Bacteriophage λ dg. *J. Mol. Biol.* **2**:392–415.

Kato, K.-I., J. M. Concalves, G. E. Hots, and F. J. Bollum. 1967. Deoxynucleo-tide-Polymerizing Enzymes of Calf Thymus Gland. II. Properties of the Terminal Deoxynucleotidyltransferase. *J. Biol. Chem.* **242**:2780–2789.

Nathans, D., and H. O. Smith. 1975. Restriction Endonucleases in the Analysis and Restructuring of DNA Molecules. *Ann. Rev. Biochem.* **44**:273-293.

Olivera, B. M., and I. R. Lehman. 1967. Linkage of Polynucleotides Through Phosphodiester Bonds by an Enzyme from *Escherichia coli*. *Natl. Acad. Sci. (USA) Proc.* **57**:1426–1433.

Recombinant DNA Research Guidelines. 1976. *Fed. Reg.* **41**:27907–27943.

Sharp, P. A., B. Sugden, and J. Sambrook. 1973. Detection of Two Restriction Endonuclease Activities in *Haemophilus parainfluenza* Using Ana-lytical Agarose Ethidium Bromide Electrophoresis. *Biochemistry* **12**:3055–3063.

Sinsheimer, R. L. 1977. Recombinant DNA. *Ann. Rev. Biochem.* **46**:415–438.

Smith, H. O., and K. W. Wilcox. 1970. A Restriction Enzyme from *Haemophilus influenzae*. I. Purification and General Properties. *J. Mol. Biol.* **51**:379–391.

Weiss, B., and C. C. Richardson. 1967. Enzymatic Breakage and Joining of Deoxyribonucleic Acid. I. Repair of Single-strand Breaks in DNA by an Enzyme System from *Escherichia coli* Infected with T4 Bacterio-phage. *Natl. Acad. Sci. (USA) Proc.* **57**:1021–1028.

Wensink, P. C., D. J. Finnegan, J. E. Donelson, and D. S. Hogness. 1974. A System for Mapping DNA Sequences in the Chromosomes of *Drosophila melanogaster*. *Cell* **3**:315–325.

Zimmerman, S. B., J. W. Little, C. K. Oshinsky, and M. Gellert. 1967. *Enzymatic Joining of DNA Strands: A Novel Reaction of Diphos-phopyridine Nucleotide*. *Natl. Acad. Sci. (USA) Proc.* **57**:1841–1848.

1

Reprinted from *J. Mol. Biol.* **78**:453–471 (1973)

Enzymatic End-to-end Joining of DNA Molecules

Peter E. Lobban† and A. D. Kaiser

Department of Biochemistry, Stanford University
School of Medicine, Palo Alto, Calif. 94305, U.S.A.

A way to join naturally occurring DNA molecules, independent of their base sequence, is proposed, based upon the presumed ability of the calf thymus enzyme terminal deoxynucleotidyltransferase to add homopolymer blocks to the ends of double-stranded DNA. To test the proposal, covalently closed dimer circles of the DNA of bacteriophage P22 were produced from linear monomers. It is found that P22 DNA as isolated will prime the terminal transferase reaction, but not in a satisfactory manner. Pre-treatment of the DNA with λ exonuclease, however, improves its priming ability. Terminal transferase can then be used to add oligo(dA) blocks to the ends of one population of P22 DNA molecules and oligo(dT) blocks to the ends of a second population, which enables the two DNAs to anneal to one another to form dimer circles. Subsequent treatment with a system of DNA repair enzymes converts the circles to covalently closed molecules at high efficiency. It is demonstrated that the success of the joining system does not depend upon any obvious unique property of the P22 DNA.

The joining system yields several classes of by-products, among them closed circular molecules with branches. Their creation can be explained on the basis of the properties of terminal transferase and the DNA repair enzymes.

1. Introduction

Given a single deoxyribonucleoside triphosphate as a substrate, calf thymus terminal deoxynucleotidyltransferase (terminal transferase) will attach homopolymer blocks to pre-existing "primer" DNA molecules by catalyzing the step-wise addition of nucleotide residues to their 3′-termini (Kato *et al.*, 1967; Chang & Bollum, 1971). Primers commonly used with the transferase are single-stranded oligonucleotides and heat-denatured DNA (Yoneda & Bollum, 1965; Kato *et al.*, 1967). If the enzyme will also accept double-stranded primers, a method for the joining of naturally occurring DNA molecules end-to-end can be proposed. First, terminal transferase would be used to add homopolymer blocks to the ends of one population of DNA molecules and complementary blocks to the ends of the other population. Next, the two DNAs would be mixed and annealed. Finally, the joined molecules, held together by hydrogen bonds between the bases in the homopolymers, would be exposed to a DNA repair system to render the junctions covalently continuous.

The subject of this paper is a test of the proposed joining system: the creation of covalently closed circular dimers of the DNA of bacteriophage P22, chosen as a typical double-stranded molecule with homogeneous physical properties (Rhoades

---

† Present address: Department of Medical Cell Biology, Medical Sciences Building, University of Toronto, Toronto 181, Ontario, Canada.

et al., 1968). Oligo(dA) and oligo(dT) were selected as the homopolymer blocks to mediate joining because they form a double helix known to be recognized by *Escherichia coli* DNA ligase (Olivera & Lehman, 1968), which was to be a component of the DNA repair system. Circular dimers were preferred to the other possible products of joining, long linear and circular oligomers, because they would be uniform in size and readily detected by their characteristic buoyant density in the presence of ethidium bromide (Radloff *et al.*, 1967) or their rapid rate of sedimentation in alkaline media (Vinograd *et al.*, 1965). Also, the production of molecules having no remaining single-strand interruptions constitutes a rigorous test of the final step of joining.

Jackson *et al.* (1972) have tested a similar joining system with the DNA of an animal virus.

2. Materials and Methods

(a) *Reagents*

Reagents and their sources were CsCl, optical grade, Harshaw Chemical Co.; ethidium bromide, Calbiochem; deoxyribonucleosides, Calbiochem; deoxyribonucleoside triphosphates, P-L Biochemicals; and Sarkosyl, Geigy Chemical Corp. Deoxyribonucleoside triphosphates labeled with ^{32}P in the α-position were prepared by the methods of Symons (1968,1969) and checked for radiochemical purity according to Gilliland *et al.* (1966). Tris/EDTA buffer is 10 mM-Tris·HCl, 1 mM-EDTA, pH as indicated.

(b) *Scintillation counting*

Samples for scintillation counting were dried onto glass filters (Whatman GF/C, 2·4 cm) and counted in a scintillation fluid consisting of 4 g 2,5-diphenyloxazole/l and 50 mg 1,4-bis-2-(4-methyl-5-phenyloxazolyl)-benzene/l in toluene.

(c) *Acid-precipitable radioactivity*

A sample to be assayed for acid-precipitable radioactivity was pipetted in a volume of 0·075 ml or less to a chilled 0·2-ml sample of 0·1 mg carrier DNA/ml in 0·1 M-sodium pyrophosphate, and 1·2 ml of 0·4 M-trichloroacetic acid, 0·02 M-sodium pyrophosphate was added. After sitting 10 min at 0°C, the mixture was poured through a glass filter, which was then washed with 40 ml 1 M-HCl, 0·1 M-sodium pyrophosphate; 40 ml 1 M-HCl, 0·1 M-sodium phosphate; and 20 ml 1 M-HCl. The filter was then wet with ethanol, dried, and counted.

(d) *Acid-soluble radioactivity*

A sample to be assayed for acid-soluble radioactivity was pipetted in a volume of 0·2 ml into a centrifuge tube containing 0·075 ml of 2·5 mg carrier DNA/ml and 0·01 ml of 0·1 M-EDTA at 0°C. After addition of 0·015 ml concentrated HCl, the mixture was left at 0°C for 10 min and then centrifuged at 6000 *g* for 10 min. A sample of the supernate was transferred to a glass filter, and the filter was dried and counted as described.

(e) *Sucrose gradients*

Sedimentation of DNA through sucrose gradients was done in cellulose nitrate tubes in the Spinco SW40 rotor. Neutral gradients were made by pouring 5% to 20% sucrose gradients in 0·3 M-NaCl, 25 mM-Tris·HCl, 2 mM-EDTA, pH 7·5, into tubes containing 1-ml "cushions" of 60% by weight CsCl in 20% sucrose. Samples were layered on in 0·2 to 0·3 ml. Sedimentation conditions are described elsewhere. Fractions were collected directly onto glass filters, which were then dried, wet with 4 drops of 1 M-trichloroacetic acid, 0·05 M-sodium pyrophosphate, washed with 20 ml 1 M-HCl on a suction filter, washed with ethanol, dried, and counted. Recovery of label was quantitative.

Alkaline gradients were 5% to 20% sucrose gradients in 0·3 M-NaOH. Fractions were collected onto glass filters which had been wet with 0·1 ml of 1 M-NaH$_2$PO$_4$ and dried before use.

14

(f) *Extraction of ethidium bromide*

Ethidium bromide was removed from DNA solutions by extraction with 5 or more successive equal volumes of cold *n*-butanol, a method suggested by J. C. Wang (personal communication).

(g) *Phenol extraction*

Phenol extraction was used to terminate preparative terminal transferase and λ exonuclease reactions and to obtain P22 DNA from the phage. The phenol was redistilled, stored under nitrogen at −20°C, and equilibrated with buffer shortly before use.

(h) *Bacteriophage P22 DNA*

P22 *tsc*$_2$^{29}, a thermally-inducible mutant of P22 phage, was supplied by M. Levine. *thyA57*, a thymine-requiring strain of *Salmonella typhimurium* LT2, came from the strain collection of K. Sanderson *via* E. Lederberg. The lysogen *thyA57* (P22 *tsc*$_2$^{29}) was prepared in this laboratory.

Tritium-labeled P22 phage were obtained by thermal induction of *thyA57* (P22 *tsc*$_2$^{29}) in the presence of tritiated thymidine. Phage were concentrated from lysates by the method of Yamamoto *et al.* (1970) and purified by sedimenting them into pre-formed CsCl step gradients and then banding them twice to equilibrium in CsCl solutions.

Phage P22 DNA was obtained from purified phage by diluting them to an absorbance at 260 nm of 25 and phenol-extracting. The DNA was then dialyzed into Tris/EDTA buffer, pH 7·9. Its absorbancy ratio (A_{260}/A_{280}) was 1·95 to 1·99, and it had a specific activity of about 4000 cts/min/μg.

The molecular weight of the sodium salt of phage P22 DNA was taken as $28·0 \times 10^6$ by correcting the data of Rhoades *et al.* (1968) to the molecular weight of phage T7 DNA given by Freifelder (1970). The absorbance at 260 nm of a solution of P22 DNA that is 1 mM in DNA-phosphorus was assumed to be 6·75 as for phage T7 DNA (Richardson, 1966).

Tritiated P22 DNA with a ^{32}P label in its terminal 5′-phosphoryl groups was prepared according to Weiss *et al.* (1968).

(i) *Enzymes*

Lambda exonuclease, $8·8 \times 10^4$ units/mg, was prepared according to Little *et al.* (1967) and had no detectable endonucleolytic activity under the conditions of use. Calf thymus terminal deoxynucleotidyltransferase at 6700 dATP units/mg (Kato *et al.*, 1967; Yoneda & Bollum, 1965) was the gift of R. L. Ratliff and J. L. Hanners and was devoid of exonucleolytic activity. DNA polymerase I of *E. coli* was obtained from A. Kornberg at 12,000 units/mg as assayed with activated calf thymus DNA (Richardson *et al.*, 1964*b*; Jovin *et al.*, 1969). *E. coli* exonuclease III at $1·3 \times 10^5$ units/mg (Jovin *et al.*, 1969) and a homogeneous preparation of *E. coli* DNA ligase at 10^4 units/mg (Modrich & Lehman, 1970, and unpublished results) were the gifts of P. Modrich.

(j) *Electron microscopy*

DNA samples were prepared for electron microscopy by a modification of the aqueous technique of Davis *et al.* (1971), in which Tris·HCl buffers at the same concentrations and pH are substituted for ammonium acetate in both the spreading solution and the hypophase. Molecules were measured on micrographs by projecting them from behind onto a translucent screen and tracing them with a map measure. Preparations of the open circular form of bacteriophage PM2 DNA (Espejo *et al.*, 1969), used as a length standard, were the gifts of F. Schachat and J. Mertz.

3. Results

(a) *Addition of homopolymers to bacteriophage P22 DNA*

Figure 1 shows that P22 DNA will prime the synthesis of oligo(dA) and oligo(dT) homopolymers by terminal transferase. However, both reactions accelerate in the beginning, suggesting that the enzyme may be initiating asynchronously. Since asynchronous initiation would tend to broaden the size distribution of the homopolymer blocks made, a way was developed for modifying P22 DNA so that it will accept nucleotide residues without an acceleration phase. (Alternatively, the acceleration may be due to a progressive improvement in the priming capacity of the P22 strands as they are elongated. In that case, circumventing the acceleration phase would still allow the transferase to synthesize homopolymers of a given average length in a shorter time, thus limiting the exposure of the DNA to deleterious contaminants in the enzyme (see below).)

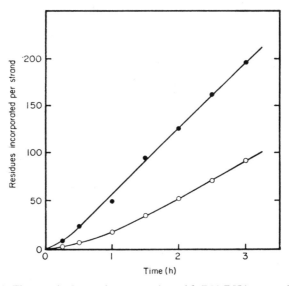

FIG. 1. The terminal transferase reaction with P22 DNA as a primer.

The reaction mixtures contained 74 μg P22 DNA/ml and 8·8 μg terminal transferase/ml in 0·1 M-potassium cacodylate, 7·5 mM-potassium phosphate, 8 mM-MgCl$_2$, 2·1 mM-β-mercaptoethanol, pH 7·0. The substrates were d-[α-^{32}P]ATP at 0·25 mM (—●—●—) and d-[α-^{32}P]TTP at 0·3 mM (—○—○—). The reactions were incubated at 37°C, and samples were withdrawn at intervals to assay for ^{32}P made acid-insoluble.

The modification chosen consisted of treating the DNA with λ exonuclease to remove a small number of residues from its 5'-ends (Little, 1967) so that the 3'-ends are no longer base-paired. The experiment of Figure 2 demonstrates that such a treatment is possible. Tritiated P22 DNA bearing a ^{32}P label in its terminal 5'-phosphoryl groups was exposed to an excess of the exonuclease at 0°C. The ^{32}P was removed rapidly and quantitatively, showing that all termini were attacked in synchrony. Subsequent digestion proceeded at the rate of 64 residues per strand per hour, slowly enough to allow the reaction to be terminated after only a limited number of residues have been removed. We designate P22 DNA treated with λ exonuclease to remove an average of x residues per strand by "P22$_{-x}$." Values of x ranged from 25 to 40.

When P22$_{-x}$ primes terminal transferase, there is no acceleration phase (Fig. 3). Therefore, treatment of the P22 DNA with λ exonuclease was incorporated into the

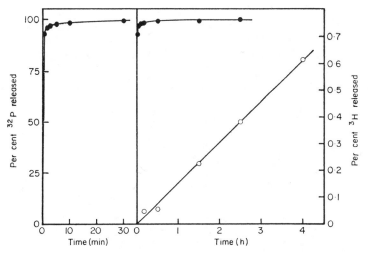

FIG. 2. The action of λ exonuclease on P22 DNA at 0°C.

The reaction mixture contained tritiated P22 DNA with a ^{32}P label in its 5′-phosphoryl groups at 60·3 μg/ml and λ exonuclease at 4·52 μg/ml in 0·067 M-potassium glycinate, 4 mM-MgCl$_2$, pH 9·4 (measured at room temperature). Incubation was at 0°C in an ice–water bath, and samples for the assay of acid-soluble radioactivity were taken with chilled pipettes. —●—●—, ^{32}P label and —○—○—, ^{3}H label, whose rate of release corresponds to the removal of 64 residues per strand/h.

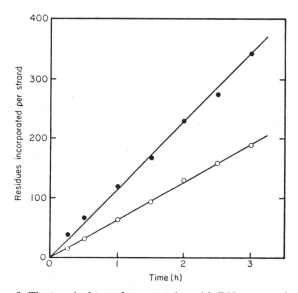

FIG. 3. The terminal transferase reaction with P22$_{-x}$ as a primer.

The reaction mixtures were identical to those of Fig. 1 except that P22$_{-29}$ at 67 μg/ml replaced the untreated P22 DNA. The symbols are the same as in Fig. 1.

joining procedure. Preparative terminal transferase reactions were done with $P22_{-x}$ by scaling up the reactions of Figure 3 and terminating by phenol extraction. The products are symbolized by "$dA_y-P22_{-x}-dA_y$" and "$dT_y-P22_{-x}-dT_y$," with y signifying the number of residues incorporated by the transferase per strand of $P22_{-x}$. The shorter symbols dA–P22–dA and dT–P22–dT serve when x and y are immaterial.

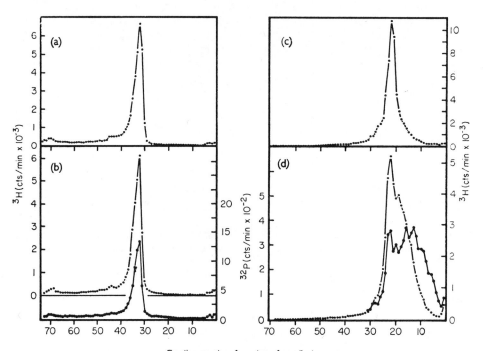

Fraction number from top of gradient

FIG. 4. Sedimentation of dA-P22-dA and dT-P22-dT.

(a) 5 μg of ^{3}H labeled $P22_{-29}$ was layered onto a neutral sucrose gradient in 0·2 ml and sedimented at 20°C for 3 h at 37,000 revs/min.

(b) 5 μg of dT_{66}-$P22_{-29}$-dT_{66}, labeled with ^{3}H in the $P22_{-29}$ (--●--●--) and with ^{32}P in the homopolymer (—●—●—), was sedimented as in (a).

(c) 9 μg of tritiated P22 DNA was layered onto an alkaline sucrose gradient in 0·2 ml and sedimented at 37,000 revs/min for 3 h at 20°C.

(d) 9·5 μg of dA_{72}-$P22_{-29}$-dA_{72}, labeled as for dT_{66}-$P22_{-29}$-dT_{66}, was sedimented as in (c). The symbols are those used in (b). ^{32}P counts from the lower portion of the gradient were near background and were not plotted.

The neutral and alkaline sucrose gradients of Figure 4 show that the homopolymer blocks (^{32}P label) synthesized by terminal transferase in the presence of $P22_{-x}$ co-sediment with the DNA (^{3}H label) and are therefore covalently attached to it. As some of the ^{32}P label sediments with full-length single strands of $P22_{-x}$ (gradient (d) Fig. 4), at least some of the homopolymers are attached at the DNA termini. However, comparison of gradients (c) and (d) in Figure 4 shows that $P22_{-x}$ is extensively nicked during treatment with terminal transferase, probably because of contaminating endonuclease activity. Moreover, the ratio of ^{32}P to ^{3}H in the region of gradient (d) (Fig. 4) where short single strands appear is greater than would be expected if the transferase acted only at termini. Thus, addition of homopolymers to

18

nicks may be occurring. Direct evidence for that possibility has been obtained by Drs David Jackson and Paul Berg (personal communication), who have shown that terminal transferase accepts nicked circular DNA as a primer, but not closed circles.

(b) *Annealing of dA–P22–dA to dT–P22–dT*

When either $P22_{-x}$, dA–P22–dA, or dT–P22–dT is self-annealed, it sediments in neutral sucrose gradients at the same rate as untreated P22 DNA (Fig. 5(b) to (e): the small peak in fractions 23 and 24 of all four gradients is apparently artifactual,

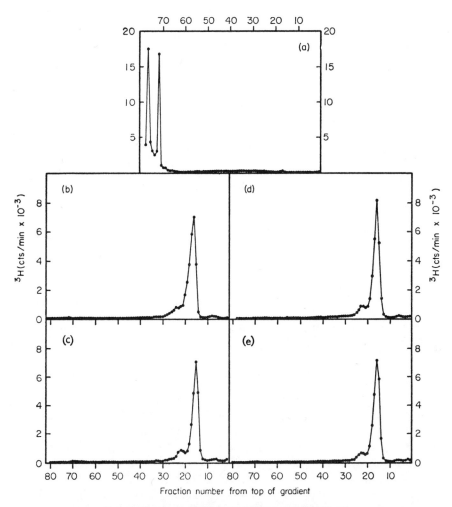

FIG. 5. The annealing of dA-P22-dA to dT-P22-dT.
The following DNA samples were sedimented through neutral sucrose gradients at 20°C for 90 min at 37,000 revs/min: (a) dA_{72}-$P22_{-29}$-dA_{72} * dT_{66}-$P22_{-29}$-dT_{66}, 10 μg; (b) P22 DNA, 6·8 μg; (c) $P22_{-29}$, 5 μg; (d) dA_{72}-$P22_{-29}$-dA_{72}, 5 μg; and (e) dT_{66}-$P22_{-29}$-dT_{66}, 5 μg. Before sedimentation, the DNAs for gradients (a) and (c) to (e) were annealed at 49 μg/ml in 0·5 M-NaCl, Tris/EDTA buffer, pH 7·8, for 10 min at 65°C, 45 min at 45°C at 45 min cooling from 45°C to 39°C. The sample for gradient (b) was heated at 34 μg/ml to 70°C for 10 min in 0·5 M-NaCl, Tris/EDTA buffer, pH 7·8, and quenched by swirling the incubation tube in an ice–water bath. The peak in fraction 72 of (a) is on the high-density cushion at the bottom of the gradient.

19

as it is not present in parallel gradients in which the 0·3 M-NaCl is omitted, and DNA from those fractions is all of unit length as judged by electron microscopy (data not shown). In contrast, when dA–P22–dA and dT–P22–dT are mixed in equimolar proportions and annealed (symbolized by dA–P22–dA * dT–P22–dT), most of the DNA sediments several times more rapidly than P22 DNA (Fig. 5(a)). We conclude that the homopolymer blocks allow molecules of dA–P22–dA to join to molecules of dT–P22–dT in a specific manner. As the aggregates in dA–P22–dA * dT–P22–dT must contain many P22 moieties in order to sediment so rapidly, it also follows that most molecules of dA–P22–dA and dT–P22–dT bear at least two sites capable of mediating joining.

The sedimentation properties of dA_y–$P22_{-x}$–dA_y * dT_z–$P22_{-x}$–dT_z are independent of the values of y and z so long as both exceed about 40 residues per strand. When y and z are less than 25, on the other hand, much of the DNA sediments as linear monomers and short aggregates (data not shown). Accordingly, values of y and z in the range of 50 to 80 residues per strand were used for joining.

TABLE 1

Structures of annealed DNA molecules

DNA preparation	Linear molecules	Branched linear molecules	Circular molecules	Branched circular molecules	Unscorable molecules
dA-P22-dA	60	0	0	0	13
dT-P22-dT	52	0	0	0	11
dA-P22-dA * dT-P22-dT	98	1	28	13	62

DNA samples were prepared for electron microscopy after annealing as described for Fig. 6. Molecules seen were scored according to structure. Unscorable molecules were those whose contours could not be followed unambiguously; in the case of dA-P22-dA * dT-P22-dT, that class included some very large, multiply branched forms. Linear fragments, which constituted about 5% of each DNA preparation, were not included in the scoring.

The structure of the aggregates in dA–P22–dA * dT–P22–dT was investigated using the electron microscope. For that purpose, the DNA was annealed at low concentration (5 μg/ml) to favor the production of dimer circles over longer forms (Wang & Davidson, 1966). Molecules were scored according to structure (Table 1), and measured (Fig. 6). The data support the following generalizations: (1) the predominant species in self-annealed dA–P22–dA or dT–P22–dT is the linear monomer of P22 DNA, and no circles are seen; (2) among the linear molecules in dA–P22–dA * dT–P22–dT are some which are dimeric or trimeric; and (3) there are circles in dA–P22–dA * dT–P22–dT, most of which are dimers. It follows that the homopolymer blocks attached to $P22_{-x}$ by terminal transferase can mediate an end-to-end joining reaction.

The joining of dA–P22–dA to dT–P22–dT also yields branched DNA molecules (Table 1). The most frequent types seen were "σ-forms", which were circles with a single branch, and "θ-forms", in which three DNA threads began at one point and converged on a second point (Plate I). There were also some more highly branched forms too complex for classification.

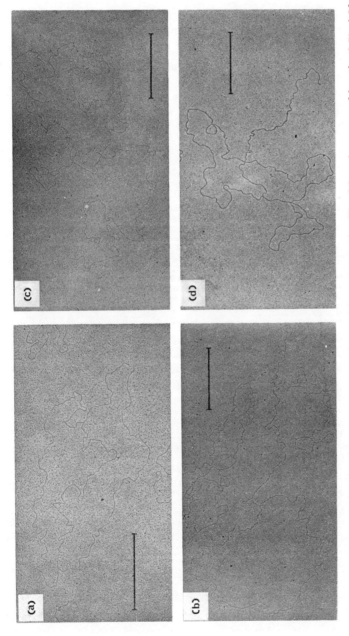

PLATE I. Shown are electron micrographs of one θ-form (a) and three σ-forms ((b) to (d)). The θ-form has arms of lengths 0·70, 0·56 and 0·71 P22 unit. The circular parts of the σ-forms are, respectively, 1·26, 2·01 and 1·49 units long, and the linear parts are of lengths 0·75, 1·02 and 0·49 units. Each calibration mark shows 2 μm.

21

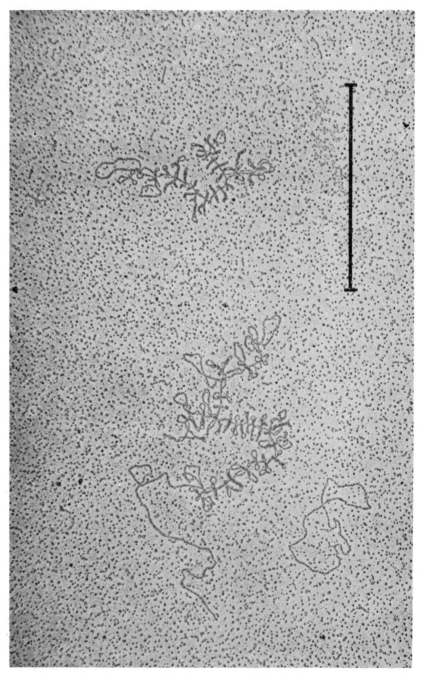

PLATE II. This electron micrograph shows DNA from region I spread without removal of the ethidium bromide. The 3 molecules are a supercoil, a supercoil with a branch and an open PM2 circle. The calibration mark is 2 μm long.

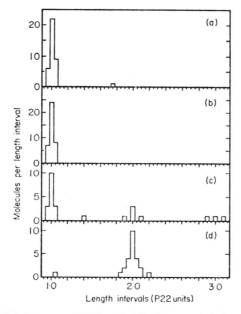

FIG. 6. Measurement of DNA molecules in annealed preparations.

DNA samples were annealed at 5 μg/ml in 0·1 M-NaCl, 25 mM-Tris·HCl, 0·5 mM-EDTA (pH 7·5), for 15 min at 65°C, 2 h at 45°C, and 3 h cooling from 45°C to 32°C. Electron micrographs were taken at random, and the molecules were measured using P22 monomers present as length standards. The samples were (a) dA-P22-dA; (b) dT-P22-dT; (c) dA-P22-dA * dT-P22-dT, linear molecules only; and (d) dA-P22-dA * dT-P22-dT, circular molecules only. One of the linear trimers in (c) was a Y-shaped molecule whose arm lengths were 0·69, 0·98 and 1·26 P22 units; all other linear forms were unbranched.

Most θ-forms are dimers with non-integral arm lengths (Table 2). The lengths of the circular parts of the majority of σ-forms are non-integral in the range from one to two P22 units (Fig. 7, filled and open circles), and most of those molecules are dimeric or trimeric in over-all length (Fig. 7, filled circles). Those data imply that most branches occur where the homopolymer block at an end of one DNA molecule

TABLE 2

Measurements of θ-forms

Arm 1	Arm 2	Arm 3	Total length
0·22	0·50	1·35	2·07
0·28	0·82	0·98	2·08
0·56	0·70	0·71	1·97
0·18	0·50	1·37	2·05
0·42	0·79	0·82	2·03
0·36	0·76	0·84	1·96
0·08	0·09	1·92	2·09
0·38	0·62	0·67	1·67

The arm lengths (in P22 units) of several θ-forms in dA-P22-dA * dT-P22-dT are tabulated here in order of increasing magnitude.

23

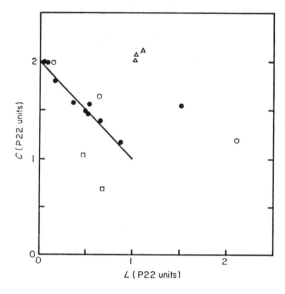

Fig. 7. Measurement of σ-forms.

The lengths L of the linear portions and C of the circular portions of several randomly selected σ-forms in dA-P22-dA * dT-P22-dT are plotted above. Four classes are distinguished: dimers and trimers with $1 < C < 2$ (●), molecules with $1 < C < 2$ whose over-all lengths are non-integral (○), molecules with $C = 2$ and $L = 1$ (△), and other types (□). The line shows the theoretical curve for dimers of the first class.

has annealed to an internal site on a second molecule, probably at a nick where terminal transferase has attached a complementary homopolymer. Only a minority of σ-forms, those consisting of a dimer circle with a monomeric branch (Fig. 7, triangles), can have originated from the annealing of two homopolymer blocks of one type to a single complementary block. Several of the σ-forms have non-integral over-all lengths, suggesting that DNA fragments were involved in their formation.

(c) *Covalent closure of the joined molecules*

The postulated structure of the dimer circles in dA–P22–dA * dT–P22–dT is shown diagrammatically in Figure 8. Each molecule contains four gaps bounded by 3'-hydroxyl and 5'-phosphoryl groups at the points where the P22 moieties are joined. One or more nicks acquired during the incubation with terminal transferase may also be present, and it has been shown that our preparations of the transferase contain an activity that attacks DNA in such a way as to leave 3'-phosphoryl groups (D. Brutlag, personal communication). Accordingly, the system chosen for the covalent closure of the dimer circles consisted of the *E. coli* enzymes DNA polymerase I, DNA ligase and exonuclease III. The former two enzymes can repair single-strand interruptions with either type of 5'-boundary so long as the 3'-boundaries bear free hydroxyl groups (Richardson *et al.*, 1963,1964*a*; Goulian & Kornberg, 1967; Masamune *et al.*, 1971). Since exonuclease III can remove 3'-phosphoryl groups from DNA (Richardson & Kornberg, 1964), its participation in the closure reaction should enable interruptions with phosphorylated 3'-boundaries to be repaired as well. In addition,

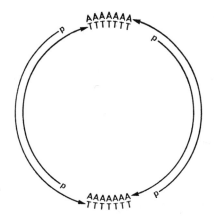

Fig. 8. The postulated structure of the dimer circle.

The strings of A's and T's represent the homopolymer blocks, the p's stand for phosphoryl groups, and the arrows are directed toward the 3'-ends of the respective strands. The end groups are shown as terminal transferase (Kato *et al.*, 1967) and λ exonuclease (Little, 1967) would leave them.

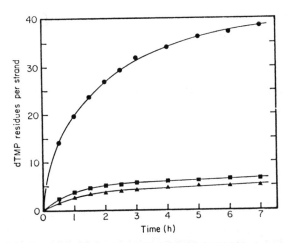

Fig. 9. Incorporation by DNA polymerase during closure.

DNA was prepared for closure by annealing it as for Fig. 6 and adding $(NH_4)_2SO_4$ to 10 mM; $MgCl_2$ to 5 mM; β-mercaptoethanol to 0·5 mM; dATP, dGTP, and dCTP to 45 μM each; NAD+ to 75 μM; bovine plasma albumin to 5 μg/ml; and d-[α-^{32}P]TTP to 8 μM. The mixture was then chilled to 0°C, and the enzymes were added: ligase, 0·75 units/ml; polymerase, 1·2 μg/ml; and exonuclease III, 0·12 units/ml. The reaction was started by placing the mixture at 15°C, and samples were taken at intervals to measure acid-insoluble ^{32}P. A supplement of polymerase (0·8 μg/ml) was added at 135 min and of ligase (0·2 units/ml) at 165 min and at 330 min. The substrates for closure were dA-P22-dA (—▲—▲—), dT-P22-dT (—■—■—), and dA-P22-dA * dT-P22-dT (—●—●—).

the 3′-to-5′ exonuclease of DNA polymerase I (Brutlag & Kornberg, 1972) will remove any residues added to the 3′-boundaries of nicks by terminal transferase, so that repair can then occur; any unpaired residues that may be present at the 3′-terminus of a homopolymer block which has annealed to a shorter complementary block should be subject to removal in a similar manner.

Figure 9 shows the kinetics of incorporation of dTTP when dA–P22–dA, dT–P22–dT, and dA–P22–dA * dT–P22–dT were each subjected to closure. As expected, dA–P22–dA * dT–P22–dT supported much more incorporation than either of the other two, indicating that repair was occurring at the gaps formed by the annealed homopolymers. The over-all incorporation is not great in any of the three reactions, so very little of the DNA present in the products of closure is newly synthesized.

Samples taken from closure reactions and sedimented to equilibrium in ethidium bromide/CsCl gradients gave the profiles of Figure 10. When dA–P22–dA * dT–P22–dT was the substrate for closure, the products appeared in four distinct regions of the gradient: I, a discrete, high-density peak as expected for closed circles (Radloff et al., 1967); II, a polydisperse population whose densities ranged from that of closed

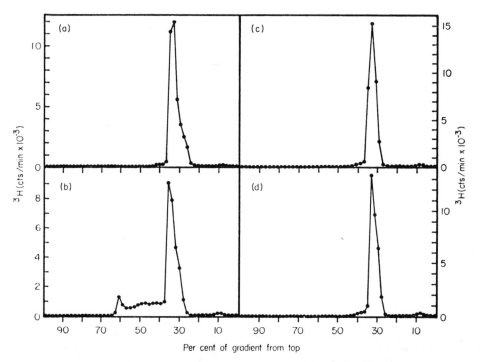

FIG. 10. The closure products in ethidium bromide/CsCl gradients.

Closure reactions were done as for Fig. 9 except that the dTTP was unlabeled. At specified times, samples of 1·8 ml were removed, mixed with an ethidium bromide/CsCl solution (0·1 ml 10% Sarkosyl; 0·25 ml 0·2 M-EDTA; 0·4 ml 1 M-Tris·HCl, pH 8·0; 2·88 ml water; 0·25 ml 10 mg ethidium bromide/ml; and 5·117 g CsCl), and centrifuged at 7°C for 40 h at 40,000 revs/min in the Spinco 65 rotor. Fractions were collected and treated as for sucrose gradients. The DNA samples were (a) dA-P22-dA * dT-P22-dT after zero min of closure (enzymes present), (b) dA-P22-dA * dT-P22-dA after 390 min of closure, (c) dA-P22-dA after 390 min of closure, and (d) dT-P22-dT after 390 min of closure. Region I of gradient (b), centered at 60% from the top, contained 6·0% of the DNA recovered from the gradient.

circles to that of linear molecules; III, a discrete peak at the density of the starting material; and IV, a small, broad peak to the low-density side of region III. In contrast, the products of exposing dA–P22–dA or dT–P22–dT to the closure enzymes banded only in regions III and IV, as did dA–P22–dA * dT–P22–dT taken from the closure reaction at zero time.

TABLE 3

Effects of various modifications to the closure reaction on the distribution of the product DNA in ethidium bromide/CsCl gradients

Sampling time (min)	Ligase present?	Exo III concentration (units/ml)	Region I (%)	Region II (%)	Region III (%)	Region IV (%)
0	+	0·12	<0·1	0·0	98·0	2·0
390	−	0·12	<0·1	0·0	97·8	2·2
390	+	0·0	0·9	6·9	89·8	2·4
390	+	0·012	1·3	8·2	88·0	2·5
390	+	0·12	4·0	13·0	80·2	2·8
390	+	0·36	5·3	16·2	75·3	3·2
390	+	1·2	7·3	15·9	73·6	3·2
90	+	0·36	5·0	15·3	76·1	3·6
210	+	0·36	5·2	20·1	71·6	3·1
390	+	0·36	7·6	20·4	69·0	3·0

The closure reaction was run with dA-P22-dA * dT-P22-dT as the substrate according to the methods of Fig. 10 except for the indicated modifications. Samples were taken at the times shown for analysis on ethidium bromide/CsCl gradients.

Table 3 shows how modifications to the closure system affect the nature of the product DNA. Ligase is absolutely required for material to appear in region I, and exonuclease III has a stimulatory effect, suggesting that most but not all of the dimer circles in dA–P22–dA * dT–P22–dT contain 3′-phosphoryl groups that block repair unless removed. The data also show that the closure reaction is rapid and that the closed molecules are stable in the reaction mixture once formed.

DNA molecules taken from regions I and II of an ethidium bromide/CsCl gradient were scored for structure after removal of the ethidium bromide and CsCl (Table 4).

TABLE 4

Structures of DNA molecules from an ethidium bromide/CsCl gradient

Region	Circles	θ-forms	σ-forms	Multiply branched circles
I	74(74%)	11(11%)	12(12%)	3(3%)
II	4(8%)	4(8%)	23(46%)	19(38%)

Fractions from a given region of a preparative ethidium bromide/CsCl gradient were pooled, freed of ethidium bromide and CsCl, and prepared for electron microscopy. Molecules seen were scored for structure. The most common type of multiply branched circle was like a σ-form except that it had more than one branch. There were also θ-forms with branches and molecules resembling two σ-forms joined at the distal ends of their branches. Region III was too heterogeneous to score.

27

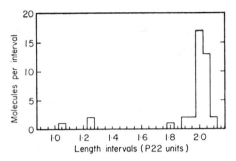

FIG. 11. Lengths of region I circles.

Forty randomly selected circles from region I of an ethidium bromide/CsCl gradient like that of Fig. 10(b) were measured with respect to the open-circular form of PM2 DNA as a length standard. The ratio of the length of P22 monomer to the length of PM2 DNA was determined separately to be 4·21 (±1·4%, 16 observations); the corresponding ratio for the dimer circles was 8·48 (±1·9%, 35 observations), making their average length 2·01 P22 units. The circle in length interval 1·05 was 1·06 units long and therefore not monomeric; and one of the circles in interval 1·90 was too short to be a dimer (1·91 units).

Approximately three-quarters of the molecules in region I were circles. Forty such circles were measured to give the histogram of Figure 11, which shows that 35 of the molecules were dimeric, with a mean length of 2·01 P22 units. The other five circles had non-integral lengths ranging from one to two units, suggesting that they resulted from the joining of one P22 molecule to a DNA fragment. Thus, closed dimer circles were produced by the closure reaction in an over-all yield of about 4%; 6% of the DNA banded in region I, 74% of that DNA was circular, and a fraction 35/40 of the circles were dimeric.

Closed circles could also be demonstrated among the products of closure by sedimentation of the DNA through alkaline sucrose gradients (Fig. 12). The peak in fractions 45 to 50 of gradient (c), displaying the characteristic rapid sedimentation rate of closed forms (Vinograd et al., 1965), constituted 4·4% of the DNA recovered, in good agreement with the amount of DNA from the same closure reaction that banded in region I of an ethidium bromide/CsCl gradient (6·1%, remembering that only three-quarters of the DNA in region I is circular). The slowly sedimenting peak in gradient (c) of Figure 12 appears three fractions further down the gradient than do full-length single strands of P22 DNA (gradient (a) Fig. 12), suggesting that efficient covalent joining of DNA strands has occurred even in those products of closure that are not closed circles.

(d) Closed branched molecules

The data of Table 4 show that among the products of closure are branched circular molecules appearing in regions I and II of ethidium bromide/CsCl gradients. The buoyant densities of those molecules indicate that they are "closed"; that is, a constraint operates in at least some portion of each molecule to prevent the two strands from rotating with respect to one another about the axis of the double helix, so that only as much ethidium can be bound there as in a closed circle of equivalent size (Radloff et al., 1967). We would like to suggest that closed σ and θ-forms have the structures shown diagrammatically in Figure 13. Structures (b) and (c) are closed by covalent bonds, while (a) is closed topologically: one strand of the circular portion

28

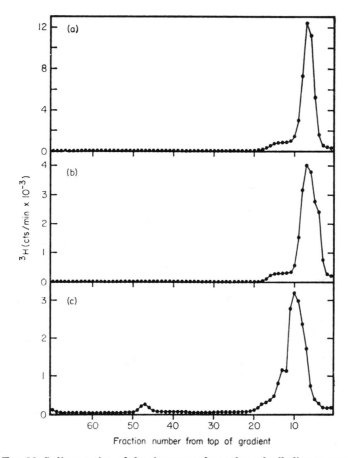

FIG. 12. Sedimentation of the closure products through alkaline sucrose.

Samples were taken from a closure reaction at zero time (b) and after 390 min of incubation (c). Each was made 0·38% in Sarkosyl and 12·5 mM in EDTA, concentrated by dialysis against solid polyethylene glycol 6000, and then dialyzed into 0·2% Sarkosyl, 20 mM·Tris·HCl, 1 mM·EDTA, pH 7·5. A sample of 0·3 ml of each was then layered onto an alkaline sucrose gradient and centrifuged at 20°C for 75 min at 31,000 revs/min. Gradient (a) shows a control sample of P22 DNA.

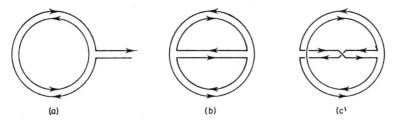

FIG. 13. The postulated structures of closed σ and θ-forms.

Shown are the structures attributed to closed σ-forms (a) and to closed θ-forms ((b) and (c)). The arrows indicate strand polarities.

29

cannot rotate with respect to the other without breaking all of the base pairs in the branch. Because the branch of a closed σ-form is under no such constraint, it should bind as much ethidium as a linear molecule of the same length; therefore, the whole molecule can have any buoyant density between that of a closed circle and that of an open form, depending upon the ratio of the lengths of its two parts. Similarly, a θ-form closed except for a nick in one arm should also have an intermediate density because only that arm will bind the full amount of ethidium. In agreement with this

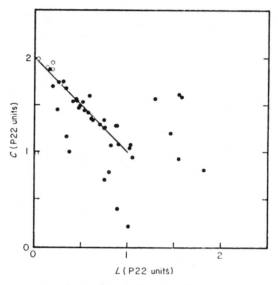

FIG. 14. Measurements of closed σ-forms.
The graph is plotted in the manner of Fig. 7 for σ-forms found in region I (○) and region II (●) of an ethidium bromide/CsCl gradient.

TABLE 5

Measurements of closed θ-forms

Arm 1	Arm 2	Arm 3	Total length
0·53	0·68	0·78	1·99
0·40	0·62	1·04	2·06
0·49	0·52	0·96	1·97
0·50	0·67	0·79	1·96
0·18	0·77	0·96	1·91
0·48	0·68	0·81	1·97
0·24	0·84	0·93	2·01
0·06	0·31	1·71	2·08
0·13	0·34	1·60	2·07
0·49	0·57	1·10	2·16
0·37	0·66	0·96	1·99
0·55	0·76	0·76	2·07
0·07	0·25	1·71	2·03
0·10	0·96	1·46	2·52

Closed θ-forms from regions I and II of a preparative ethidium bromide/CsCl gradient were measured and the results tabulated as for Table 2.

reasoning, the circular part of a closed σ-form mounted for electron microscopy in the presence of ethidium bromide displays the twisted configuration of a supercoil (Vinograd *et al.*, 1965), while its branch has the extended appearance typical of linear DNA (Plate II).

The most probable source of the closed branch forms is the population of open branched circles present in dA–P22–dA * dT–P22–dT before closure. Closed σ and θ-forms obey the same constraints on the lengths of their various parts as do the corresponding open forms (Fig. 14 and Table 5). Note that closed σ-forms from region I tend to have short branches, as expected.

4. Discussion

The experiments presented here show that P22 DNA molecules can be covalently joined to each other. Since the joining is observed only when molecules bearing both kinds of homopolymer blocks are present, and since the circular products of joining are exclusively dimeric, it follows that the reaction is mediated by the homopolymers and not by some unique structural feature of the P22 DNA (for example, its terminally repetitious regions (Rhoades *et al.*, 1968)). Thus it is likely that the joining method can be applied to any pair of double-stranded DNAs. Indeed, Jackson *et al.* (1972) have developed a similar joining system and have used it to create dimer circles of the DNA and to join Simian virus 40 and λ*dvgal* DNAs together. Also, Jensen *et al.* (1971) have reported progress toward the joining of phage T7 DNA molecules by analogous methods.

The creation of closed branched molecules by the closure enzymes merits further discussion. The probable structure of a σ-form in dA–P22–dA * dT–P22–dT is diagrammed in Figure 15. The molecule has three gaps which are identical to those in a dimer circle and therefore present no problem for repair. The fourth gap, whose 3'-boundary is the end of strand 1, is different: soon after initiating synthesis there, DNA polymerase will encounter the branch point rather than the 5'-end of another strand. At that time the enzyme can continue to use strand 2 as its template by displacing strand 3 (Kelly *et al.*, 1970); after each insertion of a base residue, the structure of the DNA will be the same as in the diagram except that the single-stranded portion of the strand being displaced will have grown. If at any time, however, the

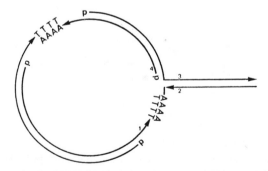

Fig. 15. The postulated structure of a σ-form in dA-P22-dA * dT-P22-dT.

The structure shown is based upon the evidence that most branch points form where terminal transferase has added a homopolymer block to the 3'-boundary of a nick. The symbols are the same as in Fig. 8.

31

enzyme begins to use strand 3 as its template, it will move toward the 5'-end of strand 4 and eventually create a structure in which ligase can join that end to the growing end of strand 1 to give a closed σ-form as depicted in Figure 13(a). Similar events could close a θ-form. The ability of DNA polymerase I to switch template strands during strand displacement has been previously implicated in the production of branched DNA (Schildkraut et al., 1964; Masamune & Richardson, 1971). It is also possible that the closing of branched forms could be catalyzed by ligase if that enzyme were able to join the 3'-end of strand 1 to the 5'-end of strand 4 as soon as the polymerase brought the former into juxtaposition with the latter at the branch point. We know of no precedent for such a reaction, however.

A different method for joining DNA molecules arises from the ability of the ligase of bacteriophage T4 to catalyze the formation of phosphodiester bonds between the ends of fully double-stranded DNA molecules (Sgaramella et al., 1970). Linear dimers and trimers of P22 DNA have been made with that enzyme (Sgaramella, 1972). Joining by T4 ligase differs from the method described here in that there is no apparent way to direct the enzyme to join molecules of one type only to molecules of another type and not to each other. The same is true of a third joining method dependent upon the ability of the R_I restriction endonuclease to generate small cohesive ends when it hydrolyzes DNA (Mertz & Davis, 1972; Hedgpeth et al., 1972).

The apparent generality of the method described here for the joining of any pair of double-stranded DNA molecules to each other may provide a new way to study eukaryotic genes and their controlling elements. Any block of genes from any organism could be inserted into the genome of a temperate bacteriophage to generate a specialized transducing phage. The genes borne by the phage could then perhaps be studied in bacteria, possibly even at the level of fine-structure genetic analysis. Alternatively, the phage could be used to produce many copies of the genes to be used to detect specific DNA-binding macromolecules in the donor organism. In short, it may become possible to apply the techniques of bacterial genetics and biochemistry to genes from eukaryotic cells.

We would like to acknowledge the fruitful discussions and free exchange of ideas we enjoyed with Drs David Jackson, Robert Symons and Paul Berg, who worked concurrently on a similar project. This work was supported by a research grant (AI04509) from the National Institutes of Health. One of us (P. E. L.) was a pre-doctoral fellow of the National Science Foundation while part of this work was in progress.

REFERENCES

Brutlag, D. & Kornberg, A. (1972). J. Biol. Chem. 247, 241.
Chang, L. M. S. & Bollum, F. J. (1971). Biochemistry, 10, 536.
Davis, R. W., Simon, M. & Davidson, N. (1971). In Methods in Enzymology (Grossman, L. & Moldave, K., eds), vol. 21, p. 413, Academic Press, New York.
Espejo, R. T., Canelo, E. S. & Sinsheimer, R. L. (1969). Proc. Nat. Acad. Sci., U.S.A. 63, 1164.
Freifelder, D. (1970). J. Mol. Biol. 54, 567.
Gilliland, J. M., Langman, R. E. & Symons, R. H. (1966). Virology, 30, 716.
Goulian, M. & Kornberg, A. (1967). Proc. Nat. Acad. Sci., U.S.A. 58, 1723.
Hedgpeth, J., Goodman, H. M. & Boyer, H. W. (1972). Proc. Nat. Acad. Sci., U.S.A. 69, 3448.
Jackson, D. A., Symons, R. H. & Berg, P. (1972). Proc. Nat. Acad. Sci., U.S.A. 69, 2904.

Jensen, R. H., Wodzinski, R. J. & Rogoff, M. H. (1971). *Biochem. Biophys. Res. Commun.* **43**, 384.

Jovin, T. M., Englund, P. T. & Bertsch, L. L. (1969). *J. Biol. Chem.* **244**, 2996.

Kato, K.-I., Gonçalves, J. M., Houts, G. E. & Bollum, F. J. (1967). *J. Biol. Chem.* **242**, 2780.

Kelly, R. B., Cozzarelli, N. R., Deutscher, M. P., Lehman, I. R. & Kornberg, A. (1970). *J. Biol. Chem.* **245**, 39.

Little, J. W. (1967). *J. Biol. Chem.* **242**, 679.

Little, J. W., Lehman, I. R. & Kaiser, A. D. (1967). *J. Biol. Chem.* **242**, 672.

Masamune, Y., Fleischman, R. A. & Richardson, C. C. (1971). *J. Biol. Chem.* **246**, 2680.

Masamune, Y. & Richardson, C. C. (1971). *J. Biol. Chem.* **246**, 2692.

Mertz, J. E. & Davis, R. W. (1972). *Proc. Nat. Acad. Sci., U.S.A.* **69**, 3370.

Modrich, P. & Lehman, I. R. (1970). *J. Biol. Chem.* **245**, 3626.

Olivera, B. M. & Lehman, I. R. (1968). *J. Mol. Biol.* **36**, 261.

Radloff, R., Bauer, W. & Vinograd, J. (1967). *Proc. Nat. Acad. Sci., U.S.A.* **57**, 1514.

Rhoades, M., MacHattie, L. A. & Thomas, C. A., Jr. (1968). *J. Mol. Biol.* **37**, 21.

Richardson, C. C. (1966). *J. Mol. Biol.* **15**, 49.

Richardson, C. C., Schildkraut, C. L., Aposhian, H. V., Kornberg, A., Bodmer, W. & Lederberg, J. (1963). In *Symposium on Informational Macromolecules* (Vogel, H., ed.), p. 13, Academic Press, New York.

Richardson, C. C. & Kornberg, A. (1964). *J. Biol. Chem.* **239**, 242.

Richardson, C. C., Inman, R. B. & Kornberg, A. (1964a). *J. Mol. Biol.* **9**, 46.

Richardson, C. C., Schildkraut, C. L., Aposhian, H. V. & Kornberg, A. (1964b). *J. Biol. Chem.* **239**, 222.

Schildkraut, C. L., Richardson, C. C. & Kornberg, A. (1964). *J. Mol. Biol.* **9**, 24.

Sgaramella, V. (1972). *Proc. Nat. Acad. Sci., U.S.A.* **69**, 3389.

Sgaramella, V., van de Sande, J. H. & Khorana, H. G. (1970). *Proc. Nat. Acad. Sci., U.S.A.* **67**, 1468.

Symons, R. H. (1968). *Biochim. Biophys. Acta,* **155**, 609.

Symons, R. H. (1969). *Biochim. Biophys. Acta,* **190**, 548.

Vinograd, J., Lebowitz, J., Radloff, R., Watson, R. & Laipis, P. (1965). *Proc. Nat. Acad. Sci., U.S.A.* **53**, 1104.

Wang, J. C. & Davidson, N. (1966). *J. Mol. Biol.* **19**, 469.

Weiss, B., Live, T. R. & Richardson, C. C. (1968). *J. Biol. Chem.* **243**, 4530.

Yamamoto, K. R., Alberts, B. M., Benzinger, R., Lawhorne, L. & Treiber, G. (1970). *Virology,* **40**, 734.

Yoneda, M. & Bollum, F. J. (1965). *J. Biol. Chem.* **240**, 3385.

2

Reprinted from *Natl. Acad. Sci. (USA) Proc.* **69**:2904–2909 (1972)

Biochemical Method for Inserting New Genetic Information into DNA of Simian Virus 40: Circular SV40 DNA Molecules Containing Lambda Phage Genes and the Galactose Operon of *Escherichia coli*

(molecular hybrids/DNA joining/viral transformation/genetic transfer)

DAVID A. JACKSON*, ROBERT H. SYMONS†, AND PAUL BERG

Department of Biochemistry, Stanford University Medical Center, Stanford, California 94305

Contributed by Paul Berg, July 31, 1972

ABSTRACT We have developed methods for covalently joining duplex DNA molecules to one another and have used these techniques to construct circular dimers of SV40 DNA and to insert a DNA segment containing lambda phage genes and the galactose operon of *E. coli* into SV40 DNA. The method involves: (*a*) converting circular SV40 DNA to a linear form, (*b*) adding single-stranded homodeoxypolymeric extensions of defined composition and length to the 3′ ends of one of the DNA strands with the enzyme terminal deoxynucleotidyl transferase (*c*) adding complementary homodeoxypolymeric extensions to the other DNA strand, (*d*) annealing the two DNA molecules to form a circular duplex structure, and (*e*) filling the gaps and sealing nicks in this structure with *E. coli* DNA polymerase and DNA ligase to form a covalently closed-circular DNA molecule.

Our goal is to develop a method by which new, functionally defined segments of genetic information can be introduced into mammalian cells. It is known that the DNA of the transforming virus SV40 can enter into a stable, heritable, and presumably covalent association with the genomes of various mammalian cells (1, 2). Since purified SV40 DNA can also transform cells (although with reduced efficiency), it seemed possible that SV40 DNA molecules, into which a segment of functionally defined, nonviral DNA had been covalently integrated, could serve as vectors to transport and stabilize these nonviral DNA sequences in the cell genome. Accordingly, we have developed biochemical techniques that are generally applicable for joining covalently any two DNA molecules.‡ Using these techniques, we have constructed circular dimers of SV40 DNA; moreover, a DNA segment containing λ phage genes and the galactose operon of *Escherichia coli* has been covalently integrated into the circular SV40 DNA molecule. Such hybrid DNA molecules and others like them can be tested for their capacity to transduce foreign DNA sequences into mammalian cells, and can be used to determine whether these new nonviral genes can be expressed in a novel environment.

---

* Present address: Department of Microbiology, University of Michigan Medical Center, Ann Arbor, Mich. 48104.

† Present address: Department of Biochemistry, University of Adelaide, Adelaide, South Australia, 5001 Australia.

‡ Drs. Peter Lobban and A. D. Kaiser of this department have performed experiments similar to ours and have obtained similar results using bacteriophage P22 DNA (Lobban, P. and Kaiser, A. D., in preparation).

MATERIALS AND METHODS

DNA. (*a*) Covalently closed-circular duplex SV40 DNA [SV40(I)] (labeled with [³H]dT, 5 × 10⁴ cpm/μg), free from SV40 linear or oligomeric molecules [but containing 3–5% of nicked double-stranded circles—SV40(II)] was purified from SV40-infected CV-1 cells (Jackson, D., & Berg, P., in preparation). (*b*) Closed-circular duplex λ*dvgal* DNA labeled with [³H]dT (2.5 × 10⁴ cpm/μg), was isolated from an *E. coli* strain containing this DNA as an autonomously replicating plasmid (see ref. 3) by equilibrium sedimentation in CsCl–ethidium bromide gradients (4) after lysis of the cells with detergent. A more detailed characterization of this DNA will be published later. Present information indicates that the λ*dvgal* (λ*dv–120*) DNA is a circular dimer containing tandem duplications of a sequence of several λ phage genes (including C_I, O, and P) joined to the entire galactose operon of *E. coli* (Berg, D., Mertz, J., & Jackson, D., in preparation). DNA concentrations are given as molecular concentrations.

Enzymes. The circular SV40 and λ*dvgal* DNA molecules were cleaved with the bacterial restriction endonuclease RI (Yoshimori and Boyer, unpublished; the enzyme was generously made available to us by these workers). Phage λ-exonuclease (given to us by Peter Lobban) was prepared according to Little *et al.* (5), calf-thymus deoxynucleotidyl terminal transferase (terminal transferase), prepared according to Kato *et al.* (6), was generously sent to us by F. N. Hayes; *E. coli* DNA polymerase I Fraction VII (7) was a gift of Douglas Brutlag; and *E. coli* DNA ligase (8) and exonuclease III (9) were kindly supplied by Paul Modrich.

Substrates. [α-³²P]deoxynucleoside triphosphates (specific activities 5–10 Ci/μmol) were synthesized by the method of Symons (10). All other reagents were obtained from commercial sources.

Centrifugations. Alkaline sucrose gradients were formed by diffusion from equal volumes of 5, 10, 15, and 20% sucrose solutions with 2 mM EDTA containing, respectively, 0.2, 0.4, 0.6, and 0.8 M NaOH, and 0.8, 0.6, 0.4, 0.2 M NaCl. 100-μl samples were run on 3.8-ml gradients in a Beckman SW56 Ti rotor in a Beckman L2-65B ultracentrifuge at 4° and 55,000 rpm for the indicated times. 2- to 10-drop fractions were collected onto 2.5-cm diameter Whatman 3MM discs, dried without washing, and counted in PPO–dimethyl POPOP–toluene scintillator in a Nuclear Chicago Mark II scintillation spectrometer. An overlap of 0.4% of ^{32}P into the ^{3}H channel was not corrected for.

CsCl–ethidium bromide equilibrium centrifugation was performed in a Beckman Type 50 rotor at 4° and 37,000 rpm for 48 hr. SV40 DNA in 10 mM Tris·HCl (pH 8.1)–1 mM Na EDTA–10 mM NaCl was adjusted to 1.566 g/ml of CsCl and 350 μg/ml of ethidium bromide. 30-Drop fractions were collected and aliquots were precipitated on Whatman GF/C filters with cold 2 N HCl; the filters were washed and counted.

Electron Microscopy. DNA was spread for electron microscopy by the aqueous method of Davis *et al.* (11) and photographed in a Phillips EM 300. Projections of the molecules were traced on paper and measured with a Keuffel and Esser map measurer. Plaque-purified SV40(II) DNA was used as an internal length standard.

Conversion of SV40(I) DNA to Unit Length Linear DNA [SV40(L_{RI})] with R_I Endonuclease. [^{3}H]SV40(I) DNA (18.7 nM) in 100 mM Tris·HCl buffer (pH 7.5)–10 mM MgCl$_2$–2 mM 2-mercaptoethanol was incubated for 30 min at 37° with an amount of R_I previously determined to convert 1.5 times this amount of SV40(I) to linear molecules [SV40(L_{RI})]; Na EDTA (30 mM) was added to stop the reaction, and the DNA was precipitated in 67% ethanol.

Removal of 5'-Terminal Regions from SV40(L_{RI}) with λ Exonuclease. [^{3}H]SV40(L_{RI}) (15 nM) in 67 mM K-glycinate (pH 9.5), 4 mM MgCl$_2$, 0.1 mM EDTA was incubated at 0° with λ-exonuclease (20 μg/ml) to yield [^{3}H]SV40(L_{RI}exo) DNA. Release of [^{3}H]dTMP was measured by chromatographing aliquots of the reaction on polyethyleneimine thin-layer sheets (Brinkmann) in 0.6 M NH$_4$HCO$_3$ and counting the dTMP spot and the origin (undegraded DNA).

Addition of Homopolymeric Extensions to SV40(L_{RI}exo) with Terminal Transferase. [^{3}H]SV40(L_{RI}exo) (50 nM) in 100 mM K-cacodylate (pH 7.0), 8 mM MgCl$_2$, 2 mM 2-mercaptoethanol, 150 μg/ml of bovine serum albumin, [α-^{32}P]dNTP (0.2 mM for dATP, 0.4 mM for dTTP) was incubated with terminal transferase (30–60 μg/ml) at 37°. Addition of [^{32}P]dNMP residues to SV40 DNA was measured by spotting aliquots of the reaction mixture on DEAE-paper discs (Whatman DE-81), washing each disc by suction with 50 ml (each) of 0.3 M NH$_4$-formate (pH 7.8) and 0.25 M NH$_4$HCO$_3$, and then with 20 ml of ethanol. To determine the proportion of SV40 linear DNA molecules that had acquired at least one "functional" (dA)$_n$ tail, we measured the amount of SV40 DNA (^{3}H counts) that could be bound to a Whatman GF/C filter (2.4-cm diameter) to which 150 μg of polyuridylic acid had been fixed (13). 15-μl Aliquots of the reaction mixture were mixed with 5 ml of 0.70 M NaCl–0.07 M Na citrate (pH 7.0)–2% Sarkosyl, and filtered at room temperature through the poly(U) filters, at a flow rate of 3–5 ml/min. Each filter was washed by rapid suction with 50 ml of the same buffer at 0°, dried, and counted. Control experiments showed that 98–100% of [^{3}H]oligo(dA)$_{125}$ bound to the filters under these conditions. When the ratio of [^{32}P]dNMP to [^{3}H]DNA reached the value equivalent to the desired length of the extension, the reaction was stopped with EDTA (30 mM) and 2% Sarkosyl. The [^{3}H]SV40(L_{RI}exo)–[^{32}P]dA or –dT DNA was purified by neutral sucrose gradient zone sedimentation to remove unincorporated dNTP, as well as any traces of SV40(I) or SV40(II) DNA.

Formation of Hydrogen-Bonded Circular DNA Molecules. [^{32}P]dA and -dT DNAs were mixed at concentrations of 0.15 nM each in 0.1 M NaCl–10 mM Tris·HCl (pH 8.1)–1 mM EDTA. The mixture was kept at 51° for 30 min, then cooled slowly to room temperature.

Formation of Covalently Closed-Circular DNA Molecules. After annealing of the DNA, a mixture of the enzymes, substrates, and cofactors needed for closure was added to the DNA solution and the mixture was incubated at 20° for 3–5 hr. The final concentrations in the reaction mixture were: 20 mM Tris·HCl (pH 8.1), 1 mM EDTA, 6 mM MgCl$_2$, 50 μg/ml bovine-serum albumin, 10 mM NH$_4$Cl, 80 mM NaCl, 0.052 mM DPN, 0.08 mM (each), dATP, dGTP, dCTP, and dTTP, (0.4 μg/ml) *E. coli* DNA polymerase I, (15 units/ml) *E. coli* ligase, and (0.4 unit/ml) *E. coli* exonuclease III.

RESULTS

General approach

Fig. 1 outlines the general approach used to generate circular, covalently-closed DNA molecules from two separate DNAs. Since, in the present case, the units to be joined are themselves circular, the first step requires conversion of the circular structures to linear duplexes. This could be achieved by a double-strand scission at random locations (see *Discussion*) or, as we describe in this paper, at a unique site with R_I restriction endonuclease. Relatively short (50–100 nucleotides) poly(dA) or poly(dT) extensions are added on the 3'-hydroxyl termini of the linear duplexes with terminal transferase; prior

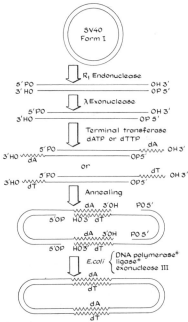

FIG. 1. General protocol for producing covalently closed SV40 dimer circles from SV40(I) DNA.

* The four deoxynucleoside triphosphates and DPN are also present for the DNA polymerase and ligase reactions, respectively.

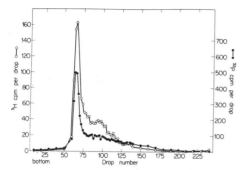

FIG. 2. Alkaline sucrose gradient sedimentation of [³H]SV40-(L$_{RI}$exo)–[³²P](dA)$_{80}$ DNA. 0.16 μg of DNA was centrifuged for 6.0 hr.

either dATP or dTTP resulted in appreciable addition of mononucleotidyl units to the DNA. But, for example, after addition of 100 residues of dA per end, only a small proportion of the modified SV40 DNA would bind to filter discs containing poly(U) (13). This result indicated that initiation of terminal nucleotidyl addition was infrequent with SV40(L$_{RI}$), but that once initiated those termini served as preferential primers for extensive homopolymer synthesis.

Lobban and Kaiser (unpublished) found that P22 phage DNA became a better primer for homopolymer synthesis after incubation of the DNA with λ exonuclease. This enzyme removes, successively, deoxymononucleotides from 5'-phosphoryl termini of double-stranded DNA (15), thereby rendering the 3'-hydroxyl termini single-stranded. We confirmed their finding with SV40(L$_{RI}$) DNA; after removal of 30–50

removal of a short sequence (30–50 nucleotides) from the 5'-phosphoryl termini by digestion with λ exonuclease facilitates the terminal transferase reaction. Linear duplexes containing (dA)$_n$ extensions are annealed to the DNA to be joined containing (dT)$_n$ extensions at relatively low concentrations. The circular structure formed contains the two DNAs, held together by two hydrogen-bonded homopolymeric regions (Fig. 1). Repair of the four gaps is mediated by *E. coli* DNA polymerase with the four deoxynucleosidetriphosphates, and covalent closure of the ring structure is effected by *E. coli* DNA ligase; *E. coli* exonuclease III removes 3'-phosphoryl residues at any nicks inadvertently introduced during the manipulations (nicks with 3'-phosphoryl ends cannot be sealed by ligase).

Principal steps in the procedure

Circular SV40 DNA Can Be Opened to Linear Duplexes by R_I Endonuclease. Digestion of SV40(I) DNA with excess R$_I$ endonuclease yields a product that sediments at 14.5 S in neutral sucrose gradients and appears as a linear duplex with the same contour length as SV40(II) DNA when examined by electron microscopy [(18); Jackson and Berg, in preparation; see Table 1]. The point of cleavage is at a unique site on the SV40 DNA, and few if any single-strand breaks are introduced elsewhere in the molecule (18); moreover, the termini at each end are 5'-phosphoryl, 3'-hydroxyl (Mertz, J., Davis, R., in preparation). Digestion of plaque-purified SV40 DNA under our conditions yields about 87% linear molecules, 10% nicked circles, and 3% residual supercoiled circles.

Addition of Oligo(dA) or -(dT) Extensions to the 3'-Hydroxyl Termini of SV40 (L$_{RI}$). Terminal transferase has been used to generate deoxyhomopolymeric extensions on the 3'-hydroxyl termini of DNA (7); once the chain is initiated, chain propagation is statistical in that each chain grows at about the same rate (12). Although the length of the extensions can be controlled by variation of either the time of incubation or the amount of substrate, we have varied the time of incubation to minimize spurious nicking of the DNA by trace amounts of endonuclease activity in the enzyme preparation; we have so far been unable to remove or selectively inhibit these nucleases (Jackson and Berg, in preparation). Incubation of SV40(L$_{RI}$) with terminal transferase and

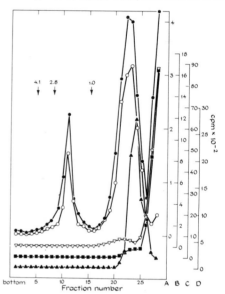

FIG. 3. Alkaline sucrose gradient sedimentation of [³H]-SV40(L$_{RI}$exo)–[³²P](dA)$_{80}$ and −(dT)$_{80}$ DNA incubated 4 hr with and without addition of *E. coli* DNA polymerase I (*P*), ligase (*L*), and exonuclease III (*III*). Conditions are described in *Methods*. 8-Drop fractions were collected. Samples *A*, *C*, and *D* were centrifuged for 60 min, sample *B* for 90 min. Line *A*, dA-ended, plus dT-ended SV40 linears, plus (P+L+III) (³²P, ●; ³H, ○); line *B*, dT-ended, SV40 omitted, plus (P+L+III) (³²P, ▼); line *C*, dA-ended SV40 omitted, plus (P+L+III) (³²P, ■); line *D*, dA-ended plus dT-ended SV40 linears, without (P+L+III) (³²P, ▲). ³H profiles are not shown for lines *B*, *C*, and *D*, but all show that the SV40 DNA sediments in its normal monomeric position. The ³²P and ³H profiles in line *A* are shifted to a faster-sedimenting position with respect to the ³²P profile in line *D* because SV40 strands are covalently linked to one another through (dA)$_{80}$ or (dT)$_{80}$ bridges in most of the molecules, whether or not covalently closed-circles are formed. Very little ³²P remains associated with the SV40 DNA in lines *B* and *C* because tails that remain single-stranded are degraded to 5'-mononucleotides by the 3'- to 5'-exonuclease activity of *E. coli* DNA polymerase I (7).

The *arrows* indicate the position in the gradient of different size supercoiled marker DNAs; the *number* is the multiple of SV40 DNA molecular size (1.0).

36

nucleotides per 5′-end (see *Methods*), the number of SV40(L_{RI}) molecules that could be bound to poly(U) filters after incubation with terminal transferase and dATP increased 5- to 6-fold. Even after separation of the strands of the SV40(L_{RI}exo)-dA, a substantial proportion of the [³H]-label in the DNA was still bound by the poly(U) filter, indicating that both 3′-hydroxy termini in the duplex DNA can serve as primers.

The weight-average length of the homopolymer extensions was 50–100 residues per end. Zone sedimentation of [³H]-SV40(L_{RI}exo)-[³²P](dA)₈₀ (this particular preparation, which is described in *Methods*, had on the average, 80 dA residues per end) in an alkaline sucrose gradient showed that (*i*) 60–70% of the SV40 DNA strands are intact, (*ii*) the [³²P](dA)₈₀ is covalently attached to the [³H]SV40 DNA, and (*iii*) the distribution of oligo(dA) chain lengths attached to the SV40 DNA is narrow, indicating that the deviation from the calculated mean length of 80 is small (Fig. 2). SV40(L_{RI}exo), having (dT)₈₀ extensions, was prepared with [³²P]dTTP and gave essentially the same results when analyzed as described above.

Hydrogen-Bonded Circular Molecules Are Formed by Annealing SV40(L_{RI}exo)-(dA)₈₀ and SV40(L_{RI}exo)-(dT)₈₀ Together. When SV40(L_{RI}exo)-(dA)₈₀ and SV40(L_{RI}exo)-(dT)₈₀ were annealed together, 30–60% of the molecules seen by electron microscopy were circular dimers; linear monomers, linear dimers, and more complex branched forms were also seen. If SV40(L_{RI}exo)-(dA)₈₀ or -(dT)₈₀ alone was annealed, no circles were found. Centrifugation of annealed preparations in neutral sucrose gradients showed that the bulk of the SV40 DNA sedimented faster than modified unit-length linears (as would be expected for circular and linear dimers, as well as for higher oligomers). Sedimentation in alkaline gradients, however, showed only unit-length single strands containing the oligonucleotide tails (as seen in Fig. 2).

Covalently Closed-Circular DNA Molecules Are Formed by Incubation of Hydrogen-Bonded Complexes with DNA Polymerase, Ligase, and Exonuclease III. The hydrogen-bonded complexes described above can be sealed by incubation with the *E. coli* enzymes DNA polymerase I, ligase, and exonuclease III, plus their substrates and cofactors. Zone sedimentation in alkaline sucrose gradients (Fig. 3) shows that 20% of the

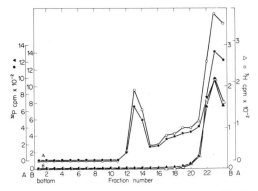

Fig. 4. CsCl–ethidium bromide equilibrium centrifugation of the products analyzed in Fig. 4. Line *A*, dA-ended, plus dT-ended SV40 linears, plus (P+L+III) (³²P, ●; ³H, ○); line *B*, the same mixture without (P+L+III) (³²P, ▲; ³H, △).

TABLE 1.   *Relative lengths of SV40 and λdvgal-120 DNA molecules*

DNA species	Length ± standard deviation in SV40 units*	Number of molecules in sample
SV40(II)†	1.00	224
SV40(L_{RI})†	1.00 ± 0.03	108
(SV40–dA dT)₂	2.06 ± 0.19	23
λ*dvgal–120*(I)	4.09 ± 0.14	65
λ*dvgal–120*(L_{RI})	2.00 ± 0.04	163
λ*dvgal*–SV40	2.95 ± 0.04	76
λ*dv*-1	2.78 ± 0.05	13

* The contour length of plaque-purified SV40(II) DNA is defined as 1.00 unit.

† Data supplied by J. Morrow.

input ³²P label derived from the oligo(dA) and -(dT) tails sediments with the ³H label present in the SV40 DNA, in the position expected of a covalently closed-circular SV40 dimer (70–75 S). About the same amount of labeled DNA bands in a CsCl–ethidium bromide gradient at a buoyant density characteristic of covalently closed-circular DNA (Fig. 4).

DNA isolated from the heavy band of the CsCl–ethidium bromide gradient contains primarily circular molecules, with a contour length twice that of SV40(II) DNA (Table 1) when viewed by electron microscopy. No covalently closed DNA is formed if either one of the linear precursors is omitted from the annealing step or if the enzymes are left out of the closure reaction. We conclude, therefore, that two unit-length linear SV40 molecules have been joined to form a covalently closed-circular dimer.

Covalent closure of the hydrogen-bonded SV40 DNA dimers is dependent on Mg^{2+}, all four deoxynucleoside triphosphates, *E. coli* DNA polymerase I, and ligase, and is inhibited by 98% if exonuclease III is omitted (Lobban and Kaiser first observed the need for exonuclease III in the joining of P22 molecules; we confirmed their finding with this system). Exonuclease III is probably needed to remove 3′-phosphate groups from 3′-phosphoryl, 5′-hydroxyl nicks introduced by the endonuclease contaminating the terminal transferase preparation. 3′-phosphoryl groups are potent inhibitors of *E. coli* DNA polymerase I (14) and termini having 5′-hydroxyl groups cannot be sealed by *E. coli* ligase (8). The 5′-hydroxyl group can be removed and replaced by a 5′-phosphoryl group by the 5′- to 3′-exonuclease activity of *E. coli* DNA polymerase I (7).

Preparation of the Galactose Operon for Insertion into SV40 DNA. The galactose operon of *E. coli* was obtained from a λ*dvgal* DNA; λ*dvgal* is a covalently closed, supercoiled DNA molecule four times as long as SV40(II) DNA (Table 1). After complete digestion of λ*dvgal* DNA with the R_I endonuclease, linear molecules two times the length of SV40(II) DNA are virtually the exclusive product (Table 1). This population has a unimodal length distribution by electron microscopy and appears to be homogeneous by ultracentrifugal criteria (Jackson and Berg, in preparation). The R_I endonuclease seems, therefore, to cut λ*dvgal* circular DNA into two equal length linear molecules. Since one R_I endonuclease cleavage per λ*dv* monomeric unit occurs in the closely related λ*dv-204* (Jackson and Berg, in preparation), it is likely that λ*dvgal* is cleaved at the

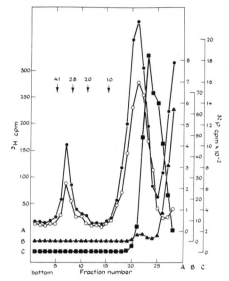

FIG. 5. Alkaline sucrose gradient sedimentation of annealed [³H]SV40(L$_{RI}$exo)–[³²P](dA)$_{80}$ and [³H]λ*dvgal–120*) (L$_{RI}$exo)–[³²P](dT)$_{80}$ incubated for 3 hr with and without (P+L+III). Centrifugation was for 60 min. Line *A*, dA-ended SV40, plus dT-ended λ*dvgal–120* linears, plus (P+L+III)(³²P, ●; ³H, ○); line *B*, dT-ended λ*dvgal–120* linears, plus dT-ended SV40 linears, plus (P+L+III) (³²P, ▲); line *C*,dA-ended SV40 linears, plus dT-ended λ*dvgal–120* linears, without (P+L+III) (³²P, ■).

The *arrows* indicate the position in the gradient of supercoiled marker DNAs having the indicated multiple of SV40 DNA molecular size.

same sites and, therefore, that each linear piece contains an intact galactose operon.

The purified λ*dvgal* (L$_{RI}$) DNA was prepared for joining to SV40 DNA by treatment with λ-exonuclease, followed by terminal transferase and [³²P]dTTP, as described for SV40-(L$_{RI}$).

Formation of Covalently Closed-Circular DNA Molecules Containing both SV40 and λdvgal DNA. Annealing of [³H]-SV40(L$_{RI}$exo)–[³²P](dA)$_{80}$ with [³H]λ*dvgal*(L$_{RI}$exo)–[³²P](dT)$_{80}$, followed by incubation with the enzymes, substrates, and cofactors needed for closure, produced a species of DNA (in about 15% yield) that sedimented rapidly in alkaline sucrose gradients (Fig. 5) and that formed a band in a CsCl–ethidium bromide gradient at the position expected for covalently closed DNA (Fig. 6). The putative λ*dvgal*–SV40 circular DNA sediments just ahead of λ*dv-1*, a supercoiled circular DNA marker [2.8 times the length of SV40(II)DNA], and behind λ*dvgal* supercoiled circles [4.1 times SV40(II)DNA] in the alkaline sucrose gradient. Electron microscopic measurements of the DNA recovered from the dense band of the CsCl–ethidium bromide gradient showed a mean cóntour length for the major species of 2.95 ± 0.04 times that of SV40(II) DNA (Table 1). Each of these measurements supports the conclusion that the newly formed, covalently closed-circular DNA contains one SV40 DNA segment and one λ*dvgal* DNA monomeric segment.

Omission of the enzymes from the reaction mixture prevents λ*dvgal*–SV40 DNA formation (Figs. 5 and 6). No covalently closed product is detectable (Fig. 5) if λ*dvgal* and SV40 linear molecules with identical, rather than complementary, tails are annealed and incubated with the enzymes. This result demonstrates directly that the formation of covalently closed DNA depends on complementarity of the homopolymeric tails.

We conclude from the experiments described above that λ*dvgal* DNA containing the intact galactose operon from *E. coli*, together with some phage λ genes, has been covalently inserted into an SV40 genome. These molecules should be useful for testing whether these bacterial genes can be introduced into a mammalian cell genome and whether they can be expressed there.

DISCUSSION

The methods described in this report for the covalent joining of two SV40 molecules and for the insertion of a segment of DNA containing the galactose operon of *E. coli* into SV40 are general and offer an approach for covalently joining any two DNA molecules together. With the exception of the fortuitous property of the R$_I$ endonuclease, which creates convenient linear DNA precursors, none of the techniques used depends upon any unique property of SV40 and/or the λ*dvgal* DNA. By the use of known enzymes and only minor modifications of the methods described here, it should be possible to join DNA molecules even if they have the wrong combination of hydroxyl and phosphoryl groups at their termini. By judicious use of generally available enzymes, even DNA duplexes with protruding 5'- or 3'-ends can be modified to become suitable substrates for the joining reaction.

One important feature of this method, which is different from some other techniques that can be used to join unrelated DNA molecules to one another (16, 19), is that here the joining is directed by the homopolymeric tails on the DNA. In our protocol, molecule A and molecule B can only be joined to each other; all AA and BB intermolecular joinings and all A and B intramolecular joinings (circularizations) are prevented. The yield of the desired product is thus increased, and subsequent purification problems are greatly reduced.

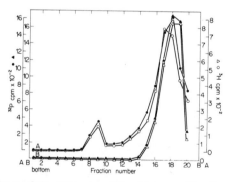

FIG. 6. CsCl–ethidium bromide equilibrium centrifugations of joined [³H]SV40(L$_{RI}$exo)–[³²P](dA)$_{80}$ and [³H]λ*dvgal–120*(L$_{RI}$-exo)–[³²P](dT)$_{80}$ DNA. The samples were those referred to in Fig. 5. Line *A*, dA-ended SV40 linears, plus dT-ended λ*dvgal–120* linears, plus (P+L+III) (³²P, ●; ³H, ○); line *B*, the same mixture without (P+L+III) (³²P, ▲; ³H, △).

For some purposes, however, it may be desirable to insert λ*dvgal* or other DNA molecules at other specific, or even random, locations in the SV40 genome. Other specific placements could be accomplished if other endonucleases could be found that cleave the SV40 circular DNA specifically. Since pancreatic DNase in the presence of Mn^{2+} produces randomly located, double-strand scissions (17) of SV40 circular DNA (Jackson and Berg, in preparation), it should be possible to insert a DNA segment at a large number of positions in the SV40 genome.

Although the λ*dvgal* DNA segment is integrated at the same location in each SV40 DNA molecule, it should be emphasized that the orientation of the two DNA segments to each other is probably not identical. This follows from the fact that each of the two strands of a duplex can be joined to *either* of the two strands of the other duplex $\left(e.g., \begin{smallmatrix} W \frown W \\ C \smile C \end{smallmatrix} \text{ or } \begin{smallmatrix} W \frown C \\ C \smile W \end{smallmatrix} \right)^{\S}$. What possible consequences this fact has on the genetic expression of these segments remains to be seen.

We have no information concerning the biological activities of the SV40 dimer or the λ*dvgal*–SV40 DNAs, but appropriate experiments are in progress. It is clear, however, that the location of the R_I break in the SV40 genome will be crucial in determining the biological potential of these molecules; preliminary evidence suggests that the break occurs in the late genes of SV40 (Morrow, Kelly, Berg, and Lewis, in preparation).

A further feature of these molecules that may bear on their usefulness is the $(dA \cdot dT)_n$ tracts that join the two DNA segments. They could be helpful (as a physical or genetic marker) or a hindrance (by making the molecule more sensitive to degradation) for their potential use as a transducer.

The λ*dvgal*–SV40 DNA produced in these experiments is, in effect, a trivalent biological reagent. It contains the genetic information to code for most of the functions of SV40, all of the functions of the *E. coli* galactose operon, and those functions of the λ bacteriophage required for autonomous replication of circular DNA molecules in *E. coli*. Each of these

sets of functions has a wide range of potential uses in studying the molecular biology of SV40 and the mammalian cells with which this virus interacts.

We are grateful to Peter Lobban for many helpful discussions. D. A. J. was a Basic Science Fellow of the National Cystic Fibrosis Research Foundation; R. H. S. was on study leave from the Department of Biochemistry, University of Adelaide, Australia and was supported in part by a grant from the USPHS. This research was supported by Grant GM-13235 from the USPHS and Grant VC-23A from the American Cancer Society.

1. Sambrook, J., Westphal, H., Srinivasan, P. R. & Dulbecco, R. (1968) *Proc. Nat. Acad. Sci. USA* **60**, 1288–1295.
2. Dulbecco, R. (1969) *Science* **166**, 962–968.
3. Matsubara, K., & Kaiser, A. D. (1968) *Cold Spring Harbor Symp. Quant. Biol.* **33**, 27–34.
4. Radloff, R., Bauer, W., Vinograd, J. (1967) *Proc. Nat. Acad. Sci. USA* **57**, 1514–1521.
5. Little, J. W., Lehman, I. R. & Kaiser, A. D. (1967) *J. Biol. Chem.* **242**, 672–678.
6. Kato, K., Goncalves, J. M., Houts, G. E., & Bollum, F. J. (1967) *J. Biol. Chem.* **242**, 2780–2789.
7. Jovin, T. M., Englund, P. T. & Kornberg, A. (1969) *J. Biol. Chem.* **244**, 2996–3008.
8. Olivera, B. M., Hall, Z. W., Anraku, Y., Chien, J. R. & Lehman, I. R. (1968) *Cold Spring Harbor Symp. Quant. Biol.* **33**, 27–34.
9. Richardson, C. C., Lehman, I. R. & Kornberg, A. (1964) *J. Biol Chem.* **239**, 251–258.
10. Symons, R. H. (1969) *Biochim. Biophys. Acta* **190**, 548–550.
11. Davis, R., Simon, M. & Davidson, N. (1971) in *Methods in Enzymology*, eds. Grossman, L. & Moldave, K. (Academic Press, New York), Vol. 21, pp. 413–428.
12. Chang, L. M. S. & Bollum, F. J. (1971) *Biochemistry* **10**, 536–542.
13. Sheldon, R., Jurale, C. & Kates, J. (1972) *Proc. Nat. Acad. Sci. USA* **69**, 417–421.
14. Richardson, C. C., Schildkraut, C. L. & Kornberg, A. (1963) *Cold Spring Harbor Symp. Quant. Biol.* **28**, 9–19.
15. Little, J. W. (1967) *J. Biol. Chem.* **242**, 679–686.
16. Sgaramella, V., van de Sande, J. H. & Khorana, H. G. (1970) *Proc. Nat. Acad. Sci. USA* **67**, 1468–1475.
17. Melgar, E. & Goldthwait, D. A. (1968) *J. Biol. Chem.* **243**, 4409–4416.
18. Morrow, J. F. & Berg, P. (1972) *Proc. Nat. Acad. Sci. USA* **69**, in press.
19. Sgaramella, V. & Lobban, P. (1972) *Nature*, in press.

§ The symbols W and C refer to one or the other complementary strands of a DNA duplex, and the "connectors" indicate how the strands can be joined in the closed-circular duplex.

3

Reprinted from *Natl. Acad. Sci. (USA) Proc.* **69**:3370–3374 (1972)

Cleavage of DNA by R_I Restriction Endonuclease Generates Cohesive Ends

(SV40/restriction site/cyclization/electron microscopy/DNA joining)

JANET E. MERTZ AND RONALD W. DAVIS

Department of Biochemistry, Stanford University Medical Center, Stanford, California 94305

Communicated by Paul Berg, September 11, 1972

ABSTRACT R_I restriction endonuclease cleaves duplex DNA at a specific sequence, probably 6 nucleotide pairs in length, by making two single-strand staggered cleavages, generating 5′-phosphoryl and 3′-hydroxyl termini. The single-strand ends produced at each break have identical and complementary sequences of 4 or 6 nucleotides in length. Therefore, the cleavage site possesses a 2-fold rotational axis of symmetry perpendicular to the helix axis. The ends of full-length linear SV40 DNA, generated by R_I endonuclease cleavage, can be joined by *Escherichia coli* ligase to regenerate duplex, fully infectious, covalently-closed circular molecules. It was further found that *all* R_I endonuclease-generated ends are identical and complementary. Therefore, any two DNA molecules with R_I sites can be "recombined" at their restriction sites by the sequential action of R_I endonuclease and DNA ligase to generate hybrid DNA molecules.

Restriction endonucleases are believed to make double-strand scissions within a specific sequence of base pairs if that sequence lacks a particular modified base (see refs. 1–3 for review of restriction–modification systems). In only one instance, the restriction endonuclease of *Hemophilus influenzae*, is the nucleotide sequence specifying the restriction site and the location of the endonuclease cleavage known (4, 5). In that case, cleavage occurs in the middle of either of two sequences of six base-pairs, generating molecules that are base paired to the ends with 5′-phosphoryl, 3′-hydroxyl termini; an interesting feature of the restriction site is its two-fold rotational axis of symmetry perpendicular to the helix axis.

After the observation that R_I restriction endonuclease (6) cleaves SV40 form-I DNA [SV40 (I)] to unique linear duplex DNA molecules (7, 8), we discovered that, in contrast to the *H. influenzae* enzyme, the R_I endonuclease makes "staggered" breaks, generating "cohesive" ends. As a consequence, the linear molecules can be recycled through intramolecular hydrogen bonds and covalently sealed by the action of *Escherichia coli* DNA ligase. Our analysis of the single-strand termini suggests that the cohesive ends generated by the R_I endonuclease are identical for all DNA substrates, are either 4 or 6 nucleotides in length, and possess the same kind of 2-fold axis of symmetry as in the *H. influenzae* site.

MATERIALS AND METHODS

DNAs and Enzymes. Monkey kidney cell lines were obtained and grown as described (7). Plaque-purified SV40 virus was obtained from J. Morrow. Purified ^{3}H-labeled SV40(I) DNA was prepared essentially as described (7). Purified, closed circular duplex λdv-120 DNA (D. Berg, J. Mertz, and D. Jackson, manuscript in preparation) was prepared and supplied by D. Jackson. The closed circular duplex DNA of the episome F_8(P17) from *E. coli* (9) was a gift of M.-T. Hsu. The R_I endonuclease was the same prepara-

tion described (7). *E. coli* DNA ligase (10), 7500 units/mg (75% pure) (11), was kindly supplied by P. Modrich.

Enzyme Reactions. (a) R_I endonuclease reactions were performed in 0.1 M Tris·HCl (pH 7.5)–0.01 M $MgCl_2$ at 37° for 15 min. To insure that the cleavage went to completion, further incubation with additional enzyme was frequently performed. The reaction was stopped by addition of Na_3-EDTA to a final concentration of 20 mM. Where indicated, the R_I endonuclease-treated DNA was purified by neutral sucrose gradient sedimentation. (b) *E. coli* DNA ligase reactions were in 20 mM Tris·HCl (pH 8.0), 1 mM EDTA, 10 mM $(NH_4)_2SO_4$, 4 mM $MgCl_2$, 0.1 M KCl (10 mM $MgCl_2$ without KCl was also used), 100 μg/ml of bovine serum albumin and 100 μM DPN (chromatopure, P-L Biochemicals) at 15° for about 24 hr with an excess of enzyme (1–5 μg). The concentration of DNA varied depending upon whether intra- or intermolecular joining was to be favored. The DNA was warmed to room temperature, then kept at 0° for 5 min or longer before the ligase was added.

Equilibrium Centrifugation. Density gradient centrifugations were performed in CsCl [containing 10 mM Tris·HCl (pH 7.5)–1 mM EDTA and, where indicated, 330 μg/ml of ethidium bromide (13)]. Samples were counted and corrected for background (<10 cpm) and a channel overlap of 0.68% ^{32}P into the ^{3}H channel.

Electron Microscopy. DNA was mounted for electron microscopy by the formamide and aqueous techniques described by Davis, Simon, and Davidson (14). To visualize hydrogen-bonded circles, aqueous spreading was performed in a suitably temperature-controlled cold room, with all solutions equilibrated to the indicated temperature while sitting in aluminum blocks. The DNAs used in experiments on the formation of hydrogen-bonded circles remained at 5° for at least 2 days before they were used. Grids were examined and photographed with a Philips EM 300 electron microscope. Length measurements were made with a Hewlett-Packard 9864A Digitizer and 9810A Calculator with a fully smoothed length calculation program, giving an accuracy of ±0.5% and greater precision on sample figures of known length.

RESULTS

SV40(L_{RI}) DNA Molecules are Infectious. Circular, covalently-closed SV40 DNA [SV40 (I)] is cleaved by R_I endonuclease at a unique site to produce full-length linear duplex molecules [SV40(L_{RI})] (7. 8). Unexpectedly, we found that even purified SV40(L_{RI}), which contained less than 0.1% circles, had about 10% of the specific plaque-forming activity of SV40(I); the progeny virus, however, contained supercoiled

SV40(I) DNA. This finding suggests that the ends of SV40 (L$_{RI}$) can be joined *in vivo* to regenerate infectious circular molecules.

SV40(L$_{RI}$) DNA Molecules Can Be Circularized In Vitro. Only 0.14% of the purified SV40(L$_{RI}$) molecules are circular when mounted for electron microscopy at 25° in 50% formamide. The same preparation spread in 0.5 M ammonium acetate at temperatures below 25° gives an appreciable number of circular monomers, and some linear and circular dimers (Fig. 1). At 5.9 ± 0.3° there are equal numbers of linear and circular monomer structures. The T_m for melting a hydrogen-bonded circular dimer to a linear dimer is 2 ± 1° (Fig. 1), and is close to that predicted for the dimer (1°) from the relation K dimer = $(1/2)^{5/2} \cdot K$ monomer (15). Clearly, under some conditions, the ends of SV40(L$_{RI}$) can be annealed to produce hydrogen-bonded circles.

SV40(L$_{RI}$) can be converted to a covalently-closed circular form by incubation with *E. coli* DNA ligase at 15°. More than 98% closed molecules have been observed by electron microscopy. These closed circular DNA molecules produce a denser band in an ethidium bromide–CsCl density gradient (Fig. 2); the enzymatically closed molecules appear at a higher buoyant density than SV40(I) extracted *in vivo* because the negative superhelical turns, normally present in native molecules, are lost under these conditions of closure (16). Incubation of SV40(L$_{RI}$) with ligase at high DNA concentrations (10 μg/ml) produces linear and circular oligomers, as well as monomer closed circles (Fig. 2). After covalent joining of the ends of SV40(L$_{RI}$) at low DNA concentration (to produce

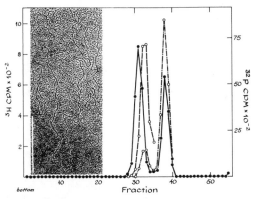

FIG. 2. Equilibrium centrifugation of ligase-treated SV40(L$_{RI}$) in a CsCl–ethidium bromide gradient. SV40(L$_{RI}$) [³H]DNA (1 μg/ml) was incubated with *E. coli* ligase for 24 hr at 18° and, after addition of EDTA, ethidium bromide (330 μg/ml), cesium chloride (final density 1.5656), and a marker of SV40 [³²P]DNA [about 17% SV40(I)–83% SV40(II)] were added. The mixture (3.0 ml) was contrifuged at 38,000 rpm and 4° for 47 hr in an SW50.1 rotor. ●——●, ³H-SV40(L$_{RI}$) DNA after ligase treatment; O–·–·–O, ³²P-SV40 marker DNA; O– – –O, the dense peak of SV40 [³²P]DNA plotted at five times the scale shown. The micrograph is a similar sample treated with ligase at 10 μg of DNA per ml.

largely monomer circles), the infectivity and R$_I$ sensitivity is restored to that of the original SV40(I) DNA. Thus, the original cleavage site is probably regenerated during joining.

We conclude from these findings that R$_I$ endonuclease cleavage of SV40(I) DNA generates short "cohesive" ends, and that these "cohesive" ends, under suitable conditions, can be paired through hydrogen-bonds to produce circular structures; treatment of these molecules with DNA ligase restores the covalently-closed structure and full infectivity.

An Estimate of the Number of Nucleotides in the Cohesive End. The equilibrium constant, K, for joining the cohesive ends of SV40(L$_{RI}$) is defined by the ratio of circular to linear monomers (Fig. 1). By Eq. **1**, the slope of the plot of ln K against $1/T$ for the monomer yields an enthalpy change, ΔH, of -50 ± 3 Cal/mol. The number of bases involved in the cy-

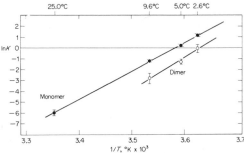

FIG. 1. The equilibrium between linear and circular forms of SV40(L$_{RI}$) as a function of temperature. A single sample of SV40(L$_{RI}$) was mounted for microscopy by the aqueous technique at the temperatures indicated. Monomers or hydrogen-bonded dimers were scored as linear (*L*) or circular (*C*) in random fields, with about 1% being unscorable because of tangling, uranyl oxide crystals from staining, or because they were residual supercoiled SV40(I). The background number of circles, obtained from a sample spread in 50% formamide–0.1 M Tris·HCl buffer (pH 8) (equivalent to 75°), was 10 of 7000 monomers scored. The data for the graph are shown below; standard error in K = $K(1/C + 1/L)^{1/2}$.

No. of molecules

Temp, °C	Monomers C	Monomers L	Dimers C	Dimers L	(Ratio of circles to linears) K(monomers)	K(dimers)
2.6	394	119	16	16	3.30 ± 0.35	1.00 ± 0.35
5.0	533	415	21	72	1.28 ± 0.08	0.29 ± 0.07
9.6	271	875	5	75	0.307 ± 0.02	0.068 ± 0.03
25.0	24	5973	0	3	0.0026 ± 0.0007	

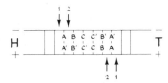

FIG. 3. Model for R$_I$ endonuclease recognition site and possible modes of cleavage. The letters *A*, *B*, and *C* signify DNA bases, and the *prime symbol* indicates the complement of that base. The *numbered arrows* indicate two alternate sites of cleavage that would generate either 6- or 4-base single-strand ends. *Plus* and *minus* distinguishes the two complementary strands of the duplex. *H* and *T* are arbitrary designation for head and tail.

clization, n, can be estimated from Eqs. **1** and **2** (17):

$$\frac{d \ln K}{d(1/T)} = -\Delta H/R \quad [1]; \qquad n\Delta h = \Delta H + 7 \quad [2]$$

where Δh is the enthalpy change per mol base-pair [-8 to -9 Cal/mol (18)] and -7 Cal/mol is assumed to be the contribution from an additional stacking interaction accompanying pairing of the ends (17). Substituting, one obtains n, the length of the single-strand ends $= 5 \pm 1$ nucleotides.

Another method of estimating the number of nucleotides in the single-strand ends created by R_I endonuclease is that used for determining the length of the cohesive ends of coliphage lambda DNA from the melting temperature of the hydrogen-bonded circles (19). This approach introduces an empirical term, which accounts for additional effects on the stability of the cohesive joint; for example, the steric and electrostatic effect of two phosphodiester breaks (end effect). By use of a value for that term, that proved applicable to the cohesive ends of lambda and 186 DNA (17, 20), and the T_m for hydrogen-bonded SV40 monomeric circles ($5.9 \pm 0.3°$), the estimated length of the SV40 cohesive ends ranges between 6 AT pairs and 4 GC pairs. However, since the R_I endonuclease-generated ends may differ significantly in their base composition, sequence, and length from those of lambda DNA, and since K for SV40(L_{RI}) DNA may be affected by

electron microscopic mounting, the estimate of the length of the R_I endonuclease cohesive ends is approximate.

The R_I Endonuclease Cleavage Site Possesses a 2-Fold Axis of Symmetry. It is possible that the R_I endonuclease restriction site has a 2-fold rotational axis of symmetry perpendicular to the helix axis and that the cleavages occur in a staggered

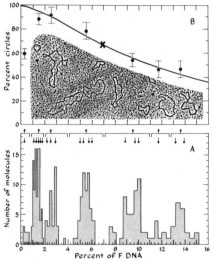

Fig. 5. (A) Fragments produced by R_I endonuclease treatment of F_8(P17) DNA. Circular F_8(P17) DNA was reacted to completion with R_I endonuclease. After aqueous spreading at 25° of the cleaved DNA (0.5 µg/ml), random fields were photographed, and the lengths of 286 fragments were determined relative to circular lambda DNA present on the same grid. 10 Molecules that were greater than 16% of the length of F_8(P17) were discarded, on the assumption that they resulted from incomplete cleavage by R_I. The small divisions on the abscissa are lengths plotted in standard deviation intervals calculated from the equation $\sigma = 0.1 \,[\% \text{ length of } F_8(P17)]^{1/2}$ (14). 19 Size classes (denoted in their approximate locations by *downward arrows*) containing about 15 molecules each were obtained. The length of the smallest molecule (about 200 nucleotides) has a larger standard deviation than expected, probably due to measurement error. Whole F_8(P17) DNA is 1.584 the length of lambda DNA and contains about 74,000 base pairs. (B) Cyclization of the different size fragments in panel A: theoretical curve and experimental points. The R_I endonuclease cleavage fragments of F_8(P17) DNA (panel A) were mounted for electron microscopy by incubation of a 50-µl drop of the DNA (in 0.45 M ammonium acetate containing 0.5 µg of DNA/ml and 70 µg/ml of cytochrome c) on a Teflon block in a 3.0° cold room; a parlodion-coated grid was touched to it. The above conditions are equivalent to 0.5 M ammonium acetate at 3.7° (27). The lengths of 300 random molecules were measured, put into one of seven distinguishable size classes (indicated by *upward arrows*), scored as circular or linear, and plotted at the position corresponding to the mean length of the molecules within each size class. The *error bars* are calculated as given in Fig. 1. An additional 86 molecules were scored for the smallest size class. Very little polymerization occurred at this low temperatue and concentration, since the number average molecular lengths of the fragments at 25° and 3.7° were 5.1 and 5.8% of the F_8(P17) DNA, respectively. The micrograph shows a field of hydrogen-bonded fragments at a magnification as shown on the *abscissa* in panel A.

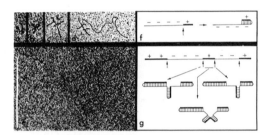

Fig. 4. Covalent joining of plus and minus strands to produce "snap-back" molecules. The covalently joined linear (and circular) oligomers of SV40(L_{RI}) were prepared by treatment of the DNA sample at high concentration (15 µg/ml) with DNA ligase. About $1/3$ of the mass of this sample was linear oligomers before denaturation, with the longest oligomer observed being 33 monomer units. The DNA (0.01 µg) was denatured with NaOH (pH 13) and quickly neutralized with Tris·HCl (total volume = 50 µl). The sample, containing "snap-back" linear oligomers and undenatured closed circles, was then quickly mounted for electron microscopy by aqueous spreading and stained with uranyl acetate. Panels a, b, and c are electron micrographs of single-stranded monomer, dimer, and trimer SV40 DNA molecules, respectively. The molecular weight of the "bushes" that result from aqueous spreading of single-stranded DNA can be estimated from the periphery length of a "bush" (25). The interpretation of the molecules in panels d and e is diagrammed in panels f and g, respectively. The diagram in panel g depicts several possible structures that the 9-mer containing joined plus and minus strands could assume after "branch migration" (26); the middle figure most closely resembles the photograph in panel e. Our estimate of the ratio of single-strand to double-strand mass in the 5-mer (panel d) and 9-mer (panel e) (characteristic of those tabulated in Table 2) are 1.5 and 0.125, respectively. The *white bar* is the length of one SV40 duplex.

TABLE 1. *Ratios of single- to double-strand mass in denatured (snap-back) linear oligomers of SV40(L$_{RI}$)*

Oligomer size (n, n + 1)	Total number of monomer units scored	Ratio of single- to double-strand mass	
		Experimental*	Predicted†
2, 3	612	1.00	1.00
4, 5	315	0.57	0.60
6, 7	175	0.57	0.46
8, 9	91	0.14	0.38
10, 11	30	0.25	0.25
12, 13	12	0.20	0.22
14, 15	14	0.17	0.21
16, 17	16	0.00	0.20
	Weight average	0.723	0.736

* See legend to Fig. 4.

† Calculated from the binomial coefficients on the assumption of equal probability of joining "head-to-head" and "head-to-tail." For example, a tetramer can join as (4+); (3+, 1−); (2+, 2−); (1+, 3−); (4−), in the respective ratio 1:4:6:4:1. After denaturation, the average single- to double-strand mass ratio is 0.60. The calculated ratio is the same for n and $n + 1$, where n is any even integer.

fashion about the site, as shown by the numbered arrows in Fig. 3. The base sequence of the two cohesive ends generated from such a symmetrical site must be identical, as well as complementary.

If they are identical, heads and tails can join at random, and, if two heads or two tails are covalently joined, then a plus strand will be joined to its complementary minus strand. After denaturation a single strand of such a molecule would instantly renature (snap-back) to form a "hairpin" duplex.

This possibility was tested by denaturing, in alkali, covalently-joined linear oligomers of SV40(L$_{RI}$) (produced by treatment with DNA ligase at high DNA concentration), quickly neutralizing the pH, and mounting the DNA for electron microscopy by the aqueous technique. The unrenatured single strands, appearing as bushes (Fig. 4a–c), are readily distinguished from the renatured (snap-back) double strands (d and e). Since the DNA before denaturation is entirely duplex and a DNA sample not treated with ligase yields only single strands after denaturation, it is clear from Fig. 4 that the ligase has joined plus and minus strands. As shown in Table 1 the observed ratio of single- to double-strand mass in all size classes of all "snap-back" molecules corresponds closely to the predicted ratios for random covalent joining of heads and tails of SV40(L$_{RI}$) molecules. Therefore, the two cohesive ends of SV40(L$_{RI}$) have identical sequences, and it follows that the R$_I$ endonuclease cleavage site must have a 2-fold axis of symmetry. Also, the single-strand ends must contain an even number of nucleotides; with an odd number of nucleotides there would be at least a one base-pair mismatch in the head-to-head or tail-to-tail joined molecules. This one mismatch would lower the T_m by 15–20° (21) (assuming ligase treatment has no effect on T_m) and would thus lower the probability of this type of joining at least 200-fold.

All R$_I$ Endonuclease-Generated Ends Are Probably Identical in Their Nucleotide Sequence. Although the cohesive ends produced by R$_I$ endonuclease cleavage of SV40 DNA are the same, one may ask whether identical ends are generated by

TABLE 2. *Number of R$_I$ endonuclease cleavages in various DNAs*

DNA (Molecular weight × 10⁻⁶)		Number of cleavages	Average number of base-pairs/cleavage
SV40	3.4	1	5,100
Polyoma*	3	1	5,000
PM2	6.6	0	>10,000
Mouse mitochondrial*	10.3	2	7,800
Adenovirus 2†	24.8	5	7,500
F$_8$(P17)	48.7	19 ± 1	3,900
	Weighted average		5,260

* J. F. Morrow and D. L. Robberson, (unpublished).

† C. Mulder, U. Petterson, H. Delius, and P. A. Sharp, (unpublished).

cleavages in other DNAs. To answer this question, fragments generated by R$_I$ endonuclease cleavage of the large episomal DNA, F$_8$(P17), were tested for their ability to cyclize under

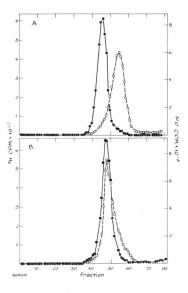

FIG. 6. Demonstration of the covalent joining of λdv-120 and SV40 DNA by CsCl density gradient centrifugation. 0.58 μg of λdv-120(L$_{RI}$) [³H]DNA (12,000 cpm/μg) and 0.25 μg of SV40(L$_{RI}$) [³²P]DNA (about 70,000 cpm/μg) were incubated with *E. coli* DNA ligase at a final DNA concentration of 83 μg/ml. In one mixture, each of the DNAs was treated with the ligase in separate reactions; the two mixed together after the reactions were stopped with EDTA (panel *A*). In the other reaction (panel *B*) both DNAs together were incubated with the ligase. Electron microscopy of each of the mixtures showed that 65% of the mass was contained in molecules averaging 7.7 SV40 DNA equivalents. Each reaction was added to 4.4 ml of CsCl in 10 mM Tris·HCl–1 mM EDTA (pH 7.5) (final ρ was 1.698), then centrifuged at 15°, 35,000 rpm for 46 hr in a Spinco no. 50 rotor. ●———●, λdv-120 DNA; ○———○, SV40 DNA.

the conditions described for forming hydrogen-bonded SV40 DNA circles. F_8(P17) DNA is cleaved to 19 ± 1 fragments (Fig. 5A), ranging from 200 to 10,000 nucleotide pairs in length. If R_I endonuclease produces only one type of single-stranded end; all fragments will form hydrogen-bonded circles. On the other hand, if a fragment has a terminal sequence that differs by even one base, <1% hydrogen-bonded circles will be formed under these conditions. Accordingly, the limit digest of R_I endonuclease on F_8 (P17) was incubated and prepared for electron microscopy at 3°; the % circles in various size classes (Fig. 5A) was then scored (Fig. 5B). If all R_I-produced single-strand ends are identical, the expected % circles in any size class would be as indicated by the curve in Fig. 5B. The % circles as a function of length is obtained from the relation K for any fragment = K for SV40(L_{RI}) $\times$ (length SV40/length F_8(P17) fragment)$^{3/2}$ (22). K for SV40, under these conditions, is 2.0 (from Fig. 1).

Except for the two smallest size classes, the experimental points are within one standard error of the theoretical curve. If a given fragment within a size class could not be cyclized, the deviation of the experimental point would be more than two standard errors. Conceivably, one fragment, occurring in the second-smallest size class, could have had nonidentical ends. But, if one fragment had nonidentical ends, a second noncyclizable fragment also should have occurred, but none are evident. The deviation in the frequency of cyclization seen for the smallest fragment is probably due to the stiffness of a DNA helix, since the smallest fragment is considerably smaller than one statistical segment length (23). Therefore, we conclude that all the R_I-generated ends in F_8(P17) DNA are very likely identical.

How Many Nucleotides Are Needed to Specify the R_I Endonuclease Restriction Site? Assume, as a first approximation, that the nucleotide order in DNA is random and that the number and sequence of nucleotides specifying the R_I restriction site is unique; then, the probability that a given sequence of length n will occur in a long random sequence composed of 4 bases is $1/4^n$. For n equal to 4, 5, 6, and 7, the statistical frequency of the number of restriction sites is of the order of one in 256, 1024, 4100, and 16,400 base-pairs, respectively. Although the sample is limited, the number of R_I endonuclease cleavages produced in several different sizes of DNAs (Table 2) [coliphage DNAs have been omitted because of a possible selection against sequences that sensitize them to restriction (1)] gives values that fit closest to $n = 6$.

R_I Endonuclease-Generated Fragments from Different DNAs Can Be Covalently Joined. If R_I endonuclease generates identical cohesive ends at the cleavage sites of all DNAs, it should be possible to covalently join DNA segments generated in this way. This has been accomplished with SV40(L_{RI}) DNA and the linear monomer duplexes [λdv-120(L_{RI})] formed from λdv-120 supercoiled dimer circles with R_I endonuclease. When the two DNAs were incubated with *E. coli* DNA ligase separately, then mixed and centrifuged to equilibrium in a CsCl gradient, SV40 DNA formed a band at a lower density than the λdv-120 DNA (Fig. 6A). When the two DNAs were mixed together, incubated with ligase, then centrifuged in the same way, the SV40 and λdv-120 DNAs banded together at an intermediate density. This experiment proves that a cohesive end of SV40(L_{RI}) can be covalently joined to a cohesive end of λdv-120(L_{RI}). Quite likely, any two R_I endonuclease fragments can be covalently joined.

DISCUSSION

R_I endonuclease, in conjunction with DNA ligase, provides a means for *in vitro*, site-specific recombination: Any two DNAs with R_I endonuclease cleavage sites can be "recombined" at their restriction sites by the sequential action of R_I endonuclease and DNA ligase. These hybrid DNAs can be cleaved by R_I endonuclease to regenerate the original DNAs. It is more difficult to limit the mode of joining of two different DNA duplexes by this approach than can be done with the protocol developed by Jackson and Berg (24) and by Lobban and Kaiser (manuscript in preparation). Nevertheless, what the joining of R_I endonuclease-cut fragments lacks in specificity is compensated for by the ease and efficiency of the reaction; quite possibly, with appropriately chosen concentrations and molecular species of DNA one may, in this simple way, be able to generate specifically oriented recombinant DNA molecules *in vitro*.

We thank P. Berg for advice and help in the preparation of this manuscript, N. Davidson for his helpful suggestions, and R. Yoshimori and H. Boyer for R_I endonuclease. This work was supported in part by research grants from the U.S. Public Health Service (National Institutes of Health, GM-13235, GM-19177 and TI GM-0196) and the American Cancer Society (VC 23A). J. E. M. was a U.S. Public Health Service Trainee.

1. Arber, W. & Linn, S. (1969) *Annu. Rev. Biochem.* 38, 467-500.
2. Boyer, H. W. (1971) *Annu. Rev. Microbiol.* 25, 153-176.
3. Meselson, M., Yuan, R. & Heywood, J. (1972) *Annu. Rev. Biochem.* 51, 447-466.
4. Smith, H. O. & Wilcox, K. W. (1970) *J. Mol. Biol.* 51, 379-391.
5. Kelly, T. J., Jr. & Smith, H. O. (1970) *J. Mol. Biol.* 51, 393-409.
6. Yoshimori, R. N. (1971) Ph.D. Thesis, University of California, San Francisco Medical Center.
7. Morrow, J. & Berg, P. (1972) *Proc. Nat. Acad. Sci. USA* 69, 3365-3369.
8. Mulder, C. & Delius, H. (1972) *Proc. Nat. Acad. Sci. USA* 69, 3215-3219.
9. Sharp, P. A., Hsu, M. T., Ohtsubo, N. & Davidson, N. (1972) *J. Mol. Biol.*, in press.
10. Richardson, C. C. (1969) *Annu. Rev. Biochem.* 38, 795-840.
11. Modrich, P. & Lehman, I. R. (1972) *Fed. Proc.* 31, 441.
12. Howard, B. V., Estes, M. K. & Pagano, J. S. (1971) *Biochim. Biophys. Acta* 228, 105-116.
13. Radloff, R., Bauer, W. & Vinograd, J. (1967) *Proc. Nat. Acad. Sci. USA* 57, 1514-1521.
14. Davis, R., Simon, M. & Davidson, N. (1971) in *Methods in Enzymology*, eds. Grossman, L. & Moldave, K. (Academic Press, New York), Vol. 21, pp. 413-428.
15. Wang, J. & Davidson, N. (1966) *J. Mol. Biol.* 19, 469-482.
16. Gray, H. B., Jr., Upholt, W. B. & Vinograd, J. (1971) *J. Mol. Biol.* 62, 1-19.
17. Wang, J. & Davidson, N. (1966) *J. Mol. Biol.* 15, 111-123.
18. Scheffler, I. & Sturtevant, J. (1969) *J. Mol. Biol.* 42, 577-580.
19. Davidson, N. & Szybalski, W. (1971) in *The Bacteriophage Lambda*, ed. Hershey, A. D. (Cold Spring Harbor Laboratories, New York), p. 60.
20. Wang, J. C. (1967) *J. Mol. Biol.* 28, 403-411.
21. Uhlenbeck, O. C., Martin, F. H. & Doty, P. (1971) *J. Mol. Biol.* 57, 217-229.
22. Jacobsen, H. & Stockmayer, W. H. (1950) *J. Chem. Phys.* 18, 1600-1606.
23. Gray, H. B., Jr. & Hearst, J. E. (1968) *J. Mol. Biol.* 35, 111-129.
24. Jackson, D. A. & Berg, P. (1972) *Proc. Nat. Acad. Sci. USA* 69, 2904-2909.
25. Davis, R. W. & Davidson, N. (1968) *Proc. Nat. Acad. Sci. USA* 60, 243-250.
26. Lee, C. S., Davis, R. W. & Davidson, N. (1970) *J. Mol. Biol.* 48, 1-22.
27. Schildkraut, C. & Lifson, S. (1965) *Biopolymers* 3, 195.

4

Reprinted from *Natl. Acad. Sci. (USA) Proc.* **69**:3389–3393 (1972)

Enzymatic Oligomerization of Bacteriophage P22 DNA and of Linear Simian Virus 40 DNA

(T4 and *E. coli* ligase/terminal and cohesive joining/zone sedimentation/electron microscopy)

VITTORIO SGARAMELLA

Department of Genetics, Lt. Joseph P. Kennedy Jr. Laboratories for Molecular Medicine, Stanford University Medical School, Stanford, California 94305

Communicated by J. Lederberg, September 5, 1972

ABSTRACT Linear double-stranded molecules of the circularly permuted and terminally redundant DNA of *Salmonella* bacteriophage P22 have been converted to oligomeric products in the presence of polynucleotide ligase coded for by the coliphage T4. The reaction has been monitored by sucrose density-gradient centrifugation and electron microscopy. It goes slowly and gives yields of 30–40%. The products are mainly dimers and trimers, but higher oligomers are also present.

DNA ligase extracted from uninfected *Escherichia coli* seems unable to perform a similar reaction, which is concluded to involve the fully base-paired termini.

Linear double-stranded molecules of simian virus(SV) 40 DNA, produced by the action of the bacterial restriction endonuclease R₁, are oligomerized by either ligase; therefore, this reaction seems to involve single-stranded cohesive ends. No mixed products could be found when P22 DNA and linear SV 40 DNA were exposed together to the T4 ligase.

The class of enzymes called DNA ligases was discovered by their property of repair of nicks (interruptions of one strand of a DNA duplex) and joining of DNA molecules annealed through single-stranded cohesive ends (1). In addition, bihelical DNA with fully base-paired termini can be joined in the presence of T4 polynucleotide ligase (2). A simplified scheme for the alternative modes of end-to-end joining is given in Fig. 1.

The terminal joining reaction has been studied with synthetic DNA duplexes (3, 4) corresponding to segments of the structural gene for an alanine transfer RNA of yeast (5).

Among naturally available products, DNA of the *Salmonella* bacteriophage P22 (6) is an advantageous substrate since: its linear double-helical molecules are circularly permuted and have redundant base-paired termini and they are short enough (13.5 μm) for solutions to be handled with little danger of shear degradation.

The present paper describes the oligomerization of P22 DNA in the presence of T4 ligase. The products have been characterized by zone sedimentation and electron microscopy, and evidence is presented that the reaction involves the apposition and the covalent joining of fully base-paired termini. The ligase extracted from uninfected *Escherichia coli* cells failed to catalyze a similar reaction. However, both enzymes oligomerize linear SV40 DNA, previously cleaved by the restriction endonuclease R₁ (7) of the *E. coli* resistance transfer factor RTF-I (H. Boyer, personal communication). Thus, R₁ linear SV40 DNA is inferred to have single-stranded cohesive ends, which cannot be joined to the base-paired

ends of P22 DNA to form interspecific "molecular translocations."

MATERIALS AND METHODS

DNA. Phage P22 DNA was extracted (8) from P22 *tsc₂*²⁹, a temperature-inducible mutant grown on ThyA 57, a thymine-requiring strain of *Salmonella typhimurium* LT. R₁ linear SV40 DNA was a gift of J. F. Morrow. The synthetic DNAs [oligo(dT), poly(dA), poly(d[A-T])] were generously donated by P. Modrich.

Enzymes. The T4 ligase was prepared according to Weiss *et al.* (9) and contained 700 units/ml, as defined by Gupta *et al.* (10). The *E. coli* exonuclease III and DNA ligase were gifts of P. Modrich and I. R. Lehman (11). Bacterial phosphatase was purchased from Worthington.

Oligomerization Reaction. A typical reaction mixture with the T4 ligase (12) contained (in 20 μl): 13.5 nM P22 [³H]DNA (expressed as duplex molecules); 20 mM Tris·HCl (pH 7.6) 10 mM MgCl₂; 35 μM ATP; 10 mM dithiothreitol; 1 mM EDTA; and 0.7–1.4 units of T4 ligase. The mixtures for the *E. coli* ligase contained in the same volume the same amount of DNA, plus 20 mM Tris·HCl (pH 8.0); 3 mM MgCl₂; 1 mM EDTA; 50 mg/ml bovine-serum albumin; 10 mM (NH₄)₂-SO₄; 27 μM DPN; and 1–2 units of ligase. Incubations were at 20° for several hours, as indicated, and were stopped by addition of 0.3 M NaCl–10 mM Tris·HCl–1 mM EDTA (pH 8.0) to a final volume of 200 μl, and heating at 65° for 3 min. The mixtures were then layered on a sucrose gradient (5–20%) with a cushion of 60% sucrose dissolved in the same buffer used to stop the reaction. Centrifugation was at 20° in a Beckman L2-65B ultracentrifuge with the SW40 rotor, or in an International B 60 with the SB283 rotor. 10-Drop fractions were collected from the tube bottom on filter paper discs and counted in Liquifluor (New England Nuclear Co.). The recovery of counts was generally more than 75%.

Electron Microscopy. Samples of DNA were prepared as described (8).

RESULTS

T4 Ligase Catalyzes the Oligomerization of P22 DNA. A mixture containing about 300 μg/ml of P22 [³H]DNA was incubated at 20° for 6 hr in the presence of the ligase extracted from T4-infected *E. coli* cells. The mixture was then diluted as described in *Methods*, heated at 65° for 3 min, and sedi-

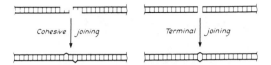

FIG. 1.   Alternative modes for the end-to-end joining of DNA molecules. The term "cohesive" joining is proposed for the reaction (*left half*) involving the interaction of single-stranded complementary sequences (often called cohesive ends). "Terminal" joining describes the alternative reaction, which takes place at completely base-paired, apposed termini (*right half*). By analogy with chromosomal rearrangements, "molecular translocations" is proposed as an apt term for the products resulting from either mode of joining of different DNA molecules.

mented through a neutral sucrose gradient. The resulting pattern is shown in Fig. 2a. The main band, which has gone through about 40% of the gradient, is unreacted DNA, as demonstrated by the pattern given in c, where the sedimentation profile of the original P22 is represented. In a, another peak is present ahead of the main band, (at about 56% of the gradient): according to the equation of Burgi and Hershey (13), its position is close to that expected for linear dimers. A third band can be seen at about 63% of the gradient: its

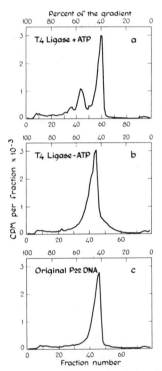

FIG. 2.   Zone sedimentation profile of the oligomerization mixture of P22 [³H]DNA in the presence of the T4 ligase. Two 20-μl reaction mixtures were kept at 20° for 6 hr: a, with ATP, b, without ATP. The original untreated P22 DNA sedimented as shown in c. Centrifugation was at 38,000 rpm for 3 hr at 20°.

position suggests it may be composed mainly of linear trimers. When ATP was omitted from the reaction mixture, thus depriving the T4 ligase of its cofactor, centrifugation gave the pattern shown in Fig. 2b. No fast-sedimenting material was present; on the contrary, some degradation products can be noticed at the right-hand side of the main peak, which exhibited the same sedimentation velocity as the original P22 DNA.

Additional evidence for the occurrence of the terminal joining reaction in the presence of the viral enzyme was obtained by sedimenting a reaction mixture similar to that described in Fig. 2a through an alkaline sucrose gradient. Two distinct peaks appeared in the T4 ligase-catalyzed reaction, and only one in the untreated and zero-time samples, as well as in the E. coli ligase-treated mixture.

The rate of the oligomerization reaction was monitored by withdrawing aliquots from a reaction mixture and sedimenting them through neutral sucrose gradients. The dimer fraction appears to increase monotonically for at least 20 hr (Fig. 3). Little can be said about the rate of appearance of the trimers, since the corresponding peak barely emerged from the background. The conversion of monomers into dimers and trimers seems to proceed at a biphasic rate, with about 50% of the total effect attained in less than 20% of the period analyzed.

Inability of the E. coli Ligase to Oligomerize P22 DNA. When P22 DNA was challenged with the E. coli ligase in the presence of the appropriate cofactor, DPN, no effect could be detected by sedimentation analysis. The results of a typical experiment are summarized in Fig. 4. Panel a gives the pattern obtained after incubation of the DNA with the viral enzyme and ATP: at least three distinct peaks can be observed. In contrast, a single band was obtained from the reaction mixture incubated with the bacterial enzyme and DPN. Panel c shows that the product sedimented as monomeric P22 DNA.

This negative result was not due to inhibition or inactivation of the E. coli ligase, as shown by two controls. When the reaction mixture contained, in addition to P22 DNA, oligo(dT)·poly(dA) with the 5′-phosphoryl termini of the

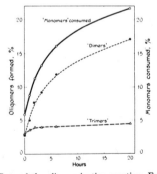

FIG. 3.   Rate of the oligomerization reaction. From an 80-μl reaction mixture, 10-μl aliquots were withdrawn before and 0, 1, 2, 3, 6, and 20 hr after the addition of ligase. The samples were diluted, heated, kept at 0°, and then centrifuged. Of each gradient, only the regions corresponding to the monomers, dimers, and trimers have been considered for the calculations. The values at 0 time are due to base-line noise.

oligo(dT) segments labeled, almost 65% of the ³²P had been converted to phosphatase-resistance after the reaction, as found in the absence of P22 DNA. In another control, an aliquot of the usual P22 DNA reaction mixture was assayed for *E. coli* ligase activity at the end of the incubation; with [³H]poly[d(A-T)], the extent of conversion to exonuclease III-resistant circles (11) was within 5% of the level obtained with the same amount of fresh *E. coli* enzyme in a parallel experiment. Experiments performed and analyzed at 5° also showed no oligomerization.

In consideration of the relative impurity of the viral enzyme, as compared to the purified bacterial ligase, it seemed appropriate to treat the P22 DNA with the *E. coli* enzyme after, or simultaneously with, the T4 enzyme, but without ATP. After prolonged (7 hr) incubation with the two enzymes

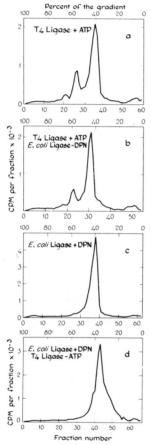

FIG. 4. Comparison of the T4 and *E. coli* ligases in the terminal oligomerization of P22 DNA. The incubation lasted 20 hr. The pattern shown in *a* was obtained with T4 ligase. In *b*, the mixture was as in *a*, plus the *E. coli* ligase (no DPN). In *c*, the mixture contained all the components required by the *E. coli* ligase. In *d*, the mixture was as in *c*, plus the T4 ligase (no ATP). Centrifugations were in an IEC ultracentrifuge, with an SB283 rotor, at 38,000 rpm and 20° for 3 hr for *a*, *b*, and *c* and 2.5 hr for *d*.

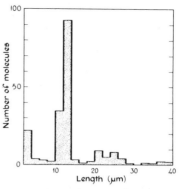

FIG. 5. Size frequency distribution in an unfractionated T4 joining mixture. 25 Photomicrographs were taken at random on grids spread with an unfractionated mixture. The length of 177 molecules was measured with a map measurer; fragments shorter than 2 μm were neglected.

and DPN, but no ATP, sedimentation analysis of the reaction mixture gave the profile shown in Fig. 4*d*: no fast-sedimenting material had been formed. Rather, the appearance of smaller fragments could again be noticed at the skewed right-hand side of the main peak, which otherwise sedimented as monomeric P22 DNA. Similar results were obtained when the DNA was first incubated with the T4 ligase (minus ATP), then treated with the *E. coli* enzyme and DPN. The *E. coli* ligase does not contain any inhibitor of the terminal joining reaction as displayed by the T4 enzyme. When the mixture containing the T4 ligase and ATP was also treated with the *E. coli* enzyme (but without DPN), the reaction went almost normally as shown in the *b* panel of Fig. 4. The extent of oligomerization was only slightly less than that obtained in the mixture containing only the T4 ligase and its cofactor ATP (Fig. 4*a*).

Electron Microscopy. After oligomerization, the unfractionated mixture, as well as the isolated and dialyzed monomers, dimers, and trimers, were spread according to a slight modification (8) of the aqueous technique of Davis *et al.* (14). About 25 micrographs were analyzed for the size distribution of the products. The proportion of dimers and trimers is close to that obtained by zone sedimentation: of 177 measurable molecules, 31 had a length between 20 and 30 μm–attributable to dimers–and 9 gave lengths equal to or higher than 30 μm–attributable to trimers and higher oligomers. The results are summarized in Fig. 5. No circles were found after more than 500 molecules were screened.

Oligomerization of Linear SV40 DNA by T4 and E. coli Ligases: Failure to Form Molecular Translocations Between SV40 DNA and P22 DNA. Restriction endonuclease R$_I$ extracted from *E. coli* carrying the resistance transfer factor RTF-I cleaves circular SV40 DNA into unique linear molecules (7). By analogy with the mode of action of other known restriction enzymes, particularly the one extracted from *H. influenzae* (15), it was thought that the *E. coli* endonuclease R$_I$ would produce fully base-paired ends capable of undergoing, like P22 DNA, terminal joining with the T4 ligase. A reaction was therefore performed in which P22 [³H]DNA

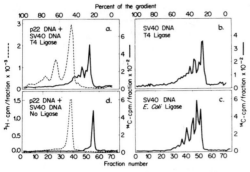

FIG. 6.   Cohesive oligomerization of linear SV40 DNA. Four 20-μl reaction mixtures were set up containing: (a) 8 μg of P22 [^{3}H]DNA and 0.4 μg of linear SV40 [^{14}C]DNA (25,000 cpm/μg), plus the components for T4 ligation; (b) the same amount of SV40 DNA, plus components for the *E. coli* ligase reaction; (c) the same amount of SV40 DNA, plus the components for T4 ligase, (d) a mixture of the two DNAs as in (a) without enzymes. Incubations were at 20° for 15 hr; centrifugation was for 4.5 hr.

was mixed with linear SV40 [^{14}C]DNA in a 2:1 molar ratio. Incubation was performed in the presence of the T4 ligase, and the products were characterized by zone sedimentation. No translocations could be detected, while both DNAs were clearly oligomerized (Fig. 6a and d), implying a difference between the termini of the two kinds of DNAs. Linear SV40 DNA alone could be oligomerized by the *E. coli* ligase (Fig. 6b), as well as by the T4 enzyme (Fig. 6c). This result indicates that the ends produced in SV40 DNA by the restriction endonuclease R$_I$ are single-stranded and cohesive (unlike those of P22 DNA) and carry 5'-phosphoryl and 3'-hydroxyl groups, as also shown by other workers (16).

DISCUSSION

The ability of the T4 ligase to join covalently two different DNA molecules at their base-paired termini has been demonstrated with short synthetic DNA duplexes carrying different radioactive labels (^{32}P and ^{33}P) at the terminal 5'-phosphoryl groups (3, 4).

The present results show that this enzyme catalyzes the oligomerization of P22 DNA molecules with formation of alkali-resistant bonds. The properties of this substrate (6) argue against a mechanism of reaction mediated by the annealing of either naturally pre-existing or accidentally introduced complementary single-stranded regions at the termini of the molecules. If that were the case, then these sites should cause the formation of hydrogen-bonded oligomers, and should be as good substrates for the *E. coli* as for the T4 ligase. Both possibilities have been investigated with negative results. Samples annealed alone, or in the presence of the T4 ligase but without the necessary cofactor, ATP, have been centrifuged in sucrose gradients without the usual heating step at the end of the incubation or at 5°. Unpublished experiments show that no faster-sedimenting material was formed. With regard to the accidental introduction of single-stranded regions, resulting from exonucleolytic contaminants of the T4 ligase, very low levels of nucleolytic activities have been detected in this preparation of the enzyme, even after

extensive incubation of both synthetic and natural DNAs. On the other hand, the removal of 5–10 nucleotides would be sufficient to stabilize the interaction of double-helical molecules by their exposed single-stranded termini, provided they are complementary (2, 10, 12). Such limited degradation even if present, could easily remain unnoticed in these experimental conditions. However, it should lead to the exposure of single-stranded regions of random sequence (given the circular permutation of the DNA), as such unable to cause the annealing of 30–40% of the molecules usually found in these experiments. Only after about 2–3% (6) of the DNA was exonucleolytically digested would complementary regions become exposed. No circles have been found by electron-microscope analysis, as expected for the DNA concentration selected, which is several-fold higher than that known to allow intramolecular annealing (17).

Both the sedimentation values of the peaks from the zone centrifugations and the absolute length measurements of spread molecules in electron micrographs are in good agreement with the values known for the monomers (6) and expected for the linear oligomers. By the two independent techniques, I found that 30–40% of the starting material has undergone terminal joining under the conditions described.

The bacterial and the viral ligases are known to differ in their cofactor requirements, as well as in their ability to use DNA–RNA substrates. (ref. 18 and O. Westergaard, D. Brutlag, and A. Kornberg, personal communication). Without excluding more complicated hypotheses, these experiments do not support the postulation of "helping" or "activating" factor(s) present in the T4 ligase preparation, unless they too are ATP-dependent. It seems therefore possible to consider the terminal-joining as an additional distinctive feature between the two enzymes.

High levels of enzyme have been used in these experiments. If a molecular weight is assumed for the ligase of 5×10^4, the enzyme to substrate ratio was close to unity; preliminary data indicate that the extent of the reaction responded to the amount of enzyme. Instability of recent preparations of the enzyme has delayed further investigation of this point. At present, I cannot rule out that the "terminal ligase" is different from the traditional sealing enzyme, and represents only a minor contaminant in the preparation, [even though preliminary experiments show that the protein species that binds [α-^{32}P]AMP gives a single band in sodium dodecyl sulfate–polyacrylamide gel electrophoresis (unpublished results of D. Uyemura, P. Modrich, and V. Sgaramella)]. If the above hypothesis is true, it could help to explain the slow rate of the reaction (Fig. 3), as compared to the cohesive joining in different experiments.

I am most grateful to Dr. J. Lederberg for continuous help and rigorous criticisms, to Dr. A. D. Kaiser for his suggestions, and to H. Bursztyn for skillful technical assistance. Dr. P. E. Lobban donated samples of P22 DNA, and useful suggestions, as did P. Modrich and F. Schachat. The Department of Biochemistry at Stanford provided facilities and expertise for the electron microscopy. The study was made possible by grants from NIH (AI-05160), NSF (GB-29024), the National Cystic Fibrosis Foundation, and from the U.S.A.–Italy Science Cooperative Program (CNR. No. 115-0308-04633 to the University of Pavia).

1. Weiss, B. & Richardson, C. C. (1967) *Proc. Nat. Acad. Sci. USA* **57**, 1021–1028; Zimmerman, S. B., Little, J. W., Oshinsky, C. K. & Gellert, M. (1967) *Proc. Nat. Acad. Sci. USA* **57**, 1841–1848; Olivera, B. M. & Lehman, I. R. (1967) *Proc.*

Nat. Acad. Sci. USA **57,** 1426–1433; Gefter, M. L., Becker, A. & Hurwitz, J. (1967) *Proc. Nat. Acad. Sci. USA* **58,** 240–247; Cozzarelli, N. R., Melechen, N. E., Jovin, T. M. & Kornberg, A. (1967) *Biochem. Biophys. Res. Commun.* **28,** 578–585.

2. Sgaramella, V. & Khorana, H. G. (1972) *J. Mol. Biol.,* in press.
3. Sgaramella, V., Van de Sande, J. H. & Khorana, H. G. (1970) *Proc. Nat. Acad. Sci. USA* **67,** 1468–1475.
4. Sgaramella, V. (1971) *Fed. Proc.* **30,** 1524.
5. Agarwal, K. L., Buchi, H., Caruthers, M. H., Gupta, N., Khorana, H. G., Kleppe, K., Kumar, A., Ohtsuka, E., RajBhandary, U. L., Van de Sande, J. H., Sgaramella, V., Weber, H. & Yamada, T. (1970) *Nature* **227,** 27–34.
6. Thomas, C. A., Jr., Kelly, T. J., Jr. & Rhoades, M. (1968) *Cold Spring Harb. Symp. Quant. Biol.* **33,** 417–424.
7. Morrow, J. F. & Berg, P. (1972) *Proc. Nat. Acad. Sci. USA* **69,** 2904–2909.
8. Lobban, P. E. (1972) Ph. D. Thesis, Stanford University.
9. Weiss, B., Jacquemin-Sablon, A., Live, T. R., Fareed, G. C. & Richardson, C. C. (1968) *J. Biol. Chem.* **243,** 4543–4555.

10. Gupta, N. K., Ohtsuka, E., Weber, H., Chang, S. H. & Khorana, H. G. (1968) *Proc. Nat. Acad. Sci. USA* **60,** 285–292.
11. Modrich, P. & Lehman, I. R. (1970) *J. Biol. Chem.* **245,** 3626–3631.
12. Sgaramella, V. & Khorana, H. G. (1972) *J. Mol. Biol.,* in press.
13. Burgi, E. & Hershey, A. D. (1963) *Biophys. J.* **3,** 309–321.
14. Davis, R. W., Simon, M. & Davidson, N. (1971) in *Methods in Enzymology,* eds. Grossman, L. & Moldave, K. (Academic Press, New York), Vol. 21, pp. 413–428.
15. Kelly, T. J., Jr. & Smith, H. O. (1970) *J. Mol. Biol.* **51,** 393–409.
16. Mertz, J. E. & Davis, R. W. (1972) *Proc. Nat. Acad. Sci. USA* **69,** 3370–3374; Hedgpeth, J., Goodman, H. & Boyer, H. (1972) *Proc. Nat. Acad. Sci. USA* **69,** in press.
17. Wang, J. C. & Davidson, N. (1968) *Cold Spring Harb. Symp. Quant. Biol.* **33,** 409–415.
18. Fareed, G. C., Wilt, E. M. & Richardson, C. C. (1971) *J. Biol. Chem.* **246,** 925–932; Kleppe, K., Van de Sande, J. H. & Khorana, H. G. (1970) *Proc. Nat. Acad. Sci. USA* **67,** 68–73.

5

Copyright © 1973 by the American Society for Microbiology
Reprinted from J. Virol. **12**:741–747 (1973)

Transfection of *Escherichia coli* Spheroplasts

III. Facilitation of Transfection and Stabilization of Spheroplasts by Different Basic Polymers

WILLIAM D. HENNER, INGRID KLEBER, AND ROLF BENZINGER

Department of Biology, University of Virginia, Charlottesville, Virginia 22901

In a previous paper (5), protamine sulfate was shown to facilitate transfection with seven different double-stranded phage DNAs by at least 300-fold. A major difficulty with this and other published *Escherichia coli* transfection systems is that high competence levels are very unstable and that only 1 in 10 spheroplast preparations gives high levels of competence (5).

Other basic polymers were now tested for their effects on transfection. When polyamines as well as protamine sulfate were added to spheroplasts, stable and highly competent preparations were reproducibly obtained.

MATERIALS AND METHODS

The strains used in this study have been previously described (5). Polymers were purchased from the following sources: heparin, calf thymus histones, arginine-rich histones, lysine-rich histones, and poly-L-lysine (mol wt 3,000–5,000) from Nutritional Biochemicals Corp.; spermine tetrahydrochloride, spermidine trihydrochloride, putrescine dihydrochloride, and cadaverine dihydrochloride from Calbiochem. 3,3′ diaminopropylamine; N', N'-diethyl-1,4-pentanediamine; 4-dodecyldiethylene triamine; tetraethylenepentamine; N, N'-bis (2-aminoethyl) 1,3-propanediamine; N, N'-bis (3-aminopropyl)-1,3-propanediamine; 1,1,4,7,7-pentamethyldiethylene-triamine; and N, N'-bis [(3-(2 aminoethylamine) propyl)] ethylene diamine from Eastman Ko-

dak. DEAE-dextran was purchased from Pharmacia, Inc.; poly-L-ornithine HBr (manufacturer's mol wt 76,000) was from Mann Research Laboratories; and poly-L-lysine HCl (mol wt 16,400; 75,000; and 200,000) and poly-L-arginine HCl (mol wt 10,000–20,000) were from Miles Laboratories.

The purification of bacteriophage and extraction of phage DNAs has been described previously (5). One unit of absorbance at 260 nm was assumed to represent 40 μg of double-stranded DNA. Standard growth medium for bacterial cultures was the following modification of the formula òf Fraser and Jerrel (7): 4.5 g of KH_2PO_4, 8.3 g of Na_2HPO_4, 15 g of Casamino Acids (Difco), 1 g of NH_4Cl, 10 ml of glycerol, 2.5 ml of 10% $MgSO_4$, and 0.3 ml of 0.1 M $CaCl_2$, per liter.

The optimal method developed for preparing spheroplasts was altered from the previous procedure (5) in the following manner: the absorbance of the overnight culture was between 8 and 12 in the Gilford spectrophotometer (at 550 nm, bacteria diluted 10-fold before measurement) when 2 ml was transferred to 400 ml of fresh medium. The culture was shaken vigorously at 37 C. When the absorbance of the culture reached 0.6, the cells were centrifuged at room temperature and resuspended in: 2.1 ml of 1.5 M sucrose, followed by 0.6 ml of 30% Povite albumin, 0.12 ml of freshly dissolved lysozyme (2 mg/ml in 0.25 M Tris buffer, pH 8.1), and 0.24 ml of unbuffered EDTA.

After 2 min, 55 ml of minimal PA medium (100 g of sucrose, 1 g of glucose, and 0.5 g of Casamino Acids [Difco] per liter of water) was added. After 12 min at room temperature without stirring, 1.2 ml of MgSO₄ and 0.15 ml of 1% protamine sulfate, as well as 0.1 ml of a fresh solution of spermine tetrahydrochloride (250 mg/ml in sterile water), were added. Other basic polymers were also added at this time; if they were added to spheroplasts at later times, their effect on competence was always weaker (no more than 10% of the control). The spheroplasts were stored on ice and assayed for competence with φX174 DNAs 3 h later, since full competence was not developed till that time (5). Such spheroplasts worked best with lambda and φX174 DNAs. For T7 DNA assays, protamine sulfate was often omitted to obtain higher competence levels. For T5 assays, spermidine trihydrochloride (250 µg/ml) or N,N'-bis (3 aminopropyl)-1,3-propanediamine (180 µg/ml) was sometimes substituted for spermine tetrahydrochloride. For lambda and φX174 DNA assays, spheroplasts were stable for 3 to 5 days. For T5 assays, they were stable for no longer than 3 days. For T7 DNA assays without protamine sulfate, spheroplasts were competent between 24 and 72 h after preparation.

For the purposes of this paper, facilitation of transfection is defined as an increase in transfection efficiency at any time after the preparation of the spheroplasts. Stabilization of transfection is defined as the maintenance of a high level of competence any time beyond the first 10 h.

RESULTS

Table 1 presents the results obtained when a variety of basic polymers was used to facilitate transfection by φX174 replicative form DNA. None of the compounds was more than 10% as efficient as protamine sulfate in facilitating transfection, not even the chemically closely related poly-L-arginine. The optimum concentration of the polymers depended inversely on the molecular weight; for example, DEAE-dextran (molecular weight 2×10^6) was most effective at 0.3 µg/ml, whereas spermine tetrahydrochloride (mol wt 348) was best at 400 µg/ml. The concentration optima observed in all experiments were relatively sharp (data not shown), as has also been demonstrated by Melechen, Hudnik-Plevnik, and Pfeifer (19) for protamine sulfate.

Since some variation in the facilitation of transfection, depending on the age of the spheroplasts, was observed (Table 1, column 4), transfection assays were performed in the presence of basic polymers at various times (Table 2). Surprisingly, spermine (which had little or no effect of transfection with 3-h-old spheroplasts [see Table 1]) dramatically increased the efficiency of transfection by T7 DNA with 48- and 72-h-old spheroplasts when protamine sulfate had already lost effect. Also, the effect of spermine on φX174 replicative form transfection was minimal, showing that facilitation of transfection by spermine depends on the DNA employed (Table 2). Most fortuitously, adding both protamine sulfate and spermine resulted in the early facilitation of transfection (which is observed with protamine sulfate alone) as well as the stabilization of the spheroplasts (which is observed with spermine alone). Addition of both protamine sulfate and a polyamine was adopted as the new standard procedure for preparing spheroplasts (see Materials and Methods).

Various substances were now tested in combination with protamine sulfate to see if they

TABLE 1. *Facilitation of φX174 RF transfection by basic polymers with E. coli W3350 spheroplasts 1 to 3 h old*

Basic polymer[a]	Manufacturer's mol wt	Average facilitation over control	Range of facilitation	Optimum concn (µg/ml)	No. of expt
Protamine sulfate	Heterogeneous	215×	2–1300×	25	18
Filtered protamine[b] sulfate	10 to 50,000	248×	38–902×	6	5
Poly-L-arginine	10 to 20,000	26×	1–145×	0.5	9
Poly-L-lysine	3 to 5,000	20×	3–73×	3	6
Poly-L-ornithine	70,000	19×	2–52×	2	4
Histones		7×	1–9×	15	6
Poly-L-lysine	16,400	5×	4–7×	0.5	4
Poly-L-lysine	75 to 200,000	4×	2–7×	0.5	5
DEAE-dextran	2,000,000	3×	2–6×	0.3	5
Spermine[c]	348	2×	1–4×	400	5

[a] All compounds were added during the preparation of the spheroplasts (see Materials and Methods) since later addition produced much weaker effects.

[b] Protamine sulfate (Eli Lilly) was passed successively through Amicon filters with molecular weight cutoffs of 10,000 and 50,000. This procedure removed 75% of the biuret-positive material.

[c] The following compounds were ineffective in facilitating transfection: heparin, spermidine trihydrochloride, putrescine dihydrochloride, cadaverine hydrochloride; 3,3'-diaminopropylamine; N'N'-diethyl-1,4-pentanediamine; 4-dodecyl-diethylene triamine; tetraethylene pentamine; N,N'-bis (2-aminoethyl)-1,3,propanediamine; N,N'-bis (3-aminopropyl 1,3-propanediamine; 1,1,4,7,7-pentamethyldiethylene triamine; N,N'-bis[(3-(2 amino ethylamine) propyl] ethylene diamine.

TABLE 2. *Dependence of facilitation of transfection on the age of the spheroplasts*

Transfecting DNA	Age of spheroplasts when assayed (h)	Infective centers found with spheroplasts containing[a]			
		No addition	Protamine sulfate (25 μg/ml)	Spermine (HCl)₄ (400 μg/ml)	Protamine sulfate (25 μg/ml) plus spermine (HCl)₄ (400 μg/ml)
φX174 RF (5 × 10⁴ molecules)	3	56	1,046	50	1035
φX174 RF (5 × 10⁴ molecules)	24	2	18	2	82
φX174 RF (5 × 10⁴ molecules)	48	0	7	2	7
T7 DNA (2.6 × 10⁹ molecules)	3	0	65	119	82
T7 DNA (2.6 × 10⁹ molecules)	24	0	14	231	285
T7 DNA (2.6 × 10⁹ molecules)	48	1	5	434	1,165
T7 DNA (2.6 × 10⁹ molecules)	72	0	0	85	
Lambda DNA (8 × 10⁶ molecules)	3	1	2,260	225	2,500
Lambda DNA (8 × 10⁶ molecules)	24	0	1,403	862	3,500
	48	5	7,000	4,300	10,000

[a] All compounds were added during the preparation of the spheroplasts since later addition produced much weaker effects.

could exert a stabilizing effect on the spheroplasts. Table 3 shows that a number of polyamines other than spermine could stabilize spheroplasts for transfection. The following conclusions are drawn from this table and similar results from other experiments. (i) Protamine sulfate gave optimal facilitative effects during the first 10 h, as previously noted (5) and older spheroplasts rapidly lost competence. (ii) For lambda and φX174 DNA transfection, spermine as well as all of the other polyamines listed in the table strongly stabilized spheroplasts for 96 h and longer in the presence of protamine sulfate. (iii) For T5 and T7 DNA transfection, only spermine, N,N'-bis (3 aminopropyl)- 1,3-propanediamine, and perhaps spermidine stabilized spheroplasts. Thus the stabilization of spheroplasts varies considerably depending on the DNA used in the transfection assay.

Table 4 presents a summary of recent work on the transfection of *E. coli* spheroplasts. Four important criteria for efficient and reproducible transfection assays are listed: maximum efficiency of transfection, the fraction of spheroplasts yielding maximum efficiencies, the variation in the competence of the spheroplasts, and the stability of spheroplasts for transfection assays.

DISCUSSION

The facilitation of transfection by protamine sulfate shows a remarkably high specificity; only spermine fully substituted for protamine sulfate with some DNAs while 11 other polyamines, DEAE-dextran, histones, and polymers of basic amino acids had little or no effect (Table 1). Poly-L-arginine, a close chemical relative of protamine sulfate, was much less effective than protamine sulfate. In other transfection systems, high specificity has been observed: Koch, Quintrell, and Bishop (12) found DEAE-dextran and poly-L-ornithine most effective in stimulating transfection by poliovirus RNA, whereas Aoki and Takebe (1) found poly-L-ornithine effective in facilitating transfection of tobacco spheroplasts by tobacco mosaic virus RNA; Kohn and Green (13) and Green (8) showed that spermine enhances transfection of *Bacillus subtilis* by SP82 phage DNA.

Each basic polymer had a sharp concentration optimum for the facilitation of transfection. Higher-molecular-weight compounds were effective at lower concentrations (Table 1). Furthermore, facilitation of transfection depended on the phage DNA used in the assay (Table 2). Hopefully, these observations will aid efforts to facilitate transfection in other assay systems.

The widespread facilitation of transfection by basic polymers observed in systems as diverse as *E. coli*, mammalian, and tobacco cells raises the possibility that protamines or polyamines are components of natural competence factors. Indeed, Leonard and Cole (16) have isolated a natural competence factor with polyamine- or protamine-like characteristics from *Streptococcus challis*. Experiments were performed to measure the facilitating effect of protamine

TABLE 3. *Stabilization of E. coli W3350 spheroplasts by polyamines in the presence of protamine sulfate*

Additions to spheroplasts[a]	φX174 RF (5 × 10⁴ molecules) with spheroplasts aged for:					Lambda DNA (1 × 10⁷ molecules) with spheroplasts aged for:					T5 DNA (2.4 × 10⁹ molecules) with spheroplasts aged for:					T4 DNA (2.6 × 10⁹ molecules) with spheroplasts aged for:				
	2 h	24 h	48 h	72 h	96 h	3 h	24 h	48 h	72 h	96 h	2 h	24 h	48 h	72 h	96 h	2 h	24 h	48 h	72 h	96 h
None	38	11	51	50	55	0	0	0	0	0	0	0	0	0	0	0	0	0	0	0
25 µg of protamine sulfate per ml	413	92	98	85	72	2,210	847	268	16	7	45	3	0	0	0	4	1	0	00	0
Protamine sulfate plus 400 µg of spermine (HCl)₄ per ml	110	712	681	684	386	1,736	3,374	3,092	3,882	4,204	170	96	97	135	234	2	8	13	14	26
Protamine sulfate plus 180 µg of N,N'-bis (3 aminopropyl)-1,3 propanediamine per ml	310	329	276	281	263	1,936	4,056	3,788	3,980	3,653	914	462	433	530	485	10	31	28	25	23
Protamine sulfate plus 250 µg of spermidine (HCl)₃ per ml	654	509	591	164	88	1,960	2,605	1,953	846	2,954	1,824	385	159	50	50	6	10	21	10	9

[a] The following compounds were nearly as effective as those listed: 260 µg of N,N'-bis [3-(2-aminoethylamine) propyl]ethylene diamine per ml; 190 µg of tetraethylene-pentamine per ml; and 160 µg of N,N'-bis (2-aminoethyl)-1,3-propane-diamine per ml. The following compounds were much less effective than those listed: 3,3' diaminodipropylamine, heparin, N¹,N¹-diethyl-1,4-pentanediamine, 4-dodecyldiethylene triamine, putrescine dihydrochloride, cadaverine dihydrochloride, and 1,1,4,7,7 pentamethyl-diethylenetriamine. All of the compounds had to be added during spheroplast preparation to be effective.

TABLE 4. *Efficiency of transfection, variations in competence, and stability of competent cells used for E. coli transfection assays*

DNAs used for transfection	Type of competent cell	Maximum efficiency of transfection	Spheroplasts giving maximum efficiency of transfection (%)	Variation in efficiency of transfection	Stability of spheroplasts (h)	References
T1	Lysozyme-EDTA spheroplasts plus protamine sulfate	5×10^{-7}				10
Lambda	Helper-infected cells plus protamine sulfate	10^{-3}		10^{-3} to 10^{-5}		17
Lambda	Calcium-shocked cells	2×10^{-4}		2×10^{-4} to 2×10^{-5}		18
φX174 replicative form	Lysozyme – EDTA spheroplasts plus protamine sulfate	10^{-1}	10	10^{-1} to 10^{-3}	10	4
Lambda	Lysozyme – EDTA spheroplasts plus protamine sulfate	10^{-3}	10	10^{-3} to 10^{-5}	10	4
fd replicative form	Lysozyme – EDTA spheroplasts plus protamine sulfate	10^{-6}	10	10^{-6} to 10^{-8}	10	4
T7	Lysozyme – EDTA spheroplasts plus protamine sulfate	3×10^{-7}	10	3×10^{-7} to 3×10^{-9}	10	4
T4	Lysozyme – EDTA spheroplasts plus protamine sulfate	10^{-5}	10	10^{-5} to 10^{-7}	10	4
T5	Lysozyme – EDTA spheroplasts plus protamine sulfate	3×10^{-6}	10	3×10^{-8} to 3×10^{-8}	10	4
P22	Lysozyme – EDTA spheroplasts plus protamine sulfate	3×10^{-9}	10	3×10^{-9} to 3×10^{-11}	10	4
T2, T4	Freeze-shocked cells	10^{-8}				11
T4	Penicillin spheroplasts	2×10^{-9}				2
Lambda	Lysozyme-EDTA spheroplasts plus protamine sulfate	1×10^{-5}	20	10^{-5} to 6×10^{-5}	48	19
Lambda	Lysozyme-EDTA spheroplasts plus protamine sulfate	4×10^{-4}			1	24
T4	Lysozyme-EDTA spheroplasts plus protamine sulfate	10^{-6}			1	24
Lambda	Lysozyme-EDTA spheroplasts plus protamine sulfate and polyamine	10^{-3}	40	10^{-3} to 1×10^{-4}	48–120	This paper
T7	Lysozyme-EDTA spheroplasts plus protamine sulfate and polyamine	3×10^{-7}	20	10^{-6} to 3×10^{-8}	24–48	This paper
T5	Lysozyme-EDTA spheroplasts plus protamine sulfate and polyamine	3×10^{-6}	20	3×10^{-6} to 10^{-7}	24–48	This paper

sulfate in the streptococcal transformation system and of the streptococcal competence factor in the *E. coli* transfection system. No facilitation was observed in either case (D. G. Leonard, personal communication; R. Benzinger, unpublished observations). The failure of these experiments might be attributed to the high specificity of transfection factors observed above and the high specificity of streptococcal competence factors (20); they might also be caused by a fundamental difference between the transfection and transformation processes (25). Further studies of the basic proteins in microorganisms should yield interesting results; Kuo and August (14) have already shown that basic proteins are a requirement for phage RNA replicase and that mammalian histones and protamine sulfate will substitute in the reaction.

Previously, our spheroplasts were stable for at most 10 h for transfection assays with large phage DNAs, and only 1 out of 10 preparations gave high competence levels (5). The combination of spermine and protamine sulfate addition to spheroplasts yields stable and highly competent spheroplasts (Table 3). This discovery makes it possible to routinely use the assay in large-scale experiments and to check the reproducibility of results with the same spheroplast preparation.

Spermine may also facilitate transfection as well as stabilize *B. subtilis* competent cells (8, 13). This function is in agreement with the data of Tabor (22), who found that the addition of spermine to spheroplasts greatly stabilized them to osmotic shock. On the other hand, Razin and Rozansky (21) noted that spermine also exerts a toxic effect on *E. coli*. This toxicity is considerably reduced for nonmetabolizing cells, and it is surely no coincidence that our spheroplasts are most stable in minimal medium, a revival of earlier procedures (3). A factor which had not been assessed in this study is the sensitivity of polyamines to oxidation by air or intracellular enzymes (23); the oxidized products may be toxic to spheroplasts. When other polyamines were tested for their stabilizing effect on spheroplasts, several besides spermine (see Table 3) were found to be active; for transfection by lambda and ϕX174 DNAs as many as six different relatives of spermine were active; however, for T5 and T7 DNA transfection, mainly N,N'-bis (3 aminepropyl)-1,3-propanediamine and spermidine but not the other polyamines were effective (Table 3). These results are also in marked contrast to those obtained in the facilitation experiments (Table 1) where only spermine could be substituted for protamine sulfate. Thus, facilitation of

transfection and stabilization of spheroplasts may be entirely different processes, especially since some compounds can only perform one of these two functions. In any case, the beneficial effects of polyamines for transfection of *E. coli* spheroplasts once again point up the importance of polyamines in living systems, as has been extensively documented in a recent book (6) and symposium (9).

ACKNOWLEDGMENTS

This investigation was supported by grant from the National Institute for Allergy and Infectious Diseases (AI-08572), by a Public Health Research Career Development Award (no. 6-K04-GM-50284 GEN) from the National Institute of General Medical Sciences, and a National Science Foundation subgrant from the Center for Advanced Studies (NSF GU-1531). We thank Charles Mitchell for the analytical ultracentrifuge runs of all the phage DNAs and C. Leonard for her generous gift of *Streptococcus challis* competence factor and helpful discussion.

LITERATURE CITED

1. Aoki, S., and I. Takebe. 1969. Infection of tobacco mesophyll spheroplasts by tobacco mosaic ribonucleic acid. Virology 39:439–448.
2. Baltz, R. H. 1971. Infectious DNA of bacteriophage T4. J. Mol. Biol. 62:425–437.
3. Benzinger, R., J. Delius, R. Jaenisch, and P. H. Hofschneider. 1967. Infectious nucleic acids of *Escherichia coli* bacteriophage. 10. Preparation and properties of *Escherichia coli* competent for infectious DNA from bacteriophages ϕX174 and M13 and RNA from bacteriophage M12. Eur. J. Biochem. 2:414–428.
4. Benzinger, R., and I. Kleber. 1971. Transfection of *E. coli* and *S. typhimurium* spheroplasts: host-controlled restriction of infective bacteriophage P22 DNA. J. Virol. 8:197–202.
5. Benzinger, R., I. Kleber, and R. Huskey. 1971. Transfection of *E. coli* spheroplasts. 1. General facilitation of double-stranded deoxyribonucleic acid infectivity by protamine sulfate. J. Virol. 7:646–650.
6. Cohen, S. S. 1971. Introduction to the polyamines. Prentice-Hall, Englewood Cliffs, New Jersey.
7. Fraser, D., and E. A. Jerrel. 1953. The amino acid composition of bacteriophage T3. J. Biol. Chem. 205:291–297.
8. Green, D. M. 1966. Intracellular inactivation of infective SP82 DNA. J. Mol. Biol. 22:1–13.
9. Herbst, E. J., and U. Bachrach (ed.) 1970. Metabolism and biological function of polyamines. Ann. N.Y. Acad. Sci. 171:691–1009.
10. Hotz, G., and R. Mauser. 1969. Infectious DNA from coliphage T1. I. Some properties of the spheroplast assay system. Mol. Gen. Genet. 104:178–194.
11. Hua, S., P. B. Mackal, B. Werninghaus, and E. A. Evans, Jr. 1971. Infectious DNA preparations from T2 and T4 bacteriophages. Virology 46:192–199.
12. Koch, G., N. Quintrell, and J. M. Bishop. 1966. An agar cell-suspension plaque assay for isolated viral RNA. Biochem. Biophys. Res. Commun. 24:304–309.
13. Kohn, K. W., and D. M. Green. 1966. Transforming activity of nitrogen mustard-crosslinked DNA. J. Mol. Biol. 19:289–302.
14. Kuo, C. H., and F. T. August. 1972. Histone or basic protein required for replication of bacteriophage RNA. Nature N. Biol. 73:105–107.
15. Lawhorne, L., I. Kleber, C. Mitchell, and R. Benzinger.

1973. Transfection of *Escherichia coli* spheroplasts. II. Relative infectivity of native, denatured, and renatured lambda, T7, T5, T4, and P22 bacteriophage DNAs. J. Virol. **12**:733–740.

16. Leonard, C. G., and R. M. Cole. 1972. Purification and properties of streptococcal competence factor isolated from chemically defined medium. J. Bacteriol. **110**:273–280.

17. Mackinlay, A. G., and A. D. Kaiser. 1969. DNA replication in head mutants of bacteriophage lambda. J. Mol. Biol. **39**:679–683.

18. Mandel, M., and A. Higa. 1970. Calcium-dependent bacteriophage lambda infection. J. Mol. Biol. **53**:159–162.

19. Melechen, N. E., T. A. Hudnik-Plevnik, and G. S. Pfeifer. 1972. Increased stability and reproducibility of *Escherichia coli* spheroplasts in the transfection assay of phage lambda DNA with polyethylene glycol instead of sucrose. Virology **47**:610–617.

20. Pakula, R. 1966. Kinetics of provoked competence in streptococcal cultures and its specificity. Int. Proc. Symp. Bacterial Transformation and Bacteriocinogeny **6**:33–42. Akademiai Kiado, Budapest.

21. Razin, S., and R. Rozansky. 1959. Mechanism of the antibacterial action of spermine. Arch. Biochem. Biophys. **81**:36–43.

22. Tabor, C. W. 1962. Stabilization of protoplasts and spheroplasts by spermine and other polyamines. J. Bacteriol. **82**:1101–1111.

23. Tabor, H., and C. W. Tabor. 1964. Spermidine, spermine, and related amines. Pharmacol. Rev. **16**:245–300.

24. Wackernagel, W. 1972. An improved spheroplast assay for lambda DNA and the influence of bacterial genotype on the transfection rate. Virology **48**:94–103.

25. Yasbin, R. C., and F. E. Young. 1972. The influence of temperate bacteriophage φ105 on transformation and transfection in *B. subtilis*. Biochem. Biophys. Res. Commun. **47**:365–371.

6

Reprinted from J. Mol. Biol. **53**:159–162 (1970)

Calcium-dependent Bacteriophage DNA Infection

M. Mandel and A. Higa

It has been known that DNA extracted by phenol treatment from temperate coliphages such as λ, 434, 186 or P2 can infect sensitive *Escherichia coli* cells in the presence of "helper phage" (Kaiser & Hogness, 1960; Mandel, 1967). However, the exact role of the helper phage is still unknown. It seems that injection of the DNA of the helper phage and the presence of the intact helper phage DNA in a cell (Takano & Watanabe, 1967) are required for the cell to become competent in incorporating free DNA.

To be infective the DNA molecule must possess at least one free cohesive end (Strack & Kaiser, 1965; Kaiser & Inman, 1965). Moreover, there is a correlation between the specificity of the cohesive ends of the helper-phage DNA and the infectious DNA and the capacity of the phage to serve as a helper for DNA infectivity (Mandel & Berg, 1968; Kaiser & Wu, 1968). The DNA infection seems to depend on the homology between cohesive ends of the infecting DNA and of the DNA of the helper phage.

Since previous work by one of the authors (Mandel, 1967) had shown that changes in cell wall permeability occurred in *E. coli* (strain C600) when made competent by infection with helper phage, we became interested in the effects of both monovalent and divalent ions on *E. coli* cell wall permeability and its correlation with DNA uptake.

During the course of this investigation we found that the DNA of temperate phages P2 and λ could infect a sensitive host in the absence of helper phage and that DNA uptake depended on the presence of calcium ions.

Bacterial strains used were *E. coli* K12 strain C600 (designated K38) as a host for λi^{434} DNA and *E. coli* C1a as a host for P2 DNA. Phage λi^{434} was obtained by ultraviolet induction of a λi^{434} lysogen and P2 by infection of sensitive cells. Phages were purified by differential centrifugation and phage DNA extracted with buffer-saturated phenol (0·01 M-Tris, pH 8·0). DNA of a streptomycin-resistant mutant of K38 was extracted by the methods suggested by Smith (1968) and Avadhani, Mehta & Rege (1969). Competent cells were prepared by inoculating supplemented P medium (Radding & Kaiser, 1963) with a 1 to 500 dilution of an overnight culture of K38 or C1a and grown with aeration at 37°C until an optical density of 0·6 was reached (1×10^9 cells/ml.). The cells were then quickly chilled, centrifuged and resuspended in 0·5 volume CaCl$_2$, kept cold for 20 minutes, then centrifuged and resuspended in 0·1 volume of cold CaCl$_2$. Chilled DNA samples, 0·1 ml. in volume, in standard saline citrate (0·15 M-NaCl, 0·015 M-sodium citrate, pH 7·0) were added to 0·2 ml. of competent cells, further chilled for 15 minutes and incubated for 20 minutes at 37°C. At the end of the incubation period, the reaction mixture was either chilled or treated with DNase for five minutes at 37°C. Dilutions of the mixture were made and plated on appropriate indicators. Under those assay conditions we obtained approximately 10^5 to 10^6 plaques per μg of DNA.

Our work shows that *E. coli* K12 and *E. coli* C grown in P medium can take up phage DNA quite readily in the presence of calcium ions. In Figure 1 we see the

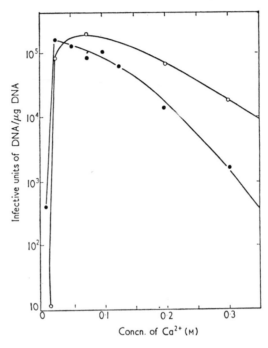

FIG. 1. DNA infectivity as a function of Ca^{2+} concentration. Experimental procedure as described in text. —○—○—, P2 DNA; —●—●—, λi^{434} DNA.

extremely rapid rise in competence of both K38 and C1a in going from 0·01 to 0·025 M-Ca^{2+} and the much slower decline in competence at molarities above 0·1. At 0·5 M-Ca^{2+} concentration the competence of K38 is reduced to practically zero while that of C1a is reduced by a factor of about 100 relative to its peak competence. This may be related to the survival of K38 and C1a cells which when incubated for 20 minutes in 0·5 M-Ca^{2+} give 0·1 and 25% survival, respectively. As a further test of changes in cell wall permeability we incubated competent K38 cells (0·1 M-Ca^{2+}) in the presence of 10 μg actinomycin D/ml. and found 25% survival compared to 100% survival for K38 in 0·01 M-Tris–0·01 M-Mg^{2+} in the presence of actinomycin.

The time-course of the interaction between λi^{434} DNA and K38 at 37°C is shown in Figure 2. The number of plaques increases very rapidly with time, reaching a maximum at about 20 minutes under the conditions of this assay. Since the competent cells and DNA are pre-chilled then mixed at cold room temperatures, and kept at 0°C for 15 minutes before starting the experiment, the plaques obtained at zero time represents DNA taken up by the cells at 0°C and protected from DNase. When the experiment is done with cells and DNA mixed at room temperature and assayed immediately, we find no DNA infectivity at zero time. In both cases the major portion of the interaction is completed in two minutes, which is quite rapid compared to the kinetics of the

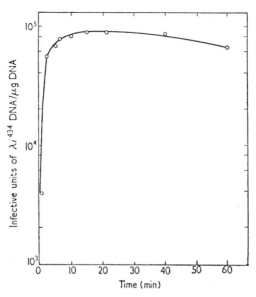

FIG. 2. Time-course of the reaction between λi^{434} DNA at a concentration of 0·1 μg/ml. and K38 cells at a concentration of 2×10^{10}/ml. in the presence of 0·05 M-Ca^{2+}. Procedure was the same as that described in the text except for proportionately larger volumes. At the times indicated on the abscissa, samples of 0·3 ml. each were removed from the incubation mixture and added to 0·1 ml. of 20 μg pancreatic DNase/ml., incubated 5 min at 37°C and plated for assay.

helper-phage DNA infection under similar conditions. The decrease of infective units after 20 minutes reflects the survival rate of K38 incubated in 0·05 M-Ca^{2+}.

The number of infectious centers obtained is linearly proportional to the concentration of phage DNA and concentration of cells. The saturation level of DNA is dependent on the cell concentration and under conditions of our assay (cell concentration~2×10^{10}ml.), 10 μg DNA/ml. was not saturating.

In contrast to helper-phage DNA infectivity, cells, competent in the presence of calcium ions, take up both linear and hydrogen-bonded circular DNA (Hershey & Burgi, 1965) with equal efficiency, as shown in Table 1.

Our attempts to transform K38 streptomycin-sensitive cells to streptomycin resistance with DNA extracted from K38 streptomycin-resistant cells met with failure.

TABLE 1

Effect of DNA form on infectivity

| | P2 DNA infectivity per μg DNA | |
	Linear DNA† (heated)	Circular DNA† (not heated)
Ca^{2+} assay	1·1 × 10^6	1·0 × 10^6
Helper-phage assay	6·0 × 10^5	1·5 × 10^5

† Circular DNA prepared as described in Hershey & Burgi (1965). The percentage of circles in DNA preparations depends on salt concentration and duration of storage. Heating at 75°C for 5 min and quick cooling to 0°C converts the closed, circular form to the open linear form.

However, since the two DNA extracts which we carefully prepared showed no hyperchromic shift on melting, the problem may well be in the technique of extracting undegraded DNA from *E. coli*.

We would like to thank Professor R. Calendar for first pointing out to us that a calcium-rich salt solution improved the helper-phage P2 DNA infectivity assay, and Leslie Jensen for her technical assistance.

This work was supported by research grant no. AI–07919 from the National Institutes of Health to one of us (M.M.).

REFERENCES

Avadhani, N.-G., Mehta, B. M. & Rege, D. V. (1969). *J. Mol. Biol.* **42**, 413.
Hershey, A. D. & Burgi, E. (1965). *Proc. Nat. Acad. Sci., Wash.* **53**, 325.
Kaiser, A. D. & Hogness, D. S. (1960). *J. Mol. Biol.* **2**, 392.
Kaiser, A. D. & Inman, R. B. (1965). *J. Mol. Biol.* **13**, 78.
Kaiser, A. D. & Wu, R. (1968). *Cold Spr. Harb. Symp. Quant. Biol.* **33**, 729.
Mandel, M. (1967). *Molec. Gen. Genetics,* **99**, 88.
Mandel, M. & Berg, A. (1968). *Proc. Nat. Acad. Sci., Wash.* **60**, 265.
Radding, C. M. & Kaiser, A. D. (1963). *J. Mol. Biol.* **7**, 225.
Smith, H. O. (1968). *Virology,* **34**, 203.
Strack, H. B. & Kaiser, A. D. (1965). *J. Mol. Biol.* **12**, 36.
Takano, T. & Watanabe, T. (1967). *Virology,* **31**, 722.

7

Reprinted from *Natl. Acad. Sci. (USA) Proc.* **69**:2110–2114 (1972)

Nonchromosomal Antibiotic Resistance in Bacteria: Genetic Transformation of *Escherichia coli* by R-Factor DNA*

(CaCl₂/extrachromosomal DNA/plasmid)

STANLEY N. COHEN, ANNIE C. Y. CHANG, AND LESLIE HSU

Division of Clinical Pharmacology, Department of Medicine, Stanford University School of Medicine, Stanford, California 94305

Communicated by A. D. Kaiser, May 15, 1972

ABSTRACT Transformation of *E. coli* cells treated with CaCl₂ to multiple antibiotic resistance by purified R-factor DNA is reported. Drug resistance is expressed in a small fraction of the recipient bacterial population almost immediately after uptake of DNA, but full genetic expression of resistance requires subsequent incubation in drug-free medium before antibiotic challenge. Transformed bacteria acquire a closed circular, transferable DNA species having the resistance, fertility, and sedimentation characteristics of the parent R factor. Covalently-closed, catenated, and open (nicked) circular forms of R-factor DNA are all effective in transformation, but denaturation and sonication abolish the transforming ability of R-factor DNA in this system.

Biochemical and genetic investigations in several different laboratories (1–9) have established that antibiotic resistance factors (R factors) of the *Enterobacteriacae* consist of autonomously replicating units of extrachromosomal DNA. Moreover, certain R factors are formed by reversible covalent linkage of separate plasmids that independently harbor either resistance or transfer functions (4–9). Recent electron microscope studies of heteroduplex formation between the DNA of various R factors, and between DNA of R factors and certain other bacterial plasmids, have indicated that these earlier conclusions about the molecular nature and structural composition of R factors are correct (ref. 10, and Sharp, Cohen & Davidson, manuscript in preparation). However, the functional interrelationships of genes located in separate regions of R-factor DNA molecules and the role of the various molecular forms of R-factor DNA in the replication, genetic expression, recombination, and transfer of these plasmids are still not well understood. Availability of a system for genetic transformation of host bacteria by purified molecular species of R-factor DNA would enable study of these and other important aspects of R-factor biology.

Although bacterial transformation has been widely investigated in *Pneumococcus, Haemophilus influenzae, Bacillus subtilis,* and certain other bacteria (11), attempts to transform *Escherichia coli* with bacterial DNA have been largely unsuccessful. In contrast, *transfection* of *E. coli* spheroplasts by ϕX174 (12), λ (13), and other bacteriophages has been accomplished with high efficiency, and transfection of intact *E. coli* cells by the DNA of lambdoid phages has been widely studied by the "helper phage" assay of Kaiser and Hogness (14). Survival of the transfected cells is not required in such

experiments, since production of viable phage particles is assayed with an appropriate strain of indicator bacteria. Recently, Mandel and Higa (15) have reported that *E. coli* cells that have been treated with calcium chloride can take up phage λ DNA and can produce viable phage particles. We report here that treatment of *E. coli* cells with calcium chloride also renders them capable of taking up molecules of purified R-factor DNA. Moreover, we find that the introduced R-factor DNA can persist in such cells as an independently replicating plasmid, and can express both the fertility and antibiotic resistance functions of the parent R factor.

MATERIALS AND METHODS

Bacterial Strains and R Factors. The I-like R factor, R64-11 (16), which specifies resistance to tetracycline (Tc) and streptomycin (Sm), was obtained from R. Curtiss. R6 (17), an F-like R factor that carries resistance to kanamycin (Km), neomycin (Nm), chloramphenicol (Cm), sulphonamide (Su), streptomycin, and tetracycline, was obtained from T. Watanabe. R6-5, a spontaneous variant of R6 that lacks tetracycline resistance, was isolated in our laboratory (18). The bacterial strains used in these experiments have been described (7, 22).

DNA Preparations. In certain instances, covalently-closed R-factor DNA was isolated and purified from *E. coli* as described (6, 7). Alternatively, a Brij-lysis procedure (19) was used for initial R-factor DNA isolation, and preparations obtained by this method were subsequently purified by centrifugation in cesium chloride–ethidium bromide gradients. The catenated, closed circular, and noncircular forms of R-factor DNA used in experiments comparing the relative transforming ability of the various R-factor DNA species were isolated from *E. coli* minicells. R-factor DNA was denatured by heating it at 98° for 5 min in 15 mM NaCl–1.5 mM Na citrate followed by rapid cooling at 0°. Sonication of R-factor DNA to about 9S fragments was done for 15 sec at 0° by a Branson model W185 D sonicator, and the size of the R-factor DNA fragments was confirmed by sucrose gradient centrifugation (7).

Transformation Reaction Mixture. Transformation was done by a variation of the procedure of Mandel and Higa (15), as modified by Lobban, Masuda, and Kaiser (personal communication). *E. coli* strain C600 was grown at 37° in H1 medium (20) to an optical density of 0.85 at 590 nm. At this point, the cells

Abbreviation: R factor, antibiotic resistance factor.

* The previous paper in this series is ref. 18.

were chilled quickly, sedimented, and washed once in 0.5 volume 10 mM NaCl. After centrifugation, bacteria were resuspended in half the original volume of chilled 0.03 M CaCl$_2$, kept at 0° for 20 min, sedimented, and then resuspended in 0.1 the original volume of 0.03 M calcium chloride solution. Chilled DNA samples in TEN buffer [0.02 M Tris (pH 8.0)– 1 mM EDTA (pH 8.0)–0.02 M NaCl] were supplemented with 0.1 M calcium chloride to a final concentration of 0.03 M.

0.2 ml of competent cells treated with CaCl$_2$ was added to 0.1 ml of DNA solution with chilled pipettes, and an additional incubation was done for 60 min at 0°. This second incubation of bacteria at 0° resulted in a 4-fold increase in transformation frequency, whether or not R-factor DNA was present. Bacteria were then subjected to a heat pulse at 42° for 2 min to enable uptake of R-factor DNA, chilled, and then either plated directly onto nutrient agar containing appropriate antibiotics or, where indicated, diluted 10 times into L broth (21) and incubated at 37° before plating. Cell survival was greater than 50% after calcium chloride treatment and heat pulse. Drug resistance was assayed on nutrient agar plates with the antibiotics indicated in specific experiments. Drug concentrations used were: neomycin (25 μg/ml), streptomycin (10 μg/ml), tetracycline (25 μg/ml), kanamycin (25 μg/ml), and chloramphenicol (25 μg/ml).

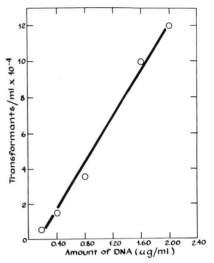

Fɪɢ. 2. Effect of concentration of R-factor DNA on transformation frequency. Various concentrations of covalently-closed R6 DNA were assayed for their ability to transform CaCl$_2$-treated *E. coli* to kanamycin resistance (see *Methods* and Fig. 1). Transformation frequency was determined after a 120-min incubation in antibiotic-free medium to allow complete expression of kanamycin resistance.

RESULTS

Expression of Antibiotic Resistance by Competent Cells. The kinetics of transformation of drug-sensitive *E. coli* to kanamycin resistance by covalently-closed R6 DNA is shown in Fig. 1. Resistance to kanamycin is expressed almost immediately by a small fraction of cells treated with CaCl$_2$ after uptake of R-factor DNA. The number of recipient bacteria exhibiting antibiotic resistance increases about 1000-fold during subsequent incubation in antibiotic-free medium and reaches a maximum in 1 hr.

Effect of DNA Concentration. The transforming ability of several different concentrations of closed circular R6 DNA is shown in Fig. 2. A linear relationship of transformation frequency to DNA concentration was observed throughout the range examined, and about 10^5 transformed bacteria were obtained per μg of R-factor DNA. Although the competence of cells treated with CaCl$_2$ to take up R-factor DNA varied in different experiments, the overall efficiency of transformation was of the same order of magnitude as the transfection efficiency reported for bacteriophage λ DNA in similarly treated cells (19) (about 2 × 10^5 transfectants/μg of λ DNA). In these experiments, bacteria were incubated in drug-free medium for 120 min after DNA uptake to allow complete expression of antibiotic resistance to occur.

Requirements for Transformation by R-Factor DNA. Since *E. coli* DNA does not appear to transform cells treated with CaCl$_2$ (15), we initially anticipated that covalent circularity of R-factor DNA might be required for its biological activity in the transformation system. However, the results of experiments presented in Table 1 indicate that closed circular, catenated, and nicked (open) circular forms of R6-5 DNA all possess transforming ability. Sonication of the R-factor DNA to 9S

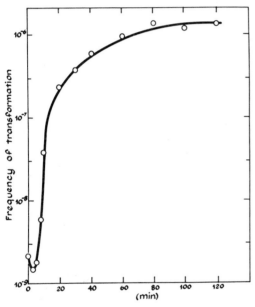

Fɪɢ. 1. Kinetics of expression of kanamycin resistance in transformed *E. coli*. After incubation of CaCl$_2$-treated cells with R6 DNA (0.6 μg/ml), bacteria were diluted 10-fold into antibiotic-free L broth. At the times shown, 0.1-ml samples of the bacterial culture were spread onto nutrient agar plates containing kanamycin and incubated overnight at 37° for determination of number of transformants. An identical sample was diluted appropriately, and plated on antibiotic-free nutrient agar to determine the total number of viable cells. Transformation frequency is expressed in terms of the number of kanamycin-resistant bacteria relative to the total number of viable cells.

fragments or denaturation of this DNA destroyed its ability to transform. Addition of DNase to reaction mixtures before the heat pulse (42°) also prevented transformation but had no effect on expression of antibiotic resistance when added after the 2-min incubation of competent bacteria with purified R-factor DNA. Moreover, phenol extraction of R-factor DNA preparations or treatment of them with either RNase or Pronase did not influence their transforming ability. The DNA of the I-like R factor, R64-11, was somewhat lower than that of the F-like R6-5 in transforming efficiency.

Expression of Different Drug Resistance Markers in Transformed Bacteria. Transformation of *E. coli* for various antibiotic resistance markers carried by R6 is shown in Table 2.

TABLE 1. *Requirements for transformation by R-factor DNA*

DNA species	Transformants/μgDNA
R6-5 (F-like)	
Closed circular	9.2×10^4
+DNase (before)	<0.3
+DNase (after)	7.7×10^4
+RNase (before)	9.6×10^4
+Pronase (before)	8.1×10^4
Phenol extraction	9.9×10^4
Isolated from transformed bacteria	7.0×10^4
Catenated	3.2×10^4
Open-circular	5.6×10^4
Denatured	<0.3
Sonicated	<0.3
No DNA	*
No bacteria	<0.3
R64-11	
Closed circular	9.4×10^3
No DNA	*
No bacteria	<0.3

Transforming ability of different molecular forms of R-factor DNA was assayed (see *Methods* and Fig. 1). All R6-5 DNA species used in this experiment were isolated from *E. coli* minicells, and DNA concentration ranged from 0.6–5 μg/ml. The sample labeled *open-circular DNA* was obtained from peak 3 of a cesium chloride–ethidium bromide gradient containing R-factor DNA isolated from minicells and was free from significant contamination by (noncircular) chromosomal DNA (22). This fraction was composed of about 85–90% nicked (open) circular R-factor DNA, and 10–15% noncircular R-factor DNA molecules (23). Transformation efficiency was determined after a 120-min incubation in antibiotic-free medium. Where indicated, R-factor DNA preparations were incubated with pancreatic DNase (10 μg/ml), pancreatic RNase (20 μg/ml, Worthington), or Pronase (100 μg/ml, Calbiochem) at 37° for 5 min before use in the transformation assay. DNase treatment was done in the presence of 10 mM $MgCl_2$; the RNase preparations were previously heated to destroy DNase activity (6). Pronase was self-digested for 37° for 1 hr and at 80° for 2 min, and then chilled rapidly before use. Phenol extraction of DNA was done as described (7). The terms *before* and *after* refer to the period of incubation of R-factor DNA with $CaCl_2$-treated cells at 42°. Since the competence of bacteria varied somewhat in different experiments, the effects of R-factor DNA structure on transformation were determined with a single batch of $CaCl_2$-treated cells.

* No colonies were observed when 10^9 bacteria were assayed in the absence of DNA.

As seen in this table, R^+ transformants could be selected with Km, Nm, Cm, or Tc, but no transformants were obtained when initial selection was carried out with Sm. Our inability to select R6 transformants with streptomycin was surprising since persistance of the Sm marker on the R factor was demonstrated by subsequent expression of streptomycin resistance in nearly all of the clones that had been initially selected with the other antibiotics (Table 2). However, the observation that streptomycin resistance is the last drug marker to be phenotypically expressed by bacteria that have received an R factor by conjugation (23, 24) suggests that expression of the Sm marker in transformed bacteria may require more extensive incubation in antibiotic-free medium than was used in these experiments.

R-Factor DNA Species Isolated from Transformed Bacteria. DNA isolated from a transformed *E. coli* clone carrying all of the antibiotic resistance determinants originally present on R6 was subjected to cesium chloride–ethidium bromide gradient centrifugation. As seen in Fig. 3 (*top*), a closed circular DNA peak (25) (peak *A*) was identified in this preparation, in addition to a peak (peak *B*) characteristic of DNA preparations isolated from the R^- *E. coli* recipient strain (7). R-factor DNA collected from peak A was treated with isopropanol to remove ethidium bromide (26), dialyzed against 0.02 M Tris (pH 8)–1 mM EDTA (pH 8)–0.02 M NaCl, and a sample of this DNA was then centrifuged in neutral sucrose in the presence of 34S λ DNA marker (7). As seen in Fig. 3 (*bottom*), this R-factor peak was composed of DNA having the same S value as the closed circular DNA of the parent R factor (75S) (7); a small amount of the 52S open circular species of R6 DNA was also observed in this preparation.

R-factor DNA isolated from transformants could transform other bacteria at an efficiency comparable to that shown by the parent R factor (Table 1). In addition, *E. coli* cells that were transformed to antibiotic resistance by R6 DNA transferred this resistance at the same frequency (about 10^{-4}) as bacteria that had received the R factor by conjugation.

DISCUSSION AND SUMMARY

The experiments we have reported here indicate that purified R-factor DNA can transform *E. coli* treated with $CaCl_2$ to

TABLE 2. *Expression of different drug resistance markers in transformed bacteria*

Selected by	No. of colonies	Number resistant to				
		Km	Nm	Cm	Tc	Sm
Km	30	30	30	30	30	27
Nm	52	52	52	52	52	48
Cm	35	35	35	35	35	33
Tc	36	36	36	36	36	36
Sm	0					

After transformation of $CaCl_2$-treated *E. coli* by R6, as indicated in Table 1, 0.1-ml samples of bacterial cultures were spread onto plates containing the antibiotic indicated, and numbers of antibiotic bacteria were determined. Separate bacterial clones isolated from each antibiotic plate were examined for the presence of other antibiotic resistance determinants by stabbing colony samples onto appropriate drug-containing plates.

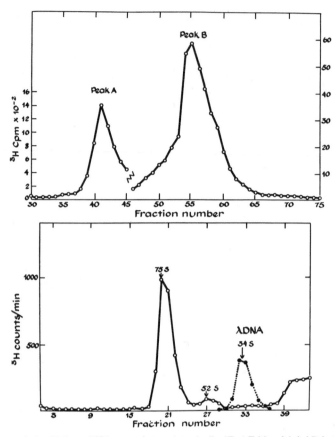

Fig. 3.   Centrifugation analysis of R-factor DNA present in transformed cells. (*Top*) Tritium-labeled R-factor DNA was prepared by growth of bacteria in modified M9 medium (2) containing [³H]thymidine (5 μCi/ml) and deoxyadenosine (250 μg/ml) as described (23). Bacterial cultures were harvested and lysed, and centrifuged for 42 hr at 39.5 rpm at 20° in cesium chloride ($\rho = 1.5570$) containing ethidium bromide (1 mg/ml) in a Spinco 50 Ti rotor. Collection of fractions and assay of a 0.01-ml sample of each fraction for radioactivity was done as described (7). Note the change in units for *peak B*, which is drawn according to the scale shown at the right of this figure. (*Bottom*) After removal of ethidium bromide from fractions 40–43 of *peak A* in Fig. 2A by equilibration with isopropyl alcohol (26), DNA was dialyzed against 0.02 M Tris (pH 8.0)–1 mM EDTA–0.02 M NaCl and a 0.1-ml aliquot was layered onto a 5–20% linear sucrose gradient containing 1mM EDTA, (pH 8.0)–1 M NaCl–0.02 M Tris(pH 7.4). The gradient was centrifuged for 50 min at 39.5 rpm in a Spinco SW50.1 rotor at 20° in the presence of ¹⁴C-labeled λ DNA marker (34S) that had previously been heated to 75° and rapidly cooled to insure its linearity. Fractions (0.12 ml) were collected, precipitated with cold 5% trichloroacetic acid, and counted (7).

multiple antibiotic resistance. R-factor DNA is rapidly taken up by competent cells, and drug resistance is expressed almost immediately in certain of the recipient bacteria. In the case of kanamycin resistance, the number of transformants continues to increase for about 1 hr after DNA uptake and subsequently forms a plateau. Since the kanamycin resistance of R6 is believed to be dependent on synthesis of the enzyme, kanamycin monophosphotransferase, in order to inactivate the drug (27), the observed rapid expression of antibiotic resistance in a fraction of the transformed cells is somewhat surprising. Presumably, uptake of growth inhibitory concentrations of kanamycin by such cells is delayed until sufficient quantities of the enzyme have been formed to allow bacterial multiplication to occur in the presence of the antibiotic.

Although prior treatment of R-factor DNA with pancreatic DNase prevents transformation, addition of DNase to transformation reaction mixtures after a 2-min incubation of competent bacteria with R-factor DNA at 42° has no effect on the frequency of transformation. Furthermore, transformation is sharply reduced by incubation of cells treated with $CaCl_2$ at 42° for 2 min before addition of the R-factor DNA. These latter results suggest that bacteria treated with $CaCl_2$ remain competent to take up R-factor DNA for only a very brief period at elevated temperature. This interpretation is consistent with the results of earlier experiments with λ DNA by Mandel and Higa (15) and by Lobban (personal communication).

Closed and open-circular forms of R-factor DNA are both

capable of transforming recipient cells. However, denaturation and sonication eliminate the transforming activity of R-factor DNA in this system. *E. coli* cells that have been transformed by R-factor DNA acquire an independently replicating closed circular DNA species indistinguishable from the parent R factor, and can transfer this DNA normally to other cells by conjugation. The general usefulness of this system for study of the biological properties of bacterial plasmids is suggested by our observation that DNA from both F-like and I-like R factors is capable of transforming *E. coli*. Moreover, the observation (van Embden and Cohen, in preparation) that DNA obtained from a *nontransmissible* plasmid coding for Tc resistance is able to transform *E. coli* indicates the additional use of this method for the investigation of nontransferable, independently replicating species of plasmid DNA.

We thank Peter Lobban and Terrie Masuda for their helpful comments regarding the calcium chloride transformation assay. These studies were supported by Grant AI 08619 from the National Institute of Allergy and Infectious Diseases, by Grant NSG GB 30581 from the National Science Foundation, and by a USPHS Career Development Award to SNC.

1. Anderson, E. S. (1968) *Annu. Rev. Microbiol.* **22**, 131–180.
2. Watanabe, T. (1971) *Ann. N.Y. Acad. Sci.* **182**, 126–140.
3. Rownd, R. (1969) *J. Mol. Biol.* **44**, 387–402.
4. Nisioka, T., Mitani, M. & Clowes, R. C. (1969) *J. Bacteriol.* **97**, 376–385.
5. Silver, R. P. & Falkow, S. (1970) *J. Bacteriol.* **104**, 331–339.
6. Cohen, S. N. & Miller, C. A. (1969) *Nature* **224**, 1273–1277.
7. Cohen, S. N. & Miller, C. A. (1970) *J. Mol. Biol.* **50**, 671–687.
8. Kopecko, D. J. & Punch, J. D. (1971) *Ann. N.Y. Acad. Sci.* **182**, 207–216.
9. Rownd, R. & Mickel, S. (1971) *Nature New Biol.* **234**, 40–43.
10. Sharp, P. A., Davidson, N. D. & Cohen, S. N. (1971) *Fed. Proc.* **30**, 1054 Abst.
11. Hotchkiss, R. D. & Gabor, M. (1970) *Annu. Rev. Genet.* **4**, 193–224.
12. Guthrie, G. D. & Sinsheimer, R. L. (1963) *Biochim. Biophys. Acta* **72**, 290–297.
13. Young, E. T. III & Sinsheimer, R. L. (1967) *J. Mol. Biol.* **30**, 147–164.
14. Kaiser, A. D. & Hogness, D. S. (1960) *J. Mol. Biol.* **2**, 392–415.
15. Mandel, M. & Higa, A. (1970) *J. Mol. Biol.* **53**, 159–162.
16. Meynell, E. & Datta, N. (1967) *Nature* **214**, 885–887.
17. Watanabe, T., Ogata, C. & Sato, S. (1964) *J. Bacteriol.* **88**, 922–928.
18. Silver, R. P. & Cohen, S. N. (1972) *J. Bacteriol.* **110**, 1082–1088.
19. Clewell, D. & Helinski, D. R. (1969) *Proc. Nat. Acad. Sci. USA* **62**, 1159–1166.
20. Radding, C. & Kaiser, A. D. (1963) *J. Mol. Biol.* **7**, 225–233.
21. Lennox, E. S. (1955) *Virology* **1**, 190–206.
22. Cohen, S. N., Silver, R. P., McCoubrey, A. E. & Sharp, P. A. (1971) *Nature New Biol.* **231**, 249–251.
23. Mitsuhashi, S. (1965) *Gumma J. Med. Sci.* **14**, 169.
24. Silver, R. P. (1970) Ph.D. Thesis, Georgetown University, Washington, D.C.
25. Radloff, R., Bauer, W. & Vinograd, J. (1967) *Proc. Nat. Acad. Sci. USA* **57**, 1514–1521.
26. Cozzarelli, N. R., Kelly, R. B. & Kornberg, A. (1968) *Proc. Nat. Acad. Sci. USA* **60**, 992–999.
27. Davies, J., Brzezinska, M. & Benveniste, M. S. (1971) *Ann. N.Y. Acad. Sci.* **182**, 226–233.

8

Reprinted from *Natl. Acad. Sci. (USA) Proc.* **70**:3240–3244 (1973)

Construction of Biologically Functional Bacterial Plasmids *In Vitro*

(R factor/restriction enzyme/transformation/endonuclease/antibiotic resistance)

STANLEY N. COHEN*, ANNIE C. Y. CHANG*, HERBERT W. BOYER†, AND ROBERT B. HELLING†

* Department of Medicine, Stanford University School of Medicine, Stanford, California 94305; and † Department of Microbiology, University of California at San Francisco, San Francisco, Calif. 94122

Communicated by Norman Davidson, July 18, 1973

ABSTRACT The construction of new plasmid DNA species by *in vitro* joining of restriction endonuclease-generated fragments of separate plasmids is described. Newly constructed plasmids that are inserted into *Escherichia coli* by transformation are shown to be biologically functional replicons that possess genetic properties and nucleotide base sequences from both of the parent DNA molecules. Functional plasmids can be obtained by reassociation of endonuclease-generated fragments of larger replicons, as well as by joining of plasmid DNA molecules of entirely different origins.

Controlled shearing of antibiotic resistance (R) factor DNA leads to formation of plasmid DNA segments that can be taken up by appropriately treated *Escherichia coli* cells and that recircularize to form new, autonomously replicating plasmids (1). One such plasmid that is formed after transformation of *E. coli* by a fragment of sheared R6-5 DNA, pSC101 (previously referred to as Tc6-5), has a molecular weight of 5.8×10^6, which represents about 10% of the genome of the parent R factor. This plasmid carries genetic information necessary for its own replication and for expression of resistance to tetracycline, but lacks the other drug resistance determinants and the fertility functions carried by R6-5 (1).

Two recently described restriction endonucleases, *Eco*RI and *Eco*RII, cleave double-stranded DNA so as to produce short overlapping single-stranded ends. The nucleotide sequences cleaved are unique and self-complementary (2–6) so that DNA fragments produced by one of these enzymes can associate by hydrogen-bonding with other fragments produced by the same enzyme. After hydrogen-bonding, the 3'-hydroxyl and 5'-phosphate ends can be joined by DNA ligase (6). Thus, these restriction endonucleases appeared to have great potential value for the construction of new plasmid species by joining DNA molecules from different sources. The *Eco*RI endonuclease seemed especially useful for this purpose, because on a random basis the sequence cleaved is expected to occur only about once for every 4,000 to 16,000 nucleotide pairs (2); thus, most *Eco*RI-generated DNA fragments should contain one or more intact genes.

We describe here the construction of new plasmid DNA species by *in vitro* association of the *Eco*RI-derived DNA fragments from separate plasmids. In one instance a new plasmid has been constructed from two DNA species of entirely different origin, while in another, a plasmid which has itself been derived from *Eco*RI-generated DNA fragments of a larger parent plasmid genome has been joined to another replicon derived independently from the same parent plasmid. Plasmids that have been constructed by the *in vitro* joining of

*Eco*RI-generated fragments have been inserted into appropriately-treated *E. coli* by transformation (7) and have been shown to form biologically functional replicons that possess genetic properties and nucleotide base sequences of both parent DNA species.

MATERIALS AND METHODS

E. coli strain W1485 containing the RSF1010 plasmid, which carries resistance to streptomycin and sulfonamide, was obtained from S. Falkow. Other bacterial strains and R factors and procedures for DNA isolation, electron microscopy, and transformation of *E. coli* by plasmid DNA have been described (1, 7, 8). Purification and use of the *Eco*RI restriction endonuclease have been described (5). Plasmid heteroduplex studies were performed as previously described (9, 10). *E. coli* DNA ligase was a gift from P. Modrich and R. L. Lehman and was used as described (11). The detailed procedures for gel electrophoresis of DNA will be described elsewhere (Helling, Goodman, and Boyer, in preparation); in brief, duplex DNA was subjected to electrophoresis in a tube-type apparatus (Hoefer Scientific Instrument) (0.6 × 15-cm gel) at about 20° in 0.7% agarose at 22.5 V with 40 mM Tris–acetate buffer (pH 8.05) containing 20 mM sodium acetate, 2 mM EDTA, and 18 mM sodium chloride. The gels were then soaked in ethidium bromide (5 μg/ml) and the DNA was visualized by fluorescence under long wavelength ultraviolet light ("black light"). The molecular weight of each fragment in the range of 1 to 200×10^6 was determined from its mobility relative to the mobilities of DNA standards of known molecular weight included in the same gel (Helling, Goodman, and Boyer, in preparation).

RESULTS

R6-5 and pSC101 plasmid DNA preparations were treated with the *Eco*RI restriction endonuclease, and the resulting DNA products were analyzed by electrophoresis in agarose gels. Photographs of the fluorescing DNA bands derived from these plasmids are presented in Fig. 1b and c. Only one band is observed after *Eco*RI endonucleolytic digestion of pSC101 DNA (Fig. 1c), suggesting that this plasmid has a single site susceptible to cleavage by the enzyme. In addition, endonuclease-treated pSC101 DNA is located at the position in the gel that would be expected if the covalently closed circular plasmid is cleaved once to form noncircular DNA of the same molecular weight. The molecular weight of the linear fragment estimated from its mobility in the gel is 5.8×10^6, in agreement with independent measurements of the size of the intact molecule (1). Because pSC101 has a single *Eco*RI cleavage site and is derived from R6-5, the equivalent DNA sequences of

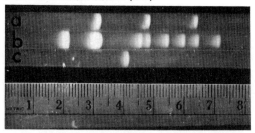

Fig. 1. Agarose-gel electrophoresis of *Eco*RI digests. (*a*) pSC102. The three fragments derived from the plasmid correspond to fragments III, V, and VIII of R6-5 (Fig. 1*b* below) as shown here and as confirmed by electrophoresis in other gels (see *text*). (*b*) R6-5. The molecular weights calculated for the fragments, as indicated in *Methods*, are (from *left* to *right*) I, 17.0; II & III (double band), 9.6 and 9.1; IV, 5.2; V, 4.9; VI, 4.3; VII, 3.8; VIII, 3.4; IX, 2.9. All molecular weight values have been multiplied by 10^{-6}. (*c*) pSC101. The calculated molecular weight of the single fragment is 5.8×10^6. Migration in all gels was from *left* (cathode) to *right;* samples were subjected to electrophoresis for 19 hr and 50 min.

the parent plasmid must be distributed in two separate *Eco*RI fragments.

The *Eco*RI endonuclease products of R6-5 plasmid DNA were separated into 12 distinct bands, eight of which are seen in the gel shown in Fig. 1*b;* the largest fragment has a molecular weight of 17×10^6, while three fragments (not shown in Fig. 1*b*) have molecular weights of less than 1×10^6, as determined by their relative mobilities in agarose gels. As seen in the figure, an increased intensity of fluorescence, of the second band suggests that this band contains two or more DNA fragments of almost equal size; when smaller amounts of *Eco*RI-treated R6-5 DNA are subjected to electrophoresis for a longer period of time, resolution of the two fragments (i.e., II and III) is narrowly attainable. Because 12 different *Eco*RI-generated DNA fragments can be identified after endonuclease treatment of covalently closed circular R6-5, there must be at least 12 substrate sites for *Eco*RI endonuclease present on this plasmid, or an average of one site for every 8000 nucleotide pairs. The molecular weight for each fragment shown is given in the caption to Fig. 1. The sum of the molecular weights of the *Eco*RI fragments of R6-5 DNA is 61.5×10^6, which is in close agreement with independent estimates for the molecular weight of the intact plasmid (7, 10).

The results of separate transformations of *E. coli* C600 by endonuclease-treated pSC101 or R6-5 DNA are shown in Table 1. As seen in the table, cleaved pSC101 DNA transforms *E. coli* C600 with a frequency about 10-fold lower than was observed with covalently closed or nicked circular (1) molecules of the same plasmid. The ability of cleaved pSC101 DNA to function in transformation suggests that plasmid DNA fragments with short cohesive endonuclease-generated termini can recircularize in *E. coli* and be ligated *in vivo;* since the denaturing temperature (T_m) for the termini generated by the *Eco*RI endonuclease is 5–6° (6) and the transformation procedure includes a 42° incubation step (7), it is unlikely that the plasmid DNA molecules enter bacterial cells with their termini already hydrogen-bonded. A corresponding observation has been made with *Eco*RI endonuclease-cleaved

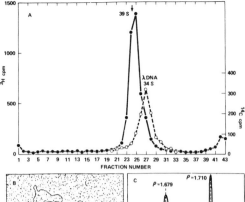

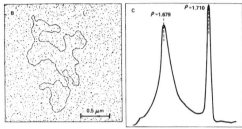

Fig. 2. Physical properties of the pSC102 plasmid derived from *Eco*RI fragments of R6-5. (*A*) Sucrose gradient centrifugation analysis (1, 8) of covalently closed circular plasmid DNA (●——●) isolated from an *E. coli* transformant clone as described in text. 34 S linear [^{14}C]DNA from λ was used as a standard (O– – –O). (*B*) Electron photomicrograph of nicked (7) pSC102 DNA. The length of this molecule is approximately 8.7 μm. (*C*) Densitometer tracing of analytical ultracentrifugation (8) photograph of pSC102 plasmid DNA. Centrifugation in CsCl ($\rho = 1.710 \text{ g/cm}^3$) was carried out in the presence of d(A-T)$_n$·-d(A-T)$_n$ density marker ($\rho = 1.679 \text{ g/cm}^3$).

SV40 DNA, which forms covalently closed circular DNA molecules in mammalian cells *in vivo* (6).

Transformation for each of the antibiotic resistance markers present on the R6-5 plasmid was also reduced after treatment of this DNA with *Eco*RI endonuclease (Table 1). Since the pSC101 (tetracycline-resistance) plasmid was derived from R6-5 by controlled shearing of R6-5 DNA (1), and no tetracycline-resistant clone was recovered after transformation by the *Eco*RI endonuclease products of R6-5, [whereas tetracycline-resistant clones are recovered after transformation with intact R6-5 DNA (1)], an *Eco*RI restriction site may separate the tetracycline resistance gene of R6-5 from its replicator locus. Our finding that the linear fragment produced by treatment of pSC101 DNA with *Eco*RI endonuclease does not correspond to any of the *Eco*RI-generated fragments of R6-5 (Fig. 1) is consistent with this interpretation.

A single clone that had been selected for resistance to kanamycin and which was found also to carry resistance to neomycin and sulfonamide, but not to tetracycline, chloramphenicol, or streptomycin after transformation of *E. coli* by *Eco*RI-generated DNA fragments of R6-5, was examined further. Closed circular DNA obtained from this isolate (plasmid designation pSC102) by CsCl–ethidium bromide gradient

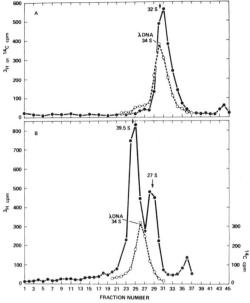

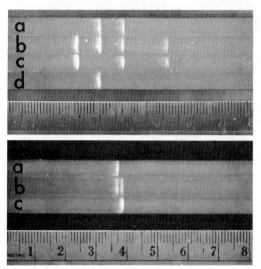

FIGS. 4 and 5.   Agarose-gel electrophoresis of *Eco*RI digests of newly constructed plasmid species.   Conditions were as described in *Methods*.

FIG. 4. (*top*) Gels were subjected to electrophoresis for 19 hr and 10 min. (*a*) pSC105 DNA. (*b*) Mixture of pSC101 and pSC102 DNA. (*c*) pSC102 DNA. (*d*) pSC101 DNA.

FIG. 5. (*bottom*) Gels were subjected to electrophoresis for 18 hr and 30 min. (*a*) pSC101 DNA. (*b*) pSC109 DNA. (*c*) RSF1010 DNA. Evidence that the single band observed in this gel represents a linear fragment of cleaved RSF1010 DNA was obtained by comparing the relative mobilities of *Eco*RI-treated DNA and untreated (covalently closed circular and nicked circular) RSF1010 DNA in gels. The molecular weight of RSF-1010 calculated from its mobility in gels is 5.5×10^6.

FIG. 3. Sucrose gradient centrifugation of DNA isolated from *E. coli* clones transformed for both tetracycline and kanamycin resistance by a mixture of pSC101 and pSC102 DNA. (*A*) The DNA mixture was treated with *Eco*RI endonuclease and was ligated prior to use in the transformation procedure. Covalently closed circular DNA isolated (7, 8) from a transformant clone carrying resistance to both tetracycline and kanamycin was examined by sedimentation in a neutral 5–20% sucrose gradient (8). (*B*) Sucrose sedimentation pattern of covalently closed circular DNA isolated from a tetracycline and kanamycin resistant clone transformed with an *untreated* mixture of pSC101 and pSC102 plasmid DNA.

centrifugation has an S value of 39.5 in neutral sucrose gradients (Fig. 2*A*) and a contour length of 8.7 μm when nicked (Fig. 2*B*). These data indicate a molecular weight

TABLE 1. *Transformation by covalently closed circular and EcoRI-treated plasmid DNA*

Plasmid DNA species	Transformants per μg DNA		
	Tetracycline	Kanamycin (neomycin)	Chloramphenicol
pSC101 covalently closed circle	3×10^5	—	—
*Eco*RI-treated	2.8×10^4	—	—
R6-5 covalently closed circle	—	1.3×10^4	1.3×10^4
*Eco*RI-treated	<5	1×10^2	4×10^1

Transformation of *E. coli* strain C600 by plasmid DNA was carried out as indicated in *Methods*. The kanamycin resistance determinant of R6-5 codes also for resistance to neomycin (15). Antibiotics used for selection were tetracycline (10 μg/ml), kanamycin (25 μg/ml) or chloramphenicol (25 μg/ml).

about 17×10^6. Isopycnic centrifugation in cesium chloride of this non-self-transmissible plasmid indicated it has a buoyant density of 1.710 g/cm³ (Fig. 2*C*). Since the nucleotide base composition of the antibiotic resistance determinant (R-determinant) segment of the parent R factor is 1.718 g/cm³ (8), the various component regions of the resistance unit must have widely different base compositions, and the pSC102 plasmid must lack a part of this unit that is rich in high buoyant density G+C nucleotide pairs. The existence of such a high buoyant density *Eco*RI fragment of R6-5 DNA was confirmed by centrifugation of *Eco*RI-treated R6-5 DNA in neutral cesium chloride gradients (Cohen and Chang, unpublished data).

Treatment of pSC102 plasmid DNA with *Eco*RI restriction endonuclease results in formation of three fragments that are separable by electrophoresis is agarose gels (Fig. 1*a*); the estimated molecular weights of these fragments determined by gel mobility total 17.4×10^6, which is in close agreement with the molecular weight of the intact pSC102 plasmid determined by sucrose gradient centrifugation and electron microscopy (Fig. 2). Comparison with the *Eco*RI-generated fragments of R6-5 indicates that the pSC102 fragments correspond to fragments III (as determined by long-term electrophoresis in gels containing smaller amounts of DNA), V, and VIII of the parent plasmid (Fig. 1*b*). These results suggest that *E. coli* cells transformed with *Eco*RI-generated DNA fragments of R6-5

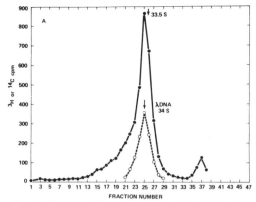

FIG. 6. Sucrose gradient sedimentation of covalently closed circular DNA representing the pSC109 plasmid derived from RSF1010 and pSC101.

TABLE 2. *Transformation of E. coli C600 by a mixture of pSC101 and pSC102 DNA*

Treatment of DNA	Transformation frequency for antibiotic resistance markers		
	Tetracycline	Kanamycin	Tetracycline + kanamycin
None	2×10^5	1×10^5	2×10^2
EcoRI	1×10^4	1.1×10^3	7×10^1
EcoRI + DNA ligase	1.2×10^4	1.3×10^3	5.7×10^2

Transformation frequency is shown in transformants per μg of DNA of each plasmid species in the mixture. Antibiotic concentrations are indicated in legend of Table 1.

can ligate reassociated DNA fragments *in vivo*, and that re-associated molecules carrying antibiotic resistance genes and capable of replication can circularize and can be recovered as functional plasmids by appropriate selection.

A mixture of pSC101 and pSC102 plasmid DNA species, which had been separately purified by dye–buoyant density centrifugation, was treated with the EcoRI endonuclease, and then was either used directly to transform *E. coli* or was ligated prior to use in the transformation procedure (Table 2). In a control experiment, a plasmid DNA mixture that had not been subjected to endonuclease digestion was employed for transformation. As seen in this table, transformants carrying resistance to both tetracycline and kanamycin were isolated in all three instances. Cotransformation of tetracycline and kanamycin resistance by the untreated DNA mixture occurred at a 500- to 1000-fold lower frequency than transformation for the individual markers. Examination of three different transformant clones derived from this DNA mixture indicated that each contained two separate covalently closed circular DNA species having the sedimentation characteristics of the pSC101 and pSC102 plasmids (Fig. 3B). The ability of two plasmids derived from the same parental plasmid (i.e., R6-5) to exist stably as separate replicons (12) in a single

bacterial host cell suggests that the parent plasmid may contain at least two distinct replicator sites. This interpretation is consistent with earlier observations which indicate that the R6 plasmid dissociates into two separate compatible replicons in *Proteus mirabilis* (8). Cotransformation of tetracycline and kanamycin resistance by the EcoRI treated DNA mixture was 10- to 100-fold lower than transformation of either tetracycline or kanamycin resistance alone, and was increased about 8-fold by treatment of the endonuclease digest with DNA ligase (Table 2). Each of four studied clones derived by transformation with the endonuclease-treated and/or ligated DNA mixture contained only a single 32S covalently closed circular DNA species (Fig. 3A) that carries resistance to both tetracycline and kanamycin, and which can transform *E. coli* for resistance to both antibiotics. One of the clones derived from the ligase-treated mixture was selected for further study, and this plasmid was designated pSC105.

When the plasmid DNA of pSC105 was digested by the EcoRI endonuclease and analyzed by electrophoresis in agarose gels, two component fragments were identified (Fig. 4); the larger fragment was indistinguishable from endonuclease-treated pSC101 DNA (Fig. 4d) while the smaller fragment corresponded to the 4.9×10^6 dalton fragment of pSC102 plasmid DNA (Fig. 4c). Two endonuclease fragments of pSC102 were lacking in the pSC105 plasmid; presumably the sulfonamide resistance determinant of pSC102 is located on one of these fragments, since pSC105 does not specify re-

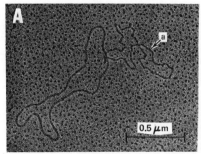
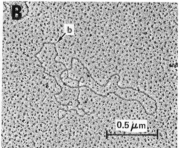

FIG. 7. (A) Heteroduplex of pSC101/pSC109. The single-stranded DNA loop marked by *a* represents the contribution of RSF1010 to the pSC109 plasmid. (B) Heteroduplex of RSF1010/pSC109. The single-stranded DNA loop marked by *b* represents the contribution of pSC101 to the pSC109 plasmid. pSC101 and RSF1010 homoduplexes served as internal standards for DNA length measurements. The scale is indicated by the bar on each electron photomicrograph.

sistance to this antibiotiç. Since kanamycin resistance *is* expressed by pSC105, we conclude that this resistance gene resides on the 4.9 × 10⁶ dalton fragment of pSC102 (fragment V of its parent, R6-5). The molecular weight of the pSC105 plasmid is estimated to be 10.5 × 10⁶ by addition of the molecular weights of its two component fragments; this value is consistent with the molecular weight determined for this recombinant plasmid by sucrose gradient centrifugation (Fig. 3*A*) and electron microscopy. The recovery of a biologically functional plasmid (i.e., pSC105) that was formed by insertion of a fragment of another plasmid fragment into pSC101 indicates that the *Eco*RI restriction site on pSC101 does not interrupt the genetic continuity of either the tetracycline resistance gene or the replicating element of this plasmid.

We also constructed new biologically functional plasmids *in vitro* by joining cohesive-ended plasmid DNA molecules of entirely different origin. RSF1010 is a streptomycin and sulfonamide resistance plasmid which has a 55% G+C nucleotide base composition (13) and which was isolated originally from *Salmonella typhimurium* (14). Like pSC101, this non-self-transmissible plasmid is cleaved at a single site by the *Eco*RI endonuclease (Fig. 5*c*). A mixture of covalently closed circular DNA containing the RSF1010 and pSC101 plasmids was treated with the *Eco*RI endonuclease, ligated, and used for transformation. A transformant clone resistant to both tetracycline and streptomycin was selected, and covalently closed circular DNA (plasmid designation pSC109) isolated from this clone by dye–buoyant density centrifugation was shown to contain a single molecular species sedimenting at 33.5 S, corresponding to an approximate molecular weight of 11.5 × 10⁶ (Fig. 6). Analysis of this DNA by agarose gel electrophoresis after *Eco*RI digestion (Fig. 5*b*) indicates that it consists of two separate DNA fragments that are indistinguishable from the *Eco*RI-treated RSF1010 and pSC101 plasmids (Fig. 5*a* and *c*).

Heteroduplexes shown in Fig. 7*A* and *B* demonstrate the existence of DNA nucleotide sequence homology between pSC109 and each of its component plasmids. As seen in this figure, the heteroduplex pSC101/pSC109 shows a double-stranded region about 3 μm in length and a slightly shorter single-stranded loop, which represents the contribution of RSF1010 to the recombinant plasmid. The heteroduplex formed between RSF1010 and pSC109 shows both a duplex region and a region of nonhomology, which contains the DNA contribution of pSC101 to pSC109.

SUMMARY AND DISCUSSION

These experiments indicate that bacterial antibiotic resistance plasmids that are constructed *in vitro* by the joining of *Eco*RI-treated plasmids or plasmid DNA fragments are bio-

logically functional when inserted into *E. coli* by transformation. The recombinant plasmids possess genetic properties and DNA nucleotide base sequences of both parent molecular species. Although ligation of reassociated *Eco*RI-treated fragments increases the efficiency of new plasmid formation, recombinant plasmids are also formed after transformation by *unligated Eco*RI-treated fragments.

The general procedure described here is potentially useful for insertion of specific sequences from prokaryotic or eukaryotic chromosomes or extrachromosomal DNA into independently replicating bacterial plasmids. The antibiotic resistance plasmid pSC101 constitutes a replicon of considerable potential usefulness for the selection of such constructed molecules, since its replication machinery and its tetracycline resistance gene are left intact after cleavage by the *Eco*RI endonuclease.

We thank P. A. Sharp and J. Sambrooke for suggesting use of ethidium bromide for staining DNA fragments in agarose gels. These studies were supported by Grants AI08619 and GM14378 from the National Institutes cf Health and by Grant GB-30581 from the National Science Foundation. S.N.C. is the recipient of a USPHS Career Development Award. R.B.H. is a USPHS Special Fellow of the Institute of General Medical Sciences on leave from the Department of Botany, University of Michigan.

1. Cohen, S. N. & Chang, A. C. Y. (1973) *Proc. Nat. Acad. Sci. USA* **70**, 1293–1297.
2. Hedgepeth, J., Goodman, H. M. & Boyer, H. W. (1972) *Proc. Nat. Acad. Sci. USA* **69**, 3448–3452.
3. Bigger, C. H., Murray, K. & Murray, N. E. (1973) *Nature New Biol.*, **224**, 7–10.
4. Boyer, H. W., Chow, L. T., Dugaiczyk, A., Hedgepeth, J. & Goodman, H. M. (1973) *Nature New Biol.*, **224**, 40–43.
5. Greene, P. J., Betlach, M. C., Goodman, H. M. & Boyer, H. W. (1973) "DNA replication and biosynthesis," in *Methods in Molecular Biology*, ed. Wickner, R. B. Marcel Dekker, Inc. New York), Vol. 9, in press.
6. Mertz, J. E. & Davis, R. W. (1972) *Proc. Nat. Acad. Sci. USA* **69**, 3370–3374.
7. Cohen, S. N., Chang, A. C. Y. & Hsu, L. (1972) *Proc. Nat. Acad. Sci. USA* **69**, 2110–2114.
8. Cohen, S. N. & Miller, C. A. (1970) *J. Mol. Biol.* **50**, 671–687.
9. Sharp, P. A., Hsu, M., Ohtsubo, E. & Davidson, N. (1972) *J. Mol. Biol.* **71**, 471–497.
10. Sharp, P. A., Cohen, S. N. & Davidson, N. (1973) *J. Mol. Biol.* **75**, 235–255.
11. Modrich, P. & Lehman, R. L. (1973) *J. Biol. Chem.*, in press.
12. Jacob, F., Brenner, S. & Cuzin, F. (1963) *Cold Spring Harbor Symp. Quant. Biol.* **23**, 329–484.
13. Guerry, P., van Embden, J., & Falkow, S. (1973) *J. Bacteriol.*, in press.
14. Anderson, E. S. & Lewis, M. J. (1965) *Nature* **208**, 843–849.
15. Davies, J., Benveniste, M. S. & Brzezinka, M. (1971) *Ann. N.Y. Acad. Sci.* **182**, 226–233.

9

Reprinted from *Natl. Acad. Sci. (USA) Proc.* **71**:1030–1034 (1974)

Genome Construction Between Bacterial Species *In Vitro:* Replication and Expression of *Staphylococcus* Plasmid Genes in *Escherichia coli*

(transformation/R plasmid/antibiotic resistance/restriction endonuclease/recombination)

ANNIE C. Y. CHANG AND STANLEY N. COHEN

Department of Medicine, Stanford University School of Medicine, Stanford, California 94305

Communicated by Joshua Lederberg, November 27, 1973

ABSTRACT Genes carried by *Eco*RI endonuclease-generated fragments of *Staphylococcus* plasmid DNA have been covalently joined to the *E. coli* antibiotic-resistance plasmid pSC101, and the resulting hybrid molecules have been introduced into *E. coli* by transformation. The newly constructed plasmids replicate as biologically functional units in *E. coli*, and express genetic information carried by both of the parent DNA molecules. In addition, electron microscope heteroduplex analysis of the recombinant plasmids indicate that they contain DNA sequences derived from *E. coli* and *Staphylococcus aureus*. Recombinant molecules can transform other *E. coli* cells for penicillin-resistance markers originally carried by the staphylococcal plasmid, and can be transferred among *E. coli* strains by conjugally proficient transfer plasmids.

Cohen *et al.* (1) have recently reported that "hybrid"* DNA molecules constructed *in vitro* by the joining of *Eco*RI (2, 3) endonuclease-generated fragments of separate plasmids can form new biologically functional replicons when inserted into *Escherichia coli* by transformation (4). Plasmid DNA species isolated from such transformed cells possess genetic properties and nucleotide base sequences from both of the parent DNA molecules.

The previously reported method for obtaining biologically functional hybrid DNA species having segments derived from diverse sources appears to be potentially applicable for the introduction of genetic material from various prokaryotic or eukaryotic organisms into *E. coli;* the antibiotic-resistance plasmid used in earlier experiments (i.e., pSC101) (1, 5) is especially useful for the selection of hybrid plasmids constructed *in vitro* from *Eco*RI-generated DNA fragments, since its replication machinery and its tetracycline-resistance gene(s) remain intact after its cleavage by the endonuclease (1).

The present report describes transformation of *E. coli* by hybrid plasmid molecules that have been constructed *in vitro* from *Eco*RI endonuclease-generated fragments of unrelated plasmid DNA species isolated from *E. coli* and *Staphylococcus aureus*. Such hybrid plasmids, which replicate autonomously in *E. coli* and which can be transferred to other bacterial cells by conjugally proficient *E. coli* plasmids, have been shown to carry DNA nucleotide sequences from *E. coli* and *Staphylococcus* and to express genetic information derived from both bacterial species.

---

* "Plasmid chimeras" might be more appropriate here, but we have not been able to establish a consensus among our colleagues for a definite terminology.

MATERIALS AND METHODS

Staphylococcus aureus strain 8325 containing the plasmid pI258 (6–8), which expresses resistance to penicillin, erythromycin, cadmium, and mercury, was obtained from R. Novick. A mutant of *E. coli* strain C600, defective in restriction and modification functions (C600 $r_K^- m_K^-$) (9), was used for initial selection of *E. coli–Staphylococcus* hybrid plasmids. Other bacterial strains, the tetracycline–resistance plasmid pSC101, and the procedures used (1, 4, 5, 10, 11, 21) for isolation of covalently-closed circular plasmid DNA from *E. coli*, transformation of *E. coli* by plasmid DNA, electron microscopy, plasmid heteroduplex studies, and agarose gel electrophoresis of *Eco*RI endonuclease-generated fragments have been described previously. The conditions used for growth of *Staphylococcus aureus*, and for isolation of staphylococcal plasmid DNA were described by Lindberg *et al.* (12). *Eco*RI restriction endonuclease and *E. coli* DNA ligase were gifts of H. Boyer or P. Modrich and I. R. Lehman, respectively, and were used as described (3, 13, 19), as indicated in Table 1.

RESULTS

Covalently-closed circular staphyloccal DNA, isolated by CsCl–ethidium bromide gradients, was cleaved by *Eco*RI endonuclease, and was examined by agarose gel electrophoresis (Fig. 1) and analytical CsCl gradient centrifugation (Fig. 2). As seen in Fig.1a, the staphylococcal plasmid pI258 is cleaved into four separate fragments having calculated molecular weights of 7.9, 4.6, 4.2, and 1.4×10^6. The sum of the molecular weights of these fragments (18.1×10^6) is in general agreement with the molecular weight reported previously for this plasmid (7, 8), and with the molecular weight calculated (14) from contour-length measurements (about 9.5 μm) of the intact molecule that we have obtained by electron microscopy (Fig. 4, and unpublished data).

The *E. coli* tetracycline-resistance plasmid, pSC101 (1, 5), was used for selection of *E. coli–Staphylococcus* hybrid plasmids. As reported (1), cleavage of pSC101 DNA by the *Eco*RI endonuclease occurs at one site, resulting in formation of a single linear fragment (Fig. 1b) having a molecular weight of 5.8×10^6 and a buoyant density in CsCl of 1.710 g/cm³.

Heterogeneity of base composition of the various component *Eco*RI-generated fragments of pI258 is evident from the buoyant density data shown in Fig. 2. In addition to a main peak banding at a buoyant density of 1.691 g/cm³, which is nearly identical to the buoyant density of *Staphylococcus* chromosomal DNA (15) and of the untreated plasmid molecule, the

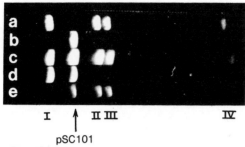

FIG. 1. Agarose gel electrophoresis (21) of EcoRI digests of plasmid DNA. The procedure of Helling, Goodman, and Boyer (in preparation) was used, as described (1). Samples were subjected to electrophoresis for 19 hr. The molecular weight of each fragment in the range of 1 to 200 $\times$ 10⁶ was determined from its mobility in the gel relative to the mobilities of DNA standards of known molecular weight. (a) pI258. The molecular weights calculated for the four fragments derived from the pI258 plasmid are (from left to right as identified in figure) I, 7.9; II, 4.6; III, 4.2; and IV, 1.4. All molecular weights have been multiplied by 10^{-6} and may be assumed to have a precision of $\pm 10\%$. (b) pSC101. Molecular weight equals 5.8×10^6 (1, 5). (c) pSC101 plus pI258. (d) pSC112. (e) pSC113. The molecular weights of the intact hybrid plasmids estimated from the weights of the component fragments are: pSC112, 13.7×10^6; pSC113, 14.6×10^6.

EcoRI-cleaved plasmid contains at least one DNA fragment, which bands at a buoyant density of 1.699 g/cm³.

Transformation of E. coli C600 $r_K^- m_K^-$ was carried out with covalently-closed circular or EcoRI-cleaved pSC101 and pI258 plasmid DNA, or a ligated mixture of the two EcoRI-

treated DNA species (Table 1). As reported (1, 5), tetracycline-resistant transformants were observed after transformation with either untreated pSC101 plasmid DNA or with the endonuclease-cleaved plasmid. No E. coli transformants for penicillin resistance or for other resistance markers carried by the Staphylococcus plasmid were observed when either covalently-closed-circular pI258 plasmid DNA or an EcoRI digest of this plasmid was used alone in the transformation assay. However, penicillin-resistant colonies and colonies that are jointly resistant to both penicillin and tetracycline were obtained after transformation with a ligated mixture of EcoRI-cleaved pSC101 and pI258 DNA (Table 1). A single clone of each type of transformant was selected for further study.

Covalently-closed circular DNA isolated from each of the selected clones was purified as described (4, 5, 10), and was examined by centrifugation in analytical CsCl gradients. As seen in Fig. 3, the plasmid DNA isolated from an E. coli transformant carrying resistance to both penicillin and tetracycline (plasmid designation pSC112) banded at a buoyant density of 1.700 g/cm³, while the plasmid DNA obtained from the clone expressing only penicillin resistance (plasmid designation pSC113) had a buoyant density of 1.703 g/cm³ in CsCl.

Treatment of both plasmid DNA species with EcoRI enzyme and subsequent CsCl gradient analysis of the fragments (Fig. 3) indicated that pSC112 contains fragments having two different buoyant densities: a DNA species banding at 1.710 g/cm³, which is identical to the buoyant density of

TABLE 1. Transformation of C600 $r_K^- m_K^-$ by pSC101 and pI258 plasmid DNA

	Transformants/μg DNA	
DNA	Tc	Pc
pSC101 closed circular	1×10^6	<3
pI258 closed circular	<3.6	<3.6
pSC101 + pI258 untreated	9.1×10^5	<5
pSC101 + pI258 EcoRI-treated	4.7×10^3	10

Transformation of E. coli strain C600 $r_K^- m_K^-$ with covalently-closed circular pSC101 (0.83 μg/ml) or pI258 (7.4 μg/ml) DNA and selection of transformants resistant to tetracycline (Tc, 25 μg/ml) or penicillin (Pc, 250 U/ml) were carried out as described (4). Covalently-closed circular pSC101 and pI258 plasmid DNA, isolated as indicated in Methods, were separately cleaved by incubation 37° for 15 min in 0.2-ml reaction mixtures containing DNA (40 μg/ml), 100 mM Tris·HCl (pH 7.4), 5 mM MgCl₂, 50 mM NaCl, and excess (2 units) EcoRI endonuclease (3) in 1-μl volume. After an additional incubation at 60° for 5 min to inactivate the endonuclease, aliquots of the two cleaved species were mixed in a ratio of 3 μg of pI258: 1 μg of pSC101 and annealed at 2–4° for 48 hr. Subsequent ligation was carried out for 6 hr at 14° (19) in 0.2-ml reaction mixtures containing 5 mM MgCl₂, 0.1 mM NAD, 100 μg/ml of bovine-serum albumin, 10 mM (NH₄)₂SO₄ (pH 7.0), and 18 U/ml of DNA ligase (13). Ligated mixtures were incubated at 37° for 5 min and chilled in ice water. Aliquots containing 3.3–6.5 μg/ml of total DNA were used directly in the transformation assay (4). Transformation frequency is expressed in terms of transformants per μg of plasmid DNA. No penicillin-resistant transformants were observed when an unligated mixture of EcoRI-cleaved pSC101 and pI258 DNA was used. Resistance of E. coli transformants to erythromycin, CaCl₂, or HgCl₂ was not observed at levels expressed by Staphylococcus aureus carrying the plasmid pI258.

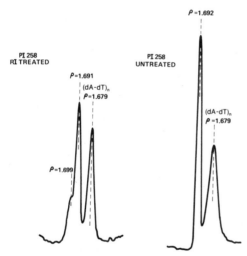

FIG. 2. Analytical ultracentrifugation (Spinco) of Staphylococcus plasmid DNA cleaved by EcoRI endonuclease (left) as described in Table 1, or untreated (right). Centrifugation in CsCl (ρ = 1.690 g/cm³) (10) was carried out for 26 hr at 44,000 rpm in the presence of a (dA-dT)ₙ (ρ = 1.679 g/cm³) density marker. Densitometer (Gilford) tracings of photographs taken during centrifugation are shown.

pSC101 DNA (1, 5), and a second species having a buoyant density ($\rho = 1.691$ g/cm³) approximately equal to that observed for the parent *Staphylococcus* plasmid (Fig. 2). pSC113 is composed of DNA fragments having three different buoyant densities in CsCl: at least two separate *Eco*RI-generated fragments of pI258 having buoyant densities of 1.691 and 1.699 g/cm³ appeared to be contained in the pSC113 hybrid plasmid, in addition to a fragment representing pSC101.

These interpretations were confirmed by electrophoresis of *Eco*RI digests of pSC112 and pSC113 DNA in agarose gels (Fig. 1d and e). Furthermore, these electrophoresis results indicate that the fragment of pI258 present in the pSC112 *E. coli–Staphylococcus* hybrid plasmid (i.e., fragment I) is absent in pSC113, while the latter plasmid contains fragments II and III of the parent staphylococcal plasmid. Since both pSC112 and pSC113 express penicillin resistance, we conclude that the *Staphylococcus* plasmid pI258 carries at least two genes capable of coding for penicillin resistance. It has not yet been determined whether these genes are duplicates.

Densitometer tracings of photographs taken during CsCl buoyant density gradient centrifugation of *Eco*RI-treated pSC113 DNA indicate that the peak banding at 1.699 g/cm³ contains less DNA than the peak at 1.691 g/cm³, and is probably fragment III (molecular weight 4.2×10^6).

The 7.9×10^6-dalton fragment of pI258 included in pSC112 necessarily carries one of the two penicillin-resistance genes identified on the *Staphylococcus* plasmid DNA. It is not yet clear which of the two *Staphylococcus* DNA fragments included in pSC113 contains the other penicillin-resistance gene. It is notable that tetracycline resistance at 25 μg/ml is expressed *only* by pSC112, although an *Eco*RI fragment comprising the tetracycline-resistance plasmid pSC101 is also included in pSC113 (Fig. 1), which failed to express tetracycline resistance at the same level as the parent plasmid. However, a tetracycline-resistance plasmid having physical and biological properties indistinguishable from those of pSC101 could be recovered after transformation of *E. coli* by an *Eco*RI digest of pSC113 DNA (Chang and Cohen, unpublished data). Although the mechanism by which expression of tetracycline resistance is reduced in pSC113 is presently unclear, nonexpression of plasmid-borne antibiotic resistance has been reported to occur by DNA insertion (5, 11).

Transformation of *E. coli* C600 and of a restriction-minus, modification-minus mutant of this strain by pSC112 or pSC113 plasmid DNA is shown in Table 2A. As seen in this table, both of these *E. coli–Staphylococcus* plasmid species are capable of transforming *E. coli* to penicillin resistance. In addition, the hybrid plasmids pSC112 and pSC113 can be mobilized to a *restriction-competent E. coli* strain (CR34) by the conjugally proficient transfer plasmid I. (Table 2B). The frequency of transformation to the $r_K^+m_K^+$ strain was about 50-fold lower than the frequency observed with C600

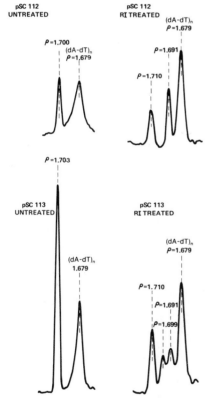

pSC 112 UNTREATED
pSC 112 RI TREATED
$(dA-dT)_n$ $\rho = 1.679$
$\rho = 1.700$
$(dA-dT)_n$ $\rho = 1.679$
$\rho = 1.691$
$\rho = 1.710$
$\rho = 1.703$

pSC 113 UNTREATED
pSC 113 RI TREATED
$(dA-dT)_n$ $\rho = 1.679$
$(dA-dT)_n$ 1.679
$\rho = 1.710$
$\rho = 1.691$
$\rho = 1.699$

FIG. 3. Analytical ultracentrifugation of DNA comprising pSC112 (*top*) and pSC113 (*bottom*). Cleavage by the *Eco*RI endonuclease (*right*) was carried out as described in Table 1. Centrifugation conditions were as indicated in legend of Fig. 2.

TABLE 2.   *Transformation and transfer of* E. coli–Staphylococcus *recombinant plasmids*

(A) Transformation (transformants/μg DNA)

Plasmid DNA	C600 $r_K^-m_K^-$		C600 $r_K^+m_K^+$	
	Pc	Tc	Pc	Tc
PSC112				
Untreated	170	180	13	34
RI treatment	7.5	170	7.5	16
PSC113				
Untreated	13,000	11	160	2.5
RI treatment	510	170	5.2	2.5

(B) Conjugal transfer

Frequency of transfer of antibiotic resistance

	Pc	Tc
SP(I) × C600 $r^-_Km^-_K$ (pSC112) × CR34N	4×10^{-9}	4×10^{-9}
SP(I) × C600 $r^-_Km^-_K$ (pSC113) × CR34N	7×10^{-5}	$<2 \times 10^{-9}$

(A) Transformation. Covalently-closed circular, *Eco*RI-cleaved pSC112 or pSC113 plasmid DNA was used to transform (4) either *E. coli* C600 $r_K^+m_K^+$ or C600 $r_K^-m_K^-$ for resistance to tetracycline or penicillin, as indicated in Table 1 and ref. 4.

(B) Conjugal transfer. pSC112 and pSC113 were mobilized by *Salmonella panama* SP (I) (16, 17) to a nalidixic acid-resistant mutant of CR34, by a modification of the procedure of Anderson and Lewis (18), Selection of the final recipient was carried out with nalidixic acid (100 μg/ml) and penicillin (250 U/ml) or tetracycline (25 μg/ml). Transfer frequency is expressed as antibiotic-resistant colonies per total number of viable cells.

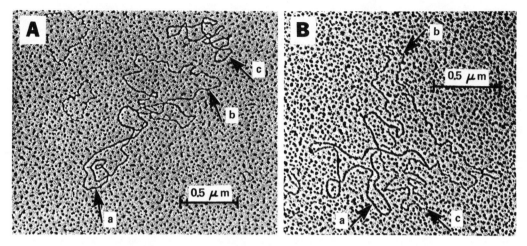

Fig. 4. (*A*) Heteroduplex of pSC101/pSC112. The region of homology is indicated by (*a*). The single-stranded loop (*b*) contains the DNA contribution of pI258 to the pSC112 hybrid plasmid. (*c*) Double-stranded molecule of pSC101 serving as internal standard for contour-length measurements. (*B*) Heteroduplex of pI258/pSC112. (*a*) Double-stranded region of homology (i.e., the segment of both plasmids containing fragment I of pI258). (*b*) Single-stranded region containing the *Eco*RI fragments of pI258 that are absent in pSC112. (*c*) Single-stranded region containing the DNA contribution of pSC101 to pSC112.

$r_K^- m_K^-$, suggesting that the staphylococcal DNA segments included in each of these plasmids contain sites susceptible to attack by the K restriction enzyme of *E. coli* (9).

Table 2 also shows that although pSC113 did not express tetracycline resistance in the clone of its origin, it is expressed in cells *transformed* by pSC113 plasmid DNA. A parallel finding was reported previously (5) for the tetracycline resistance marker carried by the R6-5 plasmid.

Heteroduplexes between pSC112 and each of its parent plasmids are shown in Fig. 4. As seen in this figure, the heteroduplex pSC101/pSC112 shows a double-stranded region approximately 3 μm in length and a much longer single-stranded region which represents the contribution of pI258 to the recombinant *E. coli–Staphylococcus* plasmid. The heteroduplex formed between pI258 and pSC112, also shows both a duplex region and two regions of nonhomology (substitution loops), which contain: (*1*) the DNA contribution of pSC101 to pSC112 and (*2*) the three *Eco*RI-generated fragments of pI258 that are absent in the hybrid plasmid.

DISCUSSION AND SUMMARY

The results of these experiments indicate that *Staphylococcus* plasmid genes can be linked to *E. coli* plasmid DNA by ligation of overlapping single-stranded ends (19, 20) of *Eco*RI endonuclease-generated fragments of both plasmids. The resulting hybrid molecules can be introduced by transformation into *E. coli* where they replicate as biologically functional units. The recombinant plasmids contain DNA nucleotide sequences derived from both *E. coli* and *Staphylococcus*, as demonstrated by electron microscope heteroduplex analysis. Moreover, genetic information carried by the *Staphylococcus* DNA is expressed in *E. coli*.

Earlier investigations (1) demonstrated that new, biologically functional plasmid species can be constructed *in vitro* with fragments of larger *E. coli* plasmids. The replication and expression of genes in *E. coli* that have been derived from a totally unrelated bacterial species (i.e., *Staphylococcus aureus*) now suggest that interspecies genetic recombination may be generally attainable. Thus, it may be practical to introduce into *E. coli* genes specifying metabolic or synthetic functions (e.g., photosynthesis, antibiotic production) indigenous to other biological classes. In addition, these results support the earlier view that antibiotic-resistance plasmid replicons such as pSC101 may be of great potential usefulness for the introduction of DNA derived from eukaryotic organisms into *E. coli*, thus enabling the application of bacterial genetic and biochemical techniques to the study of eukaryotic genes.

These studies were supported by Grant AI 08619 from the National Institute of Allergy and Infectious Disease, by Grant GB 30581 from the National Science Foundation, and by a USPHS Career Development Award to SNC. We thank H. Boyer and P. Modrich for their generous gifts of *Eco*RI endonuclease and DNA ligase respectively. We are grateful also to R. Helling for instruction in techniques of agarose gel electrophoresis.

1. Cohen, S. N., Chang, A. C. Y., Boyer, H. W. & Helling, R. B. (1973) *Proc. Nat. Acad. Sci. USA* **70**, 3240–3244.
2. Hedgepeth, J., Goodman, H. M. & Boyer, H. W. (1972) *Proc. Nat. Acad. Sci. USA* **69**, 3448–3452.
3. Greene, P. J., Betlach, M. D., Goodman, H. M. & Boyer, H. W. (1973) "DNA Replication and Biosynthesis" in *Methods in Molecular Biology*, ed. Wickner, R. B. (Marcel Dekker, Inc. New York), Vol. 9, in press.
4. Cohen, S. N., Chang, A. C. Y. & Hsu, L. (1972) *Proc. Nat. Acad. Sci. USA* **69**, 2110–2114.
5. Cohen, S. N. & Chang, A. C. Y. (1973) *Proc. Nat. Acad. Sci. USA* **70**, 1293–1297.
6. Lindberg, M. & Novick, R. P. (1973) *J. Bacteriol.* **115**, 139–145.
7. Rush, M. G., Gordon, C. N., Novick, R. N. & Warner, R. C. (1969) *Proc. Nat. Acad. Sci. USA* **63**, 1304–1310.
8. Novick, R. P. & Bouanchaud, D. (1971) *Ann. N.Y. Acad. Sci.* **182**, 279–294.
9. Meselson, M. & Yuan, R. (1968) *Nature* **217**, 1110–1114.
10. Cohen, S. N. & Miller, C. A. (1970) *J. Mol. Biol.* **50**, 671–687.

Proc. Nat. Acad. Sci. USA 71 (1974)

11. Sharp, P. A., Cohen, S. N. & Davidson, N. (1973) *J. Mol. Biol.* **75**, 235–255.
12. Lindberg, M., Sjostrom, J. E. & Johansson, T. (1972) *J. Bacteriol.* **109**, 844–847.
13. Modrich, P., Anraka, V. & Lehman, R. L. (1973) *J. Biol. Chem.* **248**, 7495–7501.
14. McHattie, L. A., Berns, K. I. & Thomas, C. A., Jr. (1965) *J. Mol. Biol.* **11**, 648–649.
15. Marmur, J. & Doty, P. (1962) *J. Mol. Biol.* **5**, 109–118.
16. Guinee, P. A. M. & Williams, A. M. C. C. (1967) *Antonie van Leeuwenhoek. J. Microbiol. Serol.* **33**, 407–412.
17. van Embden, J. & Cohen, S. N. (1973) *J. Bacteriol.* **116**, 699–709.
18. Anderson, E. S. & Lewis, M. J. (1965) *Nature* **208**, 843–849.
19. Mertz, J. E. & Davis, R. W. (1972) *Proc. Nat. Acad. Sci. USA* **69**, 3370–3374.
20. Sgaramella, V. (1972) *Proc. Nat. Acad. Sci. USA* **69**, 3389–3393.
21. Sharp, P. A., Sugden, B. & Sambrook, J. (1973) *Biochem.* **12**, 3055–3063.

75

Part II

CONSTRUCTION
OF VECTORS AND HOSTS

Editors' Comments
on Papers 10 Through 19

The phages and plasmids of *E. coli* have been studied intensively for more than three decades and have proved invaluable in establishing our current knowledge of *E. coli* as well as providing the basis for a significant portion of molecular biology. An important feature of certain phages (transducing phage) and plasmids (F' factors) was the facility with which they could pick up pieces of *E. coli* chromosome as integral, but nonessential, portions of their own genomes. Such natural recombinant DNAs were useful for fine structure genetic studies and detailed biochemical analysis (see Bukhari et al. 1977 for a detailed description of "natural" genetic enginering). It took no conceptual leap to think of phages or plasmids as vehicles for diverse DNA fragments when the DNA joining technology became available. In fact, Lobban and Kaiser (Paper 1) described the potential of their joining technique in terms of transducing bacteriophage. In Part II, we have collected some of the papers describing the concepts and construction of phages and plasmids useful in recombinant DNA work. Such self-replicating molecules have been given several names including "vehicles," "receptor chromosomes," and "vectors." The popular nomenclature at present is that any self-replicating unit designed to accept new DNA fragments is called a *vector*.

In Papers 10 and 11, the Murrays, in Edinburgh, and Thomas, Cameron, and Davis, at Stanford University, describe in detail the manipulation of wild-type coliphage lambda to produce derivatives capable of accommodating additional DNA fragments. The experiments are excellent examples of how the wealth of genetic and biochemical information on coliphage lambda was exploited for ingenious strain construction. [Concurrently, Rambach and Tiollais (1974) constructed similar phages.] Perhaps because Eco R1 was one of the first restriction enzymes available and also because the strategy described by Mertz and Davis (Paper 3) promoted Eco R1 as the restriction enzyme of choice, the first lambda phages capable of carrying foreign DNA were designed to carry Eco R1 fragments. For a variety of reasons, the terminal transferase tailing procedure was more successful with plasmids than with phages. It was established that Eco R1 cleaved lambda DNA five times to yield six specific fragments. The general problem was to eliminate all but one or two of these sites. Both groups realized that the DNA in the middle of the lambda chromosome could be removed with little effect on phage growth. This notion had been developed earlier by the isolation of many viable deletion mutations of the phage [see Hershey (1971) for a comprehensive review of phage lambda biology]. Therefore this region was ideal for insertion of new DNA fragments. Elimination of all Eco R1 sites, except those in or bracketing this nonessential block of sequences, was accomplished by ingenious use of deletion mutations as well as lambda variants carrying mutations of specific Eco R1 sites. These latter phage were isolated by employing alternate cycles of growth in *E. coli* and a similar strain that harbored the R-factor plasmid that encoded the enzyme Eco R1. Phages lacking Eco R1 sites are resistant to *in vivo* action of Eco R1 and are enriched during the cycling procedure. When Eco R1 sites were mutated and selected ("cycled out") in this way, specific Eco R1 sites could be reinstated using standard genetic crosses. Each group demonstrates the use of these newly created phage vectors.

Thomas, Cameron, and Davis (Paper 11) made use of an important concept based on the way phage lambda encapsidates its DNA. The lambda capsid is not filled simply to capacity with DNA; rather, DNA *between* the two cohesive end sites is packaged. Thus DNA molecules of varying sizes can be packaged into the lambda capsid, as evidenced by the many deletion derivatives of the phage. In addition, there is a minimum and maximum amount of DNA that must be packaged to form a viable phage. It was reasoned that if one could remove enough nonessential lambda DNA from the middle of the chromosome, a lambda variant too small to be packaged could be constructed. This would be an ideal vector because the small vector genome would not form a plaque unless its size was increased. Simple addition of any DNA fragment would enable the phage to grow and form a plaque. Such a vector, called λgt (for generalized transducer) was constructed and characterized by Thomas, Cameron, and Davis. The power of this vector is that when recombinant DNA experiments are done, every plaque obtained carries a foreign DNA fragment.

The next three papers (Papers 12, 13, and 14) were published three years later. During this time, several events occurred in recombinant DNA technology. As more restriction endonucleases were discovered and made available commercially, there was a need for phage vectors capable of accommodating fragments generated by these new enzymes. Experiments were underway exploiting several features of lambda and *E. coli* genetics to allow quick and simple detection of phages bearing new DNA fragments. There was increased awareness that recombinant DNA technology enabled scientists to isolate a gene of interest from a large, complex genome and examine it in detail. In addition, there was a growing interest in obtaining expression of foreign genes in *E. coli*. Researchers began construction of vectors capable of using well-characterized transcription systems (the *lac* system, the early promoters of phage lambda and the plasmid β-lactamase system encoding ampicillin resistance) to make RNA and perhaps protein from the inserted fragment.

Possibly the most significant change in the technology involved a series of events that had unusual impact on the nature of the new vectors and hosts as well as on all recombinant DNA research. As early as 1973, Maxine Singer and Dieter Soll sent a letter to Philip Handler, president of the National Academy of Sciences, and John R. Hogness, president of the National Institute of Medicine, voicing the concern of the participants of the Gordon Research Conference on Nucleic Acids Research (June 11–15, 1973, New Hampton, New Hampshire) about the potential hazards some recombinant DNA research might pose to the research worker and the public (Singer and Soll 1973). The National Academy of Sciences responded by asking Paul Berg to organize a committee to evaluate the potential hazards and propose a course of action. The conclusions of the Berg committee were published in *Science* (Berg et al. 1974). The letter recommended that scientists worldwide defer recombinant DNA experiments involving determinants for antibiotic resistance or bacterial toxins, animal virus DNA, and animal genomic DNA that might harbor latent tumor viruses, until such time as the potential hazard had been evaluated or appropriate containment methods devised. It further recommended that the director of the National Institutes of Health establish a Recombinant DNA Advisory Committee and that an international meeting be convened to discuss potential biohazards. The Berg committee letter resulted in a voluntary moratorium on this research that is unprecedented in the history of science. The International Conference on DNA Molecules convened in February 1975 at the Asilomar Conference Center in Pacific Grove, California. The result of that meeting was a document (Berg et al. 1975) that made specific recommendations for classification of experiments and appropriate levels of biological and physical containment. In July 1976 the National Institutes of Health released the "Guidelines for Recombinant DNA Research" (Recombinant DNA Research Guidelines 1976).

As a result of the regulations imposed on recombinant DNA research, there was a noticeable shift of emphasis in vector construction efforts as exemplified by the work in Papers 12 and 13. Specifically, there was considerable effort to improve biological containment by incorporation of mutations into the new vectors and into the *E. coli* host in hopes of obtaining enfeebled strains less likely to reproduce in nature (Curtiss 1976). A discussion of the biological containment provided naturally by phage λ and by several λ vectors is presented in Enquist and Szybalski (1978).

Enquist et al. (1976) introduced three conditionally lethal amber mutations into the λgt · C vector described in Paper 11. This new vector, λgtWES·λC, was certified as EK2 by the NIH Recombinant DNA Advisory Committee in January 1976. [Consult the NIH recombinant DNA guidelines (1976) for a precise definition of the term "EK2." It is a term to designate one class of biological containment.] It was superceded a year later by the λgtWES · λB vector (Tiemeier et al. 1976). Paper 12 by Leder, Tiemeier, and Enquist summarizes the features of the λgtWES vectors. In addition, this paper describes testing of the vectors for biological containment, including frequency of reversion of the amber mutations, ability to survive in the alimentary canal of mice, and the inability to grow on hospital isolates of *E. coli*. A similar phage, λ gt *JZvir* · λB, was designed by Donoghue and Sharp (1977). It was certified as EK2, along with λgtWES· λB, in Februrary 1977.

Paper 13 by Blattner and co-workers describes the construction and characterization of 16 lambdoid phage vectors. The phages were given the imaginative name, Charon phages, for the mythical boatsman of the river Styx. Blattner's group incorporated features of the closely related phages Φ80 and 434, lambda phages carrying *E. coli* DNA (regions of the lactose and biotin operons), and several insertion and deletion mutants of λ and Φ80 to alter the distribution of restriction sites and enhance the biological containment of these vectors.

The Charon phages vary with respect to the size and type of DNA restriction fragments they can accommodate as well as the capacity to provide transcription of the inserted fragment using promoters within the vector. Furthermore, several vectors contain readily identifiable genetic markers that facilitate detection of phages carrying DNA fragments inserted within them. For instance, some Charon phages carry a segment of *E. coli* DNA (the *lac5* substitution) encoding most of the β-galactosidase gene (*lacz*) (see also Paper 14). These phages form blue plaques on agar media containing the chromogenic substrate 5-chloro-4-bromo-3-indolyl-β-D-galactoside (XG). When the Eco R1 fragment carrying the *lac5* substitution (the indicator fragment) is removed and replaced by a foreign DNA fragment, the phage produces colorless plaques on XG containing agar. This assay and several others are described in detail in the text and footnotes of Papers 13 and 14. The complete details of the construction and restriction endonuclease mapping of 21 Charon phages can be found in Williams and Blattner (1979) and deWet et al. (1980).

A major portion of Paper 13 by Blattner and co-workers is devoted to biological containment and risk assessment. Blattner's group introduced amber mutations in two genes involved in DNA maturation and packaging (genes A and B). Three Charon phages carrying these double mutations (Ch3A, Ch4A, and Ch16A) were certified as EK2 vectors in March 1977. As a further containment measure, insertions and deletions were incorporated into the repressor genes of λ and Φ80 to ensure continued lytic growth of the vectors. Paper 13 describes a new debilitated strain of *E. coli*, DP50, constructed by Dennis Pereira in Roy Curtiss's laboratory. A derivative of DP50, DP50supF, that could be used for the λgt*WES* vectors was described by Leder and co-workers (Paper 12).

The Charon and the λ gt*WES* phages were the first vectors certified as EK2. Since that time other phage vectors have been approved. An updated list of these new vectors can be obtained from the Office of Recombinant DNA Activities (ORDA), NIH, Bethesda, MD 20205.

Murray, Brammar, and Murray (Paper 14) describe the construction of Eco R1 and Hind III cloning vectors (see also Murray and Murray 1975). Unlike the previous papers by Leder, Blattner, and their colleagues, Murray, Brammar, and Murray (Paper 14) devote little time to problems of biological containment. This most likely reflects the strength of NIH guidelines for recombinant DNA research in the United States compared to the regulations in the United Kingdom. In any event, these authors divided their new vectors into two classes: insertion vectors and replacement vectors. Insertion vectors are those containing one restriction endonuclease site for insertion of a DNA fragment. Murray and colleagues demonstrated that the presence of a foreign DNA fragment can be detected by genetic tests if the insert inactivates a phage function. Replacement vectors are those having two restriction enzyme sites flanking a dispensable DNA fragment. That fragment can be removed and replaced by a foreign DNA fragment. The authors demonstrated that if the dispensable fragment had a recognizable phenotype (that is, was an indicator fragment), its replacement by a new DNA fragment could be scored. Murray and colleagues review some of the properties of λ that are useful in controlling transcription of the cloned fragment. A more detailed discussion of transcription controls in λ and λ recombinants is presented in a review by N. Murray (1978).

A novel type of phage vector is available that enables one to propagate one strand of a duplex DNA fragment. These vectors are derived from the M13 and fd coliphages having single-stranded DNA genomes. Such vectors have utility for producing separated DNA strands needed for Sanger nucleotide sequencing methods (Wu 1978) as well as transcription studies involving hybridization or mRNA purification. Denhardt et al. (1978) is an excellent review of the biology of these unusual phages and also describes their use as vectors.

The other major type of cloning vector in use today is the plasmid (see papers in Part I for the initial work with plasmid vectors and Collins 1977). The great variety of plasmid vectors that have been constructed has been reviewed by Sutcliffe and Ausubel (1978). In Paper 15, Bolivar et al. describe pBR322, a versatile plasmid that embodies many of the properties desirable in a plasmid cloning vector. Today pBR322 is one of the most widely used plasmid vectors. It is a small plasmid with a relaxed mode of replication (that is, does not replicate coordinately with the host). It carries resistance determinants for two antibiotics, ampicillin and tetracycline, each giving a strong positive selection for cells carrying the plasmid. Furthermore, pBR322 has unique sites for insertion of DNA fragments derived from digestion with several restriction endonucleases. Many of these insertion sites inactivate one or the other of the drug resistance genes. This allows one to screen for recombinant molecules by assaying for resistance to one drug and sensitivity to the other. The plasmid is amenable for use with several different cloning techniques, including blunt-end ligation and terminal transferase tailing. Because the complete nucleotide sequence is now known (Sutcliffe and Ausubel 1978), the analysis of the inserted fragment is simplified. pBR322, when used with the enfeebled host *E. coli* χ1776 (Paper 19), is an EK2 approved cloning vector.

The final three papers on vectors describe plasmids that are novel because both are hybrids of two diverse DNA molecules. Collins and Hohn (Paper 16) describe a colE1 hybrid plasmid carrying a small portion of the phage λ genome encoding the cohesive end site (*cos*). These plasmids, termed "cosmids," can be used for cloning in the same way as any other plasmid vector. However, since they have the site encoding the λ cohesive ends, they have the additional advantage that they can be packaged *in vitro* as defective, yet infectious, phage particles (Part III, Papers 22 and 23). Even though they are defective phage particles, they are perfectly competent plasmids. They can be introduced into any λ-sensitive cell simply by infection. Once the cosmid DNA is injected into *E. coli*, it replicates as a plasmid in the relaxed mode, using the colE1 replication system. Since cosmids carry virtually no lambda DNA, save the region surrounding *cos*, and the amount of DNA required for plasmid replication is small, it is expected that cosmids could accommodate fragments as large as 40,000–50,000 base pairs.

In the next two papers, Beggs (Paper 17) and Struhl et al. (Paper 18) describe novel plasmid vectors that can "shuttle" between *E. coli* and yeast. These plasmids are composite vectors containing an *E. coli* plasmid, all or part of a small yeast plasmid (with one exception), and a specific yeast chromosomal DNA fragment encoding either leucine or histidine biosynthetic functions. Such plasmids are, in principle, similar to the λ*dvgal*SV40 hybrid described by Jackson, Symons, and Berg (Part I, Paper 2). The vectors have selectable markers for detection of the plasmid in both *E. coli* (drug resistance) and yeast (leucine or histidine biosynthetic functions encoded by the yeast chromosomal fragment). If a fragment from the yeast genome is inserted into these composite plasmids, it can be replicated in either yeast or *E. coli*. This provides the opportunity to study the functions encoded by the fragment in yeast (the homologous organism), while retaining the capacity to be propagated and manipulated in *E. coli*. In Paper 18, Struhl et al. noted that the various composite plasmids displayed different frequencies of transformation in yeast: low frequency, involving plasmids that integrated into the yeast genome; high frequency, involving plasmids that had integrated but were simultaneously carried as plasmids (presumably the same phenomenon observed by Beggs); and high frequency, involving plasmids that replicated as "minichromosomes" and did not integrate into the yeast genome. These authors present an excellent discussion of the ramifications of the three types of transformation.

The final paper in Part II by Curtiss et al. (Paper 19), describes the rationale used to construct the debilitated *E. coli* strain X1776 specifically designed for recombinant DNA experiments where biological containment was required. A similar, though less complicated, approach was used by Pereira and Curtiss to construct a λ sensitive debilitated host, DP50 (X1953), described in Blattner et al. (Paper 13). These strains of *E. coli* are important from a historical perspective. Initially, X1776 was the only approved host cell for recombinant plasmids, and DP50 was the only approved host for many of the EK2 certified phage vectors. They played an important role in recombinant DNA experiments after the first set of NIH guidelines was established. Since the requirements set forth in the current NIH guidelines for recombinant DNA research are not as stringent as the first set of rules, χ1776 and DP50 are now not always required.

REFERENCES

Berg, P., D. Baltimore, H. W. Boyer, S. N. Cohen, R. W. Davis, D. S. Hogness, D. Nathans, R. O. Roblin, J. D. Watson, S. Weissman, and N. D. Zinder. 1974. Potential Biohazards of Recombinant DNA Molecules. *Science* **185**:303.

Berg, P., D. Baltimore, S. Brenner, R. O. Roblin, and M. F. Singer. 1975. Asilomar Conference on Recombinant DNA Molecules. *Science* **188**:991–994.

Bukhari, A. I., J. A. Shapiro, and S. L. Adhya, eds. 1977. *DNA Insertion Elements, Plasmids and Episomes.* Cold Spring Harbor, N.Y.: Cold Spring Harbor Laboratory.

Collins, J. 1977. Gene Cloning with Small Plasmids. *Current Topics in Microbiol. and Immunol.* **78**:121–170.

Curtiss, R., III. 1976. Genetic Manipulation of Microorganisms: Potential Benefits and Biohazards. *Ann. Rev. Microbiol.* **30**:507–533.

Denhardt, D., D. Dressler, and D. Ray, eds. 1978. *The Single Stranded DNA Phages.* Cold Spring Harbor, N.Y.: Cold Spring Harbor Laboratory.

deWet, J. R., D. L. Daniels, J. L. Schroeder, B. G. Williams, K. Denniston-Thompson, D. D. Moore, and F. R. Blattner. 1980. Restriction maps for 21 Charon Vector Phages. *J. Virol.* **33**:401–410.

Donoghue, D. J., and P. A. Sharp. 1977. An Improved Bacteriophage Lambda Vector: Construction of Model Recombinants Coding for Kanamycin Resistance. *Gene* **1**:209–227.

Enquist, L., and W. Szybalski. 1978. Coliphage λ as a Safe Vector for Recombinant DNA Experiments. In *Viruses and the Environment,* ed. E. Kurstak and K. Maramorosch, pp. 625–652. New York: Academic Press.

Enquist, L., D. Tiemeier, P. Leder, R. Weisberg, and N. Sternberg. 1976. Safer Derivatives of Bacteriophage λgt · λC for Use in Cloning of Recombinant DNA Molecules. *Nature* **259**:596–598.

Hershey, A. D., ed. 1971. *The Bacteriophage Lambda.* Cold Spring Harbor, N.Y.: Cold Spring Harbor Laboratory.

Murray, N. E. 1978. Bacteriophage λ as a Vector in Recombinant DNA Research—Advantages and Limitations. In *Genetic Engineering,* ed. A. M. Chakrabarty, pp. 31–52. Cleveland: CRC Press.

Murray, K., and N. E. Murray. 1975. Phage Lambda Receptor Chromosomes for DNA Fragments Made with Restriction Endonuclease III of *Haemophilus influenzae* and Restriction Endonuclease I of *Escherichia coli. J. Mol. Biol.* **98**:551–564.

Rambach, A., and P. Tiollais. 1974. Bacteriophage λ Having Eco R1 Endonuclease Sites Only in the Non-essential Region of the Genome. *Natl. Acad. Sci. (USA) Proc.* **71**:3927–3930.

Recombinant DNA Research Guidelines. 1976. *Fed. Reg.* **41**:27907–27943.

Singer, M., and D. Soll. 1973. Guidelines for DNA Hybrid Molecules. *Science* **181**:1114.

Sutcliffe, J. G., and F. M. Ausubel. 1978. Plasmid Cloning Vectors. In *Genetic Engineering*, ed. A. M. Chakrabarty, pp. 83–112. Cleveland: CRC Press.

Tiemeier, D., L. Enquist, and P. Leder. 1976. Improved Derivatives of a Phage λ EK2 Vector for Cloning Recombinant DNA. *Nature* **263**: 526–527.

Williams, B. G., and F. R. Blattner. 1979. Construction and Characterization of Hybrid Bacteriophage Lambda Charon Vectors for DNA Cloning. *J. Virol.* **29**:555–575.

Wu, R. 1978. DNA Sequence Analysis. *Ann. Rev. Biochem.* **47**:607–634.

Reprinted from *Nature* **251**:476–481 (1974)

Manipulation of restriction targets in phage λ to form receptor chromosomes for DNA fragments

Noreen E. Murray & Kenneth Murray

MRC Molecular Genetics Unit and Deptartment of Molecular Biology, University of Edinburgh, King's Buildings, Mayfield Road, Edinburgh EH9 3JR, UK

RECENT advances in the biochemistry of DNA offer means of creating combinations of genetic material that cannot be achieved by genetic recombination. Biologically active DNA molecules have been reconstituted from separated, sheared halves of bacteriophage λ DNA by the successive action of λ exonuclease and polynucleotide ligase[1] and a derivative of λ DNA has been fused to an animal virus genome[2]. The highly specific restriction endonucleases[3-5] have become powerful new tools for the fragmentation of DNA molecules and the study of the organisation of these fragments within viral genomes[6,7]. Of special interest in the present context are the enzymes that make breaks a few base pairs apart in opposite strands of duplex DNA to produce fragments with short, complementary, single-stranded projections or cohesive ends[8-11]. Such DNA fragments can be joined by annealing the cohesive ends and closing the single-strand breaks with polynucleotide ligase[12].

The first restriction enzyme of this type to be described was ·isolated from a strain of *Escherichia coli* carrying the RI drug-resistance factor[9]. Some small, circular DNA molecules have only one target[12,13] for the RI restriction endonuclease, R.*Eco*RI[14], making them suitable receptors for R.*Eco*RI fragments of DNA from other sources. Insertion of a fragment of DNA into the single R.*Eco*RI site of plasmid pSC101 destroyed neither the tetracycline-resistance phenotype nor the ability of the plasmid to replicate, so that the recombinant DNA molecules were able to transform *E. coli*[13]. This type of experiment

has important potential for cloning, and hence studying, DNA from other sources. Bacteriophage λ, with its extensively studied genetics and biochemistry could be a particularly useful alternative vehicle for applications of this sort. The DNA of phage λ, however, has five targets for R.*Eco*RI[15], so our first problem was to construct derivatives of λ whose chromosomes have only one target for this enzyme. We have made a series of such phages and shown that their DNA can be broken with R.*Eco*RI to give fragments which we have joined in new combinations, thus introducing specific deletions. We have also shown that fragments of DNA produced by the action of R.*Eco*RI can be inserted between two fragments of λ DNA having respectively the right and left ends of the phage DNA. These recombinant DNA molecules were isolated as plaques following transfection of *E. coli*.

Construction of phage with single targets for R.*Eco*RI

Treatment of phage λ DNA with R.*Eco*RI gives six fragments of DNA separable by electrophoresis in polyacrylamide or agarose gels[15]. These fragments result from breakage of the DNA within the five restriction targets whose positions on the physical map of the λ chromosome have been deduced (Fig. 1)[15] (R. W. Davis, personal communication). We noted that two of these targets, designated *srl*1 and *srl*2 according to the con-

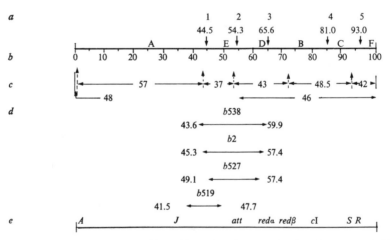

Fig. 1 Some characteristics of the λ chromosome. *a*, Positions of the targets, *srl*1–5, recognised by endonuclease R.*Eco*RI (R. W. Davies, personal communication) and location of the DNA fragments (A–F) (identified in Fig. 2) resulting from breakage of λ DNA by this enzyme. *b*, Unit length (%). *c*, The G+C content (%) of DNA fragments obtained by shearing[22]. The composition of DNA to the right of *srl*2 was calculated from that of the sheared fragments. *d*, Location of deletions[37]. *e*, Positions of some genetic markers on the physical map.[37] *A, J, reda, red*β, *cI, S* and *R* are genes; *att* is the attachment region within which recombination takes place to integrate λ into the *E. coli* chromosome.

Phage no.	Fragments							srl Targets	Restriction coefficient	Deletion	Additional genetic markers
	A H	G	B	CD	E		F				
I								1, 2, 3, 4, 5	5,000		
II								3, 4, 5	400	b538	clam
III								3, 4,	80	b538	clam
IV								3	10	b538	clam
V								0	1	b538	clam
VI								1, 2	30		clam
VII								1	10	b527	clam
I (no enzyme)											
VIII								2	3	b519	clam
IX								4	10	b538	cI857
X								5	10	b538	clam Sam7
XI								1/3a	7	srI1-3	clam
XII								1/2a	5	srI1-2	clam
I								1, 2, 3, 4, 5	5,000		

Fig. 2 Electrophoretic separation of R.EcoRI fragments of DNA from λ[+] and various derivatives. The location on the chromosome of targets for the enzyme, of the DNA fragments, and of deletions is shown in Fig. 1. The faint band visible behind the largest fragment results from adhesion of fragments by means of the natural cohesive ends of λ DNA. Superscript a denotes phage with hybrid targets generated by recombination in vitro. Although the two fragments of DNA from phages VII, VIII, and XII were not readily resolved by gel electrophoresis under the conditions used here, they have been clearly separated by equilibrium centrifugation in Cs_2SO_4–$HgCl_2$ solutions. Phage were prepared from lysates (about 2l, titre usually 5.10^{10} p.f.u. ml^{-1}) made by infection of E. coli C600 in liquid culture. After two cycles of differential centrifugation (before and after treatment with RNase and DNase at 10 µg ml^{-1}) the phage were resuspended in phage buffer (KH_2PO_4, 0.3% w/v; Na_2HPO_4, 0.7% w/v; NaCl, 0.5% w/v; $MgSO_4$, 1 mM; $CaCl_2$, 0.1 mM; gelatin 0.001% w/v), CsCl was added to 41.5% w/w and the solutions centrifuged to equilibrium. The phage band was removed with a Pasteur pipette, diluted with more CsCl in phage buffer, again centrifuged to equilibrium, and the band of phage collected after puncturing the bottom of the tube. Phage were stored in CsCl solution and DNA prepared as required by extraction of a diluted, dialysed solution with freshly distilled phenol (three extractions) followed by extensive dialysis against 0.01 M Tris-HCl (pH 8.0) 1 mM EDTA. DNA (0.3 to 1.0 µg depending upon the number and size of fragments generated) was incubated with R.EcoRI, prepared essentially as described by Yoshimori[38], in 20 µl. 0.01 M Tris-HCl (pH 7.5) 0.01 M 2-mercaptoethanol, 0.01 M $MgCl_2$ at 37° C for 30 min and the reaction stopped by addition of excess EDTA. The necessary quantity of enzyme (about 0.1 µl) was determined by titration against varying quantities of DNA and analysis by gel electrophoresis. The digests were heated in sealed capillaries at 70° C for 10 min, cooled rapidly at 0° C, mixed with 0.1% bromophenol blue in 50% w/v glycerol (2 µl) and applied to wells in 1% w/v agarose gels[40] (approx. 40 cm × 20 cm × 0.3 cm) containing 0.4 µg ml^{-1} ethidium bromide [41] and after electrophoresis for about 16 h at a constant current of 40 mA the gels were photographed under ultraviolet light on Ilford FP4 film with a 4 × red filter. For the in vitro recombination experiments, reaction mixtures of about 0.2 to 0.5 ml containing total DNA fragments, 10–30 µg ml^{-1}; Tris-HCl (pH 7.5) 66 mM; EDTA, 1 mM; $MgCl_2$, 10 mM; NaCl, 40–80 mM; dithiothreitol, 10 mM; ATP, 0.1 mM; bovine serum albumin, 0.1 mg ml^{-1}; T4 polynucleotide ligase (Miles Laboratories, Ltd), 0.5 u ml^{-1}; were incubated at 10° C for 45 min and then kept on ice for 2–10 d before sampling for transfectants.

vention of Arber and Linn[16], are within that part of the central, inessential region of the phage genome which is deleted in strain b538[17] (see Fig. 1). Should the magnitude of restriction in vivo, as defined by the restriction ratio (that is the titre of non-modified phage assayed on the propagating host, divided by the titre on the restricting host), decrease as the number of restriction targets decreases[18,19] then the deletion strain b538 should be restricted less efficiently than λ wild-type. Furthermore, phages with even fewer restriction targets should have a selective advantage on transfer from a non-modifying host to a restricting host.

Loss of the two restriction targets in the deletion strain, b538, is indeed accompanied by a fall in the restriction ratio from 5,000 to approximately 400, concomitant with which is the predicted change in the pattern of DNA fragments observed on restriction of the DNA in vitro (Fig. 2). Fragments A, E and D of the wild-type strain (phage I) were replaced by a single large fragment, while fragments B, C and F were retained in the b538 deletion strain (phage II). We attempted, therefore, to select for mutations resulting in the loss of the remaining restriction targets. Lysates of the deletion phage (phage II in Fig. 2) were made alternately on E. coli strain K and on a K strain harbouring the RI plasmid[20]. After seven such double cycles the restriction ratio had dropped from around 400 to 80 and after a total of eleven cycles the phage lysate comprised mainly restriction-resistant phages. Phages with restriction ratios of 80 (phage III) and 10 (phage IV) were isolated during the course of this enrichment and purified. Gel electrophoretic analyses of R.EcoRI digests of their DNAs (Fig. 2) showed that phage III differed from phage II in that fragments C and F had been

replaced by a larger fragment (G): we conclude that this phage had lost the right-most restriction target (srI5). DNA from the phage with a restriction ratio of 10 (phage IV) gave two large fragments: since fragments G and B were replaced by fragment H (Fig. 2) this phage must have retained only srI3 rather than srI4. The DNA of the restriction-resistant phage (phage V) migrates as a single band in a position indistinguishable from that of whole λ DNA molecules. From a cross (Table 1, cross 1) of the restriction-resistant phage (phage V), to a suitably marked strain of phage λ having a full complement of restriction targets, phages having only the two left-most restriction targets were isolated. These phages have a low restriction ratio (~ 30), and the DNA fragments resulting from in vitro restriction (phage VI, Figs 2 and 3ii) included fragment E, as expected if phage VI had acquired srI1 and srI2. Attempts to isolate spontaneous mutants lacking either srI1 or srI2 were unsuccessful because the selection favours deletions that remove both restriction targets. Phages having only srI1 (phage VII) or only srI2 (phage VIII) were therefore isolated as recombinants from crosses of phage VI (see Table 1) to the appropriately marked deletion strains b527 and b519. The agarose gel does not readily separate the two large fragments expected when the DNA of either phage VII or phage VIII is restricted (Fig. 2) but these fragments were readily separated on caesium sulphate–mercuric chloride density gradients (data not shown). Phages having only srI4 (phage IX) or only srI5 (phage X) were isolated from a cross (Table 1, cross 4) by virtue of the close linkage of srI4 to cI and of srI5 to gene S. The restricted DNA from phage IX shows two fragments including that (G) expected as the sum of fragments F and C, while the restricted DNA from phage X again shows two fragments, the smaller of which is fragment F (Fig. 2).

Separation and joining of DNA fragments

Fragments of DNA can be separated by gel electrophoresis but this method is inadequate for the preparative resolution of the larger fragments (Fig. 2). Instead, we turned to equilibrium centrifugation in caesium sulphate solutions containing mercuric chloride[21] for this gave good separation of sheared fragments of λ DNA, enabling them to be placed in order on the λ chromosome[22]. These results (Fig. 1) show that the three fragments arising from breakage of phage VI DNA with

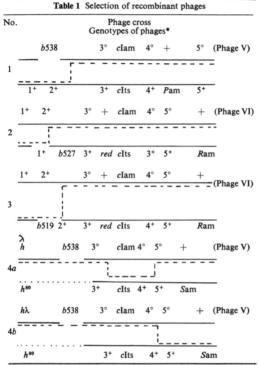

Table 1 Selection of recombinant phages

No.	Phage cross Genotypes of phages*								
	*b*538	3°	*c*Iam 4°	+	5°	(Phage V)			
1	1⁺ 2⁺	3⁺	*c*Its 4⁺	*P*am	5⁺				
	1⁺ 2⁺	3° +	*c*Iam 4° 5°	+	(Phage VI)				
2	1⁺ *b*527 3⁺	*red c*Its	3⁺ 5⁺	*R*am					
	1⁺ 2⁺	3° +	*c*Iam 4° 5°	+	(Phage VI)				
3	*b*519 2⁺	3⁺ *red c*Its	4⁺ 5⁺	*R*am					
	λ *h* *b*538	3°	*c*Iam 4° 5°	+	(Phage V)				
4*a*	*h*⁸⁰	3⁺	*c*Its 4⁺ 5⁺	*S*am					
	*h*λ *b*538	3°	*c*Iam 4° 5°	+	(Phage V)				
4*b*	*h*⁸⁰	3⁺	*c*Its 4⁺ 5⁺	*S*am					

No. 1: *srI*1⁺*srI*2⁺ (that is *b*538⁺) *c*Iam *P*⁺ recombinants were selected as phages of normal buoyant density giving clear plaques at 32° C on a suppressor-free (*sup*°) host; those with R†~30 were shown to have only targets *srI*1 and *srI*2 (Phage VI, Fig. 2). No. 2: *b*527 *red*⁺ *c*Iam *R*⁺ recombinants were selected on a *sup*° *polA* host[29] on a medium supplemented with sodium pyrophosphate (3 mM)[17]; most recombinants have R~10 and have only *srI*1 (Phage VII, Fig. 2). No. 3: *b*519 *red*⁺ *c*Iam *R*⁺ recombinants were selected on a *sup*° *polA* host[29] on a medium supplemented with sodium pyrophosphate (3 mM)[17]; most recombinants have R~4, and have only *srI*2 (Phage VIII, Fig. 2). No. 4*a*: *h*λ *b*538 *c*1857 *S*⁺ recombinants were selected as clear plaques at 39° C on a host resistant to phi80 and carrying *sup*II, which suppresses *c*Iam but not *S*am7[32]. A recombinant having R~10 was shown to have only *srI*4 (Phage IX, Fig. 2). No. 4*b*: *h*λ *b*538 *c*Iam phages were isolated as turbid plaques at 42° C (high temperature selects against phi80 host range) on a *sup*III host after enriching for those phages (*S*am7) unable to lyse a *sup*III° host. (The unlysed cells were concentrated and the phage released by chloroform.) An *S*am7 recombinant having R~10 has only *srI*5 (Phage X, Fig. 2).
*Abbreviations used: *c*Iam=*c*Iam509; *c*Its=*c*I857; *P*am=*P*am80; *R*am=*R*am5; *S*am=*S*am7; *red*=red3; 1⁺=*srI*1; 2⁺=*srI*1 etc. and 1°, 2° and so on refer to mutations conferring resistance to restriction.
†R=Restriction ratio=$\dfrac{\text{Titre on K strain}}{\text{Titre on K/RI strain}}$

The media, methods of making and assaying phage stocks and crosses were as described previously[19,31]. The standard *rec*⁺ host for phage crosses was QR47[33].

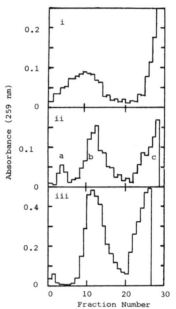

Fig. 3 Separation of DNA fragments in R.*Eco*RI digests by equilibrium centrifugation in Cs₂SO₄–HgCl₂ gradients. Density increases from right to left. (i) Phage IV, (ii) Phage VI, (iii) Phage XII. Solutions of DNA (100 μg in 1.0 ml 0.01 M Tris-HCl (*p*H 7.5) 0.01 M 2-mercaptoethanol, 0.01 M MgCl₂) were digested with R.*Eco*RI (10 μl concentrated DEAE-cellulose fraction; 37° C, 1 h), extracted with an equal volume of freshly distilled phenol and dialysed twice against 0.3 M Na₂SO₄ in 5 mM sodium borate (*p*H 9.0) and then twice against 5 mM borate (*p*H 9.0). The solutions were then heated at 65° C for 10 min, cooled in ice, weighed and their DNA concentration was determined spectrophotometrically. Mercuric chloride solution (1 mM) was then added to give an r_f of 0.22 (refs 21 and 22) and the solutions were diluted to 2.0 g with the borate buffer and Cs₂SO₄ added to 42.0% w/w. The solutions were transferred to 10 ml polypropylene centrifuge tubes, overlaid with liquid paraffin and centrifuged at 20° C at 30,000 r.p.m. for 60 h in an MSE SS65-II centrifuge, 8 × 40 Ti rotor with adaptors. Fractions of 4 drops were collected, diluted with 0.2 ml H₂O and analysed spectrophotometrically.

R.*Eco*RI would have quite different densities, as would the corresponding two fragments of phage IV DNA. Fragments of DNA from these phages were separated as predicted (Fig. 3i and ii) and recovered quantitatively.

Our first experiments on the joining of R.*Eco*RI fragments of λ DNA involved the construction of phage chromosomes with specific deletions so that the phage head would be able to accommodate additional DNA. We therefore confined our attention to phages with the targets *srI*1, *srI*2 and *srI*3 because removal of the DNA between either *srI*1 and *srI*2 or between *srI*1 and *srI*3 would affect only the inessential region of the λ chromosome. Receptor chromosomes were constructed from (*a*) the left and right fragments of phage VI DNA, and (*b*) the left fragment of phage VI DNA and the right fragment of phage IV DNA. In (*a*) the mixture of fragments in an R.*Eco*RI digest of phage VI DNA was used but in (*b*) the isolated fragments of each DNA were used. The mixtures of fragments were stored in the cold for 2 d under conditions favouring the annealing of the cohesive ends of R.*Eco*RI breaks[12] and incubated with phage T₄ polynucleotide ligase[23]. After storage at 0° C for 2–10 d, samples were taken from the reaction mixtures for transfection of *E. coli* (ref. 24 and A. E. Jacob and S. J. Hobbs, personal cummunication) and the phages from the resulting plaques were isolated and investigated.

In a second series of experiments R.*Eco*RI fragments of DNA were inserted into the genomes of our receptor phages and into that of phage IV, which possesses only *srI*3 and carries the *b*538 deletion. The central fragment of phage VI DNA, isolated from a caesium sulphate gradient (Fig. 3ii), and a DNA fragment carrying part of the *trp* operon of *E. coli*, isolated by agarose gel electrophoresis of an R.*Eco*RI digest of DNA (with a ^{32}P marker) from a λ*trp* phage (λ*trp*13, which had incorporated the bacterial *trp*D and *E* genes without deleting *srI*3)[20], were used.

Annealing of DNA fragments with cohesive ends would be expected to yield a range of products, each convertible into single DNA molecules, linear or circular, by polynucleotide ligase. SV40 DNA, for example, contains one target for R.*Eco*RI and gave monomeric and concatenated linear and circular molecules in these reations[12]. With λ DNA the situation is more complex because reactions will occur between the natural cohesive ends of the molecule[25] as well as the cohesive ends generated by R.*Eco*RI. The length and composition of the natural cohesive ends[26] relative to those created by R.*Eco*RI[9] is such that the former would cohere preferentially. The reaction products used for transfection will, therefore, comprise a complex population of DNA molecules, though many of these would not yield viable phage, either because they are too large to be packaged or because they lack essential parts of the λ genome[27]. In the experiments described here the yield of transfectants from the products of ligase reactions varied from 20 to 120 phage per µg DNA in the transfection mixture, compared with about 10^3 to 10^4 phage per µg from the corresponding DNA before restriction. (In subsequent experiments yields of transfectants from ligase reactions have been increased to about 5,000 per µg DNA and about 5×10^4 per µg from unrestricted DNA.) Omission of ligase reduced yields of transfectants from restricted DNA to about 1%, some of which appeared to result from joining of fragments *in vivo*.

Characteristics of the *in vitro* recombinants

(*a*) Receptor genomes: The three fragments from phage VI DNA could be joined to reconstitute parental phages or the two large fragments could be joined to give a phage lacking the DNA normally present between the two restriction targets (that is, ▽ *srI*1-2). Only one of twenty progeny made from the fragments of phage VI had the full complement of DNA, and could have resulted from the joining of all three fragments. The remaining 19 isolates showed an increased resistance to chelating agents[17], as expected for phages with deletions. Eighteen of these nineteen phages had a restriction ratio characteristic of phage with a single restriction target. The remaining phage was restriction-resistant and could have resulted from joining of

fragments by means other than the short cohesive ends generated by R.*Eco*RI. The deletion within this phage must extend through *srI*1 and *srI*2 and yet this and the other phages were able to integrate into the *E. coli* chromosome to give stable lysogens as shown by their ability to form stable, immune colonies[28]. (The alternative explanation that we have selected a deletion phage present in the original phage preparation seemed unlikely since the lysate had been purified by two cycles of equilibrium centrifugation in CsCl solution; furthermore most deletion mutants of phage λ are integration defective[17].) The conclusion that the essential features of the attachment region are right of *srI*2, is strengthened by the fact that the new deletion phages (▽ *srI*1-2) showed normal site-specific recombination in phage crosses (Table 2). Gel analyses of the restricted DNA of one of these ▽ *srI*1-2 phages (phage XII) confirmed that fragment E was missing: the two large fragments present were separable in a caesium sulphate–mercuric chloride gradient (Fig. 3iii). The buoyant density of phage XII indicates a deletion equivalent to 9.7% of the λ chromosome.

Phage formed by joining the left-hand fragment of phage VI and the right-hand fragment of phage IV lack the DNA normally present between targets *srI*1 and *srI*3 (that is ▽ *srI*1-3). The five phages isolated from such an experiment all had a low restriction ratio (~ 7), showed an increased resistance to chelating agents, failed to integrate into the *E. coli* chromosome and failed to form plaques on a *pol*A host. The latter property[29] is characteristic of all recombination-deficient (*red*⁻) strains of phage λ and is consistent with the deletion of at least part of the *red* genes. Crosses of the new deletion phages to appropriately marked *red*⁻ strains of phage λ (Table 3, line 4) confirm that ▽ *srI*1-3 phages are defective in the *red*α gene but not in the *red*β gene (see Fig. 1). These data provide genetic evidence that *srI*3 is either within the *red*α gene or between *red*α and *red*β. Gel electrophoreses of the restricted DNA of one of the ▽ *srI*1-3 phages (phage XI) confirm the presence of the two predicted DNA fragments and the buoyant density of phage XI indicates a deletion of 20.7% of the λ chromosome.

(*b*) Phages made from three fragments. The purified fragment E from phage VI was annealed with the restricted DNA from phage IV, which has only target *srI*3, and the mixture treated with ligase. Since phage IV has the *b*538 deletion (see Fig. 1), it lacks the DNA of fragment E and, if *srI*3 is within the *red*α gene, the insertion of DNA at this site will produce a *red*⁻ phage. We isolated 78 transfectants from the reaction mixture and detected six defined as *red*⁻ by their failure to form plaques on a *pol*A host[29]. All six *red*⁻ phages were characterised as

Table 2 Site specific recombination in phage crosses

		Cross*			Frequency of site specific recombination (Twice the frequency of turbid (c^+) $J^+ R^+$ recombinants)
Jam6	*att*⁺	*red*3	c^+	+	
					1.33%
+	*att*⁺	*red*3	*c*1857	Ram5	
Jam6	*att*⁺	*red*3	c^+	+	
					0.024%
+	*b*538 (*att* ▽)	*red*3	*c*1857	Ram5	
Jam6	*att*⁺	*red*3	c^+	+	
					1.46%, 1.37%, 1.44%**
+	*srI*1–2	*red*3	*c*1857	Ram5	

* These crosses of *red*⁻ phages were made in a *rec*A host (QR48)[33].
** Three different ▽ *srI*1–2 recombinants were used, one of which was restriction resistant and another was a recombinant made from purified fragments.

Table 3   The *red* functions of *in vitro* recombinants

| Test phage | Recombination frequency (%) between immunity and gene *R* expressed as twice the frequency of $i^{4*}R^+$ recombinants | |
	(a) Cross to *reda* i^4Ram5	(b) Cross to *redβ* i^4Ram5
λ ▽ *srI*1–2 $i^λR^+$ (*red*⁺)	7.2	2.1
λ *bio*69 $i^λR^+$ (*reda*⁻)	0.02	1.6
λ *red*3 $i^λR^+$ (*redβ*⁻ polar on *reda*)	1.4	0.02
λ ▽ *srI*1–3 $i^λR^+$ †	0.02; 0.02	2.3; 2.4
λ *b*538 *srI*3 *red*⁻R^+ ‡	0.02; 0.01; 0.01; 0.02; 0.02; 0.02.	2.8; 3.0; 2.3; 2.8; 2.9; 2.1.

* The abbreviation i^4 is used for the immunity of phage 434.
† 2 different *srI*1–3 deletion phages.
‡ The six *red* phages resulting from insertion of DNA at *srI*3.
The Red phenotypes of the phages were deduced from the frequencies of i^4R^+ recombinants generated in crosses to a *reda* phage (λ *bio*69 i^4Ram5) and to a *redβ* phage (λ *redβ*113 i^4Ram5)[39]. All crosses were made in the *recA* host, QR48[33], and the recombinants selected on a *sup*⁰ (λ $i^λ$Ram5) strain of *E. coli*.

reda⁻*redβ*⁺ (Table 3, line 5). Five of these six resembled the original phage IV in having a restriction ratio of around 10, as expected for a phage with only *one* restriction target: the remaining *red*⁻ phage (Phage XIII) had a restriction ratio of 80, suggesting two restriction targets. This phage meets the predictions for a recombinant in which the whole of fragment E had been inserted at *srI*3. Gel electrophoresis of the restriction products of the DNA from phage XIII (Fig. 4) shows two large DNA fragments like those of phage IV and a third fragment of the same size as fragment E.

One of the other five phages was investigated further. This phage (Phage XIV) had a buoyant density less than that of the receptor phage (Phage IV) and analysis by gel electrophoresis of the restricted DNA from this phage showed one fragment moving in the same position as the smaller fragment of the receptor DNA and a slower fragment moving slightly faster than the larger fragment of the receptor genome. The formation of this and the other *red*⁻ phages with only one restriction target can be explained if fragment E joined to one end of the receptor chromosome via its cohesive end and then at the other end via an illegitimate recombination event[30]. In the case of phage XIV the illegitimate recombination event was associated with the deletion of more than 10% of the phage DNA. This explanation for the origin of those *red*⁻ phages having a single restriction target is supported by a control experiment in which fragment E was not added to the receptor DNA and here the ten *red*⁻ phages isolated were all restriction resistant.

A fragment of DNA containing the *trpD* and *E* genes of the *trp* operon of *E. coli* together with their operator and promoter was isolated and purified from an $h^{80}i^λcI$ *trp*-transducing phage (λ*trpI*3)[20]. This fragment was fused to the restriction products of the receptor phage ▽ *srI*1-2 (Phage XII) and among 120 transfectants we detected[31] one $h^λ i^λ cIam$ phage having the

trpD and *E* genes. This λ*trp* phage expresses the *trp* genes from the *trp* promoter and from a λ promoter (N. E. M. and W. J. Brammar, unpublished observation). Like four of the five recombinant *red*⁻ phages described above, this phage has a low restriction ratio (R ~10) and has not incorporated the whole of the DNA fragment. In fact, the phage has become integration defective consistent with loss of DNA to the right of *srI*2. The buoyant density of the new λ*trp* phage indicated that it has about 3% more DNA than λ wild type. Had the whole fragment been incorporated into the receptor genome then a phage with at least 10% more DNA than λ wild type would be expected. At least two other phages had acquired extra DNA but did not contain the *trp* genes.

Practical implications

Our results show that the phage λ genome can be used as a receptor for genetic material that cannot be incorporated by conventional genetic techniques. The biochemical approach used provides a convenient means of cloning DNA, but more importantly it lends the extensive genetic and biochemical experience acquired with phage λ[34] to the study of the additional genetic material. This DNA can be inserted into the phage chromosome so that its expression is controlled autonomously, or by the phage regulatory systems[42], when advantage can be taken of the efficient λ promoter, P_L, and *N*-gene protein which allows RNA polymerase to read through transcriptional stops[43]. By appropriate choice of the phage genotype the yields of gene products can be greatly enhanced[35]; experiments with λ*trp* phages, for example, have shown that the five enzymes of the *trp* operon can be boosted to comprise as much as 50% of the soluble protein of the host cell (A. Moir and W. J. Brammar, personal communication). The insertion of appropriate DNA fragments into marked receptor phages offers new approaches for extending the biochemical genetics of eukaryote systems. We have shown that eukaryotic DNA can be inserted into the λ receptor phages (unpublished results). It has been established that eukaryotic DNA can be transcribed within the *E. coli* cell[36], but it is not known whether a eukaryotic messenger RNA is translated.

We expected that transfectant phages would result from covalently linked fragments of DNA as either linear or circular chromosomes[24] because a linear fragment could circularise by joining of free cohesive ends after its entry into the bacterial cell. Our results suggest that the bacterial cell may provide an alternative way of joining two non-homologous regions of DNA and in so doing generate deletions. Further experiments have proved that deletion phages are indeed created rather than merely selected (N. E. M. and K. M., unpublished observations) and they confirm the use of this method for the production of localised deletions. Illegitimate recombination events were first suspected because restriction targets were lost; those deletions generated in the region of *srI*2 have been characterised further as integration proficient, integration defective, or integration and recombination defective (K. Borck and N. E. M., unpublished observation). Lai and Nathans[44] have detected an analo-

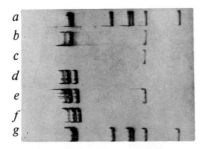

Fig. 4   Agarose gel electrophoresis of R.*Eco*RI digests of DNA used in and obtained from recombination experiments *in vitro*. Experimental details were the same as described in the legend to Fig. 2. (*a*) Phage I; (*b*) Phage VI; (*c*) Fragment E purified from phage VI (peak *a* from Fig. 3ii); (*d*) Phage IV; (*e*) Phage XIII, (*f*) Phage XIV (Phages XIII and XIV are the recombinant phages); (*g*) Phage I. The faint, slowest-moving band visible in (*b*), (*d*), (*e*) and (*f*) results from adhesion of the large fragments of DNA via their normal cohesive ends.

gous class of deletion mutants following the *in vivo* association of restriction fragments of SV40 DNA and have made elegant use of these deletions to further the genetic analysis of the SV40 virus.

Insertion of DNA fragments into a receptor chromosome *via* illegitimate recombination destroys the RI restriction target, and can therefore be a disadvantage if one is concerned with the recovery of a cloned DNA fragment. It does provide, however, a means of incorporating part of the DNA from a fragment that is too large to be accommodated within the phage genome. Removal of the DNA between *srI*1 and *srI*3 in the presence of a deletion between genes *P* and *Q* (*nin*R5)[45] eliminates 27% of the λ DNA. Since phage λ can accommodate about 10% more than its normal complement of DNA these deletions permit the insertion of DNA equivalent to about 15 genes. Although we have not used them in these experiments, powerful methods are available for selecting only those phages that have gained DNA.

We thank Sandra Fleming and Janet Brown for assistance and Dr W. J. Brammar for his constructive interest. The work was supported in part by the Science Research Council.

Received June 11, 1974.

[1] Cassuto, E., Lash, T., Sriprakash, K. S., and Radding, C. M., *Proc. natn. Acad. Sci. USA*, **68**, 1639–1643 (1971).

[2] Jackson, D. A., Symons, R. H., and Berg, P., *Proc. natn. Acad. Sci. USA*, **69**, 2904–2909 (1972).

[3] Meselson, M., Yuan, R. T., and Heywood, J., *Ann. Rev. Biochem.*, **41**, 447–466 (1972).

[4] Smith, H. O., and Wilcox, K. W., *J. molec. Biol.*, **51**, 379–391 (1970).

[5] Kelly, T. J., and Smith, H. O., *J. molec. Biol.*, **51**, 393–409 (1970).

[6] Middleton, J. H., Edgell, M. H., and Hutchison, C. A., *J. Virol.* **10**, 42–50 (1972).

[7] Danna, K. J., and Nathans, D., *Proc. natn. Acad. Sci. USA*, **68**, 2913–2917 (1971).

[8] Murray, K., and Old, R. W., in *Progress in Nucleic Acid Research Molecular Biology*, (edit. by Cohn, W. E.), **14**, 117–185, (Academic Press New York, 1974).

[9] Hedgpeth, J., Goodman, H. M., and Boyer, H. W., *Proc. natn. Acad. Sci. USA*, **69**, 3448–3452 (1972).

[10] Bigger, C. H., Murray, K., and Murray, N. E., *Nature new biol.*, **244**, 7–10 (1973).

[11] Boyer, H. W., Chow, L. T., Dugaiczyk, A., Hedgpeth, J., and Goodman, H. M., *Nature new Biol.*, **244**, 40–43 (1973).

[12] Mertz, J. E., and Davis, R. W., *Proc. natn. Acad. Sci. USA*, **69**, 3370–3374 (1972).

[13] Cohen, S. N., and Chang, A. C. Y., *Proc. natn. Acad. Sci. USA*, **70**, 1293–1297 (1973).

[14] Smith, H. O., and Nathans, D., *J. molec. Biol.*, **81**, 419–423 (1973).

[15] Allet, B., Jeppesen, P. G. N., Katagiri, K. J., and Delius, H., *Nature new Biol.*, **241**, 120–122 (1973).

[16] Arber, W., and Linn, S., *Ann. Rev. Biochem.*, **38**, 467–500 (1969).

[17] Parkinson, J. S., and Huskey, R. J., *J. molec. Biol.*, **56**, 369–384 (1971).

[18] Arber, W., and Kühnlein, U., *Pathol. Microbiol.*, **30**, 946–952 (1967).

[19] Murray, N. E., Manduca de Ritis, P., and Foster, L. A., *Mol. gen. Genet.*, **120**, 261–281 (1973).

[20] Brammar, W. J., Murray, N. E., and Winton, S., *J. molec. Biol.* (in the press).

[21] Nandi, U. S., Wang, J. C., and Davidson, N., *Biochemistry*, **4**, 1687–1696 (1965).

[22] Skalka, A. in *Methods in Enzymology*, (edit. by Grossman, L., and Moldave, K.), **21**, 341–350 (Academic Press, New York, 1971).

[23] Weiss, B., and Richardson, C. C., *Proc. natn. Acad. Sci. USA*, **57**, 1027–1028 (1967).

[24] Mandel, M., and Higa, A., *J. molec. Biol.*, **53**, 159–162 (1970).

[25] Strack, H. B., and Kaiser, A. D., *J. molec. Biol.*, **12**, 36–49 (1965).

[26] Wu, R., and Taylor, E., *J. molec. Biol.*, **57**, 491–511 (1971).

[27] Kaiser, A. D., and Inman, R. B., *J. molec. Biol.*, **13**, 78–91 (1965).

[28] Gottesman, M. E., and Yarmolinsky, M. B., *J. molec. Biol.*, **31**, 487–505 (1968).

[29] Zissler, J., Signer, E. R., and Schaefer, F., in *The Bacteriophage Lambda* (edit. by Hershey, A. D.,), 455–468 (Cold Spring Harbor Laboratories, New York, 1971).

[30] Franklin, N. C., in *The Bacteriophage Lambda* (edit. by Hershey, A. D.), 175–194 (Cold Spring Harbor Laboratories, New York, 1971).

[31] Murray, N. E., and Brammar, W. J., *J. molec. Biol.*, **77**, 615–624 (1973).

[32] Goldberg, A. R., and Howe, M., *Virology*, **38**, 200–202 (1969).

[33] Signer, E. R., and Weil, J., *J. molec. Biol.*, **34**, 261–271 (1968).

[34] *The Bacteriophage Lambda* (edit. by Hershey, A. D.), (Cold Spring Harbor Laboratories, New York, 1971).

[35] Davison, J., Brammar, W. J., and Brunel, F., *Mol. gen. Genet.*, **130**, 9–20 (1974).

[36] Morrow, J. F., Cohen, S. N., Chang, A. C. Y., Boyer, H. W., Goodman, H. M., and Helling, R. B., *Proc. natn. Acad. Sci. USA*, **71**, 1743–1747 (1974).

[37] Davidson, N., and Szybalski, W., in *The Bacteriophage Lambda* (edit. by Hershey, A. D.), 45–82 (Cold Spring Harbor, New York, 1971).

[38] Yoshimori, R. N., thesis, Univ. California (1971).

[39] Shulman, M. J., Hallick, L. M., Echols, H., and Signer, E. R., *J. molec. Biol.*, **52**, 501–520 (1970).

[40] Hayward, G. S., and Smith, M. G., *J. molec. Biol.*, **63**, 383–395 (1972).

[41] Sharp, P. A., Sugden, W., and Sambrook, J., *Biochemistry*, **12**, 3055–3063 (1973).

[42] Franklin, N. C., in *The Bacteriophage Lambda* (edit. by Hershey, A. D.,) (Cold Spring Harbor Laboratories, New York, 1971)

[43] Franklin, N. C., *J. molec. Biol.* (in the press).

[44] Lai, C.-J., and Nathans, D., *J. molec. Biol.* (in the press).

[45] Court, D., and Sato, K., *Virology*, **39**, 348–352 (1969).

11

Reprinted from *Natl. Acad. Sci. (USA) Proc.* **71**:4579–4583 (1974)

Viable Molecular Hybrids of Bacteriophage Lambda and Eukaryotic DNA

(*Eco*RI restriction endonuclease/DNA joining/calcium transfection/electron microscopy)

MARJORIE THOMAS, JOHN R. CAMERON, AND RONALD W. DAVIS

Department of Biochemistry, Stanford University School of Medicine, Stanford, California 94305

Communicated by Paul Berg, August 9, 1974

ABSTRACT A bacteriophage λ strain has been constructed and a method developed by which DNA from potentially any source can be covalently inserted through *Eco*RI cohesive ends into the middle of the λ DNA. These hybrid DNAs can infect nonrestricting *Escherichia coli* cells and can then propagate as plaque-forming phage. A unique feature of this λ strain is that extra DNA in the middle of its genome is required for plaque formation. A large number of such phages have been produced with *E. coli* DNA and *Drosophila melanogaster* DNA.

Hybrid DNA molecules can be constructed *in vitro* by joining any two or more DNA molecules through the cohesive ends that are generated by cleavage with *Eco*RI endonuclease (1). This discovery has made possible the formation of hybrid DNA molecules that can be propagated in *Escherichia coli*. Such hybrids may be formed between a vector molecule, which can be any molecule cleaved by *Eco*RI endonuclease capable of self propagation in *E. coli*, and potentially any other DNA molecule with cohesive ends generated by *Eco*RI endonuclease.

There already exist in bacteriophage λ (Fig. 1a) viable deletion mutants that lack most of the center two *Eco*RI-endonuclease-generated B and C fragments (2, 3). It was established that fragments B and C are in fact not required for lytic growth by cleaving λ DNA with *Eco*RI endonuclease, randomly rejoining the fragments with DNA ligase, and, after DNA infection, selecting viable phage that lack the B and C fragments (Abraham, Thomas, and Davis, in preparation). Therefore, it should be possible to construct viable hybrid DNA molecules by inserting foreign DNA into λ DNA in place of the B and C fragments. To facilitate such an insertion, the two rightmost *Eco*RI restriction sites were eliminated by mutation.

An advantage of λ DNA as a vector molecule is that an insertion of foreign DNA can be made to be essential for plaque formation. The shortest λ DNA molecules that produce plaques of nearly normal size are 25% deleted (J. S. Parkinson, unpublished observation). Apparently, if too much of even nonessential DNA is deleted from the λ genome, it cannot be packaged into virus particles. The *Eco*RI-B and -C fragments represent 21.1% of the λ DNA (Thomas and Davis, submitted to *J. Mol. Biol.*). Deletion of the nonessential *nin5* region (6.1% of λ DNA) (2) as well as the *Eco*RI-B and -C fragments yields a DNA that would not be expected to produce plaque-forming phage even though no essential genes have been deleted. However, this deleted λ DNA can be made to produce plaques if a new DNA segment is inserted into the remaining central *Eco*RI restriction site. This constitutes a positive selection for λ phage carrying a DNA insertion. The λ strain used

for the construction of these hybrid phages, hereafter referred to as λgt-λC, contains the λ *Eco*RI-C fragment in this central region in order to retain sufficient DNA that it may be propagated. This segment is removed after *Eco*RI endonuclease cleavage, and *Eco*RI-endonuclease-cleaved foreign DNA is inserted.

MATERIALS AND METHODS

Terminology. The term λgt (generalized transducer) designates that portion of the λ genome that is common to all of the hybrid DNA molecules and contains all of the essential genes for plaque formation. Following λgt and separated by a dash is a term referring to the origin of the DNA that is inserted. Following this term is an isolation number or capital letter(s) referring to an identified and ordered *Eco*RI-endonuclease-generated fragment(s). Prime letters indicate a fragment inserted in an inverted direction. Thus, λgt-λ*Eco*RI·C′ contains an inverted *Eco*RI-C fragment from λ DNA. For this report, the *Eco*RI term will be dropped. Thus, λgt-λ*Eco*RI·C′ becomes λgt-λC′.

Phages, Bacteria, DNAs, and Enzymes. λ *c*I857 was obtained from A. D. Kaiser, λ*bio*69*c*I857*P⁻* and λ*N⁻N⁻c*I857*nin5* from A. Campbell, and λ*b*189*c*I⁻ from J. S. Parkinson. *E. coli* strains C600, 594, PolA, and W3110 were obtained from A. D. Kaiser, C600 rK⁻mK⁻ was obtained from M. Meselson, *E. coli* rB⁺mB⁺ and *E. coli* rB⁺mB⁺ carrying the RI plasmid were obtained from J. Morrow. The DNAs and *Eco*RI endonuclease preparations have been described (refs. 5 and 7; Thomas and Davis, submitted to *J. Mol. Biol.*). *Drosophila melanogaster* DNA, bacteriophage P4 DNA, and *E. coli* DNA ligase were kindly supplied by R. Karp, R. Calendar, and P. Modrich, respectively.

Enzyme Reactions. *Eco*RI endonuclease reactions were performed in 0.1 M Tris·HCl (pH 7.5), 0.01 M MgSO₄, and 0.1 mM EDTA at 37° for 10 min as described by Mertz and Davis (1). *E. coli* DNA ligase reactions were in 20 mM Tris·HCl (pH 8.0), 1 mM EDTA, 10 mM (NH₄)₂SO₄, 4 mM MgCl₂, 0.1 M KCl, 100 μg/ml of gelatin, and 100 μM NAD at 10° for 6 hr. These conditions result in more than 50% joining of *Eco*RI ends and less than 10% joining of the λ ends. The DNA is heated to 70° for 2 min immediately before the ligase reaction to dissociate the λ cohesive ends and to inactivate excess endonuclease.

Phage Crosses and Mutagenesis. Phage crosses were performed as described by Parkinson (6). The desired recombinant phage was selected by pyrophosphate killing (7). The

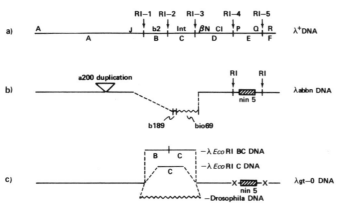

FIG. 1.   Construction of λgt-λC. (a) *Eco*RI sites in λ DNA from Thomas and Davis (submitted to *J. Mol. Biol.*). This map differs from that of Allet *et al.* (4). *Eco*RI fragments are lettered A to F below the line. λ genes are above the line. (b) Structure of λ a200 b189 *bio*69 *c*I857 *nin*5 DNA called λabbn DNA. (c) Structure of λgt-λBC, λgt-λC, and λgt-Dm DNAs.

λa200 b189 *bio*69 *c*I857 *nin*5 phage was mutagenized in 0.1 M Na acetate buffer (pH 4.6) containing 50 mM NaNO₂ for 80 min at 25°, which reduced its viability by 10³ (10).

Selection of EcoRI Restriction Site Mutants. The *Eco*RI restriction site mutants were enriched first by growth of 5 × 10⁶ mutagenized phage on an *E. coli* strain containing no restriction system and then by growth on one containing the *Eco*RI restriction systems. This process was repeated about eight times until the efficiency of plating on the *Eco*RI strain increased.

RESULTS

Elimination of EcoRI Restriction Sites 4 and 5 in λ DNA. *Eco*K restriction sites in λ DNA can be removed by genetic

selection for a nonrestricted λ phage (8). In the case of the *Eco*RI restriction sites, the selective pressure was desired against sites 4 and 5 rather than against 1, 2, or 3 (Fig. 1). It was determined that the λb189 substitution removes sites 1 and 2 and that the λ *bio*69 substitution removes site 3 without adding new restriction sites. Therefore, in the recombinant λ b189 *bio*69 *c*I857 *nin*5, sites 1, 2, and 3 should be absent and only sites 4 and 5 available for restriction (Fig. 1). But since this phage would have a 25% DNA deletion and therefore would not be readily propagated, it was necessary to select a duplication in the left arm of λ DNA (designated a200) in order to restore a suitable total DNA length (Emmons and Thomas, submitted to *J. Mol. Biol.*). The structure of this phage, illustrated in Fig. 1b, was determined by heteroduplex analysis (9)(Fig. 2a). Agarose gel electrophoresis of the *Eco*RI

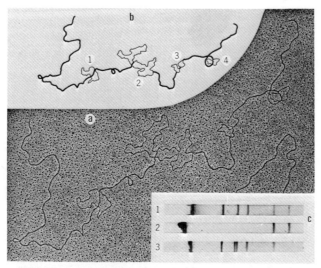

FIG. 2.   Characterization of λa200 b189 *bio*69 *c*I857 *nin*5. (a) Heteroduplex with λimm434 mounted for electron microscopy in 40% formamide and shadowed with platinum–paladium (9). DNA length measurements were performed as described by Mertz and Davis (1). (b) Illustration of micrograph in (a). Region 1 = a200 duplication; region 2 = b189 *bio*69/λ nonhomology; region 3 = imm434/λ nonhomology; region 4 = *nin*5 deletion. (c) Agarose gels of *Eco*RI endonuclease digests of (gel 3) λcI857, (gel 2) λa200 b189 *bio*69 *c*I857 *nin*5, and (gel 1) λcI857 *nin*5. The gels were prepared as described in the legend of Fig. 3.

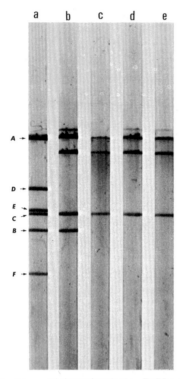

FIG. 3. Agarose gel electrophoresis of an *Eco*RI enconuclease digest of DNA from (a) λ *c*I857, (b) λgt-λBC, (c) λgt-λC, (d) λgt-λC′, and (e) mixture of λgt-λC and λgt-λC′. The λ *Eco*RI endonuclease fragments are lettered A through F. The right- and left-end fragments from the λgt strains are lettered R and L, respectively. The faint band above the left-end fragment band is from left- and right-end fragments hydrogen-bonded through the λ cohesive ends. Agarose gel electrophoresis was performed as described by Thomas and Davis (submitted to *J. Mol. Biol.*), using Tris-borate buffer and 0.5 µg/ml of ethidium bromide. Preparative gels were performed with 5 µg of DNA per tube. The DNA bands, made visible with a long-wave-length UV light (transilluminator C50 Ultra-Violet Products, Inc., San Gabriel, Calif.), were cut from the gel and eluted by electrophoresis. There was no loss in λ DNA infectivity when the DNA was treated in this manner. The gels were photographed with short-wave-length UV light.

endonuclease cleavage products of this DNA (Fig. 2c) shows that it contains only sites 4 and 5, as expected.

To select for the loss of one of two restriction sites, the efficiency of restriction must be a function of the number of restriction sites. The efficiency of plating of a number of unmodified λ deletion and substitution mutants was determined on *E. coli* cells containing only the *Eco*RI restriction system. These mutants contained from one to five *Eco*RI restriction sites. It was found that the logarithm of the efficiency of plating (E) is linearly related to the number of *Eco*RI restriction sites (R) according to the relationship $\log E = -0.88R$. Therefore, restriction sites 4 and 5 can be sequentially eliminated and do not have to be eliminated simultaneously. λa200b189*bio*69*c*I857*nin*5 was mutagenized with HNO₂ (10), and mutations in the restriction sites were selected by

infection of an *Eco*RI-restricting bacterium. Eight sequential selections were required to obtain a mutation in site 4. Five additional selections were required to obtain a further mutation in site 5.

Construction of λgt-λBC and λgt-λC. The a200b189*bio*69 region was removed from the strain lacking sites 4 and 5 and replaced with wild-type λ DNA by crossing to λN⁻ and selecting for a recombinational event between the β and the N genes. The resulting phage, called λgt-λBC, illustrated in Fig. 1c, yields 4 fragments after *Eco*RI endonuclease cleavage at *Eco*RI sites 1, 2, and 3 (Fig. 3b). The center two fragments, B and C, carry no genes essential for lytic growth. It is not necessary that both of these fragments be in the original DNA for propagation. The retention of only one of these fragments adds sufficient DNA to allow plaque formation. Therefore, the *Eco*RI-B fragment was biochemically removed. This was accomplished by first isolating the left and right DNA end fragments and the *Eco*RI-C fragment from the cleavage of λgt-λBC, by preparative electrophoresis on agarose gels (Fig. 3b). A mixture of these fragments was then covalently joined by DNA ligase and used to infect *E. coli* (nonrestricting C600 rK⁻mK⁻) cells treated with CaCl₂ (11). Several of the resulting phage were plaque-purified and characterized by heteroduplex analysis. Two types of phages were found: those that contained the *Eco*RI-C fragment oriented in the same direction as in λ⁺, called λgt-λC, and those that contained the *Eco*RI-C fragment inverted, called λgt-λC′. Agarose gels of the cleavage products of λgt-λC and λgt-λC′ are indistinguishable, as shown in Fig. 3c, d, and e. These reconstructed *Eco*RI restriction sites are still cleaved by the *Eco*RI endonuclease.

λgt-λC′ is *red*⁻ and gives a smaller plaque than λgt-λC. The *red*⁻ genotype (12) results from the inversion at *Eco*RI restriction site 3, which probably splits the *exo* gene.

Inserted DNA is Required in λgt-O for Plaque Formation. Covalent joining of the *Eco*RI-endonuclease-generated DNA end fragments from any of the λgt strains will produce a DNA molecule with all of the essential lytic genes (Fig. 1). This type of DNA molecule is termed λgt-O since it does not contain an inserted *Eco*RI DNA fragment (Fig. 1c). λgt-O DNA should not produce plaques after DNA infection since it has lost 27% of its DNA. Extra DNA must be inserted between these two *Eco*RI end fragments to produce DNA that is capable of generating plaques. These predictions have been substantiated by covalently joining the agarose-gel-purified end fragments from λgt-λBC through their *Eco*RI cohesive ends. (Substantial covalent joining was verified by electron microscopy.) The infectivity of this preparation of λgt-O DNA was less than 1% of that of uncleaved λgt-λBC DNA that had been subjected to the same process (Table 1). All 50 plaques from this infection were pooled, and the DNA from the phages was analyzed by observation of heteroduplexes with λimm434. No DNA molecules were found that did not contain a segment of DNA between the λgt end fragments. Most of the plaques resulted from λgt-λBC DNA that had not been completely digested with *Eco*RI endonuclease. The remaining plaques resulted from DNA molecules created by the insertion of contaminating *Eco*RI-B and -C fragments that were not removed by electrophoresis on agarose gels. Further evidence that λgt-O DNA does not yield plaque-forming phage is seen in the 10-fold increase in the number of

TABLE 1. *Hybrid phage formation*

Vector	DNA concentration, $\mu g/ml$	Insert	DNA concentration, $\mu g/ml$	Plaques/ plate‡ ($\times 10^{-2}$)	Plaques/ ng of vector
λgt ends	50	None	—	0.5	1
λgt ends	50	EcoRI-C	10	5	10
λgt ends	50	λCI857	50	7	13
λgt ends	50	P4	50	10	20
λgt ends	50	Drosophila	50	6	11
λgt-λBC*	—	—	—	40	400
λb2c†	—	—	—	12	1200
λgt-λC (EcoRI cleaved)	20	None	—	1	2
λgt-λC (EcoRI cleaved)	20	E. coli	20	2	4
λgt-λC (EcoRI cleaved)	20	E. coli	50	5	10
λgt-λC*	—	—	—	20	200
λb2c†	—	—	—	10	1000

* DNA used as a control was treated as in the formation of the hybrids, but no EcoRI endonuclease was added.

† λb2c DNA was used untreated as a control of transfection efficiency.

‡ The DNA infections were carried out by the procedure of Mandel and Higa (11).

plaques observed when the λ EcoRI-C fragment was added to the ligase reaction with the purified end fragments.

Although the joined λgt DNA end fragments from λgt-λC DNA are viable only if they acquire an additional segment of DNA, there is a residual infectivity from uncleaved λgt-λC DNA. A plaque produced by a hybrid phage can be identified easily in this residual background by the EMBO test for lysogeny (13) in which some of the cells from each plaque are spotted onto an EMBO (eosin and methylene blue with no sugar) plate (at 32°) overlayed with 10^9 λb2cI⁻ phage. Plaques from phage whose DNA does not contain the λ EcoRI-C fragment (the only extra λ EcoRI fragment present) must contain the foreign EcoRI-endonuclease-cleaved DNA added to the reaction mix. Phage containing the EcoRI-C fragment (λgt-λC or λgt-λC′) have the λ attachment site, the *int* and *xis* genes, and form stable lysogens which give a light-colored colony in the EMBO test. Hybrid phages containing foreign DNA in place of the EcoRI-C fragment can only form abortive lysogens, which give a dark-colored colony in the EMBO test. It is interesting to note that λgt-λC′, which has the λ *att* site and the *int* and *xis* genes inverted, still forms a stable lysogen, with the prophage presumably integrated in an inverted order. This inverted lysogen can be induced to yield phage.

Construction of Hybrid Phages with the λgt DNA End Fragments. Hybrid phages have been formed by two procedures. The first involves preparative agarose gel electrophoretic separation of the λgt DNA end fragments and the EcoRI-C fragment after complete (99%) digestion of λgt-λC DNA with EcoRI endonuclease. The addition of EcoRI-endonuclease-cleaved λ, bacteriophage P4, or *Drosophila melanogaster* DNAs to the λgt end fragments before ligase action increased

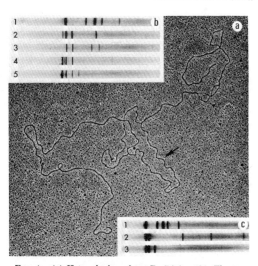

FIG. 4. (a) Heteroduplex of λgt-Dm7/λimm434. The arrow shows the inserted *Drosophila melanogaster* DNA. The DNA was prepared for electron microscopy as given in the legend of Fig. 2. (b) Agarose gels of EcoRI endonuclease digests of λ DNA carrying *Drosophila melanogaster* DNA prepared as given in the legend of Fig. 3. (Gel 1) λcI857, (gel 2) λgt-λC, (gel 3) λgt-Dm8, (gel 4) λgt-Dm9, and (gel 5) λgt-Dm7. (c) Agarose gels of EcoRI endonuclease digests of λ DNA carrying bacteriophage P4 and *Drosophila melanogaster* DNA. (Gel 1) λcI857, (gel 2) λgt-Dm3, and (gel 3) λgt-P4EcoRI·AD.

the infectivity of the product about 10 times (Table 1). These DNAs were only partially cleaved (about ⅓ of the sites) so that larger fragments could be inserted. Ninety to 95% of the plaques were judged to contain hybrid DNAs by the EMBO test. A number of these hybrids were grown and studied by heteroduplex analysis and gel electrophoresis. Fig. 4a shows a heteroduplex between λgt-Dm7 (*Drosophila melanogaster* hybrid 7) and λimm434 DNAs. Fig. 4b and c show the electrophoretic separation on agarose gels of EcoRI endonuclease digests of λgt hybrids containing *Drosophila* DNA and P4 DNA. λgt-Dm7 has a single large EcoRI DNA fragment inserted (Fig. 4b, gel 5). Also shown (Fig. 4c, gel 2) is λgt-Dm3 in which the inserted *Drosophila* DNA is cleaved to give three very small EcoRI fragments. The two EcoRI end fragments of bacteriophage P4 (72% of the P4 genome) are in λgt-P4EcoRI·AD (Fig. 4c, gel 3). This hybrid phage contains an origin of DNA replication in the P4 DNA insert as well as in the λ DNA (Goldstein, Thomas, and Davis, in preparation).

The second approach is to use the cleavage products of λgt-λC DNA directly without the removal of the λ EcoRI-C fragment. The experiment shown in the second part of Table 1 used a relatively low concentration of λgt-λC DNA with two concentrations of EcoRI-endonuclease-cleaved E. coli DNA. The E. coli DNA fragments were in large numerical excess over the λ EcoRI-C fragment (about 10 and 25 times). Twofold and 5-fold increases in the infectivity of the cleaved and rejoined λgt-λC DNA were obtained; the percentages of hybrid plaques, as measured by the EMBO test, were 50% and 80%, respectively. Higher concentrations of cleaved λgt-λC gave a smaller yield of hybrids.

DISCUSSION

This report illustrates the use of λ DNA in a generalized technique for the construction of hybrid phages containing DNA from potentially any source. In fact, hybrid phages carrying DNA from eukaryotic cells can be constructed. A plasmid DNA system for the formation and propagation of hybrid molecules has also recently been described (ref. 14; Wensink, Donelson, Finnegan, and Hogness, in preparation; Glover, White, Finnegan and Hogness, in preparation). With these λgt strains, approximately 10^4 independently generated hybrid phages containing segments of DNA up to 17,000 base pairs in length can be generated with 1 μg of foreign DNA. DNA from a eukaryotic source, *Drosophila*, gives the same yield of hybrids as DNA from *E. coli*. Thus, a sizeable proportion of *Drosophila* DNA can be propagated in *E. coli* cells. Whether there are specific eukaryotic DNA sequences that cannot be propagated in *E. coli* remains an intriguing question. At present, about 10^3 independently generated λgt-Dm phages prepared with *Drosophila melanogaster* DNA have been produced.

Insertion of DNA between the *Eco*RI endonuclease-generated λgt DNA end fragments requires two bimolecular coupling reactions. The competing nonproductive coupling reactions involving the *Eco*RI-endonuclease-generated cohesive ends are the bimolecular joining of these λgt DNA end fragments, the bimolecular oligomerization of the DNA to be inserted, and the unimolecular circularization of the DNA to be inserted. Joining of the λ DNA cohesive ends has a high activation energy and is minimized by the low temperature used for the ligase reaction. The rate of the bimolecular *Eco*RI DNA coupling reactions is determined by the concentrations of the λgt DNA end fragments and of the DNA to be inserted, while the rate of the competing unimolecular circularization reaction is determined by the length of the DNA to be inserted (2). Therefore, insertion of DNA between the *Eco*RI λgt DNA end fragments is promoted by higher concentrations and higher molecular weights of the DNA to be inserted.

A unique feature of the system described here is that it provides a selection for hybrid DNAs based upon the minimal DNA length for an infective λ particle. Also, the EMBO test readily distinguishes hybrid phage from the small background due largely to uncleaved λgt-λC DNA, since only the latter are capable of forming lysogens. Use of these phages allows easy manipulation of the hybrid DNA. For example, cells can be coinfected with two hybrids to carry out experiments involving recombination and complementation. Other possibilities for the use of lambda as a vector are being explored: for instance, a phage has been constructed that can also enter into hybrid formation but is capable of forming stable lysogens in *E. coli*. The present experimental design and its variations should, therefore, prove to be a useful tool in the study of both prokaryotic and eukaryotic DNA. Unique DNA segments from any organism can be produced in large quantities for physical and biological studies.

Note Added in Proof. The elimination of *Eco*RI restriction sites in λ DNA has been independently accomplished by N. Murray and K. Murray (15) and by A. Rambach and P. Tiollais (16).

We thank Dale Kaiser, Paul Berg, and Ray White for their helpful suggestions and Lynn Horn for help in preparation of this manuscript. This work was supported in part by Public Health Service Grant GM 19177 from the National Institutes of General Medical Sciences.

1. Mertz, J. E. & Davis, R. W. (1972) *Proc. Nat. Acad. Sci. USA* 57, 1514–1521.
2. Davidson, N. & Szybalski, W. (1971) in *The Bacteriophage Lambda*, ed. Hershey, A.D. (Cold Spring Harbor Laboratory, Cold Spring Harbor, N.Y.), pp. 45–82.
3. Davis, R. W. & Parkinson, J. S. (1971) *J. Mol. Biol.* 56, 403–423.
4. Allet, B., Jeppesen, P. G. N., Katagiri, K. J. & Delius, H. (1973) *Nature* 241, 120–122.
5. Yoshimori, R. N. (1971) Ph.D. Dissertation, University of California, San Francisco Medical Center.
6. Parkinson, J. S. (1968) *Genetics* 59, 311–325.
7. Parkinson, J. S. & Huskey, R. J. (1971) *J. Mol. Biol.* 56, 369–384.
8. Murray, N. E., Manduca de Ritis, P. & Foster, L. (1973) *Mol. Gen. Genet.* 120, 261–281.
9. Davis, R. W., Simon, M. N. & Davidson, N. (1971) in *Methods in Enzymology*, eds. Grossman, L. & Moldave, K. (Academic Press, New York), Vol. 21, pp. 413–438.
10. Miller, J. H. (1972) in *Experiments in Molecular Genetics* (Cold Spring Harbor Laboratory, Cold Spring Harbor, N.Y.), pp. 137–138.
11. Mandel, M. & Higa, A. (1970) *J. Mol. Biol.* 53, 159–162.
12. Campbell, A. (1971) in *The Bacteriophage Lambda*, ed. Hershey, A. D. (Cold Spring Harbor Laboratory, Cold Spring Harbor, N.Y.), pp. 13–44.
13. Gottesman, M. W. & Yarmolinsky, M. B. (1968) *J. Mol. Biol.* 31, 487–505.
14. Morrow, J. F., Cohen, S. N., Chang, A. C. Y., Boyer, H. W., Goodman, H. M. & Helling, R. B. (1974) *Proc. Nat. Acad. Sci. USA* 71, 1743–1747.
15. Murray, N. E. & Murray, K. (1974) *Nature* 251, 476–481.
16. Rambach, A. & Tiollais, P. (1974) *Proc. Nat. Acad. Sci. USA* 71, 3927–3930.

12

Reprinted from *Science* **196**:175–177 (1977)

EK2 DERIVATIVES OF BACTERIOPHAGE LAMBDA USEFUL IN THE CLONING OF DNA FROM HIGHER ORGANISMS: THE λgtWES SYSTEM

P. Leder, D. Tiemeier, and L. Enquist

Thomas *et al.* (*1*) have constructed a mutant strain of coliphage λ which is especially suited for cloning fragments of foreign DNA. The phage, λgt·λC, carries two Eco RI restriction sites between which Eco RI–generated DNA fragments of 1,000 to 14,000 base pairs (1 to 14 kilobase pairs) can be inserted. Because incorporation of a DNA fragment of approximately 1 kbp or more is required for phage growth, this system also provides a strong positive selection for fragment-bearing recombinants.

Recently we introduced three amber mutations—Wam403, Eam1100, and Sam100—into this phage with the aim of reducing its likelihood of encountering a susceptible host and surviving in nature (*2*). We have further modified the resulting phage, λgtWES·λC, by substituting the phenotypically inert Eco RI λ frag-

ment, λB, for λC (Fig. 1). The λC fragment contains the genes for specialized recombination and, according to its EK2 requirements, must be biochemically separated from the two larger λ fragments before in vitro recombination. The new derivative λgtWES·λB, can be used in recombination experiments directly (*3*). For testing purposes we have also constructed two model recombinants, one of which contains the genes for galactose metabolism derived from the λpgal8 recombinant described in Enquist *et al.* (*2*) (Fig. 1). The other contains a 7.6-kbp Eco RI fragment of *Escherichia coli* DNA.

These phage contain amber mutations in genes corresponding to (i) the major capsid protein of the phage, which is also required for cleavage of replicated DNA (the *E* gene), (ii) a protein needed for the

joining of phage heads and tails (the *W* gene), and (iii) a protein required for host lysis (the *S* gene). While the *W* and *E* genes are suppressed by several amber suppressors including *sup*E, the S mutation is not efficiently suppressed by *sup*E and requires a specific suppressor, *sup*F. In lysates, the frequency of wild-type revertants of each of these mutations is approximately 10^{-5} to 10^{-6} (Table 1). Wild-type revertants could not be detected in lysates of phage carrying double or triple mutations. λgtWES·λC may be propagated conveniently through lytic culture or as a lysogen since it carries the λ integration-excision system and the temperature-sensitive repressor *c*I857. λgtWES·λB can be prepared as a high-titer stock through lytic infection in liquid culture. Both phage can accept Eco RI fragments of foreign DNA as noted

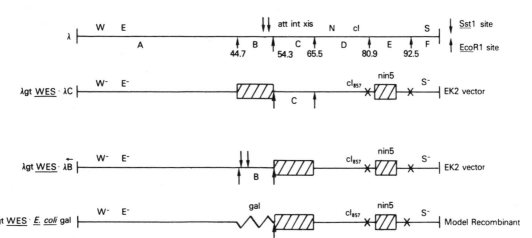

Fig. 1. Modified λ phage suitable for cloning DNA from higher organisms. The lines represent the genome of phage λ. The length of the line drawn represents the full length of the genome of wild-type bacteriophage λ. Letters over each line refer to specific λ genes. Letters under each line refer to Eco RI restriction fragments of λ with each arrow indicating an Eco RI site. The numbers under each arrow represent the position of the site as a percentage of the λ genome. Arrows over each line indicate the position of an Sst I site (note inversion as compared to wild type). Scored boxes represent deleted portions of the λ genome. X represents the point at which an Eco RI site has been eliminated by mutation. The broken line represents the location of *E. coli* DNA containing the galactose operon. Details of the construction of the two vectors have been described by Enquist *et al.* (*2*) and Tiemeier *et al.* (*3*). The model recombinant was constructed by crossing λgtSam100·λC with λWam403 Eam1100 gal8 bio256 imm434 cIts1 as described by Enquist *et al.* (*2*). The resulting phage eliminated the RI site at 54.3 percent and substituted an RI site within and close to the right end of the gal8 insertion. The internal Eco RI fragment containing a small portion of the gal8 insertion and the λ genes *int*, *xis*, and a portion of the *red* gene were deleted by in vitro recombination, yielding the phage depicted in the diagram. The position of the left end of the gal8 substitution is at 44.7 percent on the λ map (*13*), very close to the original λ Eco RI site.

Table 1. Ability of λ derivatives to grow on nonsuppressor-containing hosts. All stock lysates had titers of 1 to 2 × 10^10 per milliliter. Where possible, each mutation was tested individually in the strain used in its construction (2).

Derivative	Frequency of amber⁺ revertants
λWam403 Eam1100	< 10⁻⁸*
λWam403 Eam1100 Sam100 · λC	< 10⁻⁸*
λSam100 · λC	~ 10⁻⁵†
λWam403	~ 5 × 10⁻⁶†

*A 0.01-ml portion of undiluted stock plated on *Escherichia coli* strain 594 (*sup0*). †A 0.01-ml portion of 1000-fold diluted stock plated on *E. coli* strain 594 (*sup0*).

Table 2. Ability of phage λ, safer vector, and model recombinant to grow on environmentally encountered *E. coli* strains. Experiments on hospital samples obtained from the NIH Clinical Center Microbiology Diagnostic Laboratory were tested by plating 0.02-ml droplets of phage containing ~ 10⁸ plaque-forming units on lawns of each strain, using standard procedures (*1*). λgtWES · *E. coli* contains a 7.6 kilobase pair Eco RI fragment of *E. coli* DNA.

Strain source	Number tested	λ	Number of strains permitting plaque formation	
			λgt-WES · λC	λgt-WES · λC
Waterborne*	> 1000	0	NT	NT
Hospital samples	> 300	0	0	0

*Data supplied by R. Davis.

Table 3. Ability of model recombinant to survive in mouse alimentary tract. Three BALB/c mice were fed drinking water containing 10⁷ plaque-forming units of λgtWES · *E. coli*, a model recombinant, per milliliter. Approximately 100 ml was consumed during the experiment. Stool samples (~ five droppings) were collected on sterile paper on the days indicated and were suspended in TMG (0.01M tris, 0.01M MgSO₄, 0.01 percent gelatin, pH 7.4) containing chloroform. The suspension was tested for plaques on *E. coli* Ymel (a *supF* host).

Sample	Phage recovered (per milliliter)
Water supply, day 1	~ 10⁷
Stool, day 1	0
day 2	0
day 4	0
day 6	0
Water supply, day 7	~ 10⁷

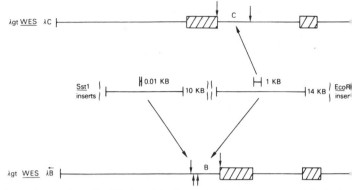

Fig. 2. Lengths of Eco RI and Sst I restriction fragments which may be inserted into λgtWES · λC and λgtWES · λB. Symbols are as indicated in the legend to Fig. 1. The arrows above the line represent Eco RI sites, and those below the line represent Sst I sites. Abbreviation: *KB*, 100 base pairs.

Table 4. Safety features of λgtWES · λC or λgtWES · λB.

1. Three conditional lethal mutations (amber)
   Eam1100; no phage heads and no cleavage of DNA to form cohesive ends
   Wam403; no joining of phage heads and tails
   Sam100; no lysis or phage release
   supF suppresses all three mutations
   *sup*E does not suppress *S*am 100 efficiently

2. *c*Its857
   Temperature-sensitive repressor; at 37°C, λ*c*Its 857 will not form a stable lysogen

3. *nin*5 deletion
   Removes a transcription stop signal; greatly reduces probability of plasmid formation

4. Recombinants are recombination defective (*red⁻*)
   λ *red⁻* phage grow less well than *red⁺* phage, conferring a selective disadvantage

5. Under permissive conditions, recombinants are capable of high-titer growth in small, easily manipulated volumes

6. Restriction barrier
   Phage can be propagated in nonmodifying hosts; unmodified DNA is destroyed on infection of a restricting host

Table 5. Test for the persistent association of cloned fragment with the permissive host used in its propagation. Routine preparative procedure: *E. coli* DP50 [*supF*, *supE*, *dap*D8, *lac*Y, Δ (*gal-uvr*B)*thy*A, *nal*Aʳ, *hsd*S] was grown overnight in tryptone broth supplemented with diaminopimelic acid (DAP, 100 μg/ml) and thymidine (50 μg/ml), and washed and resuspended in an equal volume of 0.01M MgSO₄. Approximately 1.5 × 10⁸ bacteria were incubated with 10⁶ λ Wam403 Eam 1100 *gal⁺* (*att-red*)ᴬ *c*Its857 *nin*5 *S*am100 (multiplicity of infection ≅ 10⁻²) for 10 minutes at 32°C in 0.2 ml of 0.01M MgSO₄, then diluted with 5 ml of L broth (supplemented with DAP and thymidine as above) and incubated at 38°C with vigorous shaking. Lysis occurred after 8 to 9 hours and incubation was continued for 24 hours. The phage titer was determined at the time of lysis (9 hours). Viable bacteria were determined by removing 1-ml portions, diluting and washing with 5 ml of 0.01M MgSO₄, centrifuging and washing again, and resuspending in 1 ml of 0.01M MgSO₄. Incubation was at 32°C on minimal-salts plates supplemented with DAP (100 μg/ml), thymidine (40 μg/ml), biotin (0.5 μg/ml), lysine (4 μg/ml), and glucose (0.5 percent). The *gal⁺* bacteria were determined from portions of cells treated as just described, except that plating was on medium containing 0.5 percent galactose in place of glucose. In control experiments, supplemented minimal-salts plates were checked for their ability to grow *gal⁻* and *gal⁻* derivatives of DP50 *sup*F in the presence of glucose or galactose. In the sample immediately after lysis, 4.5 × 10⁵ cells were plated on minimal-galactose plates incubated at 32°C. No *gal⁺* colonies were detected. In the sample 24 hours after lysis, 7.4 × 10⁴ cells were plated on minimal-galactose plates. No *gal⁺* colonies were detected. The validity of the detection assay was assessed by adding *gal⁺* DP50 *sup*F (λ*pgal*8 *c*Its857) cells to overgrown cells before spreading on minimal-galactose plates. Recovery of *gal⁺* cells was approximately 100 percent.

Time	Phage titer (ml⁻¹)	Bacterial titer (ml⁻¹)	*gal⁺* transductants (ml⁻¹)
At lysis	2.3 × 10¹⁰	4.8 × 10⁵	< 1
24 hours after lysis		7.4 × 10⁴	< 10

above, but λgt*WES*·λB can also accept Sst I fragments 0.01 to 10 kbp in length (Fig. 2). The presence of these Sst I sites also provides a selection against parental type Eco RI recombinants since the vector may be digested with both Sst I and Eco RI, destroying the parental λB fragment.

Bacteriophage λ has a number of fea-[Table 3 and Geier, Trigg, and Merril (*4*)].

While λgt*WES* phage are at an obvious survival disadvantage because of their suppressor requirements, they have a further disadvantage in that fragment-containing phage lack the λ general recombination functions (*red*) required for normal phage growth (*5*). The yield of such *red*-deficient phage is one-half to one-tenth that of the λ wild type. Nevertheless, sufficient recombinant phage may be prepared from 1 to 2 ml of lysates to permit detailed enzymatic, hybridization, and electron microscopic analyses (*6*). The specific safety features of the λgt*WES* system are summarized in Table 4 (*2*). The presence of each genetic feature, including the amber mutations, the temperature-sensitive repressor gene, the *nin*5 deletion, and the absence of the specialized recombination genes, can be determined by simple genetic tests (*2, 3*).

In addition to other detailed tests required for EK2 certification (*7*), we have tested the possibility that a cloned fragment could form a persistent association with a bacterial host used in its propagation. The experiments described in Table 5 represent a typical preparative phage growth using the model *gal*⁺ phage recombinant and the attenuated host strain *E. coli* K12 DP50/*sup*F (*8*). The *gal*⁺ genes carried by the model recombinant

represent a cloned DNA fragment. Any persistent association with the DP50/*sup*F host (in which the entire galactose operon is deleted) during lytic growth of the λ*WES gal*⁺ phage will give rise to *gal*⁺ colonies among the bacterial survivors in the lysate. As shown in Table 5, no *gal*⁺-containing bacteria were detected among the bacteria surviving either at the time of lysis or 24 hours thereafter.

λgt*WES*·λC was certified as an EK2 vector by the NIH Advisory Committee on Recombinant DNA Research at its January 1976 meeting, but has been superceded by λgt*WES*·λB, which was certified in February 1977. The Advisory Committee has further required that the phage be grown preparatively in the attenuated host strain *E. coli* K12 DP50/*sup*F (*8*).

We have considerable experience in the use of λgt*WES* derivatives for cloning of partially purified DNA fragments derived from total mouse genomic DNA. We generally obtain about 4500 hybrid phage per microgram of Eco RI DNA fragment (*6*). This efficiency suggests that any unique sequence purified about 1000-fold could be identified by available screening techniques (*9*). We recently cloned an Eco RI fragment containing a segment of the mouse ribosomal RNA sequences (*6*). This fragment occurs about once per 5000 mouse Eco RI fragments and is approximately 200-fold reiterated with respect to unique sequences. McClements and Skalka (*10*) cloned a similar ribosomal fragment from genomic chicken DNA using the λgt*WES*·λC vector. While the requirements of EK2 biological containment clearly reduce the flexibility of the cloning technology, the λgt*WES* system should permit cloning of

unique sequences of genomic DNA from to find any capable of supporting growth of either λ wild-type or λgt*WES* phage (Table 2). In addition, λ and the λgt*WES* derivatives are readily inactivated by passage through the mouse alimentary tract higher organisms. Other attenuated strains of λ, such as those constructed by Blattner and his associates (*11*) and Donoghue and Sharp (*12*) should be useful in this respect as well.

tures that recommend it as a safe and easily contained vector. This phage has a fastidious host range and is apparently restricted to growth in a limited number of *E. coli* strains. In fact, we have surveyed more than 300 separately isolated, human hospital strains of *E. coli* and were unable

References and Notes

1. M. Thomas, J. R. Cameron, R. W. Davis, *Proc. Natl. Acad. Sci. U.S.A.* **71**, 4579 (1974).
2. L. Enquist, D. Tiemeier, P. Leder, R. Weisberg, N. Sternberg, *Nature* **259**, 596 (1976).
3. D. Tiemeier, L. W. Enquist, P. Leder, *ibid.* **263**, 526 (1976).
4. M. R. Geier, M. E. Trigg, C. R. Merril, *ibid.* **246**, 221 (1976).
5. L. W. Enquist and A. Skalka, *J. Mol. Biol.* **75**, 185 (1973).
6. D. Tiemeier, S. Tilghman, P. Leder, in preparation.
7. P. Leder, L. Enquist, D. Tiemeier, applications to the NIH Advisory Committee on Recombinant DNA Research for EK2 certification of λgt*WES*·λC and λgt*WES*·λB, December 1976 and January 1977.
8. This strain, derived by R. Curtis, D. Pereira, and their associates at the University of Alabama (personal communication) has a number of convenient genetic features that make it useful for testing and reduce its ability to grow outside the laboratory (see Table 5). To propagate the λgt*WES* phages, we have introduced *sup*F into this strain.
9. R. A. Kramer, J. R. Cameron, R. W. Davis, *Cell* **8**, 227 (1976).
10. W. McClements and A. M. Skalka, *Science* **196**, 195 (1977).
11. F. R. Blattner *et al.*, *ibid.*, p. 161.
12. D. J. Donoghue and P. A. Sharp, *Gene*, in press.
13. W. Szybalski, personal communication.
14. We are again most grateful to C. Kunkle for her expert editorial assistance in the preparation of this manuscript.

13

Reprinted from *Science* **196**:161–169 (1977)

CHARON PHAGES: SAFER DERIVATIVES OF BACTERIOPHAGE LAMBDA FOR DNA CLONING

F. R. Blattner, B. G. Williams, A. E. Blechl, K. Denniston-Thompson, H. E. Faber, L.-A. Furlong, D. J. Grunwald, D. O. Kiefer, D. D. Moore, J. W. Schumm, E. L. Sheldon, and O. Smithies

Abstract. *The Charon λ bacteriophages have been developed as vectors for cloning. Their construction incorporates mutations that make them simple to use and also greatly increases their safety for the biological containment of cloned recombinant DNA. Three of the Charon vector phages, 3A, 4A, and 16A, have been certified for use as EK2 vector-host systems, when propagated in bulk in a special bacterial host, DP50SupF. We present here some of the data on which the safety of these systems was evaluated. DNA fragments ranging in size from 0 to 2.2 × 10⁴ base pairs can be cloned in these EK2 Charon phages.*

The desirability of using vectors for DNA cloning which have poor survival in natural environments and which require special conditions for their replication has been amply emphasized in the recent past (*1, 2*). From the point of view of containment, DNA cloning with lytic phages offers a natural advantage, since two components—phage and sensitive bacteria—must repeatedly come together for significant replication to take place. Phage vectors are also convenient because the chimeric DNA is delivered to the experimenter in a package. Since the phage and bacteria coexist only briefly, a cloned segment need not be compatible with *Escherichia coli* metabolism for ex-

tended periods of time.

Bacteriophage λ is particularly well suited for adaptation to make it useful as a lytic cloning vehicle. This is because one-third of its DNA, which forms a continuous block in the middle of the genome, can be replaced without the phage losing its ability to grow lytically (Fig. 1). Several groups of investigators, including our own, have constructed derivatives of λ specifically adapted for DNA cloning (*3–5*). The aim in all these adaptations is to construct phages so that a restriction endonuclease can make a cut (or cuts) only in the dispensable one-third of the λ DNA to permit subsequent addition of (or re-

placement by) a foreign DNA segment. Point mutations, substitutions, and deletions are, therefore, introduced into wild-type λ to alter the distribution of restriction sites and to eliminate sites in the essential regions of the phage genome. Other objectives of vector engineering include (i) cloning a variety of sizes of DNA fragments; (ii) cloning with more than one restriction enzyme; (iii) indicating by plaque type whether or not a phage has incorporated a DNA fragment; (iv) cloning with minimal manipulations; (v) controlling transcription of the cloned fragment from promoters on the vector; (vi) growing vectors and clones to a high

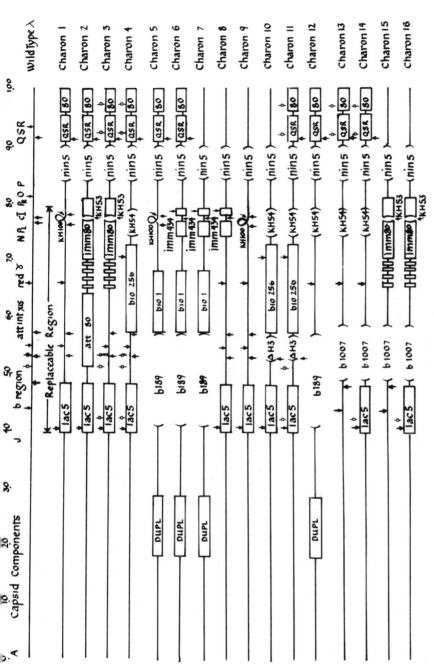

Fig. 1. Physical map of λ and the Charon phages. The phage λ map which we now estimate to be 49,400 ± 600 (S.E.M.) base pairs (34) is drawn to scale with the replaceable region indicated relative to essential genes for lytic growth. Beneath it are the Charon phage vectors with portions from λ (indicated by lines) aligned with the λ map. Boxes indicate substitutions, but the lengths of DNA substituted are not shown to scale. The length of DNA in each vector is shown in Table 1. Downward and upward arrows are Eco RI and Hind III sites, respectively. The φ symbol shows known Sst I sites; Charon 15 has none. (Sst I sites were not determined for phages 1, 5, 6, 7, 8, 9, 10, and 12.) DUPL is a duplicated piece of phage λ. *lac5*, *bio1* and *bio256* are substitutions from Lac and Bio regions of *E. coli*. The boxes labeled *att80*, *imm80*, *QSR80* are portions of phage φ80. The region shown by four small boxes at around 70 in some phages is from φ80; it is partially homologous to λ. The parenthesized b189, b1007, ΔH3, KH53, KH54, and *nin5* are deletions; KH100 is an insertion; *imm434* is a replacement for immλ [see (6)].

103

yield; (vii) readily recovering cloned DNA; and (viii) features contributing to biological containment.

The λ vectors that we have developed during the past 3 years have been designated Charon phages, after the mythological boatsman of the River Styx. They have been numbered sequentially as constructed. The structures of Charon phages 1 to 16 are shown in Fig. 1. The properties that make them useful as vectors are summarized in Table 1.

Construction of Charon phages. Most of the mutations that were incorporated into Charon phages have been described by us (6) and others (7). These were assembled into the desired arrangements by standard genetic crosses (8). We describe here only those mutations that were isolated for this project and have not previously been described.

The deletion KH53 in the immunity region of φ80 was isolated by selecting for elimination of the restriction endonuclease Eco RI site in the φ80 immunity region according to the method of Arber and Kühnlein (9). Concomitant with loss of the restriction site, the plaque morphology changed from turbid to clear. Examination of heteroduplexes (Fig. 2k) between the DNA of KH53 and of a phage with the wild-type φ80 immunity region revealed a single-stranded loop of approximately 750 base pairs near the position of the restriction site. KH53 was shown by measuring its buoyant density to be a deletion rather than an insertion; it probably deletes part of the φ80 analog of the λ cI gene. However, since little is known about the genetic organization of the immunity region of φ80, other explanations are possible. As a practical mat-

ter, KH53 is nonrevertible and drastically reduces the formation of lysogens (see safety tests).

We selected two mutations, BW1 and BW2, which eliminated the Eco RI sites just to the right of the immunity regions of λ and φ80, respectively. Another mutation isolated for this study is the duplication (DUPL) located on the left arm of Charons 5, 6, 7, and 12. This appears to be similar to duplications described by Bellett *et al.* (*10*) and has been useful for increasing the size of that arm.

The no-cut right end (NRE) of Charons 7, 8, 9, 10, 15, and 16 was specially constructed by combining the *nin*5 deletion with a hybrid isolated by Murray and Murray (5) which removes the Hind III site near gene Q, and then selecting a new point mutation (DK1) that eliminates the Eco RI site near gene S. The NRE, unlike

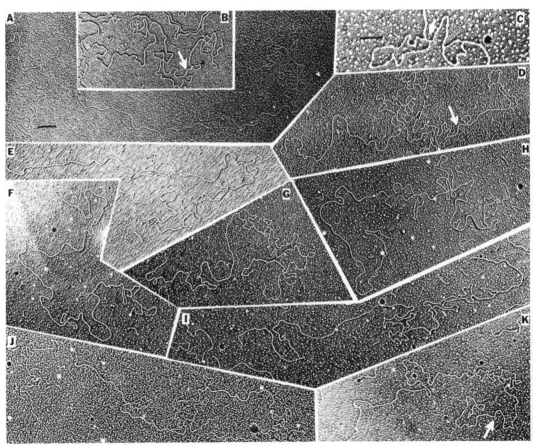

Fig. 2. Heteroduplex analysis of Charon phages. Heteroduplexes made by combining DNA strands from genetically different phages (*31*) were examined in the electron microscope. The line diagram identifies each heteroduplex and labels the loops and bifurcations where DNA strands do not pair. The calibration bar in panel (A) refers to double-stranded DNA and applies reasonably to all panels except (C). The calibration bar in panel (C) refers to single strands and applies only to that panel. The identification of phages not shown in Fig. 1 is as follows: Ch3ΔLac was made from Charon 3 by deleting its Eco RI Lac operator–containing fragment. Bio16a, panel (C), is equivalent to λ except for a small substitution of *bio* DNA to the right of *att*. 4M121 was obtained by inserting unrelated DNA from phage mu into Charon 4 (*32*). bio256nin has the same *bio* substitution and *nin*

the QSR80 substitution, has no sites to the right of *nin*5 for any of nine restriction enzymes we have tested that are potentially suitable for cloning (*11*). NRE has no visible nonhomology with λ in heteroduplexes (see Fig. 2, i and j), suggesting that the contribution of φ80 is slight.

Some of the Charon phages have had amber mutations, *A*am32 and *B*aml (*12*), introduced in their capsid genes in order to enhance biological containment. These phages are designated with an A (for example, Charon 3A is derived from Charon 3 by incorporation of these mutations).

Verification of vector structure. It is most important to verify the structure of each vector. The best way to determine whether deletions and substitutions are present and to detect unexpected structural changes is by heteroduplex mapping

(*31*). All 16 vectors were analyzed by this technique. The forms observed in the electron micrographs were completely consistent with the structures of the phages shown in Fig. 1. Electron micrographs of the heteroduplexes between DNA from wild-type λ and from eight of the vectors are shown in Fig. 2, together with three other informative heteroduplexes.

Heteroduplex analysis has also proved valuable for determining the amount of bacterial DNA remaining in some of the vectors after they have been used for cloning with Eco RI. An example of this is shown in Fig. 2c, where the bacterial DNA remaining from *lac*5 after the dispensable portion of Charon 3 had been excised with Eco RI is displayed directly as a single strand; it is less than 200 base pairs in length. The bacterial DNA re-

maining after Eco RI cloning with vectors that contain *bio*256 was similarly determined by subtracting the duplex lengths indicated by arrows in Fig. 2, d and b; this measured 980 ± 40 base pairs.

Genetic characteristics of all the vectors have been verified by plating on suitable indicator strains. Characteristics examined included the immunity type, the presence of the *lac*5 and *bio*256 substitutions, and the presence of amber mutations (*13*).

Each vector DNA was also digested with restriction enzymes, and the sizes of DNA fragments were determined on agarose gels. By analyzing these, and other related data, the positions of the restriction sites indicated in Fig. 1 were determined. The Sst I sites of Charon phages 1, 2, 5, 6, 7, 8, 9, 10, and 12 have not yet been determined.

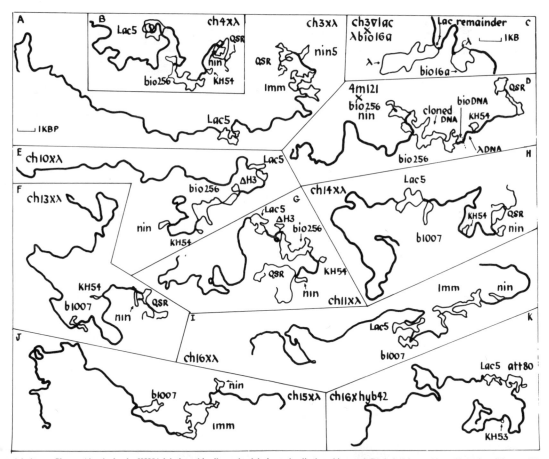

deletion as Charon 4 but lacks the KH54 deletion; this allows the deletion to be displayed in panel (D). hyb42 is λ*att*80*imm*80*nin*; its wild-type φ80 immunity region permitted display of the KH53 deletion. The arrows in the photograph panels (D) and (B) indicate duplex segments next to the KH54 marker; subtraction of their lengths provided a measurement of *E. coli* DNA remaining in Charon 4 clones; this residual *bio* DNA is indicated in the tracing of panel (D). The arrow in the photograph panel (C) indicates *E. coli* DNA remaining in Charon 3 clones. The arrow in photograph panel (D) shows the KH53 deletion.

Vectors for cloning a variety of sizes of DNA fragments. Since λ capsids cannot accommodate DNA molecules outside the size range 38 to 53 kilobase pairs (kbp) *(4, 10, 14, 34)*, each λ phage vector has a maximum and some have a minimum size of DNA fragment that can be cloned with a given restriction enzyme. The size limits for each of the Charon phages are listed in Table 1. These ranges depend not only on the positions of restriction cuts but also on the sizes of deletions and substitutions in the essential region of the vector. Any decrease in the size of the vector genome results in a corresponding increase in the maximum fragment that can be cloned, and may result in a greater minimum size. Charons 1, 4, 8, 9, 10, 11, 13, and 14 are all suitable for cloning large fragments. Charons 8

and 9 have the largest capacities, and approach the theoretical maximum for a nondefective λ vector. Among the EK2-certified Charon phages, Charon 4A can be used for cloning large Eco RI fragments.

Vectors for cloning with more than one restriction enzyme. When small DNA fragments are to be cloned, a single restriction site in a nonessential portion of the phage DNA is sufficient. Several of the Charon vectors can be used with more than one restriction enzyme in this way (Table 1). If only one site occurs for each enzyme, the heterogeneously terminated fragments obtained by digestion with all possible pairs of the enzymes can be cloned. Heterogeneously terminated vectors and fragments prevent self-cyclization reactions. The EK2-certified

phage Charon 16A can be used in this way with Eco RI and Sst I.

For cloning large fragments with multiple enzymes, a "magic" arrangement would be to have a series of different restriction sites arranged in the same order at each end of the dispensable region. With such an arrangement, restriction enzymes could be used either singly or in any possible paired combinations to clone large fragments. Charon 9 is the closest approximation to such a vector in our series.

Vector functions indicating successful cloning. A powerful indicator of successful cloning is available in Charon phages 1, 4, 8, 9, 10, 11, 13, and 14, which all have a lower limit to the size of cloned DNA they can accept (see Table 1). In these phages, when the dispensable portion of

Table 1. Characteristics of Charon cloning phages. The DNA cloning vectors and their genotypes are presented. The individual mutations are discussed in the text and in (6). Typical growth titers obtained by the PDS method are given, and they do not differ significantly with either the A series or the clone-bearing derivatives of the vectors. The total DNA length of each vector is given as a percentage of wild-type λ, determined from buoyant densities. These vectors can be used with various restriction enzymes. For each of the useful vector-enzyme combinations, the left- and right-end fragment lengths are given as percentage of λ, as determined by agarose gel electrophoresis or electron microscopy (or both). The corresponding cloning capacity is given in kilobase pairs, assuming that the packaging limits on the genome are from 38 to 53 kbp. The symbols p_L and p_{Lac}, under the heading "promoter control," indicate whether or not these major vector promoters would be expected to transcribe the cloned fragment. Other promoters ($p_{R'}$, p_{re}, or p_{rm}) might also be useful in some cases. Suggested genetic tests to indicate successful cloning are: (a) colorless plaques on Lac⁻ cells provide definite indication of foreign DNA insertion; (b) colorless plaques on Lac⁺ cells provide definite indication of foreign DNA insertion without reinsertion of dispensable vector fragment; (c) clear plaques provide definite indication of foreign DNA insertion; (d) combination of colorless plaques in Lac⁺ cells with Bio⁻ phenotype provides definite indication of foreign DNA substitution without reinsertion of either dispensable fragment. Bio⁺ phenotype is detected by a ring of bacterial growth around plaques on Bio⁻ cells on biotin-deficient plates; (e) lack of ability to plate on polA⁻ cells provides definite indication of foreign DNA insertion; (f) colorless plaques on Lac⁺ cells indicate removal of dispensable fragment but does not ensure that insertion of foreign DNA has taken place; (g) plaque formation indicates DNA insertion but does not ensure that reincorporation of dispensable fragments has not taken place; and (h) no particularly useful tests currently available. Not all genetic tests have been verified in all cases.

Vector	λ genotype	Typical growth	Total DNA length (%λ)	Fragment lengths in percent λ			Cloning capacity (kbp)	Promoter control	Genetic tests
					Left end	Right end			
Charon 1	lac5 KH100 BW1 nin5 QSR80	3 × 10¹⁰	100.5	Eco RI:	39.5	20.6	9.1–23.2		(g)
Charon 2	lac5 att80 imm80 KH53 BW2 nin5 QSR80	2 × 10¹⁰	95.1	Eco RI:	39.5	55.6	0–5.8	p_{Lac}	(a)
Charon 3	lac5 imm80 KH53 BW2 nin5 QSR80	6 × 10¹⁰	98.7	Eco RI:	39.5	42.6	0–12.3	p_L	(f)
Charon 4	lac5 bio256 KH54 BW1 nin5 QSR80	1 × 10¹⁰	93.6	Eco RI:	39.5	22.5	8.2–22.2	p_L	(d)
Charon 5	DUPL b189 bio1 KH100 BW1 nin5 QSR80	2 × 10¹⁰	93.5	Eco RI:	72.7	20.8	0–7.1		(h)
Charon 6	DUPL b189 bio1 imm434 BW1 nin5 QSR80	1 × 10¹⁰	88.6	Eco RI:	68.3	19.9	0–9.2		(c)
Charon 7	DUPL b189 bio1 imm434 BW1 nin5 NRE	2 × 10¹⁰	86.2	Eco RI:	68.3	17.6	0–9.9		(c)
				Hind III:	67.3	18.6	0–9.9		(c)
				Eco RI/Hind III:	67.3	17.6	0–10.4		(c)
Charon 8	lac5 imm434 BW1 nin5 NRE	5 × 10¹⁰	92.8	Eco RI:	39.5	17.6	10.6–24.6		(g)
				Hind III:	48.5	18.5	5.7–19.8		(g)
Charon 9	lac5 KH100 BW1 nin5 NRE	2 × 10¹¹	97.6	Eco RI:	39.5	18.3	10.2–24.3		(g)
				Hind III:	48.5	17.6	6.2–20.2		(g)
				Eco RI/Hind III:	39.5	17.6	10.6–24.6		(g)
				Sst:	42.0		≥ 13.2	p_L	(h)
				Sst/Hind III:	42.0	17.6	9.3–23.4		(g)
Charon 10	lac5 ΔH3 bio256 KH54 BW1 nin5 NRE	3 × 10¹⁰	85.3	Eco RI:	39.5	20.2	9.2–23.4	p_L	(b)
				Hind III:	52.2	33.1	0–10.7	p_L	(h)
				Sst:	42.0	34.0	1.3–15.3	p_L	(b)
				Hind III/Sst:	42.0	33.1	1.7–15.7	p_L	(b)
Charon 11	lac 5 ΔH3 bio 256 KH54 BW1 nin 5 QSR80	5 × 10¹⁰	89.0	Eco RI:	39.5	22.5	8.2–22.2	p_L	(b)
Charon 12	DUPL b189 KH54 BW1 nin5 QSR80	1 × 10¹¹	91.0	Eco RI:	64.2	26.8	0–7.9	p_L	(e)
Charon 13	b1007 KH54 BW1 nin5 QSR80	1 × 10¹¹	83.5	Eco RI:	44.5	26.8	2.4–16.7	p_L	(g)
Charon 14	lac5 b1007 KH54 BW1 nin5 QSR80	2 × 10¹⁰	83.7	Eco RI:	39.5	26.8	5.0–19.1	p_L	(b)
Charon 15	b1007 imm80 KH53 BW2 nin5 NRE	1 × 10¹¹	86.0	Eco RI:	44.5	41.5	0–10.4	p_L	(h)
				Hind III:	48.5	37.5	0–10.4	p_L	(h)
				Eco RI/Hind III:	44.5	37.5	0–12.3	p_L	(h)
Charon 16	lac5 b1007 imm80 KH53 BW2 nin5 NRE	3 × 10¹⁰	86.0	Eco RI:	39.5	46.5	0–10.4	p_{Lac}	(a)
				Sst:	42.0	44.0	0–10.4	p_{Lac}	(a)
				Eco RI/Sst:	39.5	44.0	0–11.6	p_{Lac}	(f)

DNA has been removed, the resulting vector DNA falls below 38 kbp and therefore is too small to be packaged. Only when an additional piece of DNA is inserted into the vector will a viable phage be produced (*4*). The ability to form a plaque is, therefore, a strong indicator of successful cloning. Reinsertion of a fragment or fragments from the dispensable portion will, of course, also yield viable phage. Since it is difficult to remove these dispensable fragments physically, it is advantageous to be able to detect their reinsertion genetically. It is also desirable to reduce the number and kinds of fragments from the dispensable region, so as to reduce the number of genetic tests and to reduce competition during cloning between the undesirable reinsertion fragments and the desired foreign DNA fragments. Charons 10 and 11 have a single dispensable Eco RI fragment that can be detected, when inserted in either orientation, either by plaque color on XG (defined below) plates or by tests for *bio* function.

For cloning small fragments, vectors must be used that do not have a minimum size limit. Indicators of successful cloning in these vectors can be obtained by arranging for the cloning site to be within a nonessential gene whose function can be monitored. Insertional inactivation of the nonessential gene by the cloned fragment will then indicate success. Three such indicator systems are potentially available for Charon phages. Insertion into Charon phages 6 and 7 should make the plaques clear. Insertion into the *red* gene of Charon 12 should make the phage unable to plate on *polA*⁻ cells. Insertion into the *lac*z gene of Charons 2 or 16 should render the plaques colorless on Lac⁻ cells on XG plates.

Use of lac*5 as an indicator function.* As was discussed above, several of the Charon vectors, including all three EK2-certified phages, contain the *lac*5 substitution, which carries the *E. coli* gene for β-galactosidase (*lac*z). When the chromogenic substrate 5-chloro-4-bromo-3-indolyl-β-D-galactoside (XG) is included in the plating medium, phage carrying *lac*5 give vivid blue (indigo) plaques. In order to use this reaction for indication of potential success in cloning, it was necessary to gain some understanding of the mechanism of formation of blue plaques. In *lac*5 the promoter, operator, and *z* gene are oriented leftward, and an Eco RI site is located within the *z* gene (*15*). This cut is within 200 base pairs of the left end of the substitution. To determine the effect of an interrupted *lac*z gene on plaque color, phages were constructed from Charon 3 and Charon 11 in which the Lac region

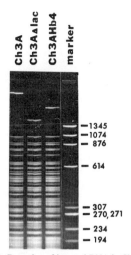

Fig. 3. Detection of inserted DNA in Charon 3A by Hha I digestion. Charon 3AHb4 was made by replacing the Eco RI Lac operator–containing fragment of Charon 3A with mouse globin cDNA. Charon 3AΔLac is Charon 3A without the Eco RI fragment. DNA from each of these phages was digested with Hha I and compared by 6 percent polyacrylamide gel electrophoresis. A Hae III digest of φX174 DNA was used as a calibration mixture. The marker sizes are indicated in base pairs. The largest fragment (> 1345 base pairs) is the piece of interest. If it is larger than that in Charon 3AΔLac, then an insert is present (as in Charon 3AHb4). If it is replaced or accompanied by new smaller pieces, then an insert with an Hha I site (or sites) is present (no example of this is shown).

had been cut out by Eco RI, turned around, and reinserted so that its orientation (see below) was changed from "n" to "u," and the *z* gene was interrupted. These phages, the parents, and other containing insertions of unrelated DNA were tested on Lac⁺ and Lac⁻ indicator bacteria on plates containing XG (80 µg/ml). We found that the development of blue plaques depended on the *lac* genotype of the plating bacteria as well as that of the phage. Dark blue plaques were always produced when the phage contained *lac*5 in the normal arrangement, and the plaques were always colorless when the fragment containing *lac*5 was removed. However, the phages with an interrupted *z* gene gave pale blue plaques on Lac⁺ cells but colorless plaques when the indicator bacteria were Lac⁻ (see cover). We conclude that a blue plaque can be produced either directly by expression of a β-galactosidase gene supplied by the infecting phage or indirectly when the many copies of the Lac operator DNA produced during phage growth bind the Lac repressor available in the cell, causing derepression of the bacterial *lac* operon.

To summarize, if we use as examples Eco RI cloning with the EK2-certified vectors, potential success with Charon 16A is indicated by loss of the Lac function detected by colorless plaques on XG plates with Lac⁻ bacteria. Potential success with Charon 4A is indicated by obtaining viable phages giving colorless plaques on XG plates with Lac⁺ bacteria. [A separate test is required for the absence of the Bio fragment (*13*).] Potential success with Charon 3A is indicated by colorless plaques on XG plates with Lac⁺ bacteria. Since Charon 3A phages that have lost the Lac fragment without insertion of foreign DNA cannot be distinguished genetically from a clone, additional tests are required, such as hybridization to a suitable probe or gel electrophoresis of a restriction enzyme digest of the phage DNA (Fig. 3).

Controlled transcription of cloned fragment. The cloning sites in many of the Charon phages can have their transcription controlled by the phage leftward promoter (p_L) or the Lac promoter. This property has already proved valuable in mapping promoters in *E. coli* fragments specifying ribosomal proteins that were cloned in Charon phages 3 and 4 (*16*).

Cloning with Charon phages. The procedure that we have developed for constructing clones differs significantly from those previously described; the conditions for ligation of fragments have been chosen to favor assimilation of all target fragments into giant concatenates rather than to attempt to limit the reaction at the dimer stage. The vector phage DNA is annealed to join its ends. (Lambdoid phages have mutually complementary single-stranded sequences on the ends of their DNA molecules.) The vector and target DNA's are digested with the desired restriction enzyme, which is then inactivated with diethyl pyrocarbonate (*17*). In order to generate recombinant DNA molecules, the vector and target DNA's are mixed. Their relative proportions are adjusted so that the molar concentration of target DNA fragments does not exceed that of the vector DNA, and the total DNA concentration is kept high (*18*) to ensure that the formation of recombinant molecules competes effectively with monomolecular cyclization reactions, particularly of short target fragments. The mixture is incubated with T4 ligase to join the molecules covalently. Viable phages can then be obtained by transfection of *E. coli* spheroplasts (*19*) or calcium-treated *E. coli* (*20*), or by in vitro encapsidation into λ capsids (*21*). We presume that λ-sized chimeric molecules are either extracted directly from concatenates by the λ packaging function, or that circles are first generated by recombination within the transfected *E. coli* cell.

Vectors with a high yield. An important safety feature for any cloning vector is its ability to be grown in high yield, so that only small volumes of culture are required for production purposes. The Charon phages and clones derived from them can usually be propagated to at least 10^{10} phage per milliliter by our PDS (preadsorb-dilute-shake) method (22). After lysis, 3 ml of chloroform is added; this immediately kills all surviving bacteria without affecting the phage. Typical yields for some of the vectors are shown in Table 1. Yields of phages containing cloned fragments have usually been in the same range.

Recovery of cloned DNA. Two convenient methods have been devised for recovering cloned DNA fragments from Charon phages. The first method is general and applies to any clones containing a single restriction enzyme fragment smaller than about 2 kbp. In this method the clone-bearing phage DNA is prepared, heated at 55°C in $2M$ NaCl to anneal the cohesive ends, dialyzed, cut with the re-

striction enzyme, and centrifuged for 4 hours at 40,000 rev/min (SW50.1 rotor) at 15°C. The small cloned fragment remains in the supernatant and can be recovered free of the large vector fragment by ethanol precipitation.

The second method is useful in selected cases, and we have used it for studying mouse globin complementary DNA (cDNA) clones that were obtained by procedures similar to those that have been used with plasmids (23). The purification method is based on the fact that the largest DNA fragment in Hha I restriction enzyme digests of Charon 3A ΔLac (see Fig. 2 legend) contains the single Eco RI site of this phage. Provided that the cloned piece of DNA inserted into this site does not contain an Hha I site, the size of the cloned DNA fragment can easily be determined from the size of the largest fragment in an Hha I digest. The position of this fragment after polyacrylamide gel electrophoresis is convenient for preparative purposes. Figure 3 illustrates this point with one of our mouse globin cDNA clones.

Hybrid phage nomenclature. Charon clones are named by appending to the

name of the vector one or more letters denoting the source of inserted DNA, followed by an isolation number. For example, Charon 3AHb4 was our fourth isolation candidate for a mouse globin clone. The orientation of an inserted fragment can be denoted "n" (natural) or "u" (unnatural), depending on whether the coordinate system assigned to the target fragment increases from left to right in agreement with λ's. In Charon 11, the Lac and Bio DNA's are in the "n" orientation since the clockwise "time-increasing" direction of the *E. coli* map corresponds to the left-right direction in λ.

Biological containment. Guidelines have been set up by the National Institutes of Health for assessing the suitability of host-vector systems for the cloning of DNA from different sources (2). When DNA from higher organisms, such as mammals, is being cloned, a host-vector system with a biological rating of EK2 is required.

The NIH guidelines specify (2) that: "For EK2 host-vector systems in which the vector is a phage, no more than one in 10^8 phage particles should be able to perpetuate itself and/or a cloned DNA fragment under nonpermissive conditions designed to represent the natural environment either (a) as a prophage or plasmid in the laboratory host used for phage propagation or (b) by surviving in natural environments and transferring itself and/ or a cloned DNA fragment to a host (or its resident lambdoid prophage) with properties common to those in the natural environment." Charon phages 3A, 4A, and 16A in conjunction with host bacteria DP50 (χ1953), or DP50SupF (χ2098) have now been certified as meeting these criteria by the director of NIH following the recommendation of his Recombinant DNA Advisory Committee.

The accumulation of data on safety and the certification process itself took more than a year and involved our submitting eight documents detailing experimental results concerning safety (24). A phage working group, appointed by the NIH Recombinant DNA Advisory Committee and headed by W. Szybalski, met three times to analyze the results. They made several useful suggestions to us for additional experiments prior to recommending that the systems be certified.

In the following, we present a brief synopsis of the safety features of the λ vectors, including (Table 2) the results of three safety experiments that were considered particularly germane by the phage working group.

General safety features of λ phages. Foreign DNA cloned in a λ phage might be propagated outside the laboratory in

Table 2. Gene transfer between Charon A phages and bacteria. Exchange of genes in both directions between Charon A phages and bacteria was estimated by considering three cases. (Experiment 1) Gene transfer from phage to productive host: Lysates of Charon A phages were prepared and the titers for phage (at lysis) and for total bacteria (after overgrowth) were determined; the $\lambda^s lacz^- y^+ sup$ E bacterium CSH18 was used as host. Lac$^+$ bacteria (those that had stably incorporated the *lacz* gene function from the phage) were measured in the overgrown cultures. The ratio of the stable associates after overgrowth to the phage titers at lysis were calculated. (Experiment 2) Gene transfer from phage to nonproductive host: Charon A phages were adsorbed at a 1 : 1 ratio to λ^sLacz$^- y^+ sup^0$ bacteria (M96CF') and to two heteroimmune lysogens of this strain ($\phi 80imm21$ and $\lambda imm21$). Lac$^+$ bacteria produced by the transfer of *lacz* from phage to host were measured and expressed as a fraction of the phage input. (Experiment 3) Gene transfer from nonproductive host to phage: The experiment was similar to experiment 2 except that the phage produced by the infected cells were studied. Phages that had either of the safety features, immunity deletion or $A^- B^-$, replaced by genes from the host lysogen (marker rescue) were assayed. Transfer of *imm21* to Charon 4A phage was measured by titration on a λ lysogen of the *sup*F strain Ymel; for Charons 3A and 16A, a $\phi 80$ lysogen was used. Transfer of $A^+ B^+$ genes from lysogen to Charon phage was assayed by counting the number of blue plaques on the sup^0 strain W3350 on XG plates. N.T., not tested.

Vector	Ch3A	Ch4A	Ch16A
Experiment 1. Phage to productive host			
Phage/ml*	1.0×10^{11}	1.0×10^{10}	3.0×10^{10}
Bacteria/ml†	5.8×10^9	4.7×10^9	3.1×10^{10}
Lac$^+$bacteria/ml	4.9×10^4	5.2×10^4	1.5×10^4
Lac$^+$bacteria/phage	4.9×10^{-7}	5.0×10^{-6}	5.0×10^{-7}
Experiment 2. Phage to nonproductive host (Lac$^+$ cells obtained per input phage)			
Nonlysogen	8.4×10^{-9}	1.8×10^{-8}	N.T.
$\phi 80imm21$ lysogen	7.0×10^{-7}	2.5×10^{-7}	$<2.9 \times 10^{-8}$
$\lambda imm21$ lysogen	4.5×10^{-6}	1.2×10^{-6}	2.7×10^{-6}
Experiment 3. Nonproductive host to phage			
(output phage with rescued imm21 *or* A$^+$B$^+$ *per input phage)*			
Nonlysogen	$<10^{-6}$	$<10^{-6}$	$<10^{-6}$
$\phi 80imm21$ lysogen:			
imm21			1.0×10^{-5}
A^+B^+	3.6×10^{-4}	2.0×10^{-5}	1.3×10^{-5}
$\lambda imm21$ lysogen:			
imm21	$<5.7 \times 10^{-5}$	$<3.9 \times 10^{-5}$	1.2×10^{-4}
A^+B^+	1.1×10^{-3}	2.4×10^{-3}	5.9×10^{-3}

*At harvest.   †After overgrowth.

two ways: (i) lytically, if the clone-bearing phage found a suitable host in the environment and started a lytic replication cycle before it had been inactivated; or (ii) as a lysogen or plasmid, if the phage DNA containing the cloned segment became incorporated into the genome of a bacterium, or if the phage DNA became established as a defective (nonlytic) plasmid in a bacterium, and if the bacterium could then grow outside the laboratory.

In order for λ to be propagated lytically outside the laboratory, phage particles must be capable of surviving in the natural environment. We find that fewer than 1 in 10^{10} phages survive exposure to pH 3 for 2.75 hours (mimicking stomach conditions); fewer than 2 in 10^7 survive 30 minutes in detergent (1 percent sodium dode-

cyl sulfate); fewer than 2 in 10^4 survive 6 hours of drying on a laboratory bench; fewer than 3 in 10^6 survive 120 hours in raw sewage. Thus, there are initial inactivation barriers to dissemination.

In order for λ to multiply in the natural environment, the phage must repeatedly encounter sensitive bacterial hosts. Typical raw sewage contains 10^7 bacteria per milliliter. Calculations show that self-sustaining lytic replication requires at least 10^5 sensitive bacteria per milliliter, so that phages can encounter bacteria sufficiently often for replication to overcome inactivation (25). Yet, of more than 2000 strains of *E. coli* independently isolated from the wild or from hospital patients, none was λ sensitive (26). Consequently sewage is unlikely to be able to sustain a lytic replication of λ. The lack of ability of

E. coli in the wild to replicate λ phages is borne out by our measurements of decreasing titers of λ phages on shaking with raw sewage (Fig. 4). Furthermore, λ cannot inject its DNA into bacteria below 15°C (27).

With the Charon phages, it has also been possible to decrease the number of potential natural hosts by including amber mutations in the phages. Such phages require a suppressor transfer RNA (tRNA) in the host bacterium of a type not usually found in wild-type *E. coli*. A further decrease in the number of potential natural hosts is obtained by propagating the phage in the laboratory on bacteria that do not modify the phage DNA; the DNA is then vulnerable to attack by the restriction endonucleases found in many wild-type *E. coli* strains.

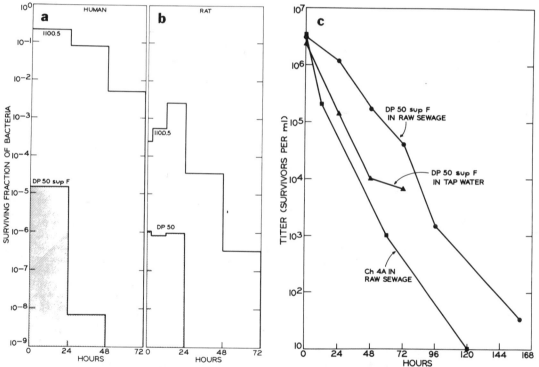

Fig. 4. Survival of host and vector in natural environments. Survival in mammals: in (a) and (b) the passage through the human and rat gut of the debilitated laboratory hosts DP50 and DP50SupF and, for comparison, a nondebilitated strain of *E. coli* K12 [1100.5 (28)] was examined. These strains are resistant to nalidixic acid, and titers can be determined on eosin–methylene blue–minimal galactose plates supplemented with vitamin-free Casamino acids and containing nalidixic acid (100 μg/ml). Such plates eliminate growth of native gut bacteria much better than plates made with yeast extract, and permit discrimination between Gal⁺ (1100.5) and Gal⁻ (DP50 and DP50SupF) bacteria. (a) A mixture of 3 × 10¹¹ DP50SupF and 9.4 × 10⁸ 1100.5 was fed in 250 ml of milk to three humans. Bar heights represent the averaged fraction of bacteria surviving gut passage during the time intervals defined by the end points of each bar. The total fraction of DP50SupF surviving is 1.4 × 10⁻⁵, which is 4.7 × 10⁻⁴ that of 1100.5. (b) Broth (0.25 ml) containing 8 × 10⁸ DP50 was placed on the tongues of three rats, and 1.6 × 10⁹ 1100.5 bacteria were fed to three other rats in the same manner. The total fraction of DP50 surviving is 2.9 × 10⁻⁶, which is 8.3 × 10⁻⁴ that of 1100.5. In neither species was there any indication of colonization by any K12 strain. Survival in sewage. In (c) City of Madison raw sewage, including University of Wisconsin effluent, was collected from a pond immediately upstream of the treatment plant; it had a titer of 10⁷ bacteria per milliliter measured on L plates. The inactivation of the EK2 vector, Charon 4A, shaken in raw sewage at 37°C, and of the debilitated host, DP50SupF, shaken in raw sewage and tap water at room temperature, was examined. The titer of Charon 4A was determined by counting blue plaques on XG plates; DP50SupF was determined as above. The initial first-order rate constants for inactivation in sewage were 1.0 day⁻¹ (bacteria) and 5.6 day⁻¹ (phage). The averages over the entire experiment were 1.7 day⁻¹ (bacteria) and 2.6 day⁻¹ (phage).

We conclude that, as far as dissemination by lytic processes is concerned, λ offers an inherently high degree of biological containment, which has been further increased in the Charon phages.

Escape of the phage-cloned DNA in a bacterial lysogen or plasmid must be considered from two aspects—the bacterium might be the host used for propagation in the laboratory, or it might be a naturally occurring one encountered by a clone-bearing phage after accidental release into the environment. In either case the self-replicating ability of bacteria, in contrast with the inability of phage to replicate without a host, makes dissemination of a cloned fragment by the bacterial route potentially serious. We have, therefore, focused much of our safety strategy on designing the vector so that its replication will be confined to the lytic mode.

The ability to form lysogens or plasmids has been reduced in Charon phages 3A, 4A, and 16A by deletions KH53 or KH54 in the immunity regions of the phages, by the *nin5* deletion (coupled with wild-type N function) which enhances lytic functions, and, in the case of Charon phages 4A and 16A, by the elimination of the gene *int* and the *att* site, which are needed for efficient insertion of the phage genome into the bacterial chromosome.

The effectiveness of these strategies can be measured by determining what proportion of bacteria surviving an encounter with a Charon phage acquire the cloned fragment of DNA in some stable association. The *lac5 E. coli* substitution carried on Charons 3A, 4A, and 16A is a convenient model cloned segment, provided that test bacteria are used which provide the *lacy* (permease) function and do not contain the *lacz* (β-galactosidase) function. Appropriate test bacteria were constructed for this purpose (*28*).

The first experiment shown in Table 2 was designed to measure the formation of stable associates with a laboratory host during productive growth. Charon phages were grown by the PDS method on Lacz⁻y⁺ cells. To mimic a worse case, the lysates were not harvested at the normal time but, instead, were allowed to overgrow until surviving bacteria had reached a high titer. Chloroform was not added. Surviving cells were separated from the free phage, diluted into broth, and shaken for 90 minutes to permit stable associates to become established and to permit unstable associates to segregate. The bacteria were then plated on XG indicator plates with lactose as the sole carbon source. Reconstruction experiments were done to verify our ability to demonstrate very low levels of Lac⁺

bacteria in lysates. The titers of phage, total bacteria, and Lac⁺ bacteria are listed in Table 2 (experiment 1). The results show that the ratio of undesirable products (Lac⁺ bacteria) to desirable products (phages) is of the order of 10^{-6}. (The majority of the bacteria surviving these lytic infections were shown to be λ-resistant.) When lysates were harvested at the proper time, these ratios were more favorable by a factor of 10^3, and chloroform is routinely used to kill all the bacteria in lysates. Nevertheless, the phage working group recommended using a host bacterium for the productive growth of phage that is able to provide an additional safety factor over that provided by *E. coli* K12. This recommendation was implemented by our using DP50, as described below.

The second experiment in Table 2 was undertaken to test the likelihood that a recombinant phage that escaped into nature might subsequently be stably integrated into a naturally occurring bacterium. Although the probability of an encounter under dilute conditions with a nonpermissive bacterium that could adsorb λ would be low, as discussed above, it might be cause for concern if the integration step turned out to be highly efficient. Again worse cases were tested in which the bacteria were λ-sensitive, lacked an amber suppressor, and carried prophages having a high homology (λ *imm*21) or moderate homology (φ80 *imm*21) with the Charon phages. We also tested nonlysogenic bacteria. In these experiments, Charon phages were adsorbed efficiently to the bacteria at a ratio of about 1 : 1 and then diluted into broth as before. The efficiency of transfer of a model cloned segment to the nonpermissive lysogen relative to the number of input phage is listed in Table 2 (experiment 2). These data show that the formation of stable associations is far from being efficient.

The third experiment in Table 2, suggested by the phage working group, provides data on marker rescue. The marker rescue experiments were designed to test the possible outcome of a phage escaping as a particle, and surviving in the natural environment long enough to find a rare bacterium (to which it can adsorb) which harbors a prophage that can remove by marker rescue the phage mutations introduced for safety. Recombination between such a prophage and the clone-bearing phage could create a new phage with some of its safety features compromised. The protocol for this experiment again involved adsorbing Charon A phages to λ-sensitive lysogens lacking an amber suppressor. The release of Lac-

containing phages that had lost either the amber mutations or the immunity deletions was then measured (Table 2, experiment 3).

The results show that the frequencies of loss by marker rescue of the amber mutations on the left side of the cloned segment are around 3×10^{-3} and of the immunity mutations on the right side are around 5×10^{-5}. Both of these events would not fully compromise the safety of the Charon system, and they would be but a small part of an extensive chain of improbable events. The results indicate that these probabilities are indeed worst case estimates because when the homology between the vector and the prophage was reduced, as would be the case in most strains from nature, the rescue frequencies were reduced.

Bacterial hosts. The bacterial hosts, DP50 and DP50SupF, were selected for the Charon phage system because of their additional contributions to the problem of biological containment. DP50 (also called χ1953) was constructed by Pereira and Curtiss [see (*29*)], University of Alabama. DP50SupF (also called χ2098) was constructed from DP50 by Leder *et al.* (*30*), and has the same genotype as DP50 except for the addition of *supF*58 (*su*III). The genotypes of the strains are: F^-, *dap*D8, *lacy*, Δ*gal-uvr*B, Δ*thyA*, *nal*A^r, *hsd*S, *supE*44 (plus *supF*58 in DP50SupF). The *dap*D8, Δ*thyA*, and Δ*gal-uvr*B mutations were introduced so that the bacteria themselves have a low survival in nature. This provides an extra margin of safety if the normal harvesting procedure, including chloroform treatment of the lysates, is not executed on time, and the entire culture including phage, bacteria, and a low frequency of plasmids or lysogens is inadvertently released into the natural environment. The survival of these bacteria during shaking in raw sewage and during intestinal passage through humans and rats is shown in Fig. 4. Relative to *E. coli* K12, the survival of these strains is reduced about a thousandfold. None of the tested bacteria colonized either the rat or human gut.

The reversion rates of the bacterial mutations important for safety, *dap*D8, Δ*thyA*, and Δ*gal-uvr*B, must be considered. The *thy* and *gal* mutations are deletions and do not revert. Dap⁺ revertants were found at about 10^{-8} in stock cultures grown from a single colony. They only accumulated to 5×10^{-8} even after 10^{12}-fold growth, so that Dap⁺ revertants do not have any appreciable selective advantage in dap-containing medium. Moreover, none of 22 tested bacteria surviving human or rat gut passage were Dap⁺. Thus, reversion of *dap*D8 is not likely to compromise the safety of the bacteria to any appreciable extent. Nevertheless, as with phage, it is advisable to frequently check the genotype of the bacteria in order to avoid contamination and mistakes as well as revertants.

Since both bacteria lack the *E. coli* K modification system, only a very small proportion, about 10^{-4}, of phage propagated on them would grow on bacteria from nature having a restriction system similar to that of *E. coli* K12. Both DP50 and DP50SupF will grow the Charon A phages efficiently.

References and Notes

1. P. Berg, D. Baltimore, S. Brenner, R. O. Rilin III, M. O. Singer, *Science* **188**, 991 (1975); W. Szybalski, in *Genetic Alteration: Impact of Recombinant Molecules on Genetic Research*, R. Beers, Ed. (Raven, New York, in press).
2. *Fed. Regist.* **41**, 27902 (1975).
3. N. E. Murray and K. Murray, *Nature (London)* **251**, 476 (1974); A. Rambach and P. Tiollais, *Proc. Natl. Acad. Sci. U.S.A.* **71**, 3927 (1974); L. Enquist, D. Tiemeier, P. Leder, R. Weisberg, N. Sternberg, *Nature (London)* **259**, 596 (1976); B. G. Williams, D. D. Moore, J. W. Schumm, D. J. Grunwald, A. E. Blechl, F. R. Blattner, in *Genetic Alteration: Impact of Recombinant Molecules in Genetic Research*, R. Beers, Ed. (Raven, New York, in press); D. J. Donoghue and P. A. Sharp, in preparation; D. Tiemeier, L. Enquist, P. Leder, *Nature (London)* **263**, 526 (1976).
4. M. Thomas, J. R. Cameron, R. W. Davis, *Proc. Natl. Acad. Sci. U.S.A.* **71**, 4579 (1974).
5. K. Murray and N. E. Murray, *J. Mol. Biol.* **98**, 551 (1975).
6. KH100 and KH54 were described by F. R. Blattner, M. Fiandt, K. K. Haas, P. A. Twoose, W. Szybalski, *Virology* **62**, 458 (1974).
7. *bio*1 was described by K. F. Manly, E. R. Signer, C. M. Radding, *ibid.* **39**, 137 (1969); *imm*434, by A. D. Kaiser and F. Jacob, *ibid.* **34**, 509 (1969); *nin*5, by D. Court and K. Sato, *ibid.* **39**, 348 (1969); *QSR*80, by N. Franklin, W. Dove, C. Yanofsky, *Biochem. Biophys. Res. Commun.* **18**, 910 (1965); *lac*5, by J. Shapiro, L. MacHattie, L. Eron, G. Ihler, K. Ippen, J. Beckwith, R. Arditti, W. Reznikoff, R. MacGilliway, *Nature (London)* **224**, 768 (1969); *b*189, by J. S. Parkinson and R. J. Huskey, *J. Mol. Biol.* **56**, 369 (1971); *b*1007, by D. Henderson and J. Weil, *Virology* **67**, 124 (1975); *imm*80, *att*80, and hybrid 42, by J. Szpirer, R. Thomas, C. M. Radding, *ibid.* **37**, 585 (1969); ΔH3 is a deletion of the *Hin*d III fragment immediately to the left of *att*, and was the gift of D. Kampf and R. Kahn.
8. B. G. Williams, thesis, University of Wisconsin, Madison (1977).
9. W. Arber and U. Kühnlein, *Pathol. Microbiol.* **30**, 946 (1967).
10. A. J. D. Bellett, H. G. Busse, R. L. Baldwin, in *The Bacteriophage Lambda*, A. D. Hershey, Ed. (Cold Spring Harbor Laboratory, Cold Spring Harbor, N.Y., 1971), p. 501.
11. These enzymes are as follows. Eco RI [M. Thomas and R. W. Davis, *J. Mol. Biol.* **91**, 315 (1975)] was purified by a modification of the method of D. Smith and J. Davies (personal communication). Hind III [P. H. Roy and H. O. Smith, *J. Mol. Biol.* **81**, 427 (1973)] was obtained from New England Bio-Labs (Beverly, Mass.). Bam HI [G. A. Wilson and F. E. Young, *J. Mol. Biol.* **97**, 123 (1975)] was obtained from New England Bio-Labs. Kpn I and Pst I [D. Smith, F. R. Blattner, J. Davies, *Nucleic Acids Res.* **3**, 343 (1976)] were both obtained from New England Bio-Labs. Sal I (J. Arand, P. Meyers, R. Roberts, personal communication) was obtained from New England Bio-Labs. Bgl II [G. A. Wilson and F. E. Young, in *Microbiology, 1976*, D. Schlessinger, Ed. (American Society of Microbiology, Washington, D.C., 1976), p. 350] was obtained from Bethesda Research Laboratories (Rockville, Md.). Xma I (R. Roberts, personal communication) was a gift from F. W. Farrelly.
12. A. Campbell, *Virology* **14**, 22 (1961).
13. The immunity specificity of the vectors was determined by plating on appropriate homoimmune and heteroimmune lysogens, and by examination of plaque type (turbid or clear) [A. D. Hershey and W. Dove, in *The Bacteriophage Lambda*, A. D. Hershey, Ed. (Cold Spring Harbor Laboratory, Cold Spring Harbor, N.Y. 1971), p. 3]. The presence or absence of amber mutations was determined by plating on hosts containing or lacking an appropriate amber suppressor (*12*). For phages with λ or 434 immunity, the presence of the *nin*5 deletion was demonstrated by the ability to plate on GroN host strains [C. P. Georgopolous, in *The Bacteriophage Lambda*, A. D. Hershey, Ed. (Cold Spring Harbor Laboratory, Cold Spring Harbor, N.Y., 1971), p. 639]. The presence or absence of the λ genes *exo*, β, and γ was determined by ability of phages to plate on RecA⁻ host strains (Fec phenotype (J. Zissler, E. Signer, F. Schaefer, in *ibid.*, p. 455) or hosts lysogenic for phage P2 [Spi phenotype (J. Zissler, E. Signer, F. Schaefer, in *ibid.*, p. 469). The presence of the *lac*5 substitution was determined by plating on XG dye indicator plates, as described in the text (*33*, p. 48)]. The presence of *bio* substitutions in phages was determined by ability to promote growth of Bio⁻ hosts on biotin-deficient plates [G. Kayajanian, *Virology* **36**, 30 (1968)].
14. N. Sternberg and R. Weisberg (personal communication); M. Shulman and M. Gottesman, in *The Bacteriophage Lambda*, A. D. Hershey, Ed. (Cold Spring Harbor Laboratory, Cold Spring Harbor, N.Y., 1971), p. 477.
15. R. B. Helling, H. M. Goodman, H. W. Boyer, *J. Virol.* **14**, 1235 (1974); D. H. Gelfand and P. O'Farrell, *Proc. Natl. Acad. Sci. U.S.A.* **73**, 3900 (1976).
16. S. R. Jaskunas, A. M. Fallon, M. Nomura, B. G. Williams, F. R. Blattner, in preparation.
17. The reaction mixture is made 0.1 percent diethyl pyrocarbonate (Calbiochem) by adding 0.1 volume of a freshly diluted 10 percent solution in ethanol. After incubation for 10 minutes at 37°C, other enzymes may then be added as desired.
18. The concentration C (micrograms per milliliter) of vector DNA, which equals the intramolecular concentration of target fragment termini, is given by $C = 46\ V/L^{3/2}$ [derived from J. C. Wang and N. Davidson, *J. Mol. Biol.* **19**, 469 (1969)], where L is the size of target molecule in kilobase pairs, and V is the length of the vector DNA as percentage of λ (including any dispensable fragments unless they are physically removed). For example, to clone a 5-kbp fragment in Charon 16, vector DNA should be at or above 354 μg/ml.
19. W. D. Henner, I. Kleber, R. Benzinger, *J. Virol.* **12**, 741 (1973).
20. M. Mandel and A. Higa, *J. Mol. Biol.* **53**, 159 (1970).
21. D. Kaiser, M. Syvanen, T. Masuda, *ibid.* **91**, 175 (1975); B. Hohn and T. Hohn, *Proc. Natl. Acad. Sci. U.S.A.* **71**, 2372 (1974); N. Sternberg, D. Tiemeier, L. Enquist, *Gene*, in press.
22. The PDS method was modified from that described by Blattner *et al.* (*6*). The phage (10^5 to 10^7 plaque-forming units) was mixed with 0.3 to 1.0 ml of stationary culture of cells plus an equal volume of 0.01*M* $MgCl_2$ and then incubated at 37°C for 10 minutes; the mixture was diluted into a liter of warmed NZY or NZYDT (see below) broth, and shaken overnight at 37°C; NZY broth contains (per liter) 10 g of NZamine (a casein hydrolyzate available from Humko-Sheffield, Linnhurst, N.J.), 5 g of yeast extract (Difco), 5 g of NaCl, and 2 g of $MgCl_2 \cdot 6\ H_2O$. NZYDT broth contains, in addition to NZY constituents, 0.04 g of thymidine and 0.1 g of diaminopimelic acid. We have found many batches of Bacto-tryptone (Difco) that do not work when substituted for NZamine. The exact ratio of phages to bacteria is generally not critical.
23. F. Rougeon, P. Kourilsky, B. Mach, *Nucleic Acids Res.* **2**, 2365 (1975); R. Higuchi, G. V. Paddock, R. Wall, W. Salser, *Proc. Natl. Acad. Sci. U.S.A.* **73**, 3146 (1976); T. Maniatis, S. G. Kee, A. Efstradiatis, F. Kafatos, *Cell* **8**, 163 (1976); T. H. Rabbitts, *Nature (London)* **260**, 221 (1976).
24. The original application, submitted 1 June 1976 is: "Application for EK2 Certification of a Host Vector System for DNA Cloning," B. G. Williams, D. D. Moore, J. W. Schumm, D. J. Grunwald, A. E. Blechl, and F. R. Blattner; Supplement I: "Part I New *dap*⁻ Bacterial Host, DP50, for Productive Growth of Charon 3A and 4A Phages. Part II Two Phages Derived from Charon 3A and 4A (16A and 14A) with Deleted Attachment Sites, and Two (15A and 13A) Which Are in Addition Devoid of Bacterial DNA," F. R. Blattner, A. E. Blechl, H. E. Faber, L. A. Furlong, D. J. Grunwald, D. O. Kiefer, E. L. Sheldon, and O. Smithies; Supplement II: "Marker Rescue Tests," F. R. Blattner, D. O. Kiefer, and D. D. Moore. Supplement III: "Additional Marker Rescue Data on Charon 3A and Charon 4A," F. R. Blattner, D. O. Kiefer, and D. D. Moore; Supplement IV: "Further Tests on DP50 (χ1953) and DP50 · SupF (χ2098)," O. Smithies, F. R. Blattner, A. Blechl, H. E. Faber, E. L. Sheldon, B. G. Williams, and L. A. Furlong; Supplement V: "Additional Data Requested by the Phage Working Group," F. R. Blattner, B. G. Williams, D. O. Kiefer; Supplement VI: "Rearmament of Charon Phages in Sewage," F. R. Blattner, D. O. Kiefer, H. E. Faber, and O. Smithies; Supplement VII: "Marker Rescue Tests on Charon Phages Having Amber Mutations in Genes W and E Instead of Genes A and B," F. R. Blattner and D. O. Kiefer.
25. For net replication to occur, $(K_2)(B)(Y)/(K_1) > 1$, where $K_1 = 5.6\ \text{day}^{-1}$ is the initial first-order rate constant for phage inactivation by City of Madison sewage, B is the concentration of λ-sensitive bacteria in sewage, $Y = 100$ is the yield of phage per infected bacterium, and $K_2 = 7.2 \times 10^{-7}$ ml per bacterium per day (J. Salstrom, personal communication) is the second-order rate constant for phage-bacteria collision and infection.
26. R. W. Davis, personal communication; P. Leder, personal communication; J. S. Parkinson, personal communication: 2000 is the sum of determinations made by the above three persons.
27. D. McKay and V. Bode, *Virology* **72**, 156 (1976).
28. CSH18 is Δ(*lac pro*) X111 *sup*E (F′ *lac* z^-y^+ *pro*A^+B^+). The F′ carries *lacz* deletion H125. CSH46 ≡ M96 and is Δ(*lac pro*) X111 sup^0 (λ*c*I857 Sts68 *h*80 *laciz*). CSH18 and CSH46 are described in (*33*, p. 16). M96CF′ was constructed by curing strain CSH46 of the prophage and introducing the F′ from CSH18. In Table 2, "λ *imm*21 lysogen" refers to λ *h*λ *att*80 *imm*21 *QSR*80 (for tests with Charon 3A and Charon 4A) and λ *h*λ *imm*21 *QSR*λ (for tests with Charon 16A) in M96CF′. "φ80 *imm*21 lysogen" refers to φ80 *h*80 *att*80 *imm*21 *QSR*80 in M96CF′. Strain 1100.5 (*nal*ʳ *end*A⁻), used in the mammalian ingestion experiments, was obtained from J. Davies.
29. As described by R. Curtiss, III, M. Inoue, D. Pereira, J. Hsu, L. Alexander, L. Rock, in *Molecular Cloning of Recombinant DNA*, W. Scott and R. Lerner, Eds. (Academic Press, New York, in press).
30. P. Leder, D. Tiemeier, L. Enquist, *Science* **196**, 175 (1977).
31. B. C. Westmoreland, W. Szybalski, H. Ris, *Science* **163**, 1343 (1969); R. W. Davis, M. Simon, N. Davidson, *Methods Enzymol.* **21**, 413 (1971).
32. D. D. Moore, J. W. Schumm, M. M. Howe, F. R. Blattner, in *DNA Insertions*, S. Adhya and A. Bukhari, Eds. (Cold Spring Harbor Laboratory, Cold Spring Harbor, N.Y., 1977).
33. J. H. Miller, *Experiments in Molecular Genetics* (Cold Spring Harbor Laboratory, Cold Spring Harbor, N.Y., 1972).
34. Measured in our laboratory by D. Daniels. Circles of PM_2 DNA (20.44 percent of λ*c*72 length, M. Fiand, personal communication) were compared to φX174 replicative form, assumed to be 5375 bp from the complete DNA sequence, as determined by F. Sanger *et al.*, *Nature (London)* **265**, 687 (1977).

35. The experiments related to cloning mouse globin cDNA were done under Asilomar guidelines; we used P3 physical containment with the (now) certified EK2 vector Charon 3A. We thank many friends who have given us samples of the phages and bacterial strains, J. Nulter for the cooperation of NIAID, and W. Szybalski and W. F. Dove for critical reading of the manuscript. The name Charon phages was suggested by Eric C. Rosenvold. We thank C. Morita, D. Stephenson and V. Farkas for technical assistance; J. Richards, S. Nitz, and A. Johnson provided help with illustrations. Supported by NIH contract AI 62506, and grants GM 21812 and GM 20069. This is Paper No. 2125 from the Laboratory of Genetics, University of Wisconsin, Madison.

14

Reprinted from *Mol. Gen. Genet.* **150**:53–61 (1977)

Lambdoid Phages that Simplify the Recovery of in vitro Recombinants

Noreen E. Murray, W.J. Brammar and K. Murray

Department of Molecular Biology, University of Edinburgh, Edinburgh, Scotland

Summary. Derivatives of phage λ are described for use as vectors for fragments of DNA generated with the *Hind*III. and *Eco*RI restriction enzymes. With some vectors, hybrid molecules are recognised by a change from a turbid to a clear plaque morphology resulting from the insertion of a fragment of DNA into the λ gene coding for the phage regressor. Other vectors contain a central, replaceable fragment of DNA which imparts a readily recognisable phenotype. This central fragment may include either a gene for a mutant transfer RNA (suppressor) or a part of the *lacZ* gene of *E. coli* able to complement a *lacZ* host. The appropriate *lacZ* host and indicator plates permit the ready distinction between recombinant and vector phages by the colour of the plaques.

1. Introduction

Fragments of DNA can be cloned very readily using plasmids or phage genomes as vector molecules. The simplest procedures use a restriction endonuclease that makes staggered breaks in DNA (e.g. Hedgpeth et al., 1972; Old et al., 1975) to generate fragments of both donor and vector DNA molecules which can then be joined by hydrogen bonding of their cohesive ends (Mertz and Davis, 1972) prior to the formation of covalent bonds with polynucleotide ligase. Plasmids, or phages, are subsequently recovered by transformation of *Escherichia coli* (Mandel and Higa, 1970).

The detection of a particular clone among many transformants is simple if an included gene can be recognised by its function; for example, a bacterial gene is readily detected by its ability to complement an appropriate, mutant strain of *E. coli* (Clark and Carbon, 1975; Borck et al., 1976). Frequently, however, the recognition of the coveted fragment of DNA

from a eukaryotic source cannot be achieved by a simple complementation test and must depend on screening of the recovered clones by other, less direct methods, such as RNA/DNA hybridisation (Grunstein and Hogness, 1975; Jones and Murray, 1975).

From the simple in vitro recombination reaction described, the majority of clones result merely from the restitution of the vector chromosome. The efficiency with which *recombinant* molecules are recovered is markedly improved by a biochemical refinement in which the synthetic addition of complementary, single-stranded projections to the donor and receptor DNAs directs the joining of the receptor chromosome via inclusion of a fragment of donor DNA (Wensink et al., 1974). This approach is convenient for use with small plasmid DNAs, but is less readily applicable to the lambda chromosome.

Phage λ vectors suitable for cloning fragments of DNA have been described by a number of authors (Murray and Murray, 1974; Rambach and Tiollais, 1974; Thomas et al., 1974; Murray and Murray, 1975; Enquist et al., 1976).

In this paper we describe genetic approaches by which phages containing donor DNA may be readily identified. Derivatives of phage λ have been made for use as vectors with the *Hind*III and the *Eco*RI systems.

2. Materials and Methods

a) Phages. The nomenclature for targets for restriction enzymes uses the abbreviation for the enzyme (Smith and Nathans, 1973) followed by the genome concerned and the number of the site within the genome. Thus sites for endo R. *Eco*RI are described by sr*I*λ1, sr*I*λ2, etc. Sites for endo R. *Hind*III should be described by shind IIIλ1, shind IIIλ2, etc., but for convenience are abbreviated to shnλ1, shnλ2, etc.

The following phages were used: Wild-type λ and derivatives with *supE*-suppressible amber mutations; *Pam*80, *Pam*3, *Qam*73

Table 1. Bacterial strains

Strain	Relevant features	Source	Reference
C600	*supE, tonA*	F.W. Stahl	Appleyard (1954)
803	*supE, hsdS⁻, met⁻*	N.C. Franklin	Wood (1966)
Ymel	*supF, λs*	A.D. Kaiser	
QR47	*supE*	E. Signer	Weil and Signer (1968)
W3350	*sup⁰*	J.R. Davison	Campbell (1961)
W3101recA	*sup⁰ recA*	N.C. Franklin	
ED8538	*sup⁰ lacZam*		
ED8792	*sup⁰ lacZM15, lac i₃*	J.G. Scaife	
ED8654	*supE, supF, hsdR⁻ M⁺ S⁺, met⁻ trpR*		
ED8739	*supE, supF, hsdS⁻, met⁻*		
ED8767	*supE, supF, hsdS⁻, met⁻ recA56*		
ED8799	*supE, supF, hsdS⁻, met⁻ lacZM15*		
ED8800	*supE, supF, hsdS⁻, met⁻ lacZM15, recA56*		

and *Nam*7 *Nam*53 (Campbell, 1961): *λ*gt. *λ*C, *Wam* 403, *Eam*1100, *Sam*100 as a donor of *Wam*403, *Eam*1100 (Enquist et al., 1976): a heat-inducible *λ*imm⁴cI857 derivative made defective in lysis by the *supF*-suppressible mutation *am*7 in gene *S* (Goldberg and Howe, 1969): a derivative of *λ*cIam509, with a deletion *b*538, of the central region of the *λ* genome (Parkinson, 1971): *λ*nin5, which has a small deletion permitting *N*-independent growth (Court and Sato, 1969) and a derivative (KH54) of *λPam* 80 having a deletion within the immunity region (Blattner et al., 1974): phage *λ* with the immunity of phage 434, *λ*imm⁴³⁴, (Kaiser and Jacob, 1957) or with the immunity of phage 21, *λ*imm²¹, (Leidke-Kulke and Kaiser, 1967): derivatives of *λ* with reduced numbers of targets for *Eco*RI (Murray and Murray, 1974) or for *Hind*III (Murray and Murray, 1975): the *lacZ*-transducing phage *λlac*5 (Ippen et al., 1971), *λsupF* and *λtrpE* phages made using endo R. *Hind*III and *λsupE* made using endo R. *Eco*RI (Borck et al., 1976). All other strains are recombinants isolated from conventional crosses or by in vitro recombination of fragmented DNAs.

b) Bacterial Strains. The bacterial strains used are listed in Table 1. Strains with 'ED' designations were constructed in this laboratory. Methods for transferring the *supF* and *hsd*_K genes have been described (Borck et al., 1976). The *recA*56 allele was introduced into strains by its linkage to *thyA* on conjugal transfer from the *recA*, Hfr donor strain JC5088 (Clark, 1967). *ThyA* mutants were isolated by selection for resistance to trimethoprim (Stacey and Simson, 1965). Generalised transduction was carried out using phage P1kc, provided by Dr.C Yanofsky, as described by Yanofsky and Lennox (1959). Transductants were routinely tested for P1-lysogeny by their ability to restrict *λ*cI, and P1-lysogenic transductants were rejected.

c) Media. The rich medium was L broth (Lennox, 1955) containing (in g/l): Difco Bacto Tryptone, 10: Difco Bacto yeast extract, 5: NaCl, 5: glucose, 1: adjusted to pH 7.2.

Phage stocks for genetic analysis were prepared on L broth agar solidified by Difco agar (10 g/1).

Phage assays were made on Baltimore Biological Laboratories Trypticase (BBL) agar, containing (in g/1): Trypticase, 10: NaCl, 5: Difco agar, 10 for plates and 6.5 for top layers (Parkinson, 1968).

Oxoid lactose MacConkey (number 3) or BBL agar containing 40 mg/l 5-bromo-4-chloro-3-indolyl-*β*-D-galactoside (XG) and 10⁻³ M isopropyl-*β*-D-thiogalactoside (IPTG) (Horwitz et al., 1964) were used as indicator media for detection of the Lac⁻ phenotype.

The minimal medium of Spizien (1958) was used with glucose (0.2% w/v) as carbon source and, for tryptophan auxotrophs, Difco

Bacto acid-hydrolysed casein (0.05% w/v) to provide all amino acids except tryptophan. Minimal plates contained New Zealand agar, 15 g/l.

d) General Techniques. The methods for preparing plating cells and phage stocks, assaying phage and carrying out phage crosses have been described (Murray et al., 1973).

e) Phage Lysates for DNA Preparations. Preparations of phage were made by infection of exponentially growing cultures of *E. coli* C600 (in L broth containing 10⁻³ M MgSO₄) at a cell density of about 2·10⁸ per ml and a multiplicity of infection of 1. Growth was followed spectrophotometrically and when the A₆₅₀ₙₘ reached a minimum (usually about 2 h after infection) lysis was completed by addition of CHCl₃ (0.5 ml/l) and 15 min later the lysate was clarified by centrifugation (10 min at 10,000 g). Phages were recovered by centrifugation, resuspended in phage buffer and treated with DNAase and RNAase (10 µg/ml each, 2 h at room temperature), pelleted, resuspended in phage buffer and recovered by equilibrium centrifugation after addition of caesium chloride to 41.5% (w/v) (Kaiser and Hogness, 1960): the caesium chloride step was repeated.

f) DNA Preparation. Phage preparations were diluted to about 5×10¹¹ to 5×10¹² plaque-forming units/ml and dialysed against 10 mM Tris-HCl (pH 8.0), 1 mM EDTA. DNA was extracted by gentle rolling with freshly distilled phenol, three times (Kaiser and Hogness, 1960) followed by dialysis against 10 mM Tris HCl (pH 8.0), 1 mM EDTA (4 changes in about 24 h). Bacterial DNA was prepared essentially as described by Marmur (1961) except that the cells were lysed with lysozyme and Triton X-100. Bacteria were grown in 2 l L-broth to a density of 2×10⁸ cells/ml, and harvested by centrifugation at 4° C. The cells were resuspended in 48 ml of ice-cold 25% (w/v) sucrose in 0.05 M Tris-HCl, pH 8.0, and 6.9 ml of a freshly prepared solution of lysozyme (10 mg/ml in 0.25 M Tris-HCl, pH 8.0) were added. The suspension was shaken gently for 30 s at 37° C, then placed on ice for 5 min. 26 ml of ice-cold 0.25 M EDTA, pH 8.0, were added, followed 5 min later by 54 ml of solution containing 2% (w/v) Triton X-100, 0.05 M Tris-HCl and 0.063 M EDTA, pH 8.0 to lyse the cells. Lysis was complete after 20 min on ice. The lysate was subsequently treated according to the Marmur procedure.

g) Enzymes and Chemicals. Pancreatic DNAase and RNAase were purchased from Worthington Biochemical Corporation, Freehold, N.J., U.S.A.: restriction endonuclease R. *Eco*RI was prepared essentially as described by Yoshimori (1971) and endo R. *Hind*III

113

was prepared as described by H.O. Smith (Old et al., 1975) or Philippsen et al. (1974). Some of the endo R. *Hind*III preparations used were generously provided by Dr. H. Cook or Dr. H.O. Smith. Lysozyme (Grade I egg-white enzyme) was obtained from Sigma Chemical Co, St. Louis, Mo. 63178, U.S.A. T4 polynucleotide ligase was purchased from Miles Laboratories Ltd, Stoke Poges, Slough, Bucks. U.K. Whenever possible chemicals used were of AR grade; caesium chloride was obtained from B.D.H. Ltd, Poole, Dorset, U.K. and agarose from Miles Laboratories Ltd, Stoke Poges, Slough, Bucks., U.K.

h) Restriction Endonuclease Digestion and Gel Electrophoresis. The quantity of the restriction endonuclease required for complete digestion (37° C, 30 min) of 1 μg of λ^+DNA was determined in trial experiments with a series of digests, which were analysed by electrophoresis in 1% (w/v) agarose gels (in 0.04 M Tris-acetate pH 8.0), containing 0.4 mg ethidium bromide/l; (Sharp et al., 1973). For analysis of the products of digestion of the various DNA preparations with the restriction endonucleases, samples of 1 to 2 μg in about 20 μl 0.01 M Tris-HCl (pH 7.5), 0.01 M MgCl$_2$, 0.01 M 2-mercaptoethanol (and for R. *Eco*RI digests, 0.1 M NaCl) were incubated with the appropriate quantity of enzyme at 37° C for 30 to 60 min, heated at 70° C for 10 min, cooled in ice, mixed with 5 μl 50% (v/v) glycerol containing bromophenol blue (about 0.1%) concentrated to about 10 μl in a vacuum desiccator and applied to wells in an agarose slab gel (40 cm × 20 cm × 0.3 cm; Sharp et al., 1973) for electrophoresis, usually for about 18 h with a constant current of 40 mA. Gels were photographed under ultraviolet light on Ilford FP4 film (4 × red filter) which was developed (9 min at 18° C) in Microphen (Ilford Ltd, Ilford, Essex, U.K.).

The efficiency of restriction reactions was usually assessed by the decrease in plaque-forming ability on transfection of a suitable *E. coli* strain made competent by starvation in CaCl$_2$ (Mandel and Higa, 1970 or Lederberg and Cohen, 1974 (see j) below).

i) Ligase Reactions. Restricted phage DNA or equal quantities of restricted donor and receptor DNAs were mixed and incubated at 30° C for 15 mins to separate pre-annealed fragments.

Reaction mixtures of 0.2 to 0.5 ml containing total DNA, 10–30 μg/ml; Tris-HCl (pH 7.5), 66 mM; EDTA, 1 mM; MgCl$_2$, 10 mM; NaCl, 100 mM; dithiothreitol, 10 mM; ATP, 0.1 mM; T4 polynucleotide ligase (Miles Laboratories Ltd), 0.2 units/ml were incubated at 10° C for 3–6 h and then kept on ice for 2 to 10 days during sampling for transfectants. The maximum yield of plaques was usually obtained after 4 to 6 days.

j) Transfection of E. coli to Recover Recombinant Phages. Cells (usually ED8654, an $hsdR_K^-$ $hsdM_K^+$ $supF$ derivative of the Met$^-$ strain 803) competent for transfection were obtained by growing in L broth and starving in calcium chloride according to the method of Lederberg and Cohen (1974).

k) Screening for Transduction of trp, sup or lacZ. Transducing phages carrying genes from *E. coli* were detected by their ability to complement mutations in appropriate bacterial hosts. Thus Trp$^+$ phages were detected by their ability to form "Trp$^+$ plaques" on a Trp$^-$ host in the absence of exogenous tryptophan (Franklin, 1971). Phages carrying amber suppressors were detected as Lac$^+$ plaques on lactose MacConkey agar, or on BBL plates containing 5-bromo-4-chloro-3-indolyl-β-D-galactoside (XG), when the Lac$^-$ indicator cells had an amber mutation in the *lacZ* gene (ED8538). Phages including either the entire *lacZ* gene, or the fragment containing most of this gene, were detected by complementation of the *lacZ* strain (ED8799). This strain has a small deletion in the *lacZ* gene but produces a β-galactosidase able to serve as an omega-donor in allelic complementation tests (Ullmann et al., 1967).

3. Results

a) Insertion Vectors. Insertion vectors have one target for the restriction enzyme. Recombinant phages are detected if insertion of a fragment of DNA inactivates a phage gene. For example, insertion of DNA at *srIλ3*, within the *redA* gene of lambda, produces a Red$^-$ phage (Murray and Murray, 1974). In principle Red$^-$ phages can be detected by their inability to grow in *polA* or ligase-deficient strains of *E. coli* (Zissler et al., 1971), but in practice these tests are demanding and laborious. An easier alternative is provided by restriction targets in the gene (*cI*) coding for the phage repressor protein. In particular, we have used λimm^{434} phages, since the *cI* gene of *imm^{434}* contains single targets for the restriction enzymes endo R. *Hind*III and endo R. *Eco*RI. The presence of donor DNA within the *cI* gene of λimm^{434} is recognised by a clear plaque morphology.

Both Red$^+$ and Red$^-$ phages have been made in which between eighteen and twenty per cent of the phage genome, including the phage attachment site, has been deleted. The Red$^+$ phages are $\lambda b538 imm^{434}$ (Fig. 1a and d) with the appropriate restriction targets removed: one Red$^-$ derivative (Fig. 1b) simply carries a mutation in the *redB* gene. In others the *red$^-$* genotype is stabilised by deletions that remove at least part of the *redA* gene. Such a Red$^-$ vector for the *Hind*III system (Fig. 1c) has a deletion of *srIλ3* generated by an illegitimate recombination event (Murray and Murray, 1974). Genetic tests (as in Murray and Murray, 1974) showed that the deletion does not extend rightwards into *redB* and analyses of the fragments resulting from digestion of the phage DNA with various restriction enzymes indicate a deletion of approximately 2 per cent of the genome. The Red$^-$ receptor for the *Eco*RI system (Fig. 1e) was derived from $\lambda srI\lambda1^0$ $srI\lambda2^+$ $srI\lambda3^+$ $imm^\lambda srI\lambda4^0$ $srI\lambda5^0$ by the removal in vitro of the DNA between *srIλ2* and *srIλ3*. This event deletes eleven per cent of the λ genome, including *att, int, xis* and part of the *redA* gene (see Fig. 1). A derivative in which the restituted restriction target has been removed by mutation (Murray and Murray, 1974) was isolated before replacing *imm$^\lambda$* with *imm^{434} srIλ4^0 nin5 srIλ5^0*; the molecular weight of the DNA of this phage is about 80% of that of λ^+ DNA.

The immunity insertion vectors for the *Eco*RI system have endo R. *Hind*III targets flanking the insertion site; those for the *Hind*III system have targets for the restriction enzyme endo R. *Bam* (Wilson and Young, 1975; Haggerty and Schlief, 1976; Perricaudet and Tiollais, 1975) on either side of the insertion site.

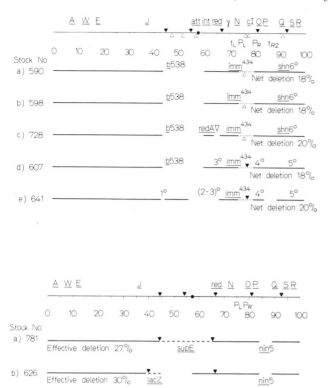

Fig. 1. Insertion vectors. The map at the top of the Figure shows some of the important genetic markers of phage λ, the scale 0 to 100 represents the length of λ^+ DNA. ▼ indicates the position of an EcoRI target; △, a HindIII target and ● the phage attachment site; P_L and P_R are the major promoters for leftward and rightward transcription, t_L and t_{R2}, two of the sites at which transcription terminates in the absence of N gene product; a third site, t_{R1}, (not shown) is located left of genes O and P whose products are necessary for DNA replication. Gene Q-product is required for transcription of the late genes, including the genes for cell lysis, S and R, and those for morphogenesis (genes A through J). Red$^-$ derivatives of phage λ result from mutations in either of two genes, redA or B; srIλ3 is within the redA gene and redB is between redA and gam. Below the phage map the Figure describes the genotypes of the phage vectors, a through e; the gaps indicate the positions of deletions and the dotted lines substitutions; shn6⁰ symbolises the loss of shnλ6 and the mutant targets for EcoRI in phages c, d and e are symbolised by 1°, 3° etc, rather than srIλ1°, srIλ3° etc

Fig. 2. Replacement vectors for DNA fragments generated by R. EcoRI. The map at the top of the Figure indicates some of the important genetic markers of phage λ, the scale 0 to 100 represents the length of λ^+ DNA, ▼ indicates the position of an EcoRI target, and ● the phage attachment site. Below the phage map the Figure describes the genotypes of the phage vectors; the gaps indicate the positions of deletions and the dotted lines substitutions. The effective deletion indicates the net loss of DNA compared to λ^+ on removal of the fragment of DNA between the two EcoRI targets. Detailed descriptions of the vectors are given in the text

b) Replacement Receptors for the EcoRI System. Replacement receptors for the RI system have two targets for endo R. EcoRI flanking a replaceable segment of the phage chromosome. The essential role of this central fragment is to fill the space between these two targets, for a phage with too little DNA, despite the presence of all necessary functions, does not form a plaque (Bellet et al., 1971; Thomas et al., 1974). All the plaques obtained on transfection after a ligase reaction using a replacement receptor result from phages that either regain their original central fragment or acquire donor DNA in its place.

The receptor phages described (Fig. 2), in contrast to previous replacement receptors (Thomas et al., 1974; Murray and Murray, 1975), contain a central fragment that imparts a readily recognisable phenotype. In one such phage this fragment of a little over twenty cent of the length of the λ^+ genome includes

a mutant transfer-RNA gene, supE, from E. coli. λsupE phages (Borck et al., 1976) are detected by the suppression of an amber mutation in the lacZ gene of a bacterial host (e.g. ED8538) either as red plaques on lactose-MacConkey agar or as blue plaques on agar containing 5-bromo-3-chloro-2-indolyl-β-D-galactoside (Davies and Jacob, 1968). In the former test immune, Lac$^+$ cells in the centre of a plaque become red; in the latter the substituted galactoside is hydrolysed in the agar to give a non-diffusing blue pigment (Smith et al., 1970). This second test is equally applicable to clear and turbid plaques. Implicit in the design of these receptors is expression of the transfer RNA gene irrespective of its orientation with in the λ genome. DNA from the λsupE phage was fragmented by endo R. EcoRI and phages recovered following incubation with polynucleotide ligase. Of 143 phages tested, 142 included a functional supE gene and we conclude that this is expressed in either orientation.

In other vectors (Fig. 2) the slightly larger space between the EcoRI target in the lacZ gene of λlac5 (Rambach and Tiollais, 1974) and srIλ3 is used. The central fragment in these derivatives of λlac5 does not contain the entire lacZ gene, but the genetic content is sufficient to complement at least some of those lacZ mutants designated as omega (ω)-donors in allelic complementation tests (Ullmann et al., 1967). On such a lacZ host, therefore, e.g. (ED8799 or ED8800, see Table 1 or Table 3) the lacZ fragment is detected by complementation irrespective of orientation. The use of imm²¹, or a deletion (KH54) in the cI gene of immλ, permits the loss of up to approximately 35% of the λ genome. The remaining phage in Figure 2 has the lacZ fragment inserted between the two phage restriction targets so that removal of this fragment leaves no residual E. coli DNA in the vector molecule (see section 3d).

All these phages derived from these vectors are recombination deficient (Red⁻) but remain plaque-forming even on a RecA host.

c) Replacement Vectors for the HindIII System. The HindIII replacement vector described previously allows the inclusion of large fragments of DNA, but its derivatives will not grow on a recA host and give very small plaques on a Rec⁺ host (see Murray and Murray, 1975). The receptors described in this paper are less extensively deleted but remain easy to handle. They were made by first incorporating a fragment of DNA into an insertion receptor (Murray and Murray, 1975) and subsequently deleting additional DNA from the phage genome. In the presence of the extra deletion, loss of the inserted fragment is tolerated only if it is substituted by an alternative fragment.

The origin of the replacement vectors from the integration-proficient, λtrpE transducing phage (Borck et al., 1976) is traced in Figure 3. The aim was to introduce a deletion extending into both the phage attachment region (att) and the recombination (red) genes. To permit the selection of such deletions the gam⁺ gene of the λtrpE phage (Fig. 3a) was replaced by the amber mutation, gam210 and imm²¹ by immλ (Fig. 3b). Red⁻, Gam⁻ (i.e. Spi⁻) derivatives were selected as plaques on a suppressor-free strain lysogenic for phage P2 (Zissler et al., 1971). The red⁻ mutations contributing to the Spi⁻ phenotype were selected as deletions by using medium supplemented with pyrophosphate (Parkinson and Huskey, 1971; Shulman and Gottesman, 1971). Those Spi⁻ phages that had retained the suppressor-sensitive gam gene were detected by their failure to give plaques on supE

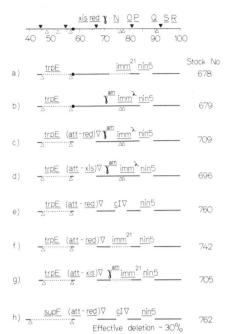

Fig. 3. The derivation of replacement vectors for DNA fragments generated by R. HindIII. The map at the top of the Figure indicates some of the important genetic markers in the right arm of phage λ, the scale 40 to 100 indicates the portion of the λ⁺ DNA shown, ▼ indicates the position of an EcoRI target, △, a HindIII target, ● the phage attachment site and γ is used as an abbreviation for gam. Below the phage map (a through h) the Figure describes the derivation of this family of vectors. All of the phages are deleted for the fragment between srIλ1 and srIλ2. More detailed descriptions are given in the text. Phage h is derived from phage e by replacing the fragment containing the trpE gene with a larger fragment including supF. Similar derivatives of the phages f and g have been made

(P2). All of the Red⁻ phages isolated were integration-deficient, as expected if they were generated by int-mediated recombination events which deleted DNA from att into or through the red genes (Davis and Parkinson, 1971; Shimada et al., 1972). One integration-defective, Red⁻, gam210 phage (Fig. 3, phage c) with reduced sensitivity to the RI-restriction system (Murray and Murray, 1974) was chosen and, as anticipated, the deletion extended through srIλ3. Since it is sometimes advantageous to have recombination-proficient transducing phages, an alternative phage was chosen which did not have the full Spi⁻ phenotype and had, therefore, remained red⁺. The deletion in this phage (Fig. 3, phage d) extends very close to redA.

The gam am210 and immλ cI⁺ alleles of the red⁻ phage (Fig. 3, phage c) were replaced by

Table 2. Vectors for cloning eukaryotic DNA

a) Immunity insertion vector for the *Eco*RI system
 $srI\lambda1^0$ $(srI\lambda2-3)\nabla^0$ imm^{434} $srI\lambda4^0$ $nin5$ $srI\lambda5^0$
 (see Fig. 1e)

b) Replacement vector for the *Eco*RI system
 $\{(srI\lambda1-3)^\nabla(lacZ)^+\}$ $imm^\lambda cI\nabla$ $srI\lambda4^0$ $nin5$ $srI\lambda5^0$
 (see Fig. 2e)

c) Replacement vector for the *Hin*dIII system
 $(srI1-2)\nabla$ $(supF)^+$ $(att-red)^\nabla$ imm^λ $cI\nabla$ $nin5$ $shn6^0$
 (see Fig. 3h)

$(srI\lambda1-3)^\nabla$ indicates that the DNA between the restriction targets $srI\lambda1$ and $srI\lambda3$ has been deleted; $(lacZ)^+$ or $(supF)^+$, the insertion of a fragment of DNA including the *lacZ*, or *supF* gene $\{(srI\lambda1-3)^\nabla (lacZ)^+\}$, the DNA between the restriction targets $srI\lambda1$ and $srI\lambda3$ has been removed and replaced by a fragment of DNA including the *lacZ* gene, or a part of the *lacZ* gene; $(srI\lambda2-3)\nabla^0$, the DNA between the restriction targets is deleted and the hybrid restriction target removed by mutation

gam^+ imm^λ $cI\nabla$ (Fig. 3, phage e), the combination of red^- and clear-plaque morphology being preferable for cloning eukaryotic DNA. The gam $am210$ imm^λ cI^+ alleles of the red^- phage were also replaced by gam^+ imm^{21} cI^+ (Fig. 3f) and the imm^λ cI^+ allele of the red^+ phage by imm^{21} cI^+ (Fig. 3g). The imm^{21} c^+ phages do not make high titre stocks quite as readily as the imm^λ $cI\nabla$ strains but the establishment of immune cells, even as absortive lysogens, facilitates the detection of transducing phages by complemention of auxotrophic bacteria.

For each of the phages (Fig. 3e, through g), the fragment containing the *trpE* gene has been replaced by one containing *supF* (Borck et al., 1976). This fragment is approximately 16 per cent of the length of the λ^+ genome and the $\lambda supF$ phages are detected in the same way as the $\lambda supE$ phages (Results, section 3b).

d) Derivatives of the Vectors. The intrinsic properties of wild-type λ can be extended by the use of mutations. Amber mutations have been incorporated into some of the vectors to facilitate studies of gene expression, to increase the efficiency of phage production or with the aim of limiting the propagation of the phage to special laboratory hosts. The following mutations in the right arm of the genome have been included in certain vectors:– *Nam*7 *Nam*53, *Pam*3, *Qam*73 and *Sam*7. (The *Sam*7 derivative of our insertion vector for the *Hin*dIII system was isolated and very kindly provided by Hamilton O. Smith.) In order to confine the λ vectors to specific permissive host strains we are uncertain whether it is preferable to block DNA replication or to permit replication, maximise killing of the host yet prevent phage production

Table 3. *E. coli* host strains for transfection

Strain	Genotype
803	*met, hsd*$_k$*S*$^-$ *R*$^+$ *M*$^+$, *supE*
ED8739	*met, hsd*$_k$*S*$^-$ *R*$^+$ *M*$^+$, *supE, supF*
ED8654	*met, hsd*$_k$*S*$^+$ *R*$^-$ *M*$^+$, *supE, supF, trpR*
ED8767	*met, hsd*$_k$*S*$^-$ *R*$^+$ *M*$^+$, *supE, supF, recA*56
ED8799	*met, hsd*$_k$*S*$^-$ *R*$^+$ *M*$^+$, *supE, supF, lacZ*M15
ED8800	*met, hsd*$_k$*S*$^-$ *R*$^+$ *M*$^+$, *supE, supF, lacZ*M15, *recA*56

(F.R. Blattner, pers. comm.). The λ vector described by Enquist et al. (1976) and certified as EK2 by a NIH advisory panel (Enquist, pers. com.) relies essentially on the ability to kill coupled with the inability to cut and package concatemeric DNA. With the same objective we are crossing amber mutations (*Eam*1100 and *Wam*403) into some of those vectors we consider most suitable for cloning eukaryotic DNA (R.D. Schmickel and N.E. Murray, unpublished) (Table 2). Since both the amber mutations in the left arm of the vector are suppressed by *supE*, the *supF*-dependent block in gene *S* (*Sam*7) has been incorporated to minimise the volumes of phage lysates handled for the preparation of DNA.

e) E. coli Hosts. The use of the λ vectors described relies on appropriate strains of *E. coli* for the efficient recovery and detection of recombinant phages. Our transfection hosts are *supF* derivatives of the restriction- and modification-deficient, *supE* strain, 803 (Wood, 1966). These hosts (see Table 3, e.g. ED8739) give good levels of competence, do not carry any known restriction system and suppress all the amber mutations in our λ strains. For some experiments we use a *K* modification-proficient derivative (Table 3, ED8654), so that the resulting phages may be tested on normal *K*-restricting *E. coli* strains. For experiments using eukaryotic DNA we choose a *recA* host (Table 3, e.g. ED8767) to minimise recombination events that might lead to changes in the DNA content of the inserted fragment. RecA$^-$ strains are also less vigorous than most laboratory strains and are much more sensitive to ultraviolet irradiation (Clark, 1967).

A special transfection host (Table 3, ED8799 or ED8800) detects recombinant derivatives of a *lacZ*-replacement vector as colourless plaques on appropriate indicator plates (Smith et al., 1970). Full advantage of the *supF* replacement vector cannot be taken in the presence of the amber mutations needed to meet the requirement of an EK2 system. Recombinant phages are easily detected by a simple screening procedure. Phages retaining the *supF* gene will grow on a *sup*^{0} or *supE* host, whilst recombinant phages require *supF* to suppress both *Sam*7 and the mutation in the left arm of the phage.

All the hosts described retain the chromosomal attachment site for phage *λ*. It can, however, be argued that this region should be deleted so that no contaminant phage is able to lysogenise the host. *TonA* derivatives of these strains offer the advantage of being resistant to phage T1, a common laboratory contaminant that is resistant to desiccation. The recombinant phages are readily propagated on either the hosts we have described or on an *rk⁻mk⁻recA* derivative of the *tonA* strain C600.

Discussion

The lambda receptors described here are designed to ease the recognition and propagation of recombinant DNA molecules in appropriate hosts. The insertion receptors are particularly easy to use and are ideally suited to the cloning of small fragments of DNA, since the detection system is not related to the size of the inserted fragment. These vectors can accommodate DNA fragments with molecular weights up to about 6×10^6. The proportion of recombinant phage normally recovered from a ligase reaction using an insertion vector is about 10%, but exceptionally this has been as high as 40% (Beggs et al., 1976). This unusually high frequency probably reflects a particularly favourable state of the donor DNA in this experiment.

The replacement vectors described have space for fragments with a molecular weight of approximately 12×10^6 and require a minimum of about 2 or 3×10^6. Although the upper limit imposed on the size of fragments that can be inserted into the lambda genome is sometimes a disadvantage, the fact that lambda has both an upper and a lower limit to the amount of DNA that it can package can be exploited, because an appropriate replacement vector selects fragments within a particular size range. In our experience, as many as 90% of the plaques retrieved using a replacement vector may result from recombined molecules, although a more usual figure is about 50%. Given this high efficiency of replacement, the direct screening of parental versus recombinant plaques may seem unnecessary, but it does provide a very quick check on the efficiency of a particular ligase reaction, as well as reducing the number of clones to be screened, for example, by hybridisation techniques.

Our original objective with the replacement receptors was to use a fragment of DNA containing the *lacZ* gene of *E. coli* (rather than amber suppressors) with both the *Eco*RI and *Hind*III systems. However, we failed to isolate a *lacZ*-transducing phage using the *Hind*III system. We assume that, when *E. coli* DNA is digested with endo R. *Hind*III, the *lacZ* gene is within a very large fragment, since the *lacZ* gene itself does not include a target for endo R. *Hind*III. In the presence of the amber mutations suggested for an EK2 vector (NIH guidelines for research on recombinant DNA, see Helinski et al., 1976) no appropriate indicator system is available for the direct detection of recombinant derivatives of the *λsupF* phages. A simple screening of individual plaques is therefore required, but in instances where the frequency of recombinant plaques is shown to be high, such screening may be unnecessary.

An advantage of phage *λ* as a cloning vector is the detailed understanding of the control of gene expression in this phage. Fragments of DNA incorporated into either type of replacement vector can be transcribed from the efficient phage promoter P_L, under the special auspices of *λ*'s *N* gene product. This protein interacts with the host DNA polymerase to permit transcription through sequences that would otherwise serve as termination signals (Franklin, 1974; Adhya et al., 1974). Phage *λ* also offers the investigator some degree of control over phage transcription, because expression from P_L may be enhanced or turned off under appropriate conditions. This, together with the use of appropriate conditional mutants, offers the opportunity for optimising the expression of genes that have been inserted into the phage chromosome (Moir and Brammar, 1976). A gene within an incorporated DNA sequence may be transcribed from its own promoter or from the phage promoter P_L. For the latter to be effective the coding sequence in the inserted DNA fragment must be on the same strand of the duplex as the strand of the phage genome that is being transcribed. Since the two strands of *λ* DNA can be separated very readily by equilibrium centrifugation in caesium chloride gradients in the presence of poly ribo UG (Szybalski et al., 1971), and the coding sequence can often be recognised by hybridisation to the appropriate messenger RNA, it is a simple matter to detect recombinant clones in which the inserted sequence is in the appropriate orientation for transcription from P_L. Experiments of this type with fragments of sea urchin DNA containing histone genes have shown that the coding sequences for the various histone fractions are in the same strand (P. Mounts, E. Southern and K. Murray, unpublished experiments).

Eukaryotic DNA characteristically contains repeated sequences. In order to preserve the identity, or integrity, of a cloned fragment of eukaryotic DNA, the use of recombination-deficient phages grown in a Rec⁻ host is suggested. Ideally, a DNA content approaching that of wild-type *λ* is desirable, for phages with too little DNA are known to be unstable (Bellet et al., 1971). The vectors chosen for cloning eukaryotic DNA (Table 2) all provide high titre lysates in the absence of all known recombination systems. Furthermore, each contains a deletion that removes the phage attachment site and either a second deletion that remove the repressor (cI) gene or, alternatively, an insertion that inactivates it. Those phages whose recombinant derivatives contain no host DNA also lack homology for *rec*-mediated integration of the phage into an *E. coli* chromosome (Gottesman et al., 1974). The presence of the *nin*5 deletion (Court and Sato, 1969) which deletes the *tr*2 termination site (see Fig. 1), ensures that even in the absence of gene *N* function, late gene functions, including those required for cell lysis, are expressed. It is highly improbable, therefore, that any of the phages listed in Table 2 could be propagated in either the prophage or plasmid states. The chance of a recombinant phage propagating lytically in a λ sensitive host is further reduced by the introduction of two amber mutations (Enquist et al., 1976) and by growing the phages in a non-modifying host strain. Unmodified phage λ plates with an efficiency of 10^{-2}–10^{-5} on host strains with restriction systems. Together these factors offer many orders of magnitude of biological containment compared with the efficiency of growth of wild-type λ in *E. coli* K12.

Acknowledgements. We are grateful to Graham Brown, Karen Brown, Sandra Bruce, Kathleen Chalmers, Susanna Dunbar and Jo Rennie for their skillful technical assistance; to the people, particularly John Scaife, Ham Smith and Fred Blattner, who generously made available bacteria or phage strains; and to Roy Schmickel, Ed Southern, Jean Beggs and George Brownlee whose interests stimulated the construction of some of these vectors and who sometimes put the vectors to test.

    This work was supported in part by grants from the Medical Research and Science Research Councils.

References

Adhya, S., Gottesman, M., de Crombrugghe, B.: Release of polarity in *E. coli* by gene *N* of phage λ: termination and antitermination of transcription. Proc. nat. Acad. Sci. (Wash.) **71**, 2534–2538 (1974)

Appleyard, R.K.: Segregation of new lysogenic types during growth of a doubly lysogenic strain derived from *Escherichia coli* K12. Genetics **39**, 440–452 (1954)

Beggs, J.D., Guerineau, M., Atkins, J.F.: A map of the restriction targets in yeast 2 micron plasmid DNA cloned on bacteriophage lambda. Molec. gen. Genet. (in press) 1976

Bellett, A.J.D., Busse, H.G., Baldwin, R.L.: Tandem genetic duplications in a derivative of phage lambda. The bacteriophage lambda, ed. by A.D. Hershey, pp. 501–514 (1971)

Blattner, F.R., Fiandt, M., Hass, K.K., Twose, P.A., Szybalski, W.: Deletions and insertions in the immunity region of coli-

phage λ: revised measurement of the promoter startpoint distance. Virology **62**, 458–471 (1974)

Borck, K., Beggs, J.D., Brammar, W.J., Hopkins, A.S., Murray, N.E.: The construction in vitro of transducing derivatives of phage lambda. Molec. gen. Genet. **146**, 199–207 (1976)

Campbell, A.: Sensitive mutants of bacteriophage λ. Virology **14**, 22–32 (1961)

Clark, A.J.: The beginning of a genetic analysis of recombination proficiency. J. cell. Phys., Suppl. to Vol. **70**, No. 2 (Part II), 165–180 (1967)

Clark, L., Carbon, J.: Biochemical construction and selection of hybrid plasmids containing specific segments of the *Escherichia coli* genome. Proc. nat. Acad. Sci. (Wash.) **72**, 4361–4365 (1975)

Court, D., Sato, K.: Studies of novel transducing variants of lambda: dispensability of genes N and Q. Virology **39**, 348–352 (1969)

Davies, J., Jacob, F.: Genetic mapping of the regulator and operator genes of the *lac* operon. J. molec. Biol. **36**, 413–416 (1968)

Davis, R.W., Parkinson, J.S.: Deletion mutants of bacteriophage lambda. III. Physical structure of *attϕ*. J. molec. Biol. **56**, 401–423 (1971)

Enquist, L., Tiemeier, D., Leder, P., Weisberg, R., Sternberg, N.: Safer derivatives of bacteriophage λgt. λC for use in cloning of recombinant DNA molecules. Nature (Lond.) **259**, 596–598 (1976)

Franklin, N.C.: The *N* operon of lambda: extent and regulation as observed in fusions to the tryptophan operon of *Escherichia coli*. In: The bacteriophage lambda (ed. A.D. Hershey), pp. 621–638. New York: Cold Spring Harbor Laboratories 1971

Franklin, N.C.: Altered reading of genetic signals fused to the *N* operon of bacteriophage λ: genetic evidence for modification of polymerase by the protein product of the *N* gene. J. molec. Biol. **89**, 33–48 (1974)

Goldberg, A.R., Howe, M.: New mutations in the *S* cistron of bacteriophage λ affecting host cell lysis. Virology **38**, 200–202 (1969)

Gottesman, M.M., Gottesman, M.E., Gottesman, S., Gellert, M.: Characterisation of bacteriophage λ reverse as an *Escherichia coli* phage carrying a unique set of host-derived recombination function. J. molec. Biol. **72**, 471–487 (1974)

Grunstein, M., Hogness, D.S.: Colony hybridisation: a method for the isolation of cloned DNAs that contain a specific gene. Proc. nat. Acad. Sci. (Wash.) **72**, 3961–3963 (1975)

Haggerty, D.M., Schlief, R.F.: Location in bacteriophage lambda DNA of cleavage sites of the site-specific endonuclease from *Bacillus amyloliquefaciens* H. J. Virol. **18**, 659–663 (1976)

Hedgpeth, J., Goodman, H.M., Boyer, H.W.: DNA nucleotide sequence restricted by the RI endonuclease. Proc. nat. Acad. Sci. (Wash.) **69**, 3448–3502 (1972)

Helinski, D.R., Falkow, S., Curtiss, R., Szybalski, W.: Guidelines for research involving recombinant DNA molecules, Appendix C. National Institutes for Health, Bethesda, Maryland, U.S.A. (1976)

Horwitz, J.P., Chua, J., Curby, R.J., Tomson, A.J., Da Rooge, M.A., Fisher, B.E., Mauricio, J., Klundt, I.: Substrate for cytochemical demonstration of enzyme activity. I. Some substituted 3-indolyl-β-D-glycopyranosides. J. med. Chem. **7**, 574–583 (1964)

Ippen, K., Shapiro, J.A., Beckwith, J.R.: Transposition of the *lac* region to the *gal* region of the *Escherichia coli* chromosome: Isolation of λ*lac* transducing bacteriophages. J. Bact. **108**, 5–9 (1971)

Jones, K.W., Murray, K.: A procedure for detection of heterologous DNA sequences in lambdoid phages by in situ hybridisation. J. molec. Biol. **96**, 455–460 (1975)

Kaiser, A.D., Hogness, D.S.: The transformation of *Escherichia coli* with Deoxyribonucleic acid isolated from bacteriophage λdg. J. molec. Biol. **2**, 392–415 (1960)

119

Kaiser. A.D.. Jacob. F.: Recombination between related temperate bacteriophage and the genetic control of immunity and prophage localisation. Virology 4. 509–521 (1957)

Lederberg. E.M.. Cohen. S.N.: Transformation of Salmonella typhimurium by plasmid deoxyribonucleic acid. J. Bact. 119. 1072–1074 (1974)

Liedke-Kulke. M.. Kaiser. A.D.: The c-region of coliphage 21. Virology 32. 475–481 (1967)

Lennox. E.S.: Transduction of linked characters of the host of bacteriophage P1. Virology 1. 190–206 (1955)

Mandel. M.. Higa. A.: Calcium-dependent bacteriophage DNA infection. J. molec. Biol. 53. 159–162 (1970)

Marmur. J.: A procedure for the isolation of Deoxyribonucleic acid from microorganisms. J. molec. Biol. 3. 208–218 (1961)

Mertz. J. E.. Davis. R.W.: Cleavage of DNA by RI restriction endonuclease generates cohesive ends. Proc. nat. Acad. Sci. (Wash.) 69. 3370–3374 (1972)

Moir. A.. Brammar. W.J.: The use of specialised transducing phages in the amplification of enzyme production. Molec. gen. Genet. (in press) (1976)

Murray. N.E.. Manduca de Ritis. P.. Foster. L.A.: DNA targets for the Escherichia coli K restriction system analysed genetically in recombinants between phages phi 80 and lambda. Molec. gen. Genet. 120. 261–281 (1973)

Murray. N.E.. Manduca de Ritis. P.. Foster. L.A.: DNA targets in phage λ to form receptor chromosomes for DNA fragments. Nature (Lond.) 251. 476–481 (1974)

Murray. K.. Murray. N.E.: Phage lambda receptor chromosomes for DNA fragments made with restriction endonuclease III of Haemophilus influenzae and restriction endonuclease I of Escherichia coli. J. molec. Biol. 98. 551–564 (1975)

Old. R.. Murray. K.. Roizes. G.: Recognition sequence of restriction endonuclease III from Haemophilus influenzae. J. molec. Biol. 92. 331–339 (1975)

Parkinson. J.S.: Genetics of the left arm of the chromosome of bacteriophage lambda. Genetics 59. 311–325 (1968)

Parkinson. J.S.: Deletion mutants of bacteriophage lambda. II. Genetic properties of att-defective mutants. J. molec. Biol. 56. 385–401 (1971)

Parkinson. J.S.. Huskey. R.J.: Deletion mutants of bacteriophage lambda. I. Isolation and initial characterisation. J. molec. Biol. 56. 369–384 (1971)

Perricaudet. M.. Tiollais. P.: Defective bacteriophage lambda chromosome. potential vector for DNA fragments obtained after cleavage by Bacillus amyloliquifaciens endonuclease (BamI). FEBS Letters 56. 7–11 (1975)

Philippsen. P.. Streeck. R.E.. Zauchau. H.G.: Defined fragments of calf, human and rat DNA produced by restriction nucleases. Europ. J. Biochem. 45. 475–488 (1974)

Rambach. A.. Tiollais. P.: Bacteriophage λ having EcoRI endonuclease sites only in the non-essential region of the genome. Proc. nat. Acad. Sci. (Wash.) 71. 3927–3930 (1974)

Sharp. P.A.. Sugden. B.. Sambrook. J.: Detection of two restriction endonuclease activities in Haemophilus parainfluenzae using analytical agarose ethidium bromide electrophoresis. Biochemistry 12. 3055–3063 (1973)

Shimada. K.. Weisberg. R.A.. Gottesman. M.E.: Prophage lambda at unusual chromosomal locations. I. Location of the secondary attachment sites and the properties of the lysogens. J. molec. Biol. 63. 483–503 (1972)

Shulman. M.. Gottesman. M.: Lambda att^2: a transducing phage capable of intra molecular int-xis Promoted Recombination. In: The bacteriophage lambda (ed. A.D. Hershey). pp. 477–488. New York: Cold Spring Harbor Laboratories 1971

Smith. H.O.. Nathans. D.: Nomenclature for restriction enzymes. J. molec. Biol. 81. 419–423 (1973)

Smith. J.D.. Barnett. L.. Brenner. S.. Russell. R.L.: More mutant tyrosine transfer RNAs. J. molec. Biol. 54. 1–14 (1970)

Spizizen. J.: Transformation of biochemically deficient strains of Bacillus subtilis by deoxyribonucleate. Proc. nat. Acad. Sci. (Wash.) 44. 1072–1078 (1958) .

Stacey. K.A.. Simson. E.: Improved method for the isolation of thymine-requiring mutants of Escherichia coli. J. Bact. 90. 554–555 (1965)

Szybalski. W.. Kubinski. H.. Hradecna. Z.. Summers. W.C.: Analytical and preparative separation of the complementary DNA strands. Mechanics in enzymology. Vol. XXI. pp. 383–413. New York: Academic Press. Inc.

Thomas. M.. Cameron. J.R.. Davis. R.W.: Viable molecular hybrids of bacteriophage λ and eukaryotic DNA. Proc. nat. Acad. Sci. (Wash.) 71. 4579–4583 (1974)

Ullmann. A.. Jacob. F.. Monod. J.: Characterisation by in vitro complementation of a peptide corresponding to an operator-proximal segment of the β-galactosidase structural gene of E. coli. J. molec. Biol. 24. 339–343 (1967)

Weil. J.. Signer. E.R.: Recombination in bacteriophage λ: II. Site-specific recombination promoted by the integration system. J. molec. Biol. 34. 273–279 (1968)

Wensink. P.C.. Finnegan. D.J.. Donelson. J.E.. Hogness. D.S.: A system for mapping DNA sequences in the chromosomes of Drosophila melanogaster. Cell 3. 315–325 (1974)

Wilson. G.A.. Young. F.E.: Isolation of a sequence-specific endonuclease (BamI) from Bacillus amyloliquefaciens. J. molec. Biol. 97. 123–125 (1975)

Wood. W.B.: Host specificity of DNA produced by E. coli: Bacterial mutations affecting the restriction and modification of DNA. J. molec. Biol. 16. 118–133 (1966)

Yanofsky. C.. Lennox. E.S.: Transduction and recombination study of linkage relationships among the genes controlling tryptophan synthesis in Escherichia coli. Virology 8. 425–447 (1959)

Yoshimori. R.N.: A genetic and biochemical analysis of the restriction and modification of DNA by resistance transfer factors. Ph. D. Thesis. University of California. (1971)

Zissler. J.. Signer. E.R.. Schaefer. F.: The role of recombination in growth of bacteriophage lambda. I. The gamma gene. The bacteriophage lambda. ed. A.D. Hershey. pp. 455–475. New York: Cold Spring Harbor Laboratories 1971

120

15

Reprinted from *Gene* **2**:95–113 (1977). Originally published by Elsevier/North-Holland Biomedical Press B.V.

CONSTRUCTION AND CHARACTERIZATION OF NEW CLONING VEHICLES
II. A MULTIPURPOSE CLONING SYSTEM

(Recombinant DNA; molecular cloning plasmid vector; EK2 host-vector system)

FRANCISCO BOLIVAR*, RAYMOND L. RODRIGUEZ*, PATRICIA J. GREENE, MARY C. BETLACH, HERBERT L. HEYNEKER*, HERBERT W. BOYER

Department of Biochemistry and Biophysics, University of California, San Francisco, Calif. 94143 (U.S.A.) and

JORGE H. CROSA and STANLEY FALKOW

Department of Microbiology, University of Washington, Seattle, Wash. 98195 (U.S.A.)

SUMMARY

In vitro recombination techniques were used to construct a new cloning vehicle, pBR322. This plasmid, derived from pBR313, is a relaxed replicating plasmid, does not produce and is sensitive to colicin E1, and carries resistance genes to the antibiotics ampicillin (Ap) and tetracycline (Tc). The antibiotic-resistant genes on pBR322 are not transposable. The vector pBR322 was constructed in order to have a plasmid with a single *Pst*I site, located in the ampicillin-resistant gene (Ap^r), in addition to four unique restriction sites, *Eco*RI, *Hin*dIII, *Bam*HI and *Sal*I. Survival of *Escherichia coli* strain X1776 containing pBR313 and pBR322 as a function of thymine and diaminopimelic acid (DAP) starvation and sensitivity to bile salts was found to be equivalent to the non-plasmid containing strain. Conjugal transfer of these plasmids in bi- and triparental matings were significantly reduced or undetectable relative to the plasmid ColE1.

---

*Present addresses: (F.B.) Departamento de Biologia Molecular, Instituto de Investigaciones Biomedicas, Universidad Nacional Autonoma de Mexico, Mexico 20, D.F., Apdo Postal 70228; (H.L.H.) Department of Molecular Genetics, University of Leiden, Wassenaarseweg 64, Leiden (The Netherlands); (R.L.R.) Department of Genetics, Briggs Hall, University of California, Davis, CA 95616 (U.S.A.).

Abbreviations: Ap^r, ampicillin-resistant; Cm^r, chloramphenicol-resistant; Col^{imm}, colicin immunity; DAP, diaminopimelic acid; DTT, dithiothreitol; Km^r, kanamycin-resistant; LB, Luria broth; Nx^r, nalidixic-resistant; SDS, sodium dodecyl sulfate; Sm^r, streptomycin-resistant; Su^r, sulfonamide-resistant; Tc^r, tetracycline-resistant.

INTRODUCTION

Bacterial plasmids and bacteriophage have a key role in recombinant DNA technology. Segments of DNA from diverse origins can be excised with the appropriate restriction endonuclease and added to plasmids or bacteriophage (Hershfield et al., 1974; Morrow et al., 1974; Cameron et al., 1975). If these new molecules contain an intact replicon, they can be propagated in a suitable host to yield large quantities of recombinant DNA and in some instances, specific gene products (Hershfield et al., 1974). Several bacterial plasmids have been used as cloning vectors: pSC101 (Cohen et al., 1973), ColE1 (Hershfield et al., 1974) and pCR1 (Covey et al., 1976). However, these plasmids and their derivatives (Hamer et al., 1975; Hershfield et al., 1976; So et al., 1976) have limited versatility in terms of genetic markers for selection of transformants and screening for recombinant plasmids.

We have described the construction of a series of plasmids containing Ap- and Tc-resistant genes derived from pRSF2124 (So et al., 1976) and pSC101 respectively in combination with replication elements of a ColE1-like plasmid (Betlach et al., 1976; Rodriguez et al., 1976). One of these plasmids, pBR313, provides single cleavage sites for the *Hind*III, *Bam*HI, *Eco*RI, *Hpa*I, *Sal*I and *Sma*I restriction endonucleases (Bolivar et al., 1977). In the case of the *Hind*III, *Bam*HI and *Sal*I endonuclease cloning sites, the insertion of DNA fragments inactivates the Tc^r gene. In this paper, we report the construction of another plasmid (pBR322) which is less than half the size of pBR313 and provides additional cloning advantages. The plasmid pBR322 contains a unique *Pst*I cleavage site located in the Ap^r gene as well as two *Hinc*II sites located in the Ap^r and Tc^r genes. The *Pst*I site can be used for molecular cloning of DNA fragments via homodeoxy polymeric extension (Lobban and Kaiser, 1973) and the *Hinc*II site for blunt-end ligation techniques (Sgaramella et al., 1970; Sugino et al., 1977). The properties of pBR313 and pBR322 in the *E. coli* strain X1776 are also presented.

MATERIAL AND METHODS

(a) Bacterial strains

E. coli K12 strain RR1 F$^-$*pro leu thi lac*Y *Str*^{r} r_k^- m_k^- was used as the recipient cells in the transformation experiments. *E. coli* B strain HB50 *pro leu try his arg met thr gal lac*Y *Str*^{r} was used to prepare unmethylated plasmid DNA for *Eco*RII digestions (Yoshimori et al., 1972). *E. coli* K12 strain X1776 F$^-$*tau*A53 *dap*D8 *mer*A1 *sup*E42 Δ40(*gal-uvr*B) λ$^-$ *min*B2 *mal*A25 *thy*A57 *met*C65 Δ29(*bio*H-*asd*) *cys*B2 *cyc*A1 *Hsd*R2 was kindly provided by R. Curtiss III.

(b) Media and buffers

For transformation RR1 was grown in either LB or M9-glucose minimal media, before CaCl$_2$ treatment. X1776 was also grown in LB supplemented

with DAP 200 μg/ml and thymine (thy) 50 μg/ml. The BSG buffer solution
used for washing X1776 in the DAP-less death experiments was 0.85% NaCl,
0.03% KH_2PO_4, 0.06% Na_2HPO_4 100 μg/ml gelatin.

(c) Preparation of plasmid DNA

Plasmid DNA was prepared by first amplifying M9-glucose-grown cultures
by the addition of 170 μg/ml of chloramphenicol during logarithmic phase
of growth (Clewell et al., 1972). Extraction and purification of plasmid
DNA was achieved by a cleared lysate technique previously described (Betlach et al., 1976).

(d) Enzymes

All the restriction enzymes used in this work, except for HpaI (BRL
laboratories) were purified according to the procedure by Greene et al. (1977)
and are itemized in Table I. Reaction conditions for the various restriction
endonucleases have been described previously (Bolivar et al., 1977). T4 DNA
ligase was purified from T4 am N82 infected E. coli B, according to the
procedure described by Panet et al. (1973). The final preparation (500 U/ml)
was homogeneous as judged by SDS-polyacrylamide gel electrophoresis.

(e) Ligation of DNA

Ligations were carried out in 66 mM Tris—HCl pH 7.6, 6.6 mM $MgCl_2$
10 mM DTT and 0.5 mM ATP at 12°C for 2—12 h. The concentration of T4
DNA ligase and of DNA termini varied to promote polymerization or circularization. When blunt-ended DNA fragments were ligated, the concentration
of ends was at least 0.2 μM and approximately 50 U of T4 DNA ligase per
ml was added to the reaction mixture (Heyneker et al., 1976). When DNA
fragments with cohesive ends were ligated, 5 U of T4 DNA ligase per ml was
sufficient and the concentration of ends was adjusted in such a way that
linear molecules were favored (Dugaiczyk et al., 1975).

(f) Agarose and acrylamide gel electrophoresis

The conditions for agarose and acrylamide electrophoresis have been previously described (Bolivar et al., 1977).

(g) Transformation of E. coli K12

E. coli RRI cells were prepared for transformation by the method described
by Cohen et al. (1972). 100 μl of DNA in 30 mM $CaCl_2$ were added to 200
μl of $CaCl_2$-treated cells (5 · 10^9 cells/ml) and the mixture was chilled in ice
for 60 min, after which it received a 75-sec, 42°C heat pulse. The pulse was
terminated with the addition of 3 ml of LB. The cells were grown for 2 h at
37°C before plating. Transformation of X1776 was achieved using the procedure described by R. Curtiss III (personal communication). An overnight
culture of X1776 in LB + DAP + thy was diluted 1/10 with 20 ml of fresh
LB + DAP + thy and incubated in a shaker et 37°C for 3 to 4 h until the

TABLE I

RESTRICTION ENDONUCLEASES

Endonucleases	Substrate site	Reference	Endonuclease	Substrate site	Reference
AluI	A G↓C T	Roberts et al., 1976	HaeIII	G G↓C C	Roberts et al., unpublished observations
BamHI	G↓G A T C C	Wilson and Young, 1975	HincII	G T Py↓PuA C	Landy et al., 1974
BglI	...	Wilson and Young, unpublished observations	HindIII	A↓A G C T T	Danna et al., 1973
EcoRI	G↓A A T T C	Greene et al., 1976	HpaI	G T T↓A A C	Gromkova and Goodgal, 1972
EcoRII	↓C C $^{A}_{T}$ G G	Yoshimori et al., 1975	PstI	C T G C A↓G	Smith et al., 1976
HaeII	PuG C G C↓Py	Roberts et al., unpublished observations	SalI	G↓T C G A C	Bolivar et al., 1977

culture reached an absorbance of 0.5 to 0.6 A_{600}. The culture was centrifuged at 7000 rpm for 10 min at 4°C, and the cells washed in 10 ml cold 10 mM NaCl. The suspension was again centrifuged as above and the pellet resuspended in 10 ml of freshly prepared cold 75 mM $CaCl_2$ (pH 8.4) and placed in ice for 25 min. Cells were centrifuged as described above and the pellet resuspended in 2 ml of 75 mM $CaCl_2$ pH 8.4, of which 200 μl was added to 100 μl of plasmid DNA in 10 mM Tris pH 8. The mixture was kept in ice for 60 min, then heated 60 sec at 42°C. Tubes were chilled for 10 min and 3 ml of LB + DAP + thy were added. The cells were incubated at 37°C for 3 h and plated in selective media. The plates were incubated 2 to 3 days at 37°C.

RESULTS

I. Construction of pBR321 and pBR322

We have described the construction of a series of cloning vehicles, one of which, pBR313, a $5.8 \cdot 10^6$ dalton Ap^r Tc^r Col^{imm} plasmid (Fig.1), has been extensively mapped using 14 restriction endonucleases (Bolivar et al., 1977). Experiments with pBR313 indicated that one of its *Pst*I sites was located in the Ap^r gene. Therefore molecular cloning into this *Pst*I site would result in recombinant molecules which could be detected by screening for Ap^s phenotypes. In order to construct a molecular cloning vector with one *Pst*I site in the Ap^r gene, it was necessary to construct two derivatives of pBR313. An $Ap^s Tc^r Col^{imm}$ plasmid, pBR318, containing one *Pst*I site was obtained by transforming *E. coli* RRI with ligated *Pst*I fragments of pBR313 and selecting for Tc^r transformants. Tc^r transformants which were Ap^s were found to carry plasmids that lack the 1.25 and $0.42 \cdot 10^6$ dalton *Pst*I fragments present in pBR313 (Fig.1). Another pBR313 derivative, pBR320, an $Ap^r Tc^s Col^s$ plasmid with a molecular weight of $1.95 \cdot 10^6$ daltons was obtained by transforming *E. coli* RRI with unligated *Eco*RII fragments of pBR313 and selecting for Ap^r transformants. Sixteen $Ap^r Tc^s$ clones were examined, and one was found to carry a plasmid, pBR320, containing only one *Pst*I site. This clone was found to be sensitive to colicin E1. Fig. 1 shows a tentative restriction endonuclease map of pBR320.

An in vitro recombination experiment using pBR318 and pBR320 was designed to restore the $Ap^r Tc^r$ markers in a single low molecular weight relaxed plasmid containing one *Pst*I substrate site. The construction of this plasmid was accomplished by the digestion of pBR318 with *Pst*I and *Hpa*I endonucleases which resulted in two pieces of DNA with molecular weights of 1.95 and $2.2 \cdot 10^6$ daltons; the smaller DNA fragment carried the Tc^r gene(s) (Rodriguez et al., 1976; Tait et al., 1976; Bolivar et al., 1977) and part of the Ap^r gene as shown in Fig. 1. The plasmid pBR320 was cleaved with the restriction enzymes *Pst*I and *Hinc*II to yield three fragments of DNA. The largest fragment, $1.15 \cdot 10^6$ daltons, carries the "origin" of replication and the remaining portion of the Ap^r gene not present in the $1.95 \cdot 10^6$ dalton fragment of pBR318.

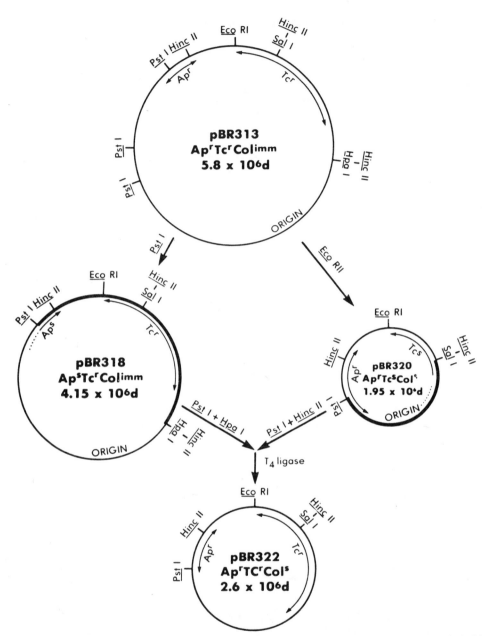

Fig. 1. Diagrammatic representation of the construction of pBR322. The parental plasmid pBR313 was used to construct pBR318 and pBR320 by *Pst*I and *Eco*RII endonuclease digestions respectively. These two plasmids were separately digested with *Pst*I and *Hpa*I endonucleases (pBR318) and with *Pst*I and *Hinc*II endonucleases (pBR320). The heavy lined regions from the *Pst*I to *Hpa*I sites in pBR318 and from *Pst*I to *Hinc*II sites (*Sal*I) in pBR320 represent the two DNA fragments that were ligated to each other to generate pBR321 and pBR322. The origins of replication in these plasmids were determined by restriction endonuclease analysis and electron microscopic examinations (unpublished observations). For detailed explanation, see the text.

126

The digested DNAs were mixed, ligated in vitro, and transformed into *E. coli* RRI. Since neither pBR320 nor the $1.95 \cdot 10^6$ dalton fragment of pBR318 carry the colicin E1 immunity gene, transformants were selected for Ap^{r}Tc^r and then screened for sensitivity to colicin E1. This transformation yielded numerous Ap^{r}Tc^{r}Col^s clones which carried plasmids (e.g. pBR 321) with a molecular weight of $3.1 \cdot 10^6$ daltons. As expected, this plasmid resulted from the addition of the $1.95 \cdot 10^6$ dalton *Hpa*I-*Pst*I fragment of pBR318 and the $1.15 \cdot 10^6$ d. *Pst*I-*Hinc*II DNA piece from pBR320.

From this transformation we obtained in one instance a smaller, $2.6 \cdot 10^6$ dalton, Ap^{r}Tc^{r}Col^s plasmid. This plasmid, possibly the result of in vivo recombination event near unligated termini, was missing $0.5 \cdot 10^6$ daltons of DNA from a region of pBR321, not associated with Ap^r, Tc^r or DNA replication (Fig. 2). Because of its lower molecular weight compared with pBR 321, pBR 322 was chosen for further characterization of number and position of restriction sites.

II. Mapping of pBR322 restriction endonuclease digestions

As determined by agarose and acrylamide gel electrophoresis of DNA digests, pBR322 was found to carry unique substrate sites for the *Bam*HI, *Eco*RI, *Hind*III, *Pst*I and *Sal*I restriction endonucleases. Double and triple endonuclease digests of the plasmid (data not shown) showed that the relative positions of these sites were identical to those mapped in pBR313. As can be seen in Figs. 2 and 3, there are only two *Hinc*II sites in pBR322, one located $0.17 \cdot 10^6$ daltons from the *Pst*I site in the Ap^r gene and the other in Tc^r gene, which is also a *Sal*I site (Bolivar et al., 1977).

The *Eco*RII restriction endonuclease was used to further characterize pBR322. As shown in Fig. 4a (slot 3), pBR322 has five *Eco*RII sites which yield fragments of 1.25, 0.64, 0.53, 0.22 and $0.04 \cdot 10^6$ daltons upon digestion. Slots 3, 4, 5, 6 and 7 (Fig. 4a) show respectively, *Eco*RI, *Hinc*II, *Sal*I, *Bam*HI and *Bgl*I digestions of *Eco*RII-digested pBR322 DNA. In the case of the *Eco*RI endonuclease digest, the largest *Eco*RII fragment when cleaved gives two new fragments of 0.08 and $1.07 \cdot 10^6$ daltons. Slots 5 and 6 show that *Sal*I and *Bam*HI endonucleases cleave the same $0.53 \cdot 10^6$ dalton *Eco*RII fragment as in pBR313 (Bolivar et al., 1977) and generates 0.29 and $0.24 \cdot 10^6$ dalton DNA fragments after the *Sal*I endonuclease digestion and 0.4 and $0.13 \cdot 10^6$ dalton fragments after *Bam*HI endonuclease digestion. Slot 4 shows that the same $0.53 \cdot 10^6$ dalton *Eco*RII piece is cleaved by *Hinc*II endonuclease, generating the same 0.29 and $0.24 \cdot 10^6$ dalton fragments that the *Sal*I endonuclease produces. The $1.25 \cdot 10^6$ dalton *Eco*RII fragment is also cleaved by *Hinc*II endonuclease into 0.82 and $0.43 \cdot 10^6$ dalton fragments. Slot 7 shows the double digestion pattern of pBR322 DNA using *Eco*RII and *Bgl*I endonucleases. Three *Eco*RII fragments are cleaved by *Bgl*I endonuclease into smaller pieces. The largest *Eco*RII fragment ($1.25 \cdot 10^6$ daltons) is cleaved into 0.57 and $0.67 \cdot 10^6$ dalton fragments. *Bgl*I endonuclease also cleaves the $0.53 \cdot 10^6$ dalton fragments into 0.47 and $0.06 \cdot 10^6$ dalton

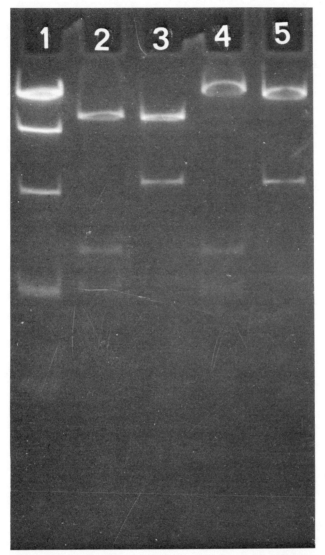

Fig. 2. *Hinc*II-*Eco*RI endonuclease analyses of pBR321 and pBR322. Molecular weight estimates are based in the seven PM2 fragments generated by the *Hind*III endonuclease (31.4, 1.34, 0.6, 0.31, 0.29, 0.14, and 0.06, the last one not seen in the gel, Wes Brown, personal communication) (slot 1). The *Hinc*II endonuclease and *Hinc*II-*Eco*RI endonuclease single and double digestions of pBR322 are shown in slots 3 and 2 respectively while the *Hinc*II endonuclease and *Hinc*II-*Eco*RI endonuclease digestions of pBR321 are shown in slots 5 and 4 respectively. It can be seen that the $0.64 \cdot 10^6$ *Hinc*II band present in pBR322 (slot 3) (see also Fig. 3) is also present in pBR321 (slot 5). This band carries the *Eco*RI site (slots 2 and 4). These data indicate that the spontaneous deletion that generates pBR322 does not extend to the Ap[r] or Tc[r] genes nor the region located in the small *Hinc*II fragment ($0.64 \cdot 10^6$ daltons) in pBR321.

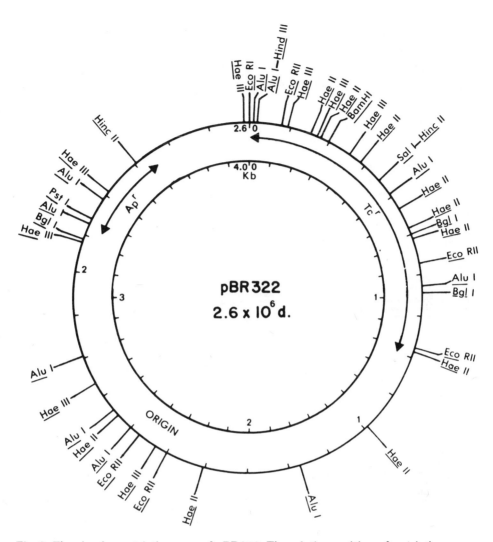

Fig. 3. The circular restriction map of pBR322. The relative position of restriction sites are drawn to scale on a circular map divided into units of $1 \cdot 10^5$ daltons (outer circle) and 0.1 kilobases (inner circle). The estimated size of the Ap^r and Tc^r genes represented in the figure were determined indirectly on the basis of the reported values for the size of the TEM β-lactamase (Datta and Richmond, 1966) and the Tc^r-associated proteins detected in the minicell system (Levy and McMurry, 1974; Tait et al., 1976). Positioning the left-hand boundary of the Tc^r gene was based on our knowledge that cloning into the *Eco*RI site of pBR313 did not affect Tc^r while cloning into the *Hind*III site did affect the Tc^r mechanism. The position and size of the Tc^r region is also consistent with the orientation of the TnA in pBR26 (Bolivar et al., 1977). Only ten out of twelve *Hae*II and *Alu*I and seven out of seventeen *Hae*III substrate sites are represented on the circular map of pBR322. The position of two of the ten *Alu*I sites located at $1.8 \cdot 10^6$ daltons and $1.86 \cdot 10^6$ daltons were mapped on a $0.7 \cdot 10^6$ dalton plasmid which encompasses this region of pBR322 (unpublished observation).

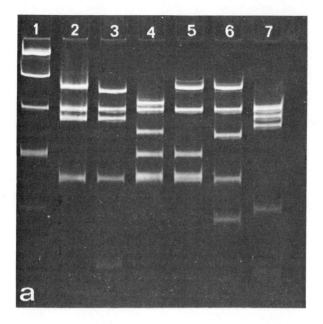

Fig. 4a. Acrylamide gel electrophoresis of EcoRII-cleaved pBR322 DNA. Analysis of
EcoRII endonuclease (slot 2) and double endonuclease EcoRII-EcoRI digestions (slot
3), EcoRII-HincII (slot 4), EcoRII-SalI (slot 5), EcoRIIBamHI (slot 6) and EcoRII-
BgII (slot 7) of pBR322 DNA. The seven HindIII-generated fragments from the DNA
of phage PM2 (slot 1) with molecular weights of 3.5, 1.34, 0.6, 0.31, 0.29, 0.15 and 0.06
(Wes Brown, personal communication) were used as molecular weight standards. For
explanation see the text.

pieces. The $0.22 \cdot 10^6$ dalton EcoRII fragment is cleaved by BgII endonuclease into 0.16 and $0.06 \cdot 10^6$ dalton DNA fragments. These data allow us
to localize four EcoRII sites in the pBR322 map (Fig. 3).

The AluI endonuclease cleaves pBR322 into approximately 12 fragments
(Fig. 4b, slot 2). Four AluI sites were mapped by analysis of the molecular
weights of DNA fragments generated by double endonuclease digestion
(Fig. 3). By comparing the AluI fragments of pBR318 and pBR320 (data
not shown), we have tentatively localized four additional AluI sites in the
map of pBR322. The AluI site located between the EcoRI and the HindIII
sites (Fig. 3) was determined by DNA sequencing (J. Shine et al., unpublished
observations).

Using the same strategy of the analysis of double digestion patterns we
were able to locate 10 out of 12 HaeII sites and the 3 BgII sites present in
the pBR322 map (Fig. 3) (data not shown).

III. Cloning properties of pBR322

(a) Cloning in the Tc^r gene. It has been previously shown (Rodriguez et al.,
1976; Bolivar et al., 1977) that the HindIII, BamHI and SalI sites are present

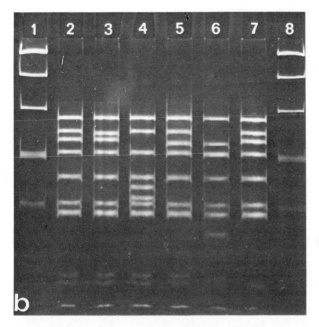

Fig. 4b. Acrylamide gel electrophoresis of *Alu*I endonuclease digested pBR322 DNA.
Analysis of *Alu*I endonuclease (slot 2) and double digestions of *Alu*I-*Eco*RI (slot 3),
*Alu*I-*Bam*HI (slot 4), *Alu*I-*Sal*I (slot 5), *Alu*I-*Hinc*II (slot 6), and *Alu*I-*Pst*I (slot 7) endonucleases of pBR322 DNA. The seven *Hin*dIII generated fragments from the DNA of
phage PM2 (slot 1 and 8) were used as molecular weight standards. It can be seen that the
second largest *Alu*I endonuclease generated fragment (0.43 ∘ 10^6 daltons) is cleaved
by the *Eco*RI endonuclease generating a 0.42 ∘ 10^6 daltons band and a 0.01 ∘ 10^6
dalton fragment not seen in the gel. *Bam*HI endonuclease cleaves the 0.38 ∘ 10^6 dalton
fragment into two fragments of 0.2 and 0.18 ∘ 10^6 daltons. *Sal*I endonuclease cleaves
the same 0.38 ∘ 10^6 dalton *Alu*I fragment generating a piece of 0.34 ∘ 10^6 daltons
and a small one of 0.04 ∘ 10^6 daltons (not seen in the gel). *Hinc*II endonuclease cleaves
two *Alu*I fragments; cleaves in the *Sal*I site generating the same 0.34 ∘ 10^6 dalton *Alu*I
fragment. The relative positions of these fragments can be seen in Fig. 3.

in the Tc^r gene(s) carried by pBR313. Since these sites are in the same relative position in pBR322, we assumed they are associated with the Tc^r gene(s)
present in this plasmid. To confirm this point, DNA fragments from *E. coli*,
Drosophila melanogaster, and *Neurospora crassa* were produced by digestion
with *Hin*dIII, *Bam*HI and *Sal*I restriction endonucleases and cloned into their
respective sites in pBR322. These recombinants were Ap^{r}Tc^s (Table II). Insertion of DNA fragments into the *Eco*RI site, as in pBR313, does not affect
the expression of the Tc^r mechanism.

   (b) Cloning in the Ap^r gene. 1. Cloning into the *Pst*I site. To confirm the
observation that the *Pst*I site in pBR322 is located in the Ap^r gene, we have
cloned several fragments of DNA in *Pst*I site of this plasmid (see Table II).
*Pst*I generated fragments of the plasmid pMB1 (Betlach et al., 1976) were

F. Bolívar et al.

ligated into the *Pst*I site of pBR322 and transformed into RR1. Transformants were selected for Tc[r] and screened for Ap[s] phenotypes. This experiment resulted in the cloning of five different *Pst*I fragments representing nearly the whole pMB1 genome (Table II).

The unique *Pst*I site in pBR322 provides two advantages for the molecular cloning of DNA by means of the homodeoxypolymeric DNA extention technique (Lobban and Kaiser, 1973). First, the *Pst*I site, which has the sequence C T G C A$\downarrow$G (Smith et al., 1976) provides a protruding 3'OH which

TABLE II

MOLECULAR CLONING OF VARIOUS DNA FRAGMENTS IN pBR322

Sources	Restriction endonuclease substrate site					
	*Eco*RI	*Hind*III	*Bam*HI	*Sal*I	*Hinc*II[a]	*Pst*I
N. crassa	2.8	2.0	4.5	2.6		
	1.1	1.8	3.2	1.1		
			2.1	0.8		
			0.5			
pMB1[b]					2.75[c]	
					2.7[d]	
pMB1						2.82[e]
						1.16
						0.7
						0.5
						0.25
pCB61[f]		1.7				
D. melanogaster						2.3[g]
						4

[a] *Hinc*II site located in the β lactamase (Ap[r]) gene.
[b] pMB1 is a clinical isolate plasmid that carries the *Eco*RI restriction and modification genes as well as the colicin E1 production and colicin E1 immunity genes (Betlach et al., 1976).
[c] pMB1 *Hinc*II fragment carrying the *Eco*RI restriction and modification genes (Greene et al., 1976, manuscript in preparation).
[d] pBM1 *Hinc*II fragment carrying colicin E1 production and colicin E1 immunity genes
[e] There are six *Pst*I fragments in pMB1. The sixth has a molecular weight of $0.046 \cdot 10^6$ daltons and was not identified in the agarose gel screening procedure.
[f] pCB61 is a pBR322 derivative plasmid carrying a $1.7 \cdot 10^6$ dalton fragment of DNA cloned in the *Hind*III site. This fragment was isolated from the TnA transposon, and possibly carries a gene(s) involved in the translocation of the ampicillin translocon (Covarrubias et al., unpublished observations).
[g] These fragments were cloned by the homopolymer extension technique in the *Pst*I site of pBR322 and can be recovered after *Pst*I digestion of the recombinant plasmid DNA (A. Dugaiczyk, personal communication).

is a direct substrate for terminal transferase. Secondly, by extending the *Pst*I site in the plasmid with poly dG and the DNA to be cloned with poly dC, it is possible to regenerate two *Pst*I sites after annealing and repairing these two species of DNA in vivo or in vitro.

By fllowing this procedure, dC tailed *Drosophila melanogaster* DNA has been successfully cloned into the dG tailed *Pst*I site of pBR322. Transformants were selected for Tc^r and screened for Ap^s phenotypes. Two out of five *Drosophila melanogaster* DNA fragments cloned in this way had restored *Pst*I sites (Table II).

(b) 2. Cloning in the *Hinc*II sites of pBR322. The *Hinc*II restriction endonuclease which generates blunt-ended DNA fragments, recognizes two sites on pBR322. One of these sites is located in the Tc^r and is also recognized by the *Sal*I endonuclease, while the other site is located in the Ap^r gene, $0.17 \cdot 10^6$ daltons from the *Pst*I site. To demonstrate the cloning of blunt-ended DNA fragments into the *Hinc*II site of the Ap^r gene, we used *Hinc*II digested pMB1 DNA which consisted of two fragments, $2.7 \cdot 10^6$ daltons and $2.75 \cdot 10^6$ daltons (P. Greene, unpublished observations). In order to preferentially cleave the *Hinc*II site in the Ap^r gene, pBR322 DNA was first cleaved with *Sal*I endonuclease followed by a *Hinc*II endonuclease digestion. These digestions generated two DNA fragments each possessing one cohesive end (*Sal*I site) and one blunt end (*Hinc*II site). The pMB1 and pBR322 DNA were ligated under conditions to promote blunt-end ligation (see MATERIALS AND METHODS) and transformed into *E. coli* RRI. After selecting Tc^r transformants and screening for Ap^s, two out of four Tc^{r}Ap^s transformants were found to contain recombinant plasmids carrying each of the two *Hinc*II fragments of pMB1. The two remaining Tc^{r}Ap^s transformants were found to carry pBR322 plasmids with no detectable change in molecular weight but only one *Hinc*II site equivalent to the *Sal*I site. We believe that these plasmids may result from in vivo recombination at the unligated *Hinc*II termini leading to the loss of both the *Hinc*II site and Ap^r.

IV. Properties of pBR322 in the E. coli strain X1776

The plasmids pBR322 and pBR313 could be used in *E. coli* X1776 as an EK2 system if it can be demonstrated that these vectors do not affect the survivability of *E. coli* 1776 (National Institutes of Health, USA, Recombinant DNA Research Guidelines, 1976). Therefore, pBR322 and pBR313 were transformed to the *E. coli* strain X1776 and examined for cell death in medium devoid of DAP, thymine and medium containing bile salts. DAP-less death of *E. coli* X1776 harboring these plasmids is as efficient as in that shown for *E. coli* X1776 alone (Fig. 5). Table III shows the plating efficiencies of the strains X1776 with and without the plasmids on media containing increasing concentrations of bile salts or lacking thymine. As in the DAP-less experiments, neither the pBR322 nor pBR313 affects the plating efficiency of X1776 under these conditions.

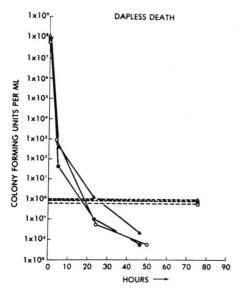

Fig. 5. DAP-less death of X1776, X1776(pBR313) and X1776(pBR322). For each of the strains the following procedure was performed. 500 ml early log cultures (~5 · 10⁷ to 1 · 10⁸ cells/ml) grown in LB + DAP (200 μg/ml) + Thy (50 μg/ml) were centrifuged. The cells were washed with BSG* and resuspended in 5 ml BSG. The resuspended cells were used to inoculate DAP⁻ medium (LB + Thy − NaCl) and DAP⁺ medium (LB + Thy + NaCl) using 0.5 ml of cells to 100 ml medium in each of 10 separate 250 ml erlenmeyer flasks. Cultures were incubated at 37°C with shaking. Samples were removed at various times from a fresh flask each time in order to reduce risk of contamination. The samples were titered on LB agar + DAP + Thy. Time points at 0.2 and 5 h were taken on the control cells grown in medium with DAP to verify their continued growth. At 24 h and later, cells were concentrated by filtering and the Millipore filters placed on LB agar. The dotted line indicates the titer at which the number of colony-forming unit have decreased by 10⁸ from the original titer. △——△, X1776 alone; ○——○, X1776 (pBR313); ●——●, X1776 (pBR322).

Data presented in Table IV show the mobilization of pBR313 and pBR 322 by de-repressed FI or FII R plasmids from "wild-type" E. coli K12 C600. In this case both plasmids are co-resident in the same cell and are mated directly with a wild-type E. coli K12 recipient cell line, SF185. As shown in Table IV, one can demonstrate that if F is coresident with wild-type ColE1, at the time of conjugation, ColE1 is mobilized at a frequency of about 0.5/ recipient cell in 24 h. When F-Km or Rldrd19 are co-resident with pBR313 or pBR322, it is clear that mobilization occurs at a very low frequency. It is also clear that, relative to ColE1, pBR313 and pBR322 are less likely to be directly mobilized from either wild-type E. coli K12 or X1776 by de-repressed F-like plasmids.

Table V shows the transfer of pBR322 and pBR313 from E. coli C600 and E. coli X1776 in triparental matings using conjugative plasmids of various

TABLE III
PLATING EFFICIENCIES OF STRAIN X1776 (pBR313) AND X1776 (pBR322) ON
MEDIA CONTAINING BILE SALTS OR WITHOUT THYMINE

Cultures were grown with shaking at 37°C in LB containing DAP (200 μg/ml) and thymine
(50 μg/ml) to about $1-2 \cdot 10^8$/ml. Cells were pelleted by centrifugation, washed with BSG
and concentrated 100-fold in BSG. Dilutions were made in growth medium and plated on
the appropriate plates. Plates were incubated at 37°C.

Medium	Plating efficiency[a]		
	X1776	X1776 (pBR313)	X1776 (pBR322)
L agar + DAP + Thy + 0.15% bile salts	$<6.25 \cdot 10^{-6}$	$<7.14 \cdot 10^{-6}$	$2.0 \cdot 10^{-7}$
L agar + DAP + Thy + 0.37% bile salts	$<6.25 \cdot 10^{-8}$	$4.29 \cdot 10^{-9}$	$2.87 \cdot 10^{-8}$
L agar + DAP + Thy + 0.75% bile salts	$<6.25 \cdot 10^{-9}$	$2.6 \cdot 10^{-9}$	$2.4 \cdot 10^{-8}$
M9CAA agar + DAP + biotin (0.5 μg/ml)	$2 \quad \cdot 10^{-9}$	$<7.14 \cdot 10^{-8}$	$6 \quad \cdot 10^{-9}$

[a] Plating efficiency was calculated using the titer of X1776, X1776 (pBR313) or X1776
(pBR322) on L agar + DAP + Thy (without bile salts).

incompatibility groups. Since the FII, I and N compatibility groups comprise
over 85% of the conjugative plasmids found in *E. coli* of healthy humans and
animals, we focused upon these and two additional conjugative plasmids.
Given the results shown in Table IV, it is not surprising that we could not
detect ($<10^8$ per final recipient cell in 24 h) any instance for the mobilization of pBR313 and pBR322 from either C600 or X1776 in triparental matings. As a control an intermediate strain carrying the non-conjugative plasmid pRSF2124 was included. pRSF2124, a ColE1 derivative to which the
Ap^r gene has been transposed, could be mobilized by F-Km at a frequency
of 10^{-4} per final recipient cell within 4 h.

DISCUSSION

A new amplifiable cloning vehicle, the plasmid pBR322, which improves
the cloning characteristics of the parental plasmid pBR313 (Bolivar et al.,
1977), has been constructed by in vitro recombination techniques.

pBR322 was constructed in order to have a plasmid cloning vector with a
single *Pst*I site in addition to those unique restriction sites already present
in pBR313. Of the three *Pst*I sites in pBR313, the one located in the Ap^r
gene provided the most useful site for cloning purposes. The advantages of
having a single *Pst*I site located in the Ap^r gene are the following: (1) molecular cloning of *Pst*I endonuclease digested DNA fragments into the *Pst*I
site will lead to the inactivation of the Ap^r gene thus allowing for the detection of cells carrying recombinant plasmids by means of their Ap^s-Tc^r phenotypes; (2) the benefits for molecular cloning by means of homodeoxypoly-

TABLE IV

MOBILIZATION OF pBR313 AND pBR322 by triparental mating[a]

E.coli K12 donor	R plasmid	Inc. group	Drug resistance	Donor cells ml(·10⁸)	C600 1h pBR313	C600 1h pBR322	C600 24h pBR313	C600 24h pBR322	C600 control 24h pRSF2124	X1776 control 4h pRSF2124	X1776 1h pBR313	X1776 1h pBR322	X1776 24h pBR313	X1776 24h pBR322
J5-3	F·Km	FI	Km^r	3	$<10^{-8}$	$<10^{-8}$	$<10^{-8}$	10^{-8}	$2\cdot10^{-4}$	N.D.[b]	$<10^{-8}$	$<10^{-8}$	$<10^{-8}$	$<10^{-8}$
J5-3	Rldrd	FII	Km^r,Su^r, Ap^r, Sm^r, Cm^r	1	$<10^{-8}$	$<10^{-8}$	$<10^{-8}$	10^{-8}	N.D.	$<10^{-8}$	$<10^{-8}$	$<10^{-8}$	$<10^{-8}$	$<10^{-8}$
J5-3	R144	I	Km^{r}Tc^r	5	$<10^{-8}$	$<10^{-8}$	$<10^{-8}$	10^{-8}	$<10^{-8}$	N.D.	$<10^{-8}$	$<10^{-8}$	$<10^{-8}$	$<10^{-8}$
J5-3	N3	N	Tc^r, Sm^r Su^r	5	$<10^{-8}$	$<10^{-8}$	$<10^{-8}$	10^{-8}	$<10^{-8}$	$<10^{-8}$	$<10^{-8}$	$<10^{-8}$	$<10^{-8}$	$<10^{-8}$
J5-3	Sa	W	Sm^r, Su^r, Cm^r, Km^r	2	$<10^{-8}$	$<10^{-8}$	$<10^{-8}$	10^{-8}	$<10^{-8}$	N.D.	$<10^{-8}$	$<10^{-8}$	$<10^{-8}$	$<10^{-8}$
J5-3	RSF1040	X	Ap^r	4.8	$<10^{-8}$	$<10^{-8}$	$<10^{-8}$	10^{-8}	$<10^{-8}$	N.D.	$<10^{-8}$	$<10^{-8}$	$<10^{-8}$	$<10^{-8}$

[a] A triparental mating is one in which the conjugative plasmid resides in one strain (donor), the nonconjugative plasmid to be tested for mobilization in another (intermediate) and a third strain carrying a chromosomal marker serves as the final recipient. Selection is made for the resistance markers present on the nonconjugative plasmid and the chromosome of the final recipient. The intermediate strains were E. coli K12 either C600 or X1776 containing the cloning vehicle pBR313 or pBR322. The final recipient was E. coli K12 W1485-1 Nx^r. Cultures were grown at 37°C in LB plus DAP and thymine. Matings were done by addition to a 250 ml flask of 5 ml each of a culture of the donor strain containing the conjugative R plasmid either C600 or X1776 containing either pBR313 or pBR322 and the final recipient W1485-1. Matings were carried out without shaking for the indicated periods of time. The mating mixtures were plated (0.1 ml) on MacConkey agar containing 20 μg/ml of either tetracycline or amphicillin. C600 (pRSF2124) was included as a control. pRSF2124 is already impaired in its mobilization as compared to ColE1.

Concentrations (cells/ml) of intermediate and final recipient strains:

	C600 (pBR313)	C600 (pBR322)	C600 (pRSF2124)	X1776 (pRSF2124)	X1776 (pBR313)	X1776 (pBR322)	W1485-1
Exp. 1	$9.2\cdot10^8$	$4.7\cdot10^8$	$3.7\cdot10^8$	$2\cdot10^8$	$2\cdot10^8$	$2\cdot10^8$	$1\cdot10^8$

[b] N.D., not done.

meric extension are two-fold. Not only is the protruding 3′OH of the cleaved *Pst*I site a direct substrate for *N*-terminal transferase, but the insertion of C-tailed DNA into a G-tailed plasmid generates two *Pst*I sites after polymerization and ligation in vitro or in vivo. This will allow for the recovery of the cloned DNA fragment after digesting the recombinant plasmid with the *Pst*I endonuclease. In addition, the main properties of pBR313 are conserved in pBR322 in that the cloning of *Hin*dIII, *Sal*I and *Bam*I endonuclease generated fragments in pBR322 continues to inactivate the Tc^r gene. Screening for Tc sensitivity and therefore recombinant plasmids can be imposed on the transformed culture. Cloning of *Pst*I fragments in the appropriate site of pBR322 also inactivates one of the antibiotic-resistance genes. In all of these cases recombinant plasmids in transformed cells possess only one functional antibiotic-resistant gene.

Although the relative positions of the *Eco*RI and *Hin*dIII sites appear to be the same in both pSC101 and pBR322, cloning into the *Eco*RI site of pBR322 does not affect the level of Tc^r as reported for pSC101 (Tait et al., 1977). Cloning of DNA fragments with one *Eco*RI terminus and the second terminus generated by one of the four endonucleases mentioned above into the appropriately digested plasmid will inactivate the Tc^r or Ap^r function.

Recent experiments of Heyneker et al. (1976) have demonstrated the usefulness of molecular cloning by blunt-end ligation. Therefore, the presence in pBR322 of only two substrate sites recognized by the *Hinc*II endonuclease, which generates blunt-ended DNA fragments, makes this plasmid a potential vector for cloning by blunt-end ligation. Cloning by blunt-end ligation has been achieved in the *Hinc*II site of the Ap^r gene when this site was preferentially cleaved by the *Hinc*II endonuclease by prior digestion of the *Hinc*II site in the Tc^r gene with *Sal*I endonuclease.

A low frequency of plasmid transmissibility has been established by the National Institutes of Health, USA, Recombinant DNA Guidelines as one of the most important safety features of a plasmid cloning vector. On the basis of the data presented in this paper, we feel that pBR313 and pBR322 within *E. coli* X1776 constitutes an improved EK2 host-vector system which can be used for the cloning of a variety of DNAs. Although the mechanism for mobilizing non-conjugative plasmids has received little attention, it is interesting to note that the frequency of mobilization of pBR313, pBR322 and pMB9 (R. Curtiss III, personal communication) has been significantly reduced with respect to wild-type ColE1. It has been recently reported (Gordon Dougan, personal communication) that a gene associated with high mobilization frequency has been mapped on the wild-type ColE1 plasmid. Therefore, ColE1 derivatives which involve deletions (pVH51) and enzymatic rearrangements (pMB9, pBR313 and pBR322) may result in the loss or alteration of mobilization. Although transposition of the Ap^r and Tc^r genes of pBR313 and pBR322 cannot be ruled out, experiments designed to test for this possibility have proven negative (data not shown).

ACKNOWLEDGEMENTS

H.W.B. is an Investigator for the Howard Hughes Medical Institute. This work was supported by grants to H.W.B. from the National Science Foundation (PCM75-10468 A01) and National Institutes of Health (5 R01 CA14026-05). H.L.H., R.L.R. and F.B. were supported by postdoctoral fellowships from the Netherlands Organization for the Advancement of Pure Research (ZWO), the A.P. Giannini Foundation for Medical Research, and CONACYT, Mexico, respectively. We would like to acknowledge Dr. Istvan Fodor, David Russel, Alejandra Covarrubias and Linda K. Luttropp for their discussion and technical assistance. We are also grateful to Patricia L. Clausen for her expert preparation of this manuscript.

REFERENCES

Betlach, M.C., Hershfield, V., Chow, L., Brown, W., Goodman, H.M. and Boyer, H.W., A restriction endonuclease analysis of the bacterial plasmid controlling the *Eco*RI restriction and modification of DNA, Fed. Proc., 35 (1976) 2037—2043.

Bolivar, F., Rodriguez, R., Betlach, M. and Boyer, H.W., Construction and characterization of new vehicles, I. Ampicillin-resistant derivatives of the plasmid pMB9, Gene, 2 (1977) 75—93.

Cameron, J.R., Panasenko, S.M., Lehmon, I.R. and Davis, R.W., In vitro construction of bacteriophage λ carrying segments of *Escherichia coli* chromosome: Selection of hybrids containing the gene for DNA ligase, Proc. Natl. Acad. Sci. USA, 72 (1975) 3416 —3420.

Clewell, D.B., Nature of ColE1 plasmid replication in the presence of chloramphenicol, J. Bacteriol., 110(1972) 667—676.

Cohen, S.N., Chang, A.C.Y., Boyer, H.W. and Helling, R., Construction of biologically functional plasmids in vitro. Proc. Natl. Acad. Sci. USA, 70 (1973) 3240—3244.

Covey, C., Richardson, D. and Carbon, J., A method for the deletion of restriction sites in bacterial plasmid deoxyribonucleic acid, Mol. Gen. Genet., 145 (1976) 155—158.

Danna, K.J., Sack, G.H. and Nathans, D., Studies of simian virus 40 DNA, VII. A cleavage map of the SV40 genome, J. Mol. Biol., 78 (1973) 363—376.

Datta, N. and Richmond, M.D., The purification and properties of a penicillinase whose synthesis is mediated by an R-factor in *Escherichia coli*, Biochem. J., 98 (1966) 204.

Dugaiczyk, A., Boyer, H.W. and Goodman, H.M., Ligation of *Eco*RI endonuclease-generated DNA fragments into linear and circular structures, J. Mol. Biol., 96 (1975) 171.

Greene, P.J., Betlach, M.C., Goodman, H.M. and Boyer, H.W., The *Eco*RI restriction endonuclease, in Wickner, R.B. (Ed.), Methods in Molecular Biology, Marcel Dekker, New York, 1974, pp. 87—111.

Greene, P.J., Heyneker, H., Betlach, M.C., Bolivar, F., Rodriguez, R., Covarrubias, A., Fodor, I. and Boyer, H.W., General method for restriction endonuclease purification, Manuscript in preparation.

Gromkova, R. and Goodgal, S.H., Action of *Haemophilus* endodeoxyribonuclease on biologically active deoxyribonucleic acid, J. Bacteriol., 109 (1972) 987.

Guerry, P., LeBlanc, D.J. and Falkow, S., General method for the isolation of plasmid deoxyribonucleic acid, J. Bacteriol., 116 (1973) 1064-1066.

Hamer, D. and Thomas, C., Molecular cloning of DNA fragments produced by restriction endonucleases *Sal*I and *Bam*I, Proc. Natl. Acad, Sci. USA, 78 (1976) 1537—1541.

Heffron, F., Bedinger, P., Champoux, J.J. and Falkow, S., Proc. Natl. Acad. Sci. USA, 74 (1977) 702—706.

Hershfield, V., Boyer, H.W., Lovett, M., Yanofsky, C. and Helinski, D., Plasmid ColE1 as a molecular vehicle for cloning and amplification of DNA, Proc. Natl. Acad. Sci. USA, 71 (1974) 3455—3461.

Hershfield, V., Boyer, H.W., Chow, L. and Helinski, D., Characterization of a mini-ColE1 plasmid, J. Bacteriol., 126 (1976) 447—453.

Heyneker, H.L., Shine, J., Goodman, H.M., Boyer, H.W., Rosenberg, J., Dickerson, R.E., Narang, S.A., Itakura, K., Lin, S. and Riggs, A.D., Synthetic lac operator DNA is functional in vivo, Nature, 263 (1976) 748—752.

Landy, A., Ruedisueli, E., Robinson, L., Foeller, C. and Ross, W., Digestion of deoxyribonucleic acids from bacteriophage T7, λ and φ80h with site-specific nucleases from Haemophilus influenzae strain Rc and strain Rd, Biochemistry, 13(1974) 2134—2141.

Levy, S. and McMurry, L., Detection of an inducible membrane protein associated with R-factor-mediated tetracycline resistance, Biochem. Biophys. Res. Commun., 56 (1974) 1060—1068.

Lobban, P. and Kaiser, A.P., Enzymatic end-to-end joining of DNA molecules, J. Mol. Biol., 78 (1973) 453.

Morrow, J.F., Cohen, S.N., Chang, A.C.Y., Boyer, H.W., Goodman, H.M. and Helling, R.B., Replication and transcription of eukaryotic DNA in Escherichia coli, Proc. Natl. Acad. Sci. USA, 71 (1974) 1743—1747.

Panet, A., Van de Sande, J.H., Loewen, P.C., Khorana, H.G., Raae, A.J., Lillehaug, J.R. and Kleppe, K., Physical characterization and simultaneous purification of bacteriophage T4 induced polynucleotide kinase, polynucleotide ligase and deoxyribonucleic acid polymerase, Biochemistry, 12 (1973) 5045—5050.

Recombinant DNA Research Guidelines, National Institutes of Health, USA, Federal Register, 41, 1976, No. 131, 27901—27943.

Roberts, R.J., Myers, P.A., Morrison, A. and Murray, K., A specific endonuclease from Arthrobacter luteus, J. Mol Biol., 102 (1976) 157—165.

Rodriguez, R.L., Bolivar, F., Goodman, H.M., Boyer, H.W. and Betlach, M.C., Construction of new cloning vehicles, in Nierlich, D.P., Rutter, W.J. and Fox, C.F. (Eds.), Molecular Mechanisms in the Control of Gene Expression, Academic Press, New York, 1976, pp. 471—477.

Sgaramella, V., Van de Sande, J.H. and Khorana, H.G., A novel joining reaction catalyzed by T4 polynucleotide ligase, Proc. Natl. Acad. Sci. USA, 67 (1970) 1468—1475.

Smith, R., Blattner, R.F. and Davis, A., The isolation and partial characterization of a new restriction endonuclease from Providencia stuartii, Nucleic Acids Res. Commun., 3 (1976) 343.

So, M., Gill, R. and Falkow, S., The generation of a ColE1-Ap^r cloning vehicle which allows detection of inserted DNA, Mol. Gen. Genet., 142 (1976) 239—249.

Sugino, A., Cozarelli, N.R., Heyneker, H.L., Shine, J., Boyer, H.W. and Goodman, H.M., Interaction of bacteriophage T4 RNA and DNA ligases in the joining of duplex DNA at base-paired ends, J. Biol. Chem. (1977) in press.

Tait, R.C., Rodriguez, R.L. and Boyer, H.W., Altered tetracycline resistance in pSC101 recombinant plasmids, Mol. Gen. Genet., 151 (1977) 327—331.

Wilson, G.A. and Young, F.F., Isolation of a sequence-specific endonuclease (BamI) from Bacillus amyloliquefaciens H, J. Mol. Biol., 97 (1975) 123—126.

Yoshimori, R., Roulland-Dussoix, D., and Boyer, H.W., R-factor controlled restriction and modification of deoxyribonucleic restriction mutants, J. Bacteriol., 112 (1972) 1275—1283.

16

Reprinted from *Natl. Acad. Sci. (USA) Proc.* **75**:4242–4246 (1978)

Cosmids: A type of plasmid gene-cloning vector that is packageable *in vitro* in bacteriophage λ heads

(bacteriophage λ morphogenesis/restriction analysis)

JOHN COLLINS* AND BARBARA HOHN[†‡]

* Gesellschaft für Biotechnologische Forschung mbH, Mascheroder Weg 1, D-3300 Braunschweig, West Germany; and [†] Biozentrum, University of Basel, Klingelbergstr. 70, CH 4056, Basel, Switzerland

Communicated by William B. Wood, June 26, 1978

ABSTRACT Evidence is presented that ColE1 hybrid plasmids carrying the cohesive-end site (*cos*) of λ can be used as gene cloning vectors in conjunction with the λ *in vitro* packaging system of Hohn and Murray [(1977) *Proc. Natl. Acad. Sci. USA* **74**, 3259–3263]. Due to the requirement for a large DNA molecule for efficient packaging, there is a direct selection for hybrids carrying large sections of foreign DNA. The small vector plasmids do not contribute a large background in the transduced population, which is therefore markedly enriched for large hybrid plasmids (over 90%). The efficiency of the *in vitro* packaging system is on the order of 10^5 hybrid clones per microgram of foreign DNA for hybrids in the 20–30 million dalton range.

The mechanism of packaging DNA into the head of *Escherichia coli* bacteriophage λ has been extensively studied through the development of *in vitro* packaging systems (1–4; for review, see ref. 5). These and studies *in vivo* (6) led to the following findings: monomeric circular DNA was not packaged; head-to-tail polymers (concatemers) of the unit-length λ DNA molecules were efficiently packaged if the cohesive-end site (*cos*), substrate for the packaging-dependent cleavage that produces the cohesive ends of mature λ DNA, was 23–33 megadaltons (MDal) apart; and only a small region in the proximity of the cleavage site was required for recognition by the packaging system (7, 8).

This information implies that *cos*-containing plasmids of less than 23 MDal would not be efficiently packaged due to the circular form of their DNA and their size, but that concatemeric derivatives with DNA inserts would be a packaging substrate. The latter DNA structure resembles a ligation mixture between a cleaved *cos*-containing plasmid and DNA to be cloned. It was expected, therefore, that cloning in a *cos*-containing plasmid in conjunction with *in vitro* packaging selects against re-ligated vector molecules but selects for hybrids in the size range of λ DNA, molecules that are recovered only poorly upon transformation.

In our present study, experiments are described in which a *cos*-containing ColE1 *rpo* plasmid (9, 10) was packaged *in vitro* after restriction and re-ligation. The results of this experiment, as well as of RI plasmid and *Pseudomonas* cloning experiments, suggest the use of packageable plasmids as a gene cloning system that is both highly efficient and selective for recovery of large hybrids.

Plasmids containing a *cos* site, which are useful as vectors for gene cloning in conjunction with the packaging system, we refer to as "cosmids."

MATERIALS AND METHODS

Plasmids and Bacteria. Preparation of plasmids pJC720 and pJC703 (Fig. 1) has been described (9, 10). The detailed mapping of these plasmids with restriction endonucleases is unpublished. *E. coli* N205, an *E. coli* K-12 strain ($r_k^+ m_k^+$ $recA^- su^-$), was from N. Sternberg; strain 5K ($r_k^- m_k^+$ $thr^- thi^-$) was from S. Glover; strain HB101 ($r_k^- m_k^-$ leu^- $pro^- recA^-$) was from H. Boyer; and strain GL1 (*pel21*; W3101) was from S. W. Emmons (11).

Packaging System. Exogenous DNA was packaged *in vitro* as described (12), with some slight modifications: single colonies of strains N205 ($\lambda imm_{434}cI_{ts}b2\ red3\ Eam4\ Sam7$)/λ and N205 ($\lambda imm_{434}cI_{ts}b2red3\ Eam\ 15\ Sam7$/λ were streaked out on LA plates (1) and grown overnight at 30°C. Controls were plated to check temperature sensitivity at 42°C. Single colonies were inoculated into warmed LB medium (1) at an OD_{600} of not more than 0.15 and incubated with shaking until an OD_{600} of 0.3 was reached. Prophages were induced by incubation of the cultures at 45°C for 15 min while standing. Thereafter they were transferred to 37°C and incubated for 3 additional hr with vigorous aeration. (A small sample of each culture, which is lysis-inhibited as a result of the mutation in gene S, was checked for induction: upon addition of a drop of chloroform the culture cleared.) The two cultures were then mixed, centrifuged at 5000 rpm for 10 min, and resuspended at 0°C in 1/500th the original culture volume in complementation buffer (40 mM Tris·HCl, pH 8.0/10 mM spermidine hydrochloride/10 mM putrescine hydrochloride/0.1% mercaptoethanol/7% dimethyl sulfoxide) which was made 1.5 mM in ATP. Biological activity of endogenous DNA can be destroyed by UV irradiation prior to concentration (12). This cell suspension was distributed in 20-μl portions in 1.5-ml Eppendorf polyallomer centrifuge tubes, frozen in liquid N_2, and stored at −60°C. When needed, a sample was transferred in liquid N_2 and put on ice. Immediately on thawing (3–4 min on ice), the DNA to be packaged (0.01–0.2 μg) was added in a volume of 1–5 μl. The DNA was usually added in the buffer in which it had just been ligated. The solutions were carefully mixed and bubbles were removed by a few seconds' centrifugation in an Eppendorf desk-top centrifuge. The mix was incubated for 30 min at 37°C. At the end of this incubation period, 20 μl of a frozen and thawed packaging mixture, which had been made 10 mM in $MgCl_2$ and to which a final 10 μg of DNase per ml was added, was mixed to each sample and incubation was continued for 20–60 min. SMC (1) buffer (0.5 ml) and a drop of chloroform were added. After mixing, denatured material was centrifuged off and the solution was used as a phage lysate.

Abbreviations: MDal, megadaltons; cosmid, plasmid containing a *cos* site.

[‡] Present address: Friedrich Miescher-Institut, P.O. Box 273, CH 4002 Basel, Switzerland.

Transduction was carried out by adding 0.4 ml of this phage suspension to 1 ml of N205 or *pel*⁻ cells from a late exponential culture (OD_{600} = 2.0) in L broth–maltose [1% Bacto-Tryptone/0.5% yeast extract (Oxoid)/0.5% NaCl/0.4% maltose]. For the experiment with *Pseudomonas* DNA, HB101 was used as recipient. After a 10-min adsorption at 30°C, the mixture was diluted 1:20 in fresh L broth and incubated for 2 hr at 30°C to allow expression of rifampicin resistance.

Transformation. Strain 5K was used. Cultures grown to an OD_{600} of 0.5 were cooled rapidly on ice, centrifuged, and resuspended in 0.5 vol of 10 mM NaCl on ice. After 30 min on ice, the cells were centrifuged and resuspended in 0.5 vol of 50 mM $CaCl_2$ and again incubated for 30 min at 0°C. After centrifugation, the cells were resuspended in 0.1 vol of 30 mM $CaCl_2$ in 20% glycerol. This competent cell preparation, divided into 1-ml aliquots, was kept frozen at −60°C until needed. For transformation the sample was thawed out on ice and 0.5 ml of 40 mM Tris, pH 8.0/40 mM NaCl/1 mM EDTA, containing the DNA for the transformation (0.1–1 μg), was added. After 30 min on ice, the mixture was heated to 42°C for 2 min and rapidly cooled on ice. The cells were diluted 1:30 in L broth and incubated for 2 hr at 37°C to allow expression of rifampicin resistance (13).

Rifampicin resistance was tested on L broth plates containing either 100 μg of rifampicin per ml, when plasmid pJC703 was used, or 30 μg/ml, when plasmid pJC720 was used. The colonies derived after transduction or transformation of pJC720 grow slowly on rifampicin, taking 2 days at 37°C to form large colonies. Since rifampicin is light sensitive, the plates must be kept dark during this prolonged incubation to prevent growth of background colonies.

Restriction and Ligation Reactions. Restriction with *Hin*dIII (Boehringer) was carried out in 30 mM Tris·HCl, pH 7.6/10 mM $MgCl_2$/10 mM NaCl to completion. Digestion with *Sal* I and *Eco*RI was carried out in the same buffer. *Sal* I and *Eco*RI were generous gifts from H. Mayer and H. Schütte, and *Bgl* II was a generous gift from E. Eichenlaub. Digestion with *Kpn* I (Bio-Labs) was in 10 mM Tris·HCl, pH 7.9/6 mM $MgCl_2$/6 mM NaCl/10 mM dithiothreitol containing 100 μg of bovine serum albumin per ml. Gel electrophoresis was in 1% agarose (14).

Ligation was carried out after heat inactivation of the restriction endonucleases at 70°C for 10 min in 6 mM $MgCl_2$, 10 mM Tris·HCl (pH 7.9), 10 mM dithiothreitol, 100 μM ATP, 100 μg of bovine serum albumin per ml, and 5×10^{-2} unit of T4 DNA ligase (Boehringer) in 100-μl aliquots. The concentration of DNA ends was between 20 and 60 pM or as described in the text. Before the ligase was added the samples were mixed, heated to 70°C for 5 min, and cooled on ice for 30 min. The ligation reaction was continued for 15 hr at 8°C. Completion of the ligation was checked by agarose gel electrophoresis: no sample was used in which linear monomers could still be detected.

Nomenclature of Plasmids. The numbering of the plasmids in the pJC series coming from John Collins will be confined to the first 500 in each thousand, with the exception of those already published, so as to avoid confusion with the collection of Alvin John Clark; pJC703 and pJC720 are two of these exceptions.

Safety Regulations. All experiments described here were carried out under P1 conditions as defined by the National Institutes of Health guidelines for recombinant DNA research.

RESULTS

Packaging of restricted and re-ligated plasmid DNA

Production of Packageable Substrate from Plasmid DNA. Plasmid pJC703 (Fig. 1) yields two *Hin*dIII restriction fragments: the 10-MDal fragment A containing the λ *cos* site and the ColE1 replicon, and the 7-MDal fragment B containing the gene for rifampicin resistance (*rpoB*). It was hoped that cleavage and re-ligation of plasmid pJC703 would produce a population of polymers (some of which are diagrammed in Fig.

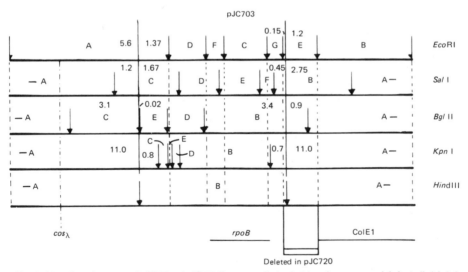

FIG. 1.   Restriction endonuclease map of pJC703 and pJC720. Fragments obtained with each enzyme are alphabetically labeled according to size. Dashed lines indicate the relative positions of the *Eco*RI cleavage site on each map, and the continuous vertical lines the positions of the *Hin*dIII cleavage sites. The distances from the *Hin*dIII sites to the nearest cleavage sites for each restriction enzyme are indicated in MDal. These values are used in the analysis of plasmids containing polymeric *Hin*dIII fragments (Table 2). The ColE1 part of these plasmids is actually derived from a freak isolate (pJC309) which contains a *Sal* I site not present in ColE1.

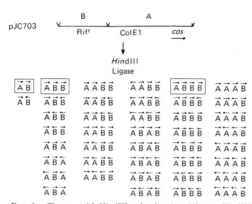

FIG. 2. Cleavage with HindIII and re-ligation of the two HindIII fragments (A and B) of plasmid pJC703 theoretically can yield a series of polymers. The diagram shows all the possible cyclic permutations that contain at least one A and one B fragment, up to the tetrameric forms. The relative orientations are indicated by the small "matchsticks" over each fragment designation. Those structures actually obtained are framed.

2) that could mimic the natural substrate for packaging. Cleavage of such molecules at the *cos* site during packaging, and recircularization subsequent to transduction, would lead to the loss of one or more entire A fragments, thus generating plasmids of the form ABB, ABBB, and AAB (or AB).

After the re-ligated HindIII fragments of pJC703 were packaged, several thousand rifampicin-resistant clones were obtained by transduction into N205. The yield of Rif^r clones is dependent on the concentration of the vector DNA during the ligation (Exp. 1 *a* and *b*, Table 1). This supports the hypothesis that efficient packaging is dependent on the formation of long polymers. In contrast, the formation of such highly polymerized chains is most detrimental to the efficiency of transformation, as has been noted elsewhere (14). The transformation data are included merely as an additional test that the ligation was successful rather than as a direct comparison of packaging and transformation, since the ligation conditions are strongly biased in favor of packaging experiments.

Structure of Packaged Plasmids. Fifty-two colonies were picked at random for further testing. They were all found to be colicin E1 resistant and colicin E2 sensitive, indicative that the plasmid coded E1-immunity carried on the HindIII A fragment was present. Small cleared lysates were made from each clone. To check for the presence and approximate size of the plasmid DNA, we electrophoresed 5-μl samples (plus 0.1% sodium dodecyl sulfate) on 0.8% agarose gels (14). Supercoiled DNA was prepared from the first 12 samples and, from the remaining 40, from those showing the presence of plasmids larger than pJC703. These DNAs (Exp. 2, Table 1), and in some cases the products of a second packaging step, were analyzed more thoroughly with the restriction enzymes Bgl II, Kpn I, Sal I, EcoRI, and HindIII (Fig. 3).

Three size classes of plasmid were obtained: 17 MDal, corresponding to the starting plasmid, and about 23 MDal and 29 MDal. To test the orientation of the fragments with respect to one another, we cleaved the plasmids with an enzyme other than HindIII. This would generate fragments overlapping the HindIII junctions and yield "junctional" fragments diagnostic of the HindIII fragment arrangement (Table 2 and Fig. 3).

Only a small number of the possible structures (Fig. 2) are found amongst the large number analyzed: $\overrightarrow{A}\overrightarrow{B}$, $\overrightarrow{A}\overrightarrow{B}\overrightarrow{B}$, and $\overrightarrow{A}\overrightarrow{B}\overrightarrow{B}\overrightarrow{B}$ (Table 2 and Fig. 3). Absent are plasmids containing duplicates of the HindIII A fragment, i.e., that fragment carrying the replicon origin and the *cos* site. Such structures would be eliminated if every *cos* site were cleaved during *in vitro* packaging. The *in vivo* probability, however, of cutting a pair of *cos* sites decreases as the amount of DNA between them decreases (6). The existence of plasmids having tandem ColE1 origins (unpublished observation) would support the argument that there is no *a priori* reason to expect plasmids containing two A fragments to be unstable. The absence of the double A combination is therefore not easily explained.

The absence of the opposed orientation of the AB fragments, namely, $\overrightarrow{A}\overleftarrow{B}$, may be due to the dependence of expression of the Rif^r *rpoB* gene on readthrough transcription from the A fragment in the correct orientation (9).

The absence of all palindromic structures (perfect inverted repeats extending to the axis of the symmetry) is also remarkable, as is the high number of deletions found (four from 52 isolates). Whether or not the elimination of palindromic

Table 1.   Transformation and packaging efficiencies of plasmids containing *cos* sites, before and after cleavage with HindIII and ligation with DNA ligase

Exp.	DNA	MDal of vector	DNA, μg/ml	Transformation		Packaging	
				Rif^r colonies/ μg DNA	% hybrids	Rif^r colonies/ μg DNA	% hybrids
1a	pJC703 × HindIII, ligated	17	500	30	NT	1.0×10^5	~50
1b	pJC703 × HindIII, ligated	17	50	1.4×10^3	NT	1.0×10^2	
2	Supercoils of pJC703,						
	configuration AB	17		1.4×10^5		1×10^2	
	ABB	23		1.7×10^4		4×10^3	
	ABBB	29		1.8×10^4		1×10^4	
3	pJC720 × HindIII	16	146				
	+	} Ligated	+	4.1×10^2	0.15	2.4×10^3	~90
	RIdrd19 × HindIII		17				
4	pJC720 × HindIII	16	330				
	+	} Ligated	+	NT		5×10^3	~80
	Pseudomonas AM1 × HindIII		75				

Transformants or (subsequent to packaging) transductants were selected for on media containing rifampicin. Yields are given as Rif^r colonies per μg of input (vector and foreign) DNA. In Exp. 1, the percentage hybrid clones refers to the percent containing more than one copy of the HindIII B fragment. About 90% of the Rif^r colonies from Exp. 3 also contained the 11.5-MDal HindIII fragment from RIdrd19, which carries ampicillin resistance. The efficiency of packaging λb2 DNA in parallel experiments was about 10^7–10^8 plaque-forming units per μg of input DNA. NT, not tested.

Biochemistry: Collins and Hohn

Proc. Natl. Acad. Sci. USA 75 (1978) 4245

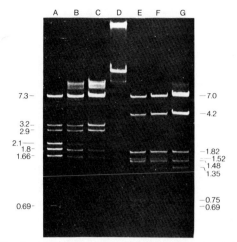

FIG. 3. Agarose gel (1%) electrophoresis of products of restriction endonuclease digestions. (Gels A, B, and C) Digests of *Sal* I endonuclease; (gels E, F, and G) digests of *Eco*RI; (gel D) uncut supercoiled pJC703. Gels A and E contain digests of a 29-MDal isolate derived from Exp. 2 in Table 1; B and F contain digests of a 23-MDal plasmid from the same experiment; C and G contain digests of pJC703. MDal are indicated.

structures, which are certainly present in the ligation mixture, takes place during the packaging step or after the transduction is not known.

Size Selectivity of Packaging. Of the 52 Rif^r clones from the experiment of packaging *Hin*dIII cleaved and re-ligated pJC703 DNA, 14 were of the 23-MDal ABB class, 7 were of the 29-MDal ABBB class, 4 were of intermediate sizes showing also aberrant fragments indicating deletions, and 27 were indistinguishable in size from pJC703. Of these latter, five were tested by restriction enzyme analysis and appeared identical to pJC703.

An even stronger size selection was obtained when Exp. 1 (Table 1) was repeated with a *pel*$^-$ host as recipient. The *pel*$^-$ mutation increases the DNA size dependency of DNA injection by lambdoid bacteriophages (15). From 29 Rif^r *pel*$^-$ clones tested, 19 had plasmids in the 24- to 25-MDal range and 8 in the 29- to 30-MDal range, with a single plasmid of 17 MDal. With the exception of three clones that had small deletions at the junction of the tandemly repeated B fragment and the A

fragment, all of the larger plasmids were found to be of the form A$\overset{\leftrightarrow}{B}$B or A$\overset{\leftrightarrow}{B}$BB.

Considering A as a cosmid vector molecule and B as foreign DNA, Exp. 1 (Table 1) can be taken as a model cloning situation. The optimum yield of "hybrid" clones in this experiment would therefore be 3×10^5 per µg of the (foreign) B fragment.

Packaging of Supercoiled DNA. Packaging of supercoiled DNAs of the plasmids (Exp. 2, Table 1) is several orders of magnitude lower than packaging of λ DNA (10^7–10^8 plaque-forming units/µg of λ DNA in parallel experiments), the efficiency for the smallest plasmid being the lowest. Moreover, their structure appears to be unaltered by the packaging-transduction step, as shown by restriction analysis of the supercoiled plasmids isolated from the transductants. Earlier studies on *in vivo* and *in vitro* packaging (summarized in ref. 5) led to the conclusion that supercoiled DNA with a single *cos* site is not packageable without a recombination step, although a low level of *in vivo* packaging of monomeric circular DNA has recently been reported (16). We do not known whether or not the low level of *in vitro* packaging of circular DNA is dependent on a low level of dimer or higher multimers in the supercoiled preparation.

Cloning of RI*drd*19 DNA

The high percentage of larger hybrids in Exp. 1 (Table 1) is evidence that a size selection is occurring in the packaging of cosmid–hybrid DNA. This size dependency was further tested by using the cosmid pJC720 (16 MDal), which contains a single *Hin*dIII site (Fig. 1), to clone fragments from the R factor RI*drd*19 (Exp. 3, Table 1). The *Hin*dIII fragments generated from this plasmid are 42.8, 11.5, 2.9, 2.0, 1.95, 1.8, 0.15, and 0.1 MDal (17). The 11.5-MDal fragment carries the gene for ampicillin resistance.

Of the Rif^r clones obtained after packaging, 90% were also found to be ampicillin resistant and therefore to be carrying at least the 11.5-MDal *Hin*dIII fragment from RI*drd*19. It would seem, therefore, that a very strong size selection had been imposed by packaging, in which the 27.5-MDal (to 30-MDal?) hybrids were produced in preference to the 16- or 19-MDal plasmids. In Exp. 3, the yield is about 2×10^4 *hybrid* clones per µg of *foreign* DNA, even though 80% of the fragments in this mixture were probably either too small (0.1–2.9 MDal) or too large (42.8-MDal) to be efficiently transduced by this method. The efficiency of packaging hybrid cosmids is therefore on the order of 10^5 per µg of foreign DNA of the correct length.

Transformation with the same DNA yielded few Rif^r Amp^r hybrids, the overall transformation efficiency being low due to the high DNA concentration used during ligation.

Cloning of *Pseudomonas* DNA

pJC720 was used to clone fragments from *Pseudomonas* AM1 chromosomal DNA partially digested with *Hin*dIII (Exp. 4, Table 1). By gel electrophoresis it was estimated that more than 80% of the *Pseudomonas* fragments were larger than 16 MDal and probably too large to be clonable by packaging with this vector. In spite of this, the efficiency of *hybrid* formation is about 3×10^4 per µg of foreign DNA. Twenty-seven of the first 32 clones tested carried new DNA fragments. The average size of the DNA insert in these 27 was 10 MDal. On this basis, a few hundred of the clones obtained should constitute a gene bank (18) of *Pseudomonas* AM1 chromosomal DNA in *E. coli*.

Table 2. Detection of different possible molecular forms in plasmids containing polymeric regions

| Molecular form | Junctional fragments Expected (MDal) | | Found |
	Sal I	*Eco*RI	
A$\overset{\leftrightarrow}{A}$	3.95	6.8	−
A$\overset{\leftrightarrow}{A}$	5.5	2.4	−
A$\overset{\leftrightarrow}{A}$	2.4	11.2	−
A$\overset{\leftrightarrow}{B}$	3.2	1.35	+
A$\overset{\leftrightarrow}{B}$	4.42	2.6	−
A$\overset{\leftrightarrow}{B}$	1.65	5.7	−
B$\overset{\leftrightarrow}{B}$	2.12	1.52	+
B$\overset{\leftrightarrow}{B}$	3.34	2.74	−
B$\overset{\leftrightarrow}{B}$	0.9	0.3	−

Plasmids containing the molecular form indicated (after the convention adopted for Fig. 2) would produce "junctional" fragments of the indicated MDal.

143

DISCUSSION

We have demonstrated that the packaging of plasmid DNA in λ bacteriophage particles can be used as a method for obtaining plasmid hybrids in the 20- to 30-MDal size range when using plasmid DNA that has been linked *in vitro* to foreign DNA fragments. The yield of clones containing these hybrids is of the order of 3×10^5, under optimal conditions, per μg of foreign DNA. Furthermore, by the use of small plasmids (less than 8 MDal) that are themselves very inefficiently packaged (unpublished results), the background of nonhybrid clones is effectively eliminated in a single step without resort to either modification of the DNA (e.g., alkaline phosphatase treatment or polynucleotide tailing) or to elaborate selection or screening procedures which are usually the most time-consuming steps in plasmid cloning experiments. In addition, the use of small cosmids will allow efficient recovery of cloned fragments in the size of up to 25–30 MDal, the selection being imposed by the requirement for packaging of a full or nearly full head.

In vitro packaging of λ cloning vectors can be made independent (12) or dependent (19) of the size of the DNA in the range of 24–30 MDal, but a lower size limit for the vector is set by the requirement for the bacteriophage genes for plaque formation. This requirement is circumvented in the cosmid cloning system, which is independent of phage genes responsible for lytic growth. The space thus provided can be taken up by DNA to be cloned.

Because of the small region required for plasmid replication it is to be expected that new derivatives for use with other restriction enzymes will be rapidly developed. Cosmid derivatives of pJC720 and pJC703 have been produced in which cloning with *Bgl* II or *Bam*HI can be carried out by using rifampicin selection, with *Sal* I, *Eco*RI, *Bgl* II, or *Bam*HI by using ampicillin selection, or with *Xma* I, *Kpn* I, and *Pst* I by using selection for colicin immunity (unpublished results). In addition, a series of cosmid vectors have been developed (unpublished results), including an 8-MDal cosmid (pJC75-58) for use with *Eco*RI, *Bam*HI, and *Bgl* II which is temperature sensitive, ampicillin resistant, and mobilization-minus. It is hoped that in conjunction with incapacitated host strains (20) this latter cosmid will provide an EKII host-vector system that will be most effective in the production of gene banks of eukaryotic DNA.

The packaging of cosmids in λ particles should allow the use of many standard genetic tricks, previously only applicable to λ, for the selection of deletions or insertions in cloned fragments. Such selection methods are based on the instability of full bacteriophage heads in chelating agents, the positive selection for large molecules on infection of *pel⁻* hosts, or the physical separations possible on the basis of density differences between full and partially filled λ particles.

We thank Hildburg Stephan for her expert technical assistance, Ulla Hartmann for preparation of the partially *Hind*III-cleaved *Pseudomonas* AM1 DNA, and Dietmar Blohm and Renate Bonewald for preparation of the R1*drd*19 plasmid DNA. Financial support was provided in part by a grant to T. Hohn from the Swiss National Fonds.

1. Hohn, B. & Hohn, T. (1974) *Proc. Natl. Acad. Sci. USA* **71**, 2372–2376.
2. Kaiser, D., Syvanen, M. & Masuda, T. (1975) *J. Mol. Biol.* **91**, 175–186. ˙
3. Becker, A., Murialdo, H. & Gold, M. (1977) *Virology* **78**, 227–290.
4. Becker, A., Marko, M. & Gold, M. (1977) *Virology* **78**, 291–305.
5. Hohn, T. & Katsura, I. (1977) *Curr. Top. Microbiol. Immunol.* **78**, 69–110.
6. Feiss, M., Fisher, R. A., Crayton, M. A. & Egner, C. (1977) *Virology* **77**, 281–293.
7. Syvanen, M. (1974) *Proc. Natl. Acad. Sci. USA* **71**, 2496–2499.
8. Hohn, B. (1975) *J. Mol. Biol.* **98**, 93–106.
9. Collins, J., Fiil, N. P., Jørgensen, P. & Friesen, J. D. (1976) in *Control of Ribosome Synthesis*, Alfred Benson Symposium 9, eds. Maaløe, O. & Kjeldgaard, V. O. (Munksgaard, Copenhagen), pp. 356–369.
10. Collins, J., Johnsen, M., Jørgensen, P., Valentin-Hansen, P., Karlström, H. O., Gautier, F., Lindenmaier, W., Mayer, H. & Sjöberg, B. M. (1978) in *Microbiology 1978*, eds. David, J. & Novik, R. (American Society of Microbiology, Washington, DC), pp. 150–153.
11. Emmons, S. W., Maccosham, V. & Baldwin, R. L. (1975) *J. Mol. Biol.* **91**, 133–146.
12. Hohn, B. & Murray, K. (1977) *Proc. Natl. Acad. Sci. USA* **74**, 3259–3263.
13. Morrison, D. A. (1977) *J. Bacteriol.* **132**, 349–351.
14. Collins, J. (1977) *Curr. Top. Microbiol. Immunol.* **78**, 121–170.
15. Emmons, S. W. (1974) *J. Mol. Biol.* **93**, 511–525.
16. Umene, K., Shimada, K. & Takagi, Y. (1978) *Mol. Gen. Genet.* **159**, 39–45.
17. Blohm, D. (1978) *Proceedings of the 2nd International Symposium on Microbeal Drug Resistance*, ed. Mitsuhashi, S. (University Press, Tokyo), Vol. 2, in press.
18. Clarke, L. & Carbon, J. (1976) *Cell* **9**, 91–99.
19. Sternberg, N., Tiemeier, D. & Enquist, L. (1977) *Gene* **1**, 255–280.
20. Curtiss, R., III, Inoue, M., Pereira, D., Hsu, J., Alexander, L. & Rock, L. (1977) in *Molecular Cloning of Recombinant DNA*, eds. Scott, W. & Lerner, R. (Academic, New York), pp. 99–114.

Transformation of yeast by a replicating hybrid plasmid

Jean D. Beggs

Department of Molecular Biology, University of Edinburgh, Edinburgh, UK and Department of Cytogenetics, Plant Breeding Institute, Trumpington, Cambridge, UK

Chimaeric plasmids have been constructed containing a yeast plasmid and fragments of yeast nuclear DNA linked to pMB9, a derivative of the ColEl plasmid from E. coli. Two plasmids were isolated which complement leuB mutations in E. coli. These plasmids have been used to develop a method for transforming a leu2 strain of S. cerevisiae to Leu⁺ with high frequency. The yeast transformants contained multiple plasmid copies which were recovered by transformation in E. coli. The yeast plasmid sequence recombined intramolecularly during propagation in yeast.

PROCEDURES for cloning DNA segments in *Escherichia coli* by insertion into a plasmid or bacteriophage genome are now well established and have been used to isolate many prokaryotic and eukaryotic DNA sequences. However, many eukaryotic genes do not have counterparts in bacteria and therefore cannot be isolated by complementation of a bacterial deficiency, and

sequences involved in regulation or development may only be usefully studied in a homologous genetic background.

A yeast cloning system would be extremely useful for the isolation of genes and regulatory sequences from yeast and other eukaryotic organisms and for the genetic manipulation of commercially important yeasts. Recently yeast cells were transformed at a frequency of 10^{-7} transformants per viable cell using a derivative of the bacterial plasmid ColEl carrying the yeast *LEU*2 gene which codes for β-isopropylmalate dehydrogenase[1]. The transforming plasmid was apparently not present in yeast clones in the free or non-integrated form but was integrated into yeast chromosomes at more than one site.

I describe here the development of a highly efficient yeast transformation system using yeast–bacterial hybrid plasmids which are able to replicate and may be selected in and recovered from both *E. coli* and *Saccharomyces cerevisiae*. The hybrid plasmids incorporate a yeast plasmid, 2 μm long (approximately 6 kilobases)[2–5] which was previously cloned in bacteriophage λ and analysed by restriction endonuclease digestion to determine its suitability as a cloning vehicle[6].

The yeast plasmid is present in many strains of *S. cerevisiae*

Fig. 1 Plasmid pMB9, the 2-μm plasmids from *S. cerevisiae*, and the hybrid plasmids used to clone nuclear DNA fragments from *S. cerevisiae*. The physical map of pMB9 is essentially that described by Bolivar *et al.*[15] and the *Hpa*I endonuclease target was mapped in this work. The tetracycline resistance gene is represented by Tc^r. The *S. cerevisiae* plasmid types A and B which are presumably interconvertible are represented as dumbells with the inverted repeat regions corresponding to the stem portion. The sizes of *Eco*RI restriction digest fragments are given as kilobases and were reported previously[6]. The structures are shown of the four hybrid plasmids chosen for further cloning experiments out of the eight possible configurations. To construct these hybrid plasmids, purified[16] yeast plasmid DNA was restricted with *Eco*RI endonuclease[19] under conditions such that digestion was incomplete and the majority of the products were in the form of full length linear molecules as determined by agarose gel electrophoresis[19]. Plasmid pMB9 DNA purified from chloramphenicol-treated cultures[20] by a cleared lysate procedure[21] was digested with *Eco*RI endonuclease and the two digested DNA samples were mixed and ligated[22]. *E. coli* strain ED8767 (*recA56 metB⁻ hsdS⁻ supE supF*; W. J. Brammar) grown in L-broth[22] was made competent by calcium chloride starvation[23] and transformed by heat shock treatment with DNA at 0.3 μg ml⁻¹. Cells were diluted threefold with L-broth and incubated at 37 °C for 40 min before plating to select for transformants on fresh L-broth agar containing tetracycline hydrochloride at 15 μg ml⁻¹. Transformant clones were screened for the presence of the yeast plasmid sequence by *in situ* hybridisation[24] with a ³²P-labelled probe prepared by nick translation[25] of DNA from a lambda clone of the yeast plasmid[6]. Clones of plasmids containing the yeast sequence were further analysed to identify those in which the complete yeast plasmid was present. The size of plasmid in each clone was estimated by gel electrophoresis of colony lysates[26]. Clones with full-size hybrid plasmids were further characterised by restriction analysis of the plasmid DNA.

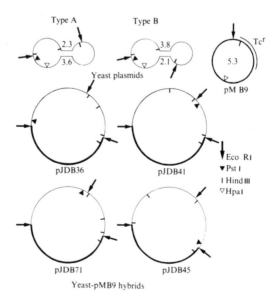

Type A   Type B

Yeast plasmids

pM B9

pJDB36   pJDB41

pJDB71   pJDB45

Yeast-pMB9 hybrids

Eco RI
Pst I
Hind III
Hpa I

Fig. 2 Cloning of *leu*B-complementing sequences from yeast nuclear DNA into hybrid plasmids. *Eco*RI endonuclease digests of (1) pMB9 DNA, (2) pJDB36 (type A hybrid plasmid) DNA, (3) pJDB41 (type B hybrid plasmid) DNA, (4) pJDB248 (type A leu$^+$ plasmid), (5) pJDB219 (type B leu$^+$ plasmid) and (6) *Eco*RI endonuclease digest of λ DNA mixed with a *Hpa*I endonuclease digest of λ DNA. DNA fragment sizes are given in kilobases and faint bands are indicated by arrows. In the diagram only yeast sequences in pJDB248 and pJDB219 are shown. The yeast plasmid sequence is represented by a continuous line and the yeast nuclear DNA inserts by discontinuous lines. Arrows in the diagram represent sites for *Eco*RI endonuclease. Nuclear DNA (a gift from C. Christiansen) from *S. cerevisiae* strain M127 (prototrophic) was sheared in a Virtis 60K homogeniser[27] such that the major fraction of sheared DNA was in fragments 6 to 15 kilobases long with a smaller proportion of fragments of length ranging to less than 1 kilobase. A mixture of the DNAs of the hybrid plasmids pJDB36, pJDB41, pJDB45 and pJDB71 was digested with *Pst*I endonuclease. Both the yeast and the plasmid DNAs were extracted with phenol, extracted with ether and extensively dialysed against 0.1 M potassium cacodylate pH 7.0. Poly(dA) tails were added to the yeast DNA and poly(dT) tails were added to the plasmid DNAs by incubation with calf thymus terminal deoxynucleotidyl transferase. The reaction conditions were as described by Lobban and Kaiser[28], except that the temperature of the poly(dT) reaction was lowered to 0 °C, and in both reactions MgCl$_2$ was replaced by 0.5 mM CoCl$_2$ (ref. 29). The reactions were stopped by the addition of EDTA when 50 nucleotides had been incorporated per 3′ terminus. The tailed DNAs (0.37 μg of each DNA ml^{-1}) were annealed in 10 mM Tris, 1 mM EDTA, 100 mM NaCl pH 8 by incubating at 65 °C for 10 min and allowing to cool to room temperature in the same water bath. The DNA was cloned in *E. coli* JA221 by selection of tetracycline-resistant transformants. A total of 21,000 clones were stored in batches of 500 at −80 °C in 12.3 g per l K$_2$HPO$_4$, 0.9 g per l sodium citrate, 0.18 g per l MgSO$_4$.7H$_2$O, 1.8 g per l (NH$_4$)$_2$SO$_4$, 3.6 g per l KH$_2$PO$_4$, 88 g per l glycerol (D. Hogness, personal communication). In addition 5,000 of these clones were maintained individually in L-broth containing 7% dimethylsulphoxide in microtitre plates at −20 °C. Clones with plasmids complementing the *leu*B6 mutation in *E. coli* JA221 were selected by growth of samples from the batches of clones on minimal selective plates (the minimal medium of Spizizen[30] was used with 0.2% w/v glucose, 1.5% Difco Bacto agar and 20 μg ml^{-1} L-tryptophan). The *leu*B6-complementing plasmids (pJDB219 and pJDB248) were analysed by digestion of the DNAs with *Eco*RI, *Hin*dIII, and *Hpa*I endonucleases and gel electrophoresis to determine the configuration of the hybrid plasmid in each case. Electrophoresis was through a 1% agarose horizontal slab gel (18 × 24 × 0.5 cm) in 40 mM Tris, 20 mM sodium acetate, 1 mM EDTA pH 8.2 at 100 V per 30 mA for 15 h. The gel was stained by soaking in a solution of ethidium bromide (0.5 μg ml^{-1}) and photographed with a UV light source below.

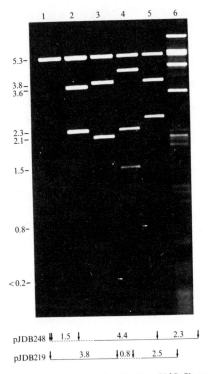

but has no known function. The plasmid has the same density as main band nuclear DNA[2,5] and replicates under nuclear control[7,8], there being 50 to 100 plasmid copies per cell[9]. It contains an inverted repeat sequence 600 base pairs long and is isolated as a mixture of two forms which differ from one another in the orientation of the two nonrepeated segments of the molecule with respect to each other. The two forms are presumably due to recombination between the inverted repeat sequences[10–12]. Plasmids isolated from several different strains of *S. cerevisiae* differ from each other by small deletions or point mutations which alter the restriction enzyme cleavage pattern of the DNAs[13,14].

Construction of chimaeric plasmids

The plasmids were constructed in two stages. First, pMB9, a derivative of the ColE1 plasmid from *E. coli*[15], and the plasmid from *S. cerevisiae* DR19/8T[16] were joined by ligation of the DNA fragments produced by *Eco*RI endonuclease (Fig. 1). The pMB9 sequence includes a tetracycline resistance determinant and tetracycline-resistant clones were selected after transforming *E. coli* with the ligated DNAs. These clones were screened for a hybrid plasmid large enough to carry the complete yeast plasmid sequence.

Hybrid yeast–pMB9 plasmids theoretically have eight possible configurations. These are determined by whether the yeast sequence is in type A or type B configuration, by which of the two *Eco*RI endonuclease targets joins the sequence to pMB9, and by the orientation of insertion into pMB9. These configurations can be differentiated by the digestion pattern of the plasmid DNA with *Hin*dIII or *Hpa*I endonucleases. Five of

these different configurations were found amongst seven complete hybrid plasmids examined. The significance of the two forms (A and B) of yeast plasmid was not known. Nor was it known whether cloning DNA fragments into either of the *Eco*RI sites of the yeast plasmid would inactivate a sequence required for the maintenance of the plasmid in yeast or for some other essential function. Therefore, four of the different hybrid plasmids, pJDB36, pJDB41, pJDB45 and pJDB71 (Fig. 1), were used in the second stage of the construction of the plasmids to be used in the transformation of yeast.

Sheared nuclear DNA from *S. cerevisiae* of average fragment length 6 to 15 kilobases was linked by means of poly(dA·dT) tails[17] to a mixture of DNAs of pJDB36, pJDB41, pJDB45 and pJDB71 which had been converted to linear molecules by digestion with *Pst*I endonuclease. Tetracycline resistant clones were selected after transformation of *E. coli* JA221 (recA1, *leu*B6, *trp*ΔE5, *hsd*R$^-$ *hsd*M$^+$ *lac*Y C600; J. Carbon) with the annealed DNAs. A total of 21,000 clones were obtained, and those which were phenotypically leu$^+$ were isolated by selection on minimal plates. Two plasmids, pJDB219 and pJDB248, were isolated which contained a sequence able to complement the *leu*B6 mutation in strains JA221 and C600. Both plasmids transformed strain JA221 to leu$^+$ or to tetracycline resistance with equal efficiency (10^5 transformants per μg DNA). The plasmids also complemented the *leu*B61 and *leu*B401 alleles[18] at the same frequency as transformation to tetracycline resistance in the same strains.

Plasmid pJDB248 was pJDB36 with an insertion of approximately 2.4 kilobases containing two sites for *Eco*RI endonuclease which were 1.5 kilobases apart (Fig. 2). Plasmid

JDB219 was formed by an insertion of approximately 1.2 kilobases (containing one target for *Eco*RI endonuclease) in JDB41.

Minor bands corresponding to DNA sequences of 3.8 and 2.9 kilobases were observed in varying amounts in *Eco*RI digests of different preparations of pJDB248 DNA from cultures produced from purified colonies. These corresponded to the fragment sizes expected if pJDB248 contained the type B yeast plasmid sequence rather than type A. Presumably the yeast plasmid sequence inverted repeat sequences recombined during propagation in *E. coli*. This recombination was not observed in other *E. coli*–yeast hybrid plasmids. However, the rate of recombination may be low relative to the growth rate of *E. coli*, and the growth of

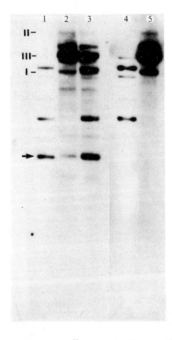

Fig. 3 Autoradiograph of ^{32}P-labelled cRNA prepared[31] to *Eco*RI digested DNA of pJDB248 (1, 2 and 3) or pJDB36 (4 and 5) hybridised with *Eco*RI digested pJDB248 DNA (1 and 4), *S. cerevisiae* DNA (2 and 5) or *S. cerevisiae* and pJDB248 DNAs together (3). The bands corresponding to the 1.5 kilobase DNA fragment are indicated by an arrow. The slower migrating bands (I, II, III) observed in the yeast DNA digests correspond to the supercoiled, nicked-circular and linear forms respectively of 2-μm yeast plasmid which was relatively resistant to digestion by *Eco*RI endonuclease although there were two targets for this enzyme in the plasmid sequence in this strain. Digestion of plasmid DNA with *Eco*RI endonuclease before the preparation of cRNA had the effect of increasing the efficiency of the RNA polymerase reaction, but the yeast sequences were then preferentially transcribed, resulting in the relatively lower intensity of the band due to pMB9 on the autoradiograph. The *S. cerevisiae* DNA used in this experiment was a different preparation from that used in the cloning experiment (and was prepared as described in the legend to Fig. 5). The possibility that the cloned DNA fragment was a contaminant in the DNA preparation is therefore unlikely. *Eco*RI digested DNAs were electrophoresed through a 1.5% agarose slab gel at 120 V per 45 mA for 13 h and transferred[32] to a nitrocellulose filter (Millipore HAWP). Hybridisation of the filters was carried out at 60 °C in 4 ml 2×SSC (0.03 M sodium citrate, 0.3 M sodium chloride, pH 7.0) containing 20 μg *E. coli* tRNA and cRNA (300,000 d.p.m.) with rapid shaking for 16 h. The filters were washed in 1×SSC at 60 °C with gentle shaking for 2 h. The sensitivity of Kodak X-Omat H film was enhanced by using Ilford fast tungstate screens and storing at -80 °C during the exposure[35]

cultures from original clones was not extended beyond the minimum required for preparation of plasmid DNA.

Figure 3 shows that DNA from *S. cerevisiae* digested with *Eco*RI endonuclease contained a 1.5 kilobase sequence which hybridised to the ^{32}P-labelled cRNA prepared to pJDB248 DNA, but not to cRNA prepared to pJDB36. This indicates that the inserted sequence corresponds to a yeast DNA fragment. The same *Eco*RI-generated DNA fragment hybridised cRNA prepared to pJDB219, indicating that the *leu*B-complementing sequences in the two independently isolated plasmids were the same.

Transformation of yeast

S. cerevisiae MC16α*leu*2-3 *his*4-712[FS] *ade*2-1 *lys*2-1 SUF2; G. R. Fink (FS denotes a suppressible mutation, SUF2 is a suppressor gene) was used as the recipient strain to investigate the uptake and replication of the plasmids and the transfer of the Leu$^+$ phenotype. Protoplasts and DNA were mixed and treated with polyethylene glycol to promote DNA uptake (see legend to Fig. 4). In different experiments between 10% and 30% of the cells thus treated regenerated the cell wall and produced colonies in supplemented minimal agar. Viable protoplasts were transformed to LEU$^+$ by pJDB219 or pJDB248 DNA with a frequency of 5×10^{-4} to 3×10^{-3} transformants per viable cell. The *leu*2 mutation in this strain reverted to *LEU*$^+$ at a frequency of only 10^{-8} to 10^{-7}. Clones transformed by pJDB248 grew as fast on selective plates as the untransformed cells grew on supplemented plates: protoplasts gave rise to colonies in 2 to 3 days. Transformants of pJDB219 required 3 to 4 days to develop colonies. [Note, efficiency of transformation of *S. cerevisiae* was not linearly proportional to DNA concentration, and a saturating DNA concentration was not found. Transformation frequencies of 10^4 to 10^5 transformants per μg DNA were obtained in the conditions used here.]

To detect the presence of the transforming plasmids in yeast clones, extracts of yeast cells were analysed by gel electrophoresis, transfer of the DNA to nitrocellulose and hybridisation with a ^{32}P-labelled cRNA prepared to pMB9 DNA. Transformants T19.1 to T19.6 (Fig. 4) produced bands hybridising to the pMB9 sequence and with similar electrophoretic mobility to the transforming plasmid pJDB219 (upper and lower bands correspond to nicked circular and supercoiled molecules respectively, slower migrating minor bands probably represent oligomeric forms of the plasmid). Transformants T48.1 to T48.6 also contained plasmids with the pMB9 sequence and with lower electrophoretic mobility than pJDB219, as expected for the larger plasmid pJDB248 (not shown in Fig. 4). There were apparently more copies per cell of pJDB219 than of pJDB248 in their respective transformant clones.

To investigate the integrity of the plasmid sequence in transformants, DNA was extracted and analysed by digestion with *Eco*RI endonuclease. Figure 5 shows the banding pattern of *Eco*RI-digested plasmid DNA isolated from two yeast clones transformed by pJDB219. The bands found in digests of pJDB219 DNA (corresponding to 5.3, 3.8, 2.5 and 0.8 kilobase DNA fragments) can be seen in the transformant DNA digests, and also the bands found in the digests of 2-μm plasmid DNA (3.8, 3.6, 2.3 and 2.1 kilobase fragments). There is also a band in the transformant DNAs corresponding to a 4.0 kilobase fragment which would be expected if the yeast plasmid in pJDB219 recombined during propagation in yeast, converting a type B to a type A plasmid (fragments of length 3.8 and 2.5 kilobases would be replaced by fragments of length 4.0 and 2.3 kilobases). Thus all the bands in the plasmid DNA of one transformant (Fig. 5, track 2) are those expected for a mixture of the 2 μm plasmid and pJDB219, however, two more bands can be seen in the plasmid DNA of the other transformant (Fig. 5, track 3) corresponding to 3.4 and 1.9 kilobase fragments. These bands cannot be readily accounted for. They may be products of recombination between plasmids[11], or of specific deletion

J. D. Beggs

events. Two examples of each type of banding pattern were found in digests of DNA from four different yeast clones transformed by pJDB219. The DNA from three yeast clones transformed by pJDB248 DNA was similarly analysed. The unexpected 3.4 and 1.9 kilobase fragments were produced from the DNA of two of these clones (not shown).

Recovery in *E. coli* of plasmids from yeast transformants

Samples of DNA (prepared as described in Fig. 5) from six pJDB219 transformants were used to transform *E. coli* JA221. Transformed cells were plated on minimal agar, minimal agar containing tetracycline (15 μg ml^{-1}) or leucine-supplemented agar containing tetracycline. No significant difference was found between the transformation frequencies for the *leu*B and the tetracycline resistance markers selected either independently or together. This suggests that the plasmids recovered from yeast transformants by complementation in *E. coli* had retained both the tetracycline resistance and the *leu*B-complementing sequences. This test gives no information about plasmids which may have retained the *leu*B-complementing sequence but lost the pMB9 sequence, as the yeast plasmid probably does not replicate in *E. coli* (J. D. B., unpublished result). Similar results were obtained in recovering DNA from pJDB248 transformants although the level of complementation was reduced probably as a result of the low copy number of plasmid in these strains (see Fig. 4).

DNA of the 'recovered' pJDB219 was isolated from eight *E. coli* transformants selected by transformation to leu$^+$ and tetra-

cycline resistance, and from two transformants selected for tetracycline resistance alone. The *Eco*RI digestion pattern of seven of these plasmid isolates was identical to that of pJDB219 DNA. Three of the recovered plasmids contained the yeast DNA sequence in the type A rather than the type B configuration. Since pJDB219 DNA used to transform yeast was in only one configuration (type B) this confirms the theory that recombination occurs in yeast between the inverted repeat sequences of the yeast plasmid. It can also be concluded that the yeast DNA sequence in pJDB219 complements the *leu*B mutation in *E. coli* regardless of the orientation of this sequence with respect to the pMB9 sequence (the inverted repeat sequences separate pMB9 from the *leu*-complementing sequence).

The results presented in this article show that *S. cerevisiae* can be readily transformed by the hybrid plasmids and that despite the generation of unexpected recombinant plasmids in some clones the unaltered plasmid can be isolated from the yeast clones directly, or purified by recloning in *E. coli* and prepared more easily from bacterial cultures.

The inheritance of the Leu$^+$ phenotype in the yeast clones containing pJDB219 or pJDB248 has been shown by tetrad analysis to be cytoplasmic (J. D. B. and Nasmyth, K., in preparation).

Yeast is a desirable organism to use in molecular cloning. It is readily and rapidly propagated. Its biochemical and physiological similarity to bacteria allows the use of techniques developed for bacterial systems. It offers the advantage to genetic studies of vegetative growth in both the haploid and the diploid states. The high frequency of transformation obtained with the system described here makes it feasible to isolate directly in yeast any

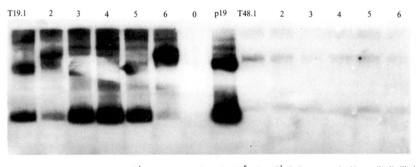

Fig. 4 Autoradiograph of ^{32}P-labelled cRNA to pMB9 DNA hybridised with undigested DNA of lysates of yeast clones transformed with pJDB219 (clones T19.1 to T19.6) or pJDB248 (clones T48.1 to T48.6) and in the lysate of the untransformed recipient strain alone (0) and with pJDB219 DNA (p19). Cultures of *S. cerevisiae* MC16 to be transformed were grown on YPDA (1% yeast extract, 2% Difco bacto-peptone, 2% glucose, 20 mg l^{-1} adenine) and collected at 10^8 cells ml^{-1}. Cells were washed in sterile distilled water, concentrated threefold in 1.2 M sorbitol, 25 mM EDTA, 50 mM dithiothreitol pH 8 and incubated at 29 °C for 10 min with gentle shaking. Cells were washed twice by centrifugation and resuspension in the original culture volume of cold, sterile 1.2 M sorbitol. Protoplasts were obtained by concentrating the cells threefold in 1.2 M sorbitol, 0.01 M EDTA, 0.1 M sodium citrate pH 5.8 containing Helicase lyophilised snail gut extract at 2 mg ml^{-1} (Industrie Biologique Francaise; 2 g powder containing 10^6 units β-glucuronidase activity were dissolved in 10 ml distilled water, filter sterilised and stored at -20 °C), and incubated at 29°C with gentle shaking for 1 h. Protoplasts were washed by centrifugation and resuspension in the original culture volume of 1.2 M sorbitol three times to remove nuclease activities present in the snail gut extract, and resuspended in 1.2 M sorbitol 10 mM CaCl$_2$ to give approximately 10^{10} protoplasts ml^{-1}. DNA was mixed with 0.1 ml or 0.2 ml protoplasts to give 10 to 20 μg DNA ml^{-1} (DNA samples were sterilised immediately before use by extraction with chloroform and extraction with ether). Protoplasts and DNA were left at room temperature for 15 min and then diluted with 10 volumes of 20% polyethylene glycol 4000, 10 mM CaCl$_2$, 10 mM Tris-HCl pH 7.5 (a modification of a protoplast fusion procedure[33]). After 15 to 20 min protoplasts were pelleted and resuspended in 100 μl 1.2 M sorbitol, 10 mM CaCl$_2$, 20 μg ml^{-1} leucine plus 50 μl YPDA containing 1.2 M sorbitol, and incubated at 29 °C for 20 min. Samples were diluted in 1.2 M sorbitol and plated by mixing in 7 ml minimal medium containing 1.2 M sorbitol and 3% agar at 45 °C (minimal medium was 0.67% Difco yeast nitrogen base without amino acids, 2% glucose, 20 mg l^{-1} adenine, lysine and histidine) and pouring onto minimal medium plates containing 1.2 M sorbitol, 2% agar. Dilutions were plated on supplemented (20 μg leucine ml^{-1}) minimal plates to measure the efficiency of protoplast regeneration. Colonies developed after 2 to 4 days incubation at 29 °C. To test for the presence of the transforming plasmid, single clones were purified on minimal selective plates. Single colonies were used as inocula to make 1 cm × 2 cm streaks on YPDA plates which were incubated at 29 °C overnight. These streaks of cells were resuspended in 0.5 ml 1.2 M sorbitol, 25 mM EDTA, 0.86 M 2–mercaptoethanol, pH 8, left at room temperature for 10 to 20 min and centrifuged. The pelleted cells were resuspended in 0.5 ml 1.2 M sorbitol, 0.1 M sodium citrate, 10 mM EDTA pH 5.8 containing Helicase snail gut extract at 2 mg ml^{-1} and incubated at 29 °C for 1.5 h. Protoplasts were washed twice by centrifugation and resuspension in 1.5 ml 1.2 M sorbitol and finally resuspended in 200 μl 10 mM Tris, 1 mM EDTA, 1% sodium dodecyl sulphate pH 8. Protoplasts were lysed by incubation at 60°C for 20 min and debris was pelleted. Lysate (40 μl) was mixed with 20 μl 20% Ficoll, 0.05% bromophenol blue and electrophoresed through 1% agarose slab gels at 120 V per 35 mA for 15 h. Gels were exposed to short wavelength ultraviolet light for 15 min to nick the supercoiled plasmids and transferred to nitrocellulose filters as usual[32], except that the denaturation step was carried out at 37 °C for 1 h to ensure plasmid denaturation and to hydrolyse RNA. The filters were hybridised with ^{32}P-labelled cRNA (300,000 d.p.m.) as described in Fig. 2.

148

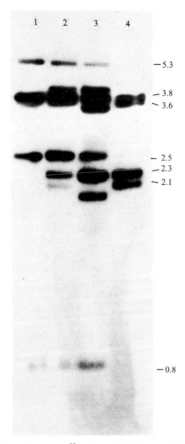

Fig. 5 Autoradiograph of ^{32}P-labelled cRNA to *Eco*RI digested DNA of pJDB219 hybridised with *Eco*RI digested DNA from yeast clones transformed by pJDB219. (1) pJDB219 DNA; (2) DNA from yeast clone T19.1 in Fig. 4; (3) DNA from yeast clone T19.2 in Fig. 4; (4) DNA from the untransformed yeast strain. DNA fragment sizes are indicated in kilobases. DNA was partially purified from cultures grown on YPDA medium by lysis of protoplasts[34] and extraction of the lysates with phenol and chloroform. Nucleic acids were ethanol precipitated, resuspended, and RNA was digested by pancreatic ribonuclease[34]. The samples were finally dialysed against 5 mM Tris, 0.5 mM EDTA *p*H 8.

yeast DNA sequence from a pool of hybrid plasmids carrying fragments representing the entire yeast genome.

All experiments using organisms containing recombinant plasmids were carried out under category I containment as recommended by the UK Genetic Manipulation Advisory Group. I thank Professor K. Murray and Dr R. B. Flavell for laboratory facilities and many materials, and J. F. Atkins and J. R. Bedbrook for discussions. Most enzymes were provided by K. Murray and J. R. Bedbrook. Calf thymus terminal transferase was a gift from the Corporate Lab., ICI. The work was supported in part by an SRC grant. An EMBO Short Term Fellowship was awarded to develop techniques for regenerating yeast protoplasts in the laboratory of Dr P. P. Slonimski. J. D. B. is a Beit Memorial Research Fellow.

Received 23 June, accepted 1 August, 1978.

1. Hinnen, A., Hicks, J. B. Fink, G. R. *Proc. natn. Acad. Sci. U.S.A.* **75**, 1929–1933 (1978).
2. Guerineau, M., Grandchamp, C., Paoletti, C. & Slonimski, P. P. *Biochem. biophys. Res. Commun.* **42**, 550–557 (1971).
3. Stevens, B. J. & Moustacchi, E. *Expl Cell Res.* **64**, 259–266 (1971).
4. Bak, A. L., Christiansen, C. & Christiansen, G. *Biochim. biophys. Acta* **269**, 527–530 (1972).
5. Clark-Walker, G. D. *Eur. J. Biochem.* **32**, 263–267 (1973).
5. Clark-Walker, G. D. *Proc. natn. Acad. Sci. U.S.A.* **69**, 388–392 (1972).
6. Beggs, J. D., Guerineau, M. & Atkins, J. F. *Molec. gen. Genet.* **148**, 287–294 (1976).
7. Petes, T. D. & Williamson, D. H. *Cell* **4**, 249–253 (1975).
8. Livingston, D. M. & Kupfer, D. M. *J. molec. Biol.* **116**, 249–260 (1977).
9. Clark-Walker, G. D. & Miklos, G. L. G. *Eur. J. Biochem.* **41**, 359–365 (1974).
10 Hollenberg, C. P., Degelmann, A., Kustermann-Kuhn, B. & Royer, H. D. *Proc. natn. Acad. Sci. U.S.A.* **73**, 2072–2076 (1976).
11. Guerineau, M., Grandchamp, C. & Slonimski, P. P. *Proc. natn. Acad. Sci. U.S.A.* **73**, 3030–3034 (1976).
12. Livingston, D. M. & Klein, H. L. *J. Bact.* **129**, 472–481 (1977).
13. Livingston, D. M. *Genetics* **86**, 73–84 (1977).
14. Cameron, J. R., Philippsen, P. & Davis, R. W. *Nucleic Acids Res.* **4**, 1429–1448 (1977).
15. Bolivar, F., Rodriguez, R. L., Betlach, M. C. & Boyer, H. W. *Gene* **2**, 75–93 (1977).
16. Guerineau, M., Slonimski, P. P. & Avner, P. R. *Biochem. biophys. Res. Commun.* **61**, 462–469 (1974).
17. Jackson, D., Symons, R. & Berg, P. *Proc. natn. Acad. Sci. U.S.A.* **69**, 2904–2909 (1972).
18. Somers, J. M., Amzallag, A. & Middleton, R. B. *J. Bact.* **113**, 1268–1272 (1973).
19. Borck, K., Beggs J. D., Brammer, W. J., Hopkins, A. S. & Murray N. E. *Molec. gen. Genet.* **146**, 199–207 (1976).
20. Clewell, D. *J. Bact.* **110**, 667–676 (1972).
21. Clewell, D. D. & Helinski, D. R. *Proc. natn. Acad. Sci. U.S.A.* **62**, 1159–1166 (1969).
22. Murray, N. E., Brammar, W. J. & Murray, K. *Molec. gen. Genet.* **150**, 53–61 (1977).
23. Lederberg, E. M. & Cohen, S. N. *J. Bact.* **119**, 1072–1074 (1974).
24. Grunstein, M. & Hogness, D. S. *Proc. natn. Acad. Sci. U.S.A.* **72**, 3961–3965 (1975).
25. Maniatis, T., Jeffrey, A. & Kleid, D. *Proc. natn. Acad. Sci. U.S.A.* **72**, 1184–1188 (1975).
26. Barnes, W. M. *Science* **195**, 393–394 (1977).
27. Davidson, E. H., Hough, B. R., Amenson, C. S. & Britten, R. J. *J. molec. Biol.* **77**, 1-23 (1973).
28. Lobban, P. E. & Kaiser, A. D. *J. molec. Biol.* **78**, 453–471 (1973).
29. Roychoudhury, R., Jay, E. & Wu, R. *Nucleic Acids Res.* **3**, 863–877 (1976).
30. Spizizen, J. *Proc. natn. Acad. Sci. U.S.A.* **44**, 1072–1078 (1958).
31. Burgess, R. R. *J. biol. Chem.* **244**, 6160–6167 (1969).
32. Southern, E. M. *J. molec. Biol.* **98**, 503–517 (1975).
33. van Solingen, P. & van der Plaat, J. B. *J. Bact.* **130**, 946–947 (1977).
34. Cryer, D. R., Eccleshall, R. & Marmur, J. *Methods in Cell Biology* **XII**, 39–44 (1975).
35. Laskey, R. A. & Mills, A. D. *FEBS Lett.* **82**, 314–316 (1977).

149

18

Reprinted from *Natl. Acad. Sci. (USA) Proc.* **76**:1035–1039 (1979)

High-frequency transformation of yeast: Autonomous replication of hybrid DNA molecules

(recombinant DNA/cloning vector/yeast plasmid/eukaryotic genetics/minichromosome)

KEVIN STRUHL, DAN T. STINCHCOMB, STEWART SCHERER, AND RONALD W. DAVIS

Department of Biochemistry, Stanford University School of Medicine, Stanford, California 94305

ABSTRACT   A set of vector DNAs (Y vectors) useful for the cloning of DNA fragments in *Saccharomyces cerevisiae* (yeast) and in *Escherichia coli* are characterized. With these vectors, three modes of yeast transformation are defined. (*i*) Vectors containing yeast chromosomal DNA sequences (YIp1, YIp5) transform yeast cells at low frequency (1–10 colonies per μg) and integrate into the genome by homologous recombination; this recombination is reversible. (*ii*) Hybrids containing endogenous yeast plasmid DNA sequences (YEp2, YEp6) transform yeast cells at much higher frequency (5000–20,000 colonies per μg). Such molecules replicate autonomously with an average copy number of 5–10 covalently closed circles per yeast cell and also replicate as a chromosomally integrated structure. This DNA may be physically isolated in intact form from either yeast or *E. coli* and used to transform either organism at high frequency. (*iii*) Vectors containing a 1.4-kilobase yeast DNA fragment that includes the centromere linked *trp1* gene (YRp7) transform yeast with an efficiency of 500–5000 colonies per μg; such molecules behave as minichromosomes because they replicate autonomously but do not integrate into the genome. The uses of Y vectors for the following genetic manipulations in yeast are discussed: isolation of genes; construction of haploid strains that are merodiploid for a particular DNA sequence; and directed alterations of the yeast genome. General methods for the selection and the analysis of these events are presented.

The molecular analysis of gene structure, function, and regulation depends upon the ability to correlate physiological, genetic, and structural data relating to a specific gene or set of genes. Many mutants of the yeast *Saccharomyces cerevisiae* have been isolated, and techniques for the manipulation and mapping of the associated genetic characteristics are routinely performed (1). Genetically defined yeast DNA sequences have been isolated in the form of viable molecular hybrids with bacteriophage λ or *Escherichia coli* plasmids (2–5). Derivatives of the cloned *his3* gene that delete DNA sequences near or in the structural gene have been isolated and physically defined (unpublished data). The demonstration by Hinnen *et al.* (5) that recombinant DNA containing cloned yeast genes can be used to transform yeast cells clearly expands the potential of molecular analysis considerably. These workers showed that yeast transformation occurred at low frequency and that it was usually accompanied by homologous recombination between the transforming DNA and the host chromosomal DNA.

The present paper reports two additional modes of yeast transformation. In both, the transformation event occurs at high frequency and is associated with autonomous replication of the transforming DNA. Yeast vectors useful for a wide variety of genetic manipulations have been constructed by combining the three mechanistically different modes of transformation with structural information of the endogenous yeast plasmid (6), the *his3* gene, the *trp1* gene, and the *ura3* gene (D. Botstein, per-

sonal communication). In particular, physically defined, cloned alterations of wild-type yeast DNA sequences may be introduced back into yeast cells in order to examine their *in vivo* phenotypic effects.

MATERIALS AND METHODS

Organisms, DNAs, and Enzymes. The following strains were used: yeast—A3617C (a *his3-532 gal2*) (3) and D13-1A (a *his3-532 trp1 gal2*); *E. coli—trpC* 9830 (7), *hisB* 463, SF8, C600 (rK^-mK^+) (2), and MB1000 ($rK^-mK^+ lac^-trp^-pyrF^-$) (D. Botstein, personal communication); phage—λgt-Sc2601 (2), λ590 (8), and λgt-Sc4104; plasmid DNAs—pMB9-Sc2601 (3), Scp1 (6), pBR322 (9), pGT2-Sc2605, pBR322-Sc2676 (unpublished data), and pMB1068 (D. Botstein, personal communication). Propagation of strains and preparation of DNAs have been described (2, 3, 6).

*Eco*RI, *E. coli* DNA ligase, and deoxynucleotidyl terminal transferase were the gifts of Marj Thomas, Robert Alazard, and Tom St. John, respectively. Other restriction endonucleases were purchased from New England BioLabs and Bethesda Research Laboratories (Rockville, MD) and used as directed. Cloning procedures have been described (2, 10).

Where appropriate p2,EK1 conditions, as described by the National Institutes of Health Guidelines fo Recombinant DNA Research, were used.

Rapid Yeast DNA Preparations. Total yeast DNA was prepared from 5-ml cultures of cells grown to the stationary phase. Yeast cells were harvested and resuspended in 0.4 ml of 0.9 M sorbitol/50 mM potassium phosphate, pH 7.5/14 mM 2-mercaptoethanol. Lyticase (25 units) (a gift from R. Schekman) was added and spheroplast formation was allowed to proceed for 30 min at 30°C. At this stage, the procedure for rapid phage DNA preparations (6) was used with two changes: the ethanol precipitation was done at room temperature, and the resulting pellet was resuspended in 50–100 μl of 10 mM Tris, pH 7.4/1 mM EDTA containing 0.5 μg of pancreatic RNase. These preparations yielded approximately 1 μg of DNA per ml of original culture. The DNA is of high molecular weight [95% is greater than 25 kilobases (kb)], is relatively un-nicked (the yeast plasmid is isolated predominantly in the closed circular form), and is cleavable by all restriction enzymes tested.

Transformation of Yeast Cells. The procedure of Hinnen *et al.* (5) was followed with some modifications. Spheroplasts were prepared by treating 100 ml of an exponentially growing culture with 300 units of lyticase for 30 min at 30°C. After treatment with polyethylene glycol, the cells were immediately plated in the regeneration agar (10^7–10^8 viable spheroplasts per plate). The relative efficiency of transformation with the different vector DNAs was simultaneously determined on individual preparations of spheroplasts.

Abbreviation: kb, kilobase(s).

Nucleic Acid Hybridization. Preparations of ^{32}P-labeled probes and hybridization to nitrocellulose filters have been described (3). Hybridizations were usually performed at 65°C in 1 M Na$^+$. More stringent conditions were achieved in 50% formamide/1 M Na$^+$ at 42°C.

RESULTS AND DISCUSSION

Yeast Transformation Vectors. All yeast transformation (Y) vectors used in this study have the following properties: the ability to replicate in yeast and *E. coli* cells; genetic characteristics selectable in yeast or *E. coli* after DNA transformation into either of these organisms, into which essentially any fragment of DNA can be inserted; and capability to isolate hybrid DNAs as covalently closed circles in at least one of these organisms. These vectors combine *E. coli* and yeast genetics into a single system, thus making it possible for cloning technology to be applied directly to yeast cells. Table 1 describes the properties of the Y vectors. Fig. 1 diagrams the physical structures of these vectors. Hybrid DNAs were enzymatically constructed *in vitro* and propagated in *E. coli* cells. Structural analysis of these DNAs was performed on material isolated from *E. coli* cells.

DNA Structural Analysis of the Yeast Transformation Event. The transformation event is operationally defined as the selected, genotypic change of a yeast cell dependent upon a particular DNA molecule. Hinnen *et al.* (5) analyzed the transformation of a *leu2$^-$* yeast strain to Leu$^+$ by a combination of genetic and molecular techniques. However, it is possible to analyze the structure of the transforming DNA in greater detail by using well-marked restriction endonuclease cleavage maps

Table 1. Properties of Y vectors

Vector*	Size, kb	Markers	Cloning sites
YIp1	9.8	*amp, his3*	*Eco*RI, *Sal* I, *Xho* I
YEp2	10.4	*tet*	*Pst* I
YIp5	5.4	*amp, tet, ura3*	*Eco*RI, *Bam*HI, *Sal* I, *Hind*III
YEp6	7.9	*amp, his3*	*Eco*RI, *Xho* I, *Sal* I
YRp7	5.7	*tet, amp, trpl*	*Bam*HI, *Sal* I

* Abbreviations: I, chromosomal integrator; E, episomal replicon; R, chromosomal replicon; p, *E. coli* plasmid.

of the relevant DNAs. Fig. 2 presents four hypothetical mechanisms to explain a given transformation event: (*i*) autonomous replication of the transforming DNA, (*ii*) integration of the transforming DNA into the yeast genome by homologous recombination, (*iii*) integration via illegitimate (nonhomologous) recombination, and (*iv*) transformation not associated with autonomous replication or stable chromosomal integration of foreign DNA. To facilitate description of the DNA sequence organization of the transformants, we define three types of DNA sequences. *Common sequences* are present in both the transforming DNA and in the chromosomal DNA of the transformed strain. *Foreign sequences* are present in the transforming DNA but not in the chromosomal DNA; because the yeast plasmid (Scp1) does not normally integrate into the chromosome (6), it is defined as a foreign sequence. *Flanking sequences* are located immediately adjacent to either side of the common sequence in the chromosome of the transformed strain.

The diagrams in Fig. 2 show that the four postulated mechanisms of transformation predict different physical organizations of the transforming DNA. The autonomous replication mechanism predicts a low molecular weight form of the transforming DNA in the yeast cell which is identical to (or a linear permutation of) the covalently closed circular form found in *E. coli*. Both integration models predict that the transforming DNA will exist in the yeast cells as part of a high molecular weight species. However, the homologous recombination model predicts a perfectly aligned, nontandem duplication of the entire common sequence. The duplication is separated by one copy of the foreign sequence. This specific organization results from the insertion of a linear permutation of the transforming DNA into any site of the chromosomal copy of the common sequence. The illegitimate recombination model predicts in-

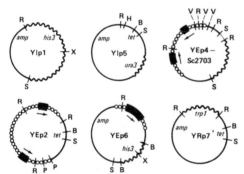

FIG. 1. Structure of Y vectors. DNA sequences diagrammed as follows: solid line, *E. coli* plasmid pBR322; wavy line, yeast chromosomal; circles, yeast plasmid Scp1; dashed line, λ; solid bar, Scp1 sequence which is repeated in an inverted orientation (arrows). Restriction endonuclease sites: R, *Eco*RI; B, *Bam*HI; S, *Sal* I; X, *Xho* I; P, *Pst* I; V, *Pvu* I; H, *Hind*III. The vectors were constructed as follows: YIp1, ligation of *Eco*RI- and *Sal* I-cleaved pBR322 and pGT2-Sc2605 DNAs; YEp2 and YEp4, joining of dG-tailed *Pst* I-cleaved pBR322 DNA and dC̄-tailed *Hpa* I-cleaved Scp1 DNA; YIp5, joining of dC-tailed *Ava* I-cleaved pBR322 DNA and dG-tailed *Hind*III-cleaved λ590-Sc2904 DNA (Sc2904 is the 1.1-kb *ura3 Hind*III fragment of pMB1068); YEp6, joining of dG-tailed *Eco*RI-cleaved pBR322-Sc2676 DNA and dC-tailed *Hpa* I-cleaved Scp1 DNA; YRp7, ligation of *Eco*RI-cleaved pBR322 and λgt-Sc4104 DNAs. YEp2 and YEp4 have the XY form of Scp1 (6) but differ by the number of *Pst* I sites regenerated by the cloning procedure (one for YEp2; two for YEp4) and by the orientation of Scp1 DNA with respect to pBR322 DNA. The structure of YEp6 DNA is most probably explained by a single deletion that removes approximately 4 kb of Scp1 DNA sequences and 0.2 kb of pBR322 sequences. The Scp1 sequences in YEp6 DNA correspond to coordinates 3700–5800 of the XY form.

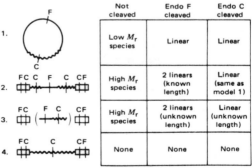

		Not cleaved	Endo F cleaved	Endo C cleaved
1.		Low M_r species	Linear	Linear
2.		High M_r species	2 linears (known length)	Linear (same as model 1)
3.		High M_r species	2 linears (unknown length)	Linear (unknown length)
4.		None	None	None

FIG. 2. Structural predictions for four transformation models. F and C, restriction endonuclease sites. Enzyme F cleaves foreign but not common sequences; enzyme C cleaves common but not foreign sequences. Predictions listed are based on hybridization of ^{32}P-labeled foreign sequence-specific probe across a gel of electrophoretically separated samples. See text for details.

tegration of the transforming DNA at an unknown genomic site. The structural organization will not be precisely predictable, and it will differ from that predicted for homologous recombination. The fourth mechanism predicts a structure indistinguishable from a nontransformed cell.

The analysis of any yeast transformant with respect to these four basic mechanisms proceeds in the following manner (see Fig. 2). DNA from a transformant is treated in three ways. One sample is uncleaved by restriction endonucleases; another sample is treated with an enzyme that cleaves at an internal cleavage site in the foreign sequence but does not cleave the common sequence (endo F); the third sample is treated with an enzyme that cleaves at least once in the common sequence or at the joint between the foreign and common sequences (endo C). DNA molecules in each sample are electrophoretically separated, transferred to a nitrocellulose filter, and challenged for hybridization with a ^{32}P-labeled probe specific for foreign DNA sequences or total transforming DNA sequences. This method is sufficiently rapid and general to be useful for the analysis of all yeast transformants.

Low-Frequency Transformation of Yeast Is Accompanied by Homologous Recombination. *his3* hybrids (such as YIp1 and pBM9-Sc2601) transform *his3*$^-$ yeast cells to His$^+$ at a frequency of 1–10 colonies per μg of DNA. Hybridization results such as those shown in Fig. 3 indicate that all 30 His$^+$ transformants examined contain one copy of the entire transforming DNA integrated at the *his3* locus. None of the low-frequency transformants contain any autonomously replicating DNA. The clearest example of the predicted integrated structure is that found in KY114, a cell transformed to His$^+$ by pMB9-Sc2683. Sc2683 contains an internal 1.4-kb deletion of the original *Eco*RI DNA fragment Sc2601 (M. Brennan and K. Struhl: unpublished data). Because the *Eco*RI fragments containing the common sequences of pMB9-Sc2683 and the host chromosome are distinguishable, the nontandem duplication is easily seen by *Eco*RI cleavage of KY114 DNA (Fig. 3, lane 7).

The nontandem duplicated structure generated upon transformation is not stable (Fig. 4). After 15 generations of growth in nonselective medium, approximately 1% of the colonies are His$^-$. This segregation is accompanied by the complete loss of the transforming DNA (Fig. 3, lane 6) and almost certainly results from a reversal of the original transformation event—i.e., excision by homologous recombination.

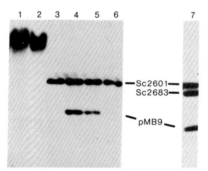

FIG. 3. Separation in 0.7% agarose; 1 μg of DNA per lane. Lanes: 1, 4, and 5, A3617C (pMB9-Sc2601); 2, A3617C (YIp1); 3, A3617C (untransformed); 6, his$^-$ segregant of A3617C (pMB9-Sc2601); 7, A3617 (pMB9-Sc2683). Samples in lanes 3–7 were cleaved with *Eco*RI. Samples in lanes 1 and 2 were uncleaved. The hybridization probe was ^{32}P-labeled pMB9-Sc2601 DNA. Appropriate size standards were present in each gel. Lane 7 is not from the same gel as lanes 1–6.

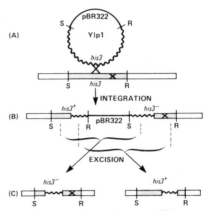

FIG. 4. (*A*) Homologous recombination at *his3* sequences between YIp1 and yeast chromosomal DNAs. (*B*) Integrated structure of a low-frequency transformant. (*C*) Possible structures after excision of transforming DNA. The *his3*-532 lesion maps in the middle of all cloned His$^+$ DNA fragments used in this paper (unpublished data).

The results indicate that low-frequency yeast transformation by *his3* hybrid DNAs is always associated with homologous recombination at the *his3* locus. Hinnen *et al.* (5) reported that most transformation events by a hybrid DNA molecule containing the yeast *leu2* gene could be accounted for by homologous recombination at the *leu2* locus. They also found transformants in which the *leu2*$^+$ character was unlinked to *leu2*$^-$ (as in model 3) and transformants that did not contain any foreign DNA sequences (as in model 4). However, the transforming DNA used in their experiments contains a sequence of yeast DNA that is present at many sites in the yeast genome (11). Therefore, *leu2*$^+$ transformants that have the transforming DNA integrated at loci other than *leu2* may be explained by homologous recombination at one of these repeated DNA sequences.

High-Frequency Transformation by Hybrid DNAs Containing Yeast Plasmid Sequences Is Accompanied by Autonomous Replication and Chromosomal Integration. Naturally occurring deletions across the unique *Hpa* I cleavage site of the endogenous yeast plasmid DNA (Scp1) do not affect replication of the plasmid in yeast cells (6). Vectors YEp2 and YEp4 contain the entire yeast plasmid sequences in functional form. A 7.1-kb *Sal* I DNA fragment (Sc2703) was inserted into YEp4. YEp4-Sc2703 DNA transforms *his3*$^-$ yeast to His$^+$ at a frequency of 5000–20,000 colonies per μg. The structural analysis of 42 of these His$^+$ transformants (3 of which are shown in Fig. 5) indicates that all 42 contain autonomously replicating, covalently closed, circular molecules of identical size to that of the transforming DNA. Based on ethidium bromide staining of the agarose gel (Fig. 5), we estimate that each cell has an average of 5–10 hybrid DNA molecules. In addition to replicating autonomously, the transforming DNA (in all cases) is integrated into the chromosome by homologous recombination at the *his3* locus; thus, these molecules are yeast episomes.

High-frequency transformants dependent upon endogenous yeast plasmid (Scp1) DNA sequences are extremely unstable with respect to the His$^+$ character. After 15 generations of growth in nonselective liquid medium, 97% of the cells are His$^-$. Fifteen of 15 His$^-$ segregants lost both the autonomously replicating and the chromosomally integrated structures. Seven of seven colonies that remained His$^+$ after this nonselective passaging contained both forms of the transforming DNA (data

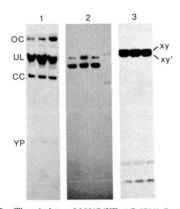

FIG. 5.   Three isolates of A3617 (YEp4-Sc2703). Samples: 1, un-cleaved DNA separated in 0.4% agarose; 2, *Pvu* I-cleaved DNA separated in 0.4% agarose [includes one lane of A3617C (YIp1) DNA]; 3, *Eco*RI-cleaved DNA separated in 0.7% agarose. The hybridization probe was ^{32}P-labeled pBR322 DNA. Abbreviations: CC, closed circles of transforming DNA; OC, open circles of transforming DNA; UL, uncleaved linear chromosomal DNA of high molecular weight; xy,xy', linear fragments containing either the XY or the XY' form of Scp1 DNA; YP, closed circular Scp1 molecules. The mobilities of these species were determined with appropriate standards present in the same gel. Scp1 DNA contains a sequence that is nontandemly repeated in an inverted orientation (6) (Fig. 1). Recombination in *E. coli* across these inverted repeat sequences has not been detected; i.e., YEp4-Sc2703 DNA is isolated from *E. coli* in only one form. This figure indicates that high-frequency transformants contain two forms of these molecules; therefore, recombination in yeast across these inverted repeat sequences does occur. Analogously, this suggests that the XY and XY' forms of Scp1 DNA are normally interchangeable in yeast. Because the transforming DNA (in one form) rapidly becomes equilibrated to roughly equimolar quantities of both forms, it is likely that the recombination across the inverted repeat sequence occurs by site-specific recombination.

not shown). Comparison of the genetic instability of transformants containing hybrids with or without Scp1 DNA sequences suggests that integration of the yeast plasmid into the chromosome may be detrimental to growth of the yeast cell. This is consistent with the observation that Scp1 does not normally integrate into the chromosome (6).

The entire sequence of Scp1 is not essential for the autonomous replication of hybrid DNA molecules in yeast. Naturally occurring deletions of Scp1 DNA replicate autonomously in yeast cells (6). YEp6 contains only 2 kb of Scp1 DNA, yet its transformation properties are indistinguishable from those of YEp2. Other fragments of Scp1 DNA, which are sufficient for autonomous replication of hybrid DNA molecules, have no sequences in common except for one copy of the inverted repeat sequence (data not shown). It is possible that the inverted repeat sequences are essential for Scp1 DNA replication.

Although high-frequency transformation is almost certainly dependent upon autonomous replication of the transforming DNA, it is striking that all 42 His$^+$ transformants tested also contain integrated structures. Because the His$^+$ character has not been separated from the presence of both the autonomous and integrated forms, it is not clear which (or both) form expresses *his3*. It is also possible that some cells of a culture of a high-frequency transformant lack either of the two forms; i.e., the transforming DNA may equilibrate rapidly between autonomous and integrated copies. In relation to *his3* expression, the intracellular location of the autonomously replicating form of the transforming DNA is significant. Although it is believed that Scp1 DNA is predominantly found in the cytoplasm (12), the observation that hybrid DNA molecules can integrate into the chromosome strongly suggests their presence in the nucleus. The possibility of nuclear copies of Scp1 DNA is also suggested by the fact that yeast cells transcribe discrete species of poly(A)-containing RNA from Scp1 DNA (J. Broach, personal communication).

In the course of these experiments, we observed that there is DNA sequence homology between the *E. coli* plasmids pBR322 and pMB9 and the yeast plasmid Scp1. This is best illustrated in Fig. 5. The degree of homology is fairly weak because hybridization was not detected under more stringent renaturation conditions.

Autonomous Replication and High-Frequency Transformation Dependent upon a Yeast Chromosomal Sequence. In other work we had isolated a 1.4-kb yeast DNA fragment (Sc4101) containing the *trp1* gene by complementation of *E. coli* mutants lacking *N*-(5'-phosphoribosyl) anthranilate isomerase. In initial transformation experiments, hybrid DNA molecules containing Sc4104 (such as YRp7) transformed *trp1*$^-$ yeast cells to Trp$^+$ at the surprisingly high frequency of 500–2000 colonies per µg. Furthermore, YRp7-Sc2605 hybrids (containing both *trp1* and *his3* genes) cotransformed *trp1*$^-$ *his3*$^-$ strains to His$^+$ and Trp$^+$ in all 175 transformants examined, demonstrating that the high transformation efficiency is dependent upon the presence of the 1.4-kb chromosomal sequence and is not due to an aberration of the *trp1* transformation. When 30 of the resulting transformants were examined (4 are shown in Fig. 6), the transforming sequences were always detected as closed circular DNA molecules. None of the hybrids was found integrated into the yeast chromosomal DNA.

The behavior of hybrid DNAs containing Sc4101 suggests that these molecules act like yeast minichromosomes. Thus, a yeast chromosomal sequence permits hybrid DNA molecules to replicate autonomously and to express structural genes (both *trp1* and *his3*) without recombining with host chromosomal sequences. This observation is striking because other modes of transformation are always accompanied by homologous recombination even when the integrated structures are highly unstable (as in Scp1 hybrid molecules). Because Sc4101 hybrids containing other yeast sequences (e.g., YRp7-Sc2605) exhibit this behavior, the failure to integrate is not explained by suppression of homologous recombination at the *trp1* locus. These transformation properties of Sc4101 hybrids are of special interest in light of the close proximity of *trp1* to the centromere of chromosome IV (1). Further studies on high-frequency transformation dependent upon Sc4101 will be presented elsewhere.

Some Uses of Yeast Transformation. Three modes of transformation of yeast have been described. Low-frequency transformation is accompanied by homologous recombination

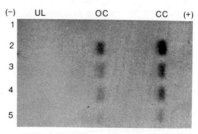

FIG. 6.   Uncleaved DNA from four isolates of D13-1A (YRp7-Sc2605) (lanes 2–5) and one isolate of D13-1A (lane 1) were separated in 0.5% agarose and probed with ^{32}P-labeled pBR322 DNA. Abbreviations as in Fig. 5.

Biochemistry: Struhl et al.

Proc. Natl. Acad. Sci. USA 76 (1979)   1039

between transforming DNA sequences and yeast chromosomal DNA. High-frequency transformation dependent upon yeast plasmid sequences is accompanied by autonomous replication and chromosomal integration of the transforming DNA. The high-frequency transformation dependent upon a yeast chromosomal sequence (Sc4101) is also accompanied by autonomous replication of the transforming DNA, but the hybrid molecules fail to integrate into the host chromosomal DNA. Although the mechanisms for the two modes of high-frequency transformation are unclear, the properties of these modes are sufficiently defined so that they may be applied to a wide range of genetic manipulations.

By using yeast transformation, a hybrid containing a desired yeast gene can be directly identified by complementation of a yeast mutation. In particular, high-frequency transformation should prove invaluable as a general method for the cloning of any yeast gene. Because all high-frequency transformation vectors are autonomous replicons in yeast and in *E. coli*, the hybrid DNA molecules are readily interchanged between these two organisms. For example, a pool of hybrid DNA molecules of YEp2 containing unselected fragments of yeast DNA could be isolated from *E. coli* cells and transformed into an appropriate yeast strain selecting for the gene of interest. The autonomously replicating DNA isolated from such a yeast transformant could then be introduced back into *E. coli*. Plasmid DNA isolated from these *E. coli* cells should contain the desired yeast gene. As a test of this system, we have cloned *Pst* I DNA fragments of yeast into YEp2. From such a hybrid pool, we have selected a hybrid that contains the predicted 2.8-kb *Pst* I DNA fragment containing the *his3* gene. In some cases, the yeast mutation to be complemented may not be directly selectable. However, because YEp6 contains the *his3* gene, transformation of a YEp6 hybrid pool into a strain that contains the mutation of interest and is also *his3⁻* allows one to select His⁺ transformants before screening for the desired marker.

An important application of yeast transformation is that of introducing physically defined alterations of a given DNA fragment back into yeast cells in order to examine the phenotypic effect of the alteration. For example, the introduction of a cloned sequence into yeast cells without radically altering the existing genome would permit detailed complementation analysis. Because yeast DNA sequences cloned in YRp7 are autonomously replicated and functionally expressed in the absence of significant recombinations with the yeast genome, this vector is ideal for such manipulations. Thus, haploid strains containing YRp7 hybrid DNA molecules are merodiploid for the cloned yeast sequences in a manner analogous to F'-containing *E. coli* strains. Such complementation analyses have been performed to study the expression of the yeast *his3* gene.

In addition, it would be useful to replace completely a functional genomic yeast sequence with a physically or genetically altered derivative. Low-frequency transformation could be used to perform these manipulations. It is essential that common DNA sequences flank both sides of the cloned alteration (e.g., deletion, insertion, altered restriction endonuclease site, genetic lesion). Integration and excision of transforming DNA containing an altered common DNA sequence generates two classes of segregants (Fig. 4). Some contain one copy of the altered common sequence at the normal chromosomal location; the others are indistinguishable from the original untransformed strain. The inclusion of a separate, functional yeast gene in the hybrid DNA makes it possible to select independently for the transformation and segregation events. Alterations of a given sequence are possible even if the sequence itself has no selectable phenotype.

The ability to isolate defined yeast genes and to introduce them back into yeast is essential for detailed studies of gene structure and function in this eukaryote. In addition, it is now possible to study the expression of other eukaryotic DNAs in yeast. Methods such as those described in this paper make it possible to genetically manipulate *S. cerevisiae* as easily as *E. coli*.

We thank Geoff Wahl for help in the construction of YEp2 and YEp4, David Botstein for MB1000 and pMB1068, and Randy Schekman for lyticase. We thank David Botstein, Jim Broach, Gerry Fink, Jim Hicks, and Randy Schekman for fruitful discussions and for communication of results prior to publication. This work was partly supported by U.S. Public Health Service Grant GM21891 from the National Institute of General Medical Sciences, by National Science Foundation Grant 77-27859, and by the Dreyfuss Foundation.

1. Mortimer, R. K. & Hawthorne, D. C. (1970) in *The Yeasts*, eds. Rose, A. H. & Harrison, J. S. (Academic, New York), Vol. 1, pp. 386–460.
2. Struhl, K., Cameron, J. R. & Davis, R. W. (1976) *Proc. Natl. Acad. Sci. USA* 73, 1471–1475.
3. Struhl, K. & Davis, R. W. (1977) *Proc. Natl. Acad. Sci. USA* 74, 5255–5259.
4. Ratzkin, B. & Carbon, J. (1977) *Proc. Natl. Acad. Sci. USA* 74, 487–491.
5. Hinnen, A., Hicks, J. B. & Fink, G. R. (1978) *Proc. Natl. Acad. Sci. USA* 75, 1929–1933.
6. Cameron, J. R., Philippsen, P. & Davis, R. W. (1977) *Nucleic Acids Res.* 4, 1429–1448.
7. Yanofsky, C., Horn, V., Bonner, M. & Stasiowski, S. (1971) *Genetics* 69, 409–433.
8. Murray, N. E., Brammer, W. J. & Murray, K. (1977) *Mol. Gen. Genet.* 150, 53–61.
9. Bolivar, R., Rodriguez, R. L., Greene, P. J., Betlach, M. C., Heyneker, H. L., Boyer, H. W., Crosa, J. H. & Falkow, S. (1977) *Gene* 2, 95–113.
10. Wensink, P. C., Finnegan, D. J., Donelson, J. E. & Hogness, D. S. (1974) *Cell* 3, 315–325.
11. Hicks, J. B. & Fink, G. R. (1977) *Nature (London)* 269, 265–267.
12. Clark-Walker, G. D. (1972) *Proc. Natl. Acad. Sci. USA* 69, 388–392.

19

Reprinted from pages 45–56 of *Recombinant Molecules: Impact on Science and Society*, edited by R. F. Beers, Jr., and E. G. Bassett, N.Y.: Raven Press, 1977

Biological Containment: The Subordination of *Escherichia coli* K-12

Roy Curtiss, III, Dennis A. Pereira, J. Charles Hsu, Sheila C. Hull, Josephine E. Clark, Larry J. Maturin, Sr., Raúl Goldschmidt, Robert Moody, Matsuhisa Inoue, and Laura Alexander

Department of Microbiology, Cancer Research and Training Center, Institute of Dental Research, University of Alabama in Birmingham, Birmingham, Alabama 35294

INTRODUCTION

The idea to manipulate genetically host strains and vectors to make them safer for recombinant DNA molecule research was strongly endorsed at the International Conference on Recombinant DNA Molecules held at the Asilomar Conference Center in Pacific Grove, California, in February 1975. The birth of this concept of biological containment as a means for augmenting the safety afforded by physical containment facilities and procedures was relatively painless as was the initial formulation of the means to manipulate hosts and vectors genetically to preclude the survival and/or transmission of cloned recombinant DNA to other hosts and/or vectors in the biosphere. The disarming of *Escherichia coli* K-12 and its plasmid and phage cloning vectors has, however, been considerably more difficult, frustrating, and time-consuming than originally anticipated.

Immediately after the completion of the Asilomar meeting, our laboratory group commenced to construct a number of *E. coli* K-12 strains that would have numerous built-in safety features for recombinant DNA research. Our specific goal was to construct strains possessing mutations that would

1. Increase usefulness for recombinant DNA molecule research.
2. Preclude colonization of and survival in the intestinal tract.
3. Preclude biosynthesis of the rigid layer of the cell wall in nonlaboratory controlled environments.
4. Lead to degradation of DNA in nonlaboratory controlled environments.
5. Permit cloning vector replication to be dependent on the host strain.

155

6. Preclude or minimize transmission of recombinant DNA to other bacteria that might be encountered in nature should the host strain escape from the confines of the laboratory.

7. Permit monitoring of the strains.

Our initial expectation was that the task could be easily attained. To the contrary, we have learned that *E. coli* has enormous resiliency and a great capacity to do the illogical and unexpected. With each passing month, our respect for the sophisticated biological mechanisms that have evolved in *E. coli* has increased. Moreover, we have learned a great deal about which mutations do and which mutations do not contribute to either the safety, utility, or both of *E. coli* strains for recombinant DNA molecule research. Toward the end of January 1976, we completed construction of a strain, designated χ1776, that seemed to possess a sufficient number of safety features to satisfy our goals and meet the standards of an EK2 host for cloning with nonconjugative plasmid vectors as set forth in the NIH Guidelines for Research Involving Recombinant DNA Molecules (16). We thus subjected this strain to an exhaustive series of tests to evaluate its safety and utility for recombinant DNA molecule experiments and to uncover new information that would hopefully permit future construction of a completely fail-safe *E. coli* strain. This chapter reviews what we have learned during our more than year-long education by *E. coli*.

RESULTS

Mutations that Increase the Usefulness of Strains for Recombinant DNA Molecule Research

As a standard necessary attribute, all strains being constructed must have *hsdR*[1] or *hsdS* alleles to abolish restriction thereby making it possible to introduce foreign DNA sequences into the strain. In studying transformability of strains, we found that transformation frequencies increased in strains with *endA* mutations (3- to 30-fold), *dapD*, and Δ[*bioH-asd*] mutations (approximately 3-fold), and with *galE* or Δ[*gal-uvrB*] mutations (4- to 10-fold). Although this information makes it possible to construct a strain that is considerably more transformable than wild-type K-12 strains, we have found that *rfb* mutations, which cause the deletion of carbohydrates from the lipopolysaccharide (LPS) core, reduce transformability 3- to 7-fold. This is unfortunate since these mutations have several desirable features in that they reduce recipient ability for many conjugative plasmid types and confer increased sensitivity to bile salts, detergents, and antibiotics. In this latter regard, strains with *dapD*, Δ[*bioH-asd*], and *rfb* mutations lyse easily, thus facilitating isolation of chimeric plasmid DNA.

---

[1] All gene symbol designations are those used and defined by Bachmann et al. (1).

Many of the strains being constructed possess mutations that lead to the continuous production of chromosome-deficient minicells during growth of the culture. Minicells result from abnormal cell divisions occurring near the polar ends of cells. Minicells are particularly useful in studies on the expression of DNA cloned on plasmid vectors, since minicells produced by plasmid-containing strains possess plasmid DNA which can be transcribed and translated (11,12,18). Consequently, such minicell-producing strains have already been useful in studying the expression of eukaryotic DNA sequences cloned on the pSC101 plasmid (5,15).

Mutations that Preclude Colonization of and Survival of Strains in the Intestinal Tract

Unpublished experiments done 10 years ago at Oak Ridge National Laboratory indicated that both *pur* and *thy* mutations caused *E. coli* strains to survive at reduced frequencies during passage through the intestinal tract of mice. We were also interested in both of these mutations since *thy* mutants should undergo thymineless death with concomitant degradation of DNA; *pur* mutations should also contribute to the further avirulence of *E. coli* K-12 since it is known that *pur* mutations in pathogenic microbial species cause avirulence because of interference with the ability of cells to grow intracellularly (3). In numerous tests in which approximately 10^{10} bacteria were suspended in milk and introduced into the esophagus of rats, we could obtain no evidence that *pur* mutations affected the intestinal titers or the duration of excretion of the *E. coli* K-12 strains before their elimination from the intestinal flora. We have thus discontinued including a purine requirement in our strains since it could impede the growth of strains in certain environments, thereby interfering with the phenotypic expression of other mutations that abolish synthesis of the rigid layer of the cell wall. Strains with mutations in the *thyA* gene, however, survive passage through the rat's intestinal tract at somewhat reduced frequencies. It is interesting to note that secondary mutations in the *deoC* gene but not the *deoB* gene usually cause a substantially lower survival and shorter duration of excretion of the *thyA* strain. Since *deoC* mutants lack deoxyriboaldolase, we can surmise that their lower survival rate in the rat intestine is owing to the presence of high concentrations of purine deoxyribonucleosides and/or thymidine, which are toxic to cells that possess such mutations (4). Further tests are underway to elucidate the basis for this behavior.

Mutations in the *rfbA* or *rfbB* cistrons confer increased sensitivity to a large number of antibiotics, drugs, ionic detergents, and, of greatest importance, bile salts. We have not yet tested the effect of an *rfb* mutation alone on colonization and survival of *E. coli* K-12 strains in the rat intestinal tract, but we know that this mutation, in conjunction with *thyA* and mutations that abolish synthesis of the rigid layer of the cell wall, has resulted in no recoverable survivors after feeding of rats. So far we have tested over 40 rats with an inoculation dose of approximately 10^{10} E. coli cells per rat. It is expected, although unconfirmed as yet, that other mutations conferring bile salts sensitivity such as lpcA, lpcB, and rfa would similarly lead to the inability of strains to survive in or colonize the intestinal tract of animals, provided that bile was produced and delivered to the duodenum.

Mutations that Preclude Biosynthesis of the Rigid Layer of the Cell Wall in Nonlaboratory Controlled Environments

Our initial goal was to construct strains that would require diaminopimelic acid (DAP), the precursor in a biosynthetic pathway unique to prokaryotic organisms leading to the synthesis of lysine. Since DAP is an unusual amino acid, we reasoned that it would not be prevalent in nature. Because DAP is an essential constituent of the murein layer of the E. coli cell wall, strains possessing defects in its synthesis would form spheroplasts in the absence of DAP and thereby lyse. Among numerous dap mutations tested, only the dapD8 allele isolated and characterized by Bukhari and Taylor (2) was found to be genetically stable (reversion frequency of approximately 10^{-9}), and thus this allele was introduced into some of our strains for initial tests. We soon found, to our dismay, that strains possessing this allele did not always undergo DAP-less death and indeed were capable of growth, not only in liquid media but also when replica plated on solid media lacking DAP. The ability to survive in the absence of DAP was dependent on the presence of certain cations (Na^+, K^+, Mg^{2+}, and/or Ca^{2+}) and was also temperature dependent with survival being noted at 37°C or lower but not at 42°C. Upon microscopic observation, surviving cells growing in DAP-deficient liquid media appeared as spheroplast-like bodies surrounded by a capsular material and colonies forming on agar medium lacking DAP were very mucoid. These colonies were also composed of spheroplast-like bodies. It was thus evident that these E. coli strains, when placed under adverse circumstances, were able to synthesize colanic acid, a polysaccharide composed of galactose, glucose, glucuronic acid, and fucose, and whose synthesis is regulated by the lon gene (14). Consequently, it was necessary to introduce one of a variety of mutations such as galE, galU, man, or non that would abolish the ability of cells to make colanic acid and thus minimize, if not preclude, survival of spheroplast forms resulting from cessation of synthesis of the rigid cell wall layer.

During these studies we also investigated properties of other mutations that would confer a Dap⁻ phenotype, such as deletions of the gene for aspartic acid semialdehyde dehydrogenase. Ultimately, a strain with the dapD8, Δ[bioH-asd], and Δ[gal-uvrB] mutations was constructed and found to undergo DAP-less death under a variety of conditions. This strain was also shown to be unable to synthesize colanic acid. Such a constellation of mutations also promoted more rapid rates of death during passage through the intestinal tract of rats. The rates of DAP-less death, however, depend on a number of parameters: cells exhibiting faster rates of protein synthesis yield more rapid rates of death, and cells inoculated from log-phase cultures die more rapidly than cells inoculated from stationary-phase cultures. The cell density at the inception of DAP-less growth is also important in that cells starting at low density die more rapidly than cells inoculated at high density. Indeed, when one starts with cell densities of $>5 \times 10^7$ cells/ml, the culture never completely dies since survivors apparently can scavenge DAP released from lysed cells.

We have also observed that the presence of other microorganisms, whether introduced into the culture deliberately or accidently, accelerates the rate of DAP-less death. Since DAP-less death is dependent on con-comitant protein synthesis and *dap* mutations confer an obligate requirement for lysine, and *asd* mutations confer an obligate requirement for both threonine and methionine in addition to DAP, DAP-less death can be achieved only when *dap asd* strains grow in the presence of threonine, methionine, and lysine. Because of this, we are currently exploring the behavior of mutations that abolish alanine racemase and D-alanyl-D-alanine ligase activities, which are necessary for the synthesis of D-alanyl-D-alanine, another unique component of the rigid layer of the *E. coli* cell wall. Since such mutants would be able to synthesize L-alanine and carry out protein synthesis in media containing a metabolizable energy source but devoid of amino acids, they should offer some safety advantages over use of strains with *dap* and/or *asd* mutations.

It is evident from our studies that strains defective in the synthesis of the rigid layer of the cell wall offer a significant safety advantage, provided that they contain additional mutations that abolish the synthesis of the extracellular capsule which balances the osmotic pressure differential be-tween the medium and the inside of the bacterial cell. It appears, therefore, that this experimental approach to safer strain construction is not only highly efficacious for *E. coli* but conceivably valid for application to a wide range of microbial species for future development of safer host-vector systems.

Mutations that Facilitate Degradation of DNA in Nonlaboratory Controlled Environments

We expected that the presence of *thyA* mutations, which lead to thymine-less death, would also lead to degradation of DNA because of the accumula-tion of unrepaired single-strand breaks (13). Although thymineless death does occur in thymine-deficient medium and this is accompanied by degra-dation of DNA, we have not observed a significant decrease in the molecular weight of single-stranded DNA as determined on alkaline sucrose gradients.

Since we have observed a loss of covalently closed circular pSC101 molecules in this strain during thymine starvation, it is possible that DNA degradation is closely associated with appearance of single-strand breaks such that we could not expect to detect the decrease in DNA molecular weights because of the formation of single-strand breaks, which have been estimated to occur at a rate of 0.98 per chromosome per minute (13). As an additional or alternative explanation, it should be noted that the strains in which DNA degradation has been investigated most extensively also possess mutations that abolish dark repair and restriction, either of which in the mutant or wild-type state might affect detection of single-strand breaks and/or subsequent DNA degradation during thymineless death. These possibilities are currently under study.

We have also investigated the behavior of strains with a *polA* (TS) allele such as *polA214* or *polA12* in combination with a *recA* allele such as *recA200* (TS) to see if these combinations of mutations in the presence of a *thyA* mutation would lead to an increased rate of DNA degradation in the presence or absence of thymidine. Although the presence of *polA* (TS) and *recA* (TS) in *thyA* strains causes an accelerated rate of DNA degradation during thymineless death at 42°C, the presence of these mutations has little noticeable effect at 37°C. The presence of a *polA* (TS) allele does, however, cause *thyA* strains to commence thymineless death immediately at 37°C with no lag in loss of colony-forming ability as is usually noted for *thyA* or *thyA recA* (TS) strains. We have not, however, observed that *polA* (TS) and/or *recA* (TS) alleles alter either the rates of death or the time for clearance of *thyA* strains from the rat intestinal tract. Based on these results, it is evident that if *polA* and *recA* mutations are to be introduced into future safer strains to accelerate the rates of DNA degradation at temperatures usually encountered by *E. coli*, it will be necessary to isolate additional temperature-sensitive alleles of the *polA* and/or *recA* genes. Since we already know what mutations eliminate survival of strains *in vivo*, it would seem judicious to isolate conditional alleles that confer cold rather than heat sensitivity. Indeed, the use of *polA* (CS) alleles should also abolish replication and transmission of ColE1 cloning vectors at temperatures below 30°C.

Mutations that Would Permit Replication of Cloning Vectors to Be Dependent on the Host

Since we have not undertaken the development of cloning vectors, no work at present has been attempted in this area other than to include a *supE* allele in all strains. The *supE* allele was maintained under the expectation that most safer cloning vectors would have some critical function that would be dependent on the presence of an amber suppressor.

Mutations that Preclude or Minimize Transmission of Recombinant DNA to Other Bacteria

Mutations that cause resistance to all of the familiar specialized and generalized transducing phages of *E. coli* are well known and indeed have been introduced into some of our strains. Nevertheless, it is highly likely that alterations in the cell surface which result in resistance to these known phages will endow the strain with sensitivity to other phages that lurk in the sewers and polluted rivers. It thus may be difficult, if not impossible, to ensure that a strain could not be infected by a potential transducing phage in nature that would act to transmit a cloned DNA fragment to some other robust organism. The probabilities for such occurrences, however, would be extremely small, especially if the disarmed host is unable to synthesize DNA because of a *thyA* mutation or is otherwise in a metabolically inactive state because of growth requirements or is in the process of dying. It should be noted in this regard that *E. coli* K-12 cells suspended in phosphate-buffered saline and then infected with T6 bacteriophage have been observed to yield a small burst of liberated phage several hours after infection, especially after prior starvation of the cells for 4 hr. Whether some of the transducing phages which are more host dependent than T6 can replicate or not is as yet unknown. This observation, coupled with our observation that *E. coli* can transfer conjugative plasmids at low frequency after long periods of starvation on nongrowth media at 37°C, underscores the necessity of ultimately introducing mutations that accelerate the rate of the death of *E. coli* under conditions of starvation when thymineless and DAP-less death do not occur.

To block the transmission of nonconjugative plasmid cloning vectors, we have been isolating and characterizing mutants with various cell surface defects that would contribute to a conjugation-deficient (Con⁻) recipient phenotype. The rationale for this approach is based on the fact that a cell possessing a nonconjugative, plasmid vector–harboring, recombinant DNA would first have to acquire a conjugative plasmid in order to transmit that cloned DNA to some other microorganism. Although a considerable amount of work has been done on the isolation and characterization of Con⁻ recipients with regard to their ability to mate with Hfr and F′ donors (8,9,17, 19), there has been no previous work on the isolation of Con⁻ mutants defective in inheriting the 18 to 20 other conjugative plasmid types found in gram-negative enteric microorganisms. Because of the diversity of plasmid types and of mutational lesions that could potentially reduce the recipient ability of safer host strains, we have spent considerable time during the past year in isolating and characterizing Con⁻ mutants. For example, we have found that the presence of deletions of the *gal* operon reduces the recipient ability 1,000-fold or more for plasmids in the M, O, W, X, and 10 incompatibility groups with smaller yet significant reductions noted for plasmids in the C, I (only some of those tested), J, L, T, and 9 incompatibility groups. The addition of an *rfb* mutation to a strain with a Δ*gal* lesion either does not alter or further reduces recipient ability for the aforementioned plasmid types but more importantly reduces recipient ability approximately 1,000-fold for plasmids in the F and N incompatibility groups, which are quite prevalent in *E. coli* strains in nature, as are I-type plasmids (10). Thus, a strain such as χ1776 with both Δ*gal* and *rfb* mutations, when mated for 90 min under optimal laboratory conditions with donors possessing repressed conjugative plasmids, gives transconjugant frequencies of 10^{-4} to 10^{-5} for L, P, and some I group plasmids; 10^{-5} to 10^{-6} for J, 9, and other I group plasmids; and less than 10^{-7} for C, FII, H, M, N, O, W, X, and 10 group plasmids. Other mutants with different cell surface defects are currently being isolated and characterized with regard to their ability to either receive or donate the various types of conjugative plasmids.

Since little is known about plasmid transfer between gram-negative microorganisms in soil, water, and sewage, we investigated conjugational plasmid transfer at temperatures likely to occur in these environments. At 27°C we were unable, even when using wild-type Con⁺ recipient strains of *E. coli,* to detect plasmid transfer at frequencies in excess of 10^{-8} except for transfer of plasmids in the T and P incompatibility groups. Thus, if we assume that laboratory conditions reflect those found in nature, it would be unlikely that a significant amount of mobilization of nonconjugative plasmid cloning vectors would occur in natural environments other than in the intestinal tract because of temperatures unsuitable for the expression of the donor and/or recipient phenotypes necessary for conjugative plasmid transmission. A more complete treatment of the probabilities of transmission of recombinant DNA contained on nonconjugative plasmid cloning vectors has been presented and discussed elsewhere (6,7).

Mutations that Permit Monitoring of Strains

We have introduced mutations that confer high-level resistance to nalidixic acid as a means to recover strains during rat feeding experiments. This particular genetic marker is suitable for such experiments and routine monitoring since resistance to nalidixic acid is not common among naturally occurring gram-negative organisms and has not been observed to be plasmid mediated and since mutations to resistance occur at a very low spontaneous frequency. In addition, nalidixic acid has a high efficiency of killing such that it can be used to detect a single Nal^r isolate in the presence of 10^8 to 10^9 Nal^s cells. Many of our strains also possess mutations to resistance to cycloserine, which also has not been observed to be plasmid mediated. It is thus possible to use double selection for resistance to cycloserine and nalidixic acid to detect very low numbers of cells when testing strains in a variety of environments in and outside of the laboratory.

One of the chief concerns expressed by many who are apprehensive about the safety of recombinant DNA molecule research is the inadvertent contamination of a culture of a disarmed strain with cells of a robust transformable strain during transformation with recombinant DNA molecules. We have found that the addition of nalidixic acid and/or cycloserine to the culture during its growth before transformation and to the selective medium for plating transformants greatly minimizes, if not precludes, transformation of such contaminants. In this regard, the use of cycloserine is somewhat preferable to nalidixic acid since the latter is difficult to dispose of in a safe way, being stable to autoclaving, whereas the former has a reasonably short half-life at neutral or acidic pHs and is rapidly destroyed in nature.

Construction, Properties, and Testing of χ1776

χ1776 was constructed in 13 steps from χ1276 (12) and possesses 15 different mutational lesions. Table 1 lists most of the phenotypic properties

TABLE 1. *Phenotypic properties of* χ1776

Phenotype	Responsible mutation(s)"
Requires diaminopimelic acid	*dapD8* Δ29[*bioH-asd*]
Requires threonine	Δ29[*bioH-asd*]
Requires methionine	*metC65* Δ29[*bioH-asd*]
Requires biotin	Δ40[*gal-uvrB*] Δ29[*bioH-asd*]
Requires thymidine	*thyA57*
Cannot use galactose for growth	Δ40[*gal-uvrB*]
Cannot use maltose for growth	Δ29[*bioH-asd*]
Cannot use glycerol for growth	Δ29[*bioH-asd*]
Cannot synthesize colanic acid	Δ40[*gal-uvrB*]
Sensitive to UV and defective in dark repair	Δ40[*gal-uvrB*]
Cannot undergo photoreactivation	Δ40[*gal-uvrB*]
Sensitive to glycerol (aerobic)	Δ29[*bioH-asd*]
Sensitive to bile salts, ionic detergents, and antibiotics	*rfb-2*
Resistant to nalidixic acid	*nalA25*
Resistant to cycloserine	*cycA1 cycB2*
Resistant to chlorate (anaerobic)	Δ40[*gal-uvrB*]
Resistant to trimethoprim	*thyA57*
Resistant to T1, T5, and φ80	*tonA53*
Resistant to λ and 21	Δ29[*bioH-asd*]
Partially resistant to P1	*rfb-2*
Conjugation defective	Δ40[*gal-uvrB*] *rfb-2*
Produces minicells	*minA1 minB2*
Temperature sensitive at 42°C	One mutation that is partially responsible for phenotype is linked to *thyA*
Cannot restrict foreign DNA	*hsdR2*
Allows for vector to be host dependent	*supE42* (amber)

" See Bachmann et al. (1).

of this strain along with the responsible mutations for the designated pheno-type. A more complete description of the properties associated with some of these mutations and data verifying its safety features will be presented elsewhere. In summary, χ1776 and its derivative χ1876, which possesses the pSC101 nonconjugative plasmid, (a) cannot survive passage through the intestinal tract of rats; (b) die, partially degrade their DNA, and lyse in growth media lacking DAP and thymidine; (c) cannot grow and die at variable rates after drying or when suspended in water or other nongrowth media; and (d) are unable to transmit genetic information contained on pSC101 to other bacteria under any of the above-described nonpermissive conditions at measurable frequencies. It should be pointed out that by fac-toring the various parameters necessary for a successful triparental mating to allow for receipt and then mobilization of pSC101 by a conjugative plasmid, we can predict that this series of events could occur in nature with a probability of approximately 10^{-22} per surviving bacterium (7).

DISCUSSION AND CONCLUDING REMARKS

Our future goals include the construction of χ1776 derivatives with addi-tional mutations that would further reduce its recipient ability in matings with donors possessing various conjugative plasmids, increase its trans-formability, and increase its rate of demise in nonlaboratory controlled environments. At the same time, we are constructing other sublines of E. coli K-12 with defects in D-alanine metabolism as a means to block cell wall biosynthesis. The latter, together with the other mutations as described in this chapter, should confer the necessary attributes required for safer and more useful E. coli strains for recombinant DNA molecule research.

Since it is generally acknowledged that the cloning of DNA in bacteria, especially E. coli, is not without potential hazard, it is imperative that investigators attempting to clone DNA in these disarmed host-vector systems undertake the responsibility to verify the relevant phenotypic properties of this system immediately after isolating a clone containing recombinant DNA. With regard to χ1776 and its derivatives, the traits to be verified should include the requirements for diaminopimelic acid and thymidine, inability to utilize or ferment galactose or maltose, inability to produce colanic acid or other capsular material, and sensitivity to UV and bile salts. These tests can be easily done by either replica plating or streak testing on appropriate agar media. Clones that are not tested or show an alteration in even one phenotypic trait should be destroyed immediately. We also believe that those who clone DNA should devote some portion of their time to more extensive tests to determine the effects of cloned DNA on the survival and transmissibility characteristics of the host-vector system. We will be more than pleased to offer advice on the design of these tests and provide any strains that might be necessary to perform them.

Soon after the discovery of recombinant DNA molecule technology using *E. coli*, it became obvious that numerous potential biohazards were associated with this research. This realization was in large part based on the facts that *E. coli* is a normal intestinal inhabitant of humans and warm-blooded animals, is sometimes a severe opportunistic pathogen, and possesses the potential to exchange genetic information, especially plasmid DNA, with representative strains of over 30 bacterial genera. It is equally obvious that the development of biological containment systems and their usage in conjunction with physical containment facilities and procedures are, therefore, imperative for continued safe recombinant DNA molecule research. Although we tend to think primarily of hazard to the human species as a direct consequence of recombinant DNA molecule research, it must be remembered that other potential biohazards in recombinant DNA molecule research can exist that could cause perturbations of the ecosystem. Also, the extent of the possible damage to the biosphere and ultimately to humans when one organism acts on another to either displace it from or interfere with its normal functioning in its ecological niche is unpredictable but potentially catastrophic. These points should therefore serve to underscore the importance for developing, testing, and using disarmed microbial host-vector systems in all experiments in which the foreign cloned DNA might contribute to such potentially biohazardous conditions. This admonition is as applicable for use of microbial hosts other than *E. coli* for recombinant DNA molecule experiments as it is when using *E. coli*. Such disarmed microbial host-vector systems should minimize if not prevent the survival and transmission of cloned DNA if the chimeric host was inadvertently released from its rigorously controlled test tube habitat.

ACKNOWLEDGMENTS

Research was supported by grants from the National Science Foundation (GB-37546) and the National Institutes of Health (DE-02670, AI-11456, and 5 P02 CA 13148) and by an NIH Postdoctoral Fellowship (F32-AI-05222) to S.C.H., an NIH Postdoctoral Traineeship (T32-GM-7090) to J.E.C., and an NIH Predoctoral Traineeship (T32-GM-07164) to D.A.P.

REFERENCES

1. Bachmann, B. J., Low, K. B., and Taylor, A. L. (1976): Recalibrated linkage map of Escherichia coli K-12. *Bacteriol. Rev.*, 40:116–167.
2. Bukhari, A. T., and Taylor, A. L. (1971): Genetic analysis of diaminopimelic acid- and lysine-requiring mutants of Escherichia coli K-12. *J. Bacteriol.*, 105:844–854.
3. Burrows, T. W. (1955): The basis of virulence for mice of Pasteurella pestis. *Symp. Soc. Gen. Microbiol.*, 5:151–175.
4. Buxton, R. S. (1975): Genetic analysis of thymidine-resistant and low-thymine-requiring mutants of Escherichia coli K-12 induced by bacteriophage Mu-1. *J. Bacteriol.*, 121:475–490.
5. Chang, A. C. Y., Lansman, R. A., Clayton, D. B., and Cohen, S. N. (1975): Studies of

mouse mitochondrial DNA in Escherichia coli: structure and function of eucaryotic-procaryotic chimeric plasmids. *Cell*, 6:231–244.

6. Curtiss, R., III (1976): Genetic manipulation of microorganisms: potential benefits and biohazards. *Annu. Rev. Microbiol.*, 30:507–533.

7. Curtiss, R., III, Clark, J. E., Goldschmidt, R., Hsu, J. C., Hull, S. C., Inoue, M., Maturin, L. J., Moody, R., and Pereira, D. A. (1976): Biohazard assessment of recombinant DNA molecule research. *Proceedings of the Third International Symposium on Antibiotic Resistance (in press)*.

8. Curtiss, R., III, Fenwick, R. G., Jr., Goldschmidt, R., and Falkinham, J. O. (1975): The mechanism of conjugation. In: *Transferable Drug Resistance Factor R*, edited by S. Mitsuhashi. University Park Press, Tokyo (*in press*).

9. Falkinham, J. O., III, and Curtiss, R., III (1976): Isolation and characterization of conjugation-deficient mutants of Escherichia coli K-12. *J. Bacteriol.*, 126:1194–1206.

10. Falkow, S. (1975): *Infectious Multiple Drug Resistance*. Pion Limited, London.

11. Frazer, A. C., and Curtiss, R., III (1973): Derepression of anthranilate synthase in purified minicells of Escherichia coli containing the Coltrp plasmid. *J. Bacteriol.*, 115:615–622.

12. Frazer, A. C., and Curtiss, R., III (1975): Production, properties and utility of bacterial minicells. *Curr. Top. Microbiol. Immunol.*, 69:1–84.

13. Freifelder, D. (1969): Single-strand breaks in bacterial DNA associated with thymine starvation. *J. Mol. Biol.*, 45:1–7.

14. Markovitz, A. (1976): Genetics and regulation of bacterial capsular polysaccharide biosynthesis and radiation sensitivity. In: *Surface Carbohydrates of Procaryotic Cells*, edited by W. Sutherland. Academic Press, New York.

15. Morrow, J. F., Cohen, S. N., Chang, A. C. Y., Boyer, H. W., Goodman, H. M., and Helling, R. B. (1974): Replication and transcription of eukaryotic DNA in Escherichia coli. *Proc. Natl. Acad. Sci. U.S.A.*, 71:1743–1747.

16. National Institutes of Health Guidelines for Research Involving Recombinant DNA Molecules (1976): U.S. Department of Health, Education and Welfare, Public Health Service, National Institutes of Health, Bethesda, Maryland.

17. Reiner, A. M. (1974): Escherichia coli females defective in conjugation and in absorption of a single-stranded deoxyribonucleic acid phage. *J. Bacteriol.*, 119:183–191.

18. Roozen, K. J., Fenwick, R. G., Jr., and Curtiss, R., III (1971): Synthesis of ribonucleic acid and protein in plasmid-containing minicells of Escherichia coli K-12. *J. Bacteriol.*, 107:21–33.

19. Skurray, R. A., Hancock, R. E. W., and Reeves, P. (1974): Con mutants: class of mutants in Escherichia coli K-12 lacking a major cell wall protein and defective in conjugation and adsorption of bacteriophage. *J. Bacteriol.*, 119:726–735.

Part III

BASIC TECHNIQUES

Editors' Comments
on Papers 20 Through 29

29   SMITH AND BIRNSTIEL

A Simple Method for DNA Restriction Site Mapping

By 1974 the stage had been set for the *in vitro* generation of recombinants between any two DNA molecules. The technology for joining two molecules was well established; plasmid and phage vectors were available to carry and amplify the DNA fragments; and methods of introducing DNA into *E. coli* existed. The next several years saw refinement of these techniques to improve the efficiency of each step, as evidenced by the newer vectors described by Blattner, Bolivar, Leder, and Murray and their colleagues (Part II). During this period a new journal called *Gene* was created "devoted to gene cloning and recombinant nucleic acids." It was an exciting time because new techniques were being developed for making and detecting recombinants and introducing them into the cell. A powerful technology was developing, and two important new techniques were those of Maxam and Gilbert (1977) and Sanger's group (Sanger and Coulson 1975; Sanger et al. 1977) who described rapid methods for determining the nucleotide sequence of DNA.

In Part III, we have included some of the many excellent "methods" papers that appeared during the three years after 1974. Paper 20 describes the study of ligation conditions to determine those DNA concentrations that would optimize joining of the target fragment to the vector fragment, while minimizing "self-closure" and concatemer formation. This study is based on the previous work of two polymer chemists, H. Jacobson and W. H. Stockmayer, that had been applied to the problem of circularization of λ DNA by the cohesive ends (Davidson and Szybalski 1971; Wang and Davidson 1966).

Dugaiczyk and co-workers recognized that joining of ends may be intramolecular or intermolecular. The rate of each reaction depends on the concentration and length of the DNA molecules in the following ways. Intermolecular joining (concatemer formation) is favored by high concentrations of DNA; circularization is favored by low concentrations. Since the desired result of ligation is the joining of one vector molecule to one target molecule, the ideal situation is to have equal amounts of target and vector ends. This condition optimizes the probability that one end of a vector molecule will meet one end of a target molecule with the same frequency that the two vector ends will meet each other. This is a length dependent phenomenon because given an equal mass of large and small fragments, the large fragment will have a lower concentration of ends. Dugaiczyk et al. have derived equations whereby these theoretical optimal concentrations can be calculated, and they experimentally confirmed their predictions.

Paper 21 by Bahl et al. introduces the application of chemically synthesized oligonucleotides to the field of recombinant DNA. Such oligonucleotides had been synthesized before, but the idea of synthesizing restriction endonuclease recognition sites and ligating them to fragments that were not, for one reason or another, able to be inserted into vectors was revolutionary and opened up a whole new realm of recombinant DNA research. For example, it was now possible to clone fragments that had no appropriate restriction sites. Such methodology helped to generalize the development of "gene banks" (see Part IV, Paper 32) and "genomic libraries" (Maniatis et al. 1978).

Several groups had calculated that an entire mammalian genome could be represented in about 10^6–10^7 recombinant. phages if each phage carried a 20,000 base pair fragment. It is difficult to obtain this number of independent recombinant phages because of the combined inefficiencies of ligation and calcium or spheroplast transfection. A major advance that permitted scientists to obtain this number of recombinant phage without using enormous amounts of reagents was to abandon transfection methods and use techniques involving packaging recombinant phage genomes *in vitro* (Papers 22 and 23).

The methodology of *in vitro* packaging was derived from basic research on the mechanism by which λ encapsidates its DNA (Kaiser and Masuda 1973). It increased the efficiency of recovery of recombinant phages some 10–100-fold over calcium or spheroplast transfection methods. In this technique recombinant DNA is made, as before, with DNA fragments and a λ vector, but the next step involves addition of this DNA mixture to crude extracts containing partially assembled λ virions. The added DNA is encapsidated or packaged *in vitro* into the phage head and then tails are joined to the filled capsids giving rise to infectious particles. The packaging extracts are crude lysates made from two induced λ lysogens. The prophages in both lysogens carry several mutations and are defective. Paper 22 by Sternberg, Tiemeier, and Enquist and Paper 23 by Hohn and Murray present two somewhat different protocols for *in vitro* packaging. Sternberg et al. modified the procedure of Becker and Gold (1975) to generate a simple protocol that subsequently proved amenable to the safety requirements of recombinant DNA research. A surprising feature of this system was that larger DNA molecules were packaged more efficiently than smaller ones, permitting one to select for larger, recombinant phages. The paper contains several methods for screening recombinant phages using standard phage techniques.

Hohn and Murray used a different variation of the Becker and Gold (1975) procedure. They discovered that if both spermidine and putrescine were included in the packaging reaction, the size bias noted by Sternberg and colleagues was eliminated (see also Paper 33, Part IV).

A major difference between the two protocols was the method used to eliminate packageable endogenous λ DNA (derived from the prophages of the packaging lysogens) as well as recombination between the exogenous recombinant phage DNA and endogenous λ DNA. Sternberg and associates introduced a deletion that blocked excision of the λ prophages present in the strains used to make the extracts. In addition, they incorporated missence mutations in the packaging strains inactivating both prophage and *E. coli* homologous recombination systems. Hohn and Murray relied on use of UV-irradiation of the packaging mixture to destroy endogenous DNA.

Sternberg and co-workers also describe a novel λ vector that contains a single amber mutation in the D gene (a capsid gene). This vector, called λ D*am*, facilitated recognition of insert size simply by inspection of plaque morphology. The authors demonstrated how the λD*am* vector could be useful for selection of deletion mutants of the cloned fragment.

The next group of papers concerns the question of identification: How does one determine which phage or plasmid vector in a population of recombinant molecules carries the desired DNA fragment? Unless the starting DNA was a purified fragment, it would still be necessary to screen many recombinant candidates to determine which contains the proper DNA fragment. Therein lies the beauty of the procedures described in Papers 24 and 25. So widespread is the use of these techniques that the authors' names have become synonymous with the techniques themselves. Although each of these techniques has been modified and improved over the years, these papers stand as classics in the field.

Although phage plaque hybridization techniques had existed for some time, they involved manipulation of individual plaques (Cramer et al. 1976; Jones and Murray 1975). The elegance of the Benton and Davis plaque hybridization procedure (Paper 24) is that of classical replica plating for selection of bacterial mutants. By making an imprint of a plate containing plaques onto nitrocelluose paper, enough DNA is transferred to detect the desired plaque(s) directly by hybridization with the appropriate probe. Because the imprinting procedure does not destroy the original plaques, it is a simple matter then to isolate the recombinant phage by standard phage techniques.

The Grunstein and Hogness colony hybridization procedure (Paper 25) preceded Benton and Davis by two years. In this procedure, cells bearing plasmids are plated on nitrocellulose paper that is resting on the surface of an agar plate. After replica plating, the cells on the filter are lysed, the DNA denatured, and the desired colonies detected by hybridization with a specific probe. Although each of these techniques may now seem quite straightforward, their impact on recombinant DNA research was substantial.

Another technique that has become synonymous with the author's name is that described by Southern in 1975 (Paper 26) for the transfer of DNA fragments from agarose gels onto nitrocellulose paper for subsequent hybridization with specific probes. Although a related procedure involving direct hybridization within the dehydrated gel was published by Shinnick et al. in 1975, the Southern technique has remained the method of choice. The technique subsequently has been modified in individual laboratories, but the so-called Southern blot or Southern transfer remains one of the most powerful tools of the molecular biologist. The technique can be used with either RNA or DNA probes; it can be used to determine the size of the desired fragment (even if it is single copy) in a restriction fragment mixture derived from a complex genome; when those fragments are fractionated by preparative gel electrophoresis (Polsky et al. 1978), the technique can be used to detect those fractions to be used in the recombinant DNA experiments; and it can be used to examine and verify the structure of the inserted fragment once propagated by the vector system.

Another technique for characterization of DNA fragments was developed by Thomas, White, and Davis at Stanford University (Paper 27). This procedure involves examination by electron microscopy of hybrids formed between DNA sequences complementary to specific RNAs. The technique, called "R-loop mapping," is based on observations of the stability of DNA-DNA and DNA-RNA hybrids. The authors show that RNA can hybridize to double-stranded DNA in the presence of 70-percent formamide by displacing the identical DNA strand. The resulting structure is called an "R-loop." When R-loops are observed in the electron microscope, they appear in a long double-stranded molecule as a "bubble" where the DNA has been displaced by RNA. This technique enables one to compare the structure of messenger RNA made from a particular gene to the primary structure of the gene itself. On first reflection, this sounds less than remarkable, but this technique has played an integral part in deciphering one of the most amazing discoveries of this decade, the presence of intervening sequences within eukaryotic genes (see, for example, Paper 30 and Abelson 1979).

The final two papers in this section concern restriction enzymes. Paper 28 by Roberts is included, in his words, "to extend current awareness of the enzymes now available." Paper 29 by Smith and Birnstiel describes a useful technique for locating restriction sites on a DNA fragment. Although such mapping is part of routine laboratory procedures, this method simplified what often was laborious using standard techniques.

REFERENCES

Abelson, J. 1979. RNA Processing and the Intervening Sequence Problem. *Ann. Rev. Biochem.* **48**:1035–1069.

Becker, A., and M. Gold. 1975. Isolation of the Bacteriophage λ A Gene Protein. *Natl. Acad. Sci. (USA) Proc.* **72**:581–585.

Cramer, R. A., J. R. Cameron, and R. W. Davis. 1976. Isolation of Bacteriophage λ Containing Yeast Ribosomal Genes: Screening by *in situ* RNA Hybridization to plaques. *Cell* **8**:227–232.

Davidson, N., and W. Szybalski. 1971. Physical and Chemical Characteristics of Lambda DNA. In *The Bacteriophage Lambda*, ed. A. D. Hershey, pp. 45–82. Cold Spring Harbor, N. Y.: Cold Spring Harbor Laboratory.

Jones, K. W., and K. Murray. 1975. A Procedure for Detection of Heterologous DNA Sequences in Lambdoid Phage By *in situ* Hybridization. *J. Mol. Biol.* **96**:455–460.

Kaiser, A. D., and T. Masuda. 1973. *In vitro* Assembly of a Bacteriophage Lambda. *Natl. Acad. Sci. (USA) Proc.* **70**:260–264.

Maniatis, T., R. C. Hardison, E. Lacy, J. Lauer, C. O'Connell, D. Quon, G. K. Sim, and A. Efstratiadis. 1978. The Isolation of Structural Genes from Libraries of Eukaryotic DNA. *Cell* **15**:687–701.

Maxam, A., and W. Gilbert. 1977. A New Method for Sequencing DNA. *Natl. Acad. Sci. (USA) Proc.* **74**:560–564.

Polsky, F., M. H. Edgell, J. G. Seidman, P. Leder. 1978. High Capacity Gel Preparative Electrophoresis for Purification of Fragments of Genomic DNA. *Anal. Biochem.* **87**:397–410.

Sanger, F., and A. R. Coulson. 1975. A Rapid Method for Determing Sequences in DNA by Primed Synthesis with DNA Polymerase. *J. Mol. Biol.* **94**:441–448.

Sanger, F., S. Nicklen, and A. R. Coulson. 1977. DNA Sequencing with Chain-Termination Inhibitors. *Natl. Acad. Sci. (USA) Proc.* **74**:5463–5467.

Shinnick, T. M., E. Lund, O. Smithies, and F. R. Blattner. 1975. Hybridization of Labeled RNA to DNA Agarose Gels. *Nucl. Acids Res.* **2**:1911–1929.

Wang, J. C., and N. Davidson. 1966. On the Probability of Ring Closure of λ DNA. *J. Mol. Biol.* **19**:469–482.

20

Reprinted from *J. Mol. Biol.* **96**:171–184 (1975)

Ligation of *Eco*RI Endonuclease-generated DNA Fragments into Linear and Circular Structures

ACHILLES DUGAICZYK[1], HERBERT W. BOYER[2] AND HOWARD M. GOODMAN[1]

Departments of Biochemistry and Biophysics[1], and Microbiology[2]
University of California School of Medicine, San Francisco
Calif. 94143, U.S.A.

Double-stranded DNA fragments terminated at their 5′-ends by the single-stranded sequence pA-A-T-T-, generated by digestion of DNA with *Eco*RI restriction endonuclease, were ligated with *Escherichia coli* polynucleotide ligase under various conditions of temperature, concentration and time. The linear and circular products of ligation were separated by electrophoresis in agarose gel and quantitated by densitometry. The rate of ligation of (*Eco*RI-cleaved) simian virus (SV40) DNA at a concentration of 100 μg/ml increased from 0°C to 5°C to 10°C (6-fold increase overall); raising the temperature to 15°C did not further increase the rate of ligation. At the appropriate DNA concentrations, the predominant products of ligation are either linear concatemers that are integral multimers of the starting DNA fragment, or covalently closed circular structures of the monomeric DNA fragment. Ligating a mixture of two different length DNA fragments gives rise to all of the possible expected recombinant molecules.

Linear or circular products of ligation were predicted by consideration of the total concentration of DNA termini, i, and the local concentration of one terminus in the neighborhood of the other on the same DNA molecule, j. The parameter j is a function of the length of a DNA molecule, providing this length is greater than the random coil segment of DNA. Experimentally it was found that circular structures are formed in significant amounts only under conditions when the value of j is several times greater than that of i. When $j = i$, equal amounts of linear and circular products would be expected, but most of the molecules were ligated into linear concatemers. No circular structure of a DNA fragment whose contour length l (6×10^{-2} μm) is smaller than the random coil segment value b ($7 \cdot 17 \times 10^{-2}$ μm) was observed, while circular structures of the dimer of the same molecule (12×10^{-2} μm) were detected.

1. Introduction

Restriction endonucleases have become increasingly useful for cleaving DNA into specific fragments containing one or a few genes (for review, see Boyer, 1974; Nathans *et al.*, 1974). Some of the endonucleases introduce staggered breaks in double-stranded DNA, generating short single-stranded cohesive ends, which are substrates for polynucleotide ligase on duplex formation with a complementary single-stranded end. Ligated mixtures of *Eco*RI endonuclease-generated DNA fragments have been used to transform *Escherichia coli* and replicative recombinant plasmids selected on the basis of genetic properties associated with one or more of the DNA fragments (Cohen *et al.*, 1973; Chang & Cohen, 1974; Hershfield *et al.*, 1974; Morrow *et al.*, 1974).

Recombinant plasmid DNA molecules were recovered from clones of transformed cells and characterized.

The present study was undertaken to determine the optimum conditions for ligation of defined DNA fragments and to determine the parameters governing their mode of polycondensation leading to linear or circular structures. In this work we analyze the relation between the size and concentration of DNA fragments, and the distribution of molecular species obtained in the course of ligation of these fragments.

2. On the Probability of Ring Closure of Linear DNA

In the theory of the cyclization of DNA a factor j is involved (Jacobson & Stockmayer, 1950), which is the effective concentration of one end in the neighborhood or volume of the other end of the same molecule:

$$j = \left(\frac{3}{2\pi \, lb}\right)^{3/2} \text{ (ends/ml)}, \tag{1}$$

where l is the contour length and b the random coil segment length of a DNA molecule. The value of b, taken from the interpretation of sedimentation coefficients of DNAs, is $7 \cdot 17 \times 10^{-2}$ μm (Hearst & Stockmayer, 1962) and l_λ is $13 \cdot 2$ μm (MacHattie & Thomas, 1964). The calculated value of j_λ from equation (1) is then $3 \cdot 6 \times 10^{11}$ ends/ml for phage λ DNA, which agrees well with the experimental value ($3 \cdot 4 \times 10^{11}$ ends/ml) reported by Wang & Davidson (1966).

Equation (1) can be written so that j for any DNA molecule can be calculated from the j value for phage λ DNA and the molecular weight (MW) of the DNA in question, i.e.

$$j = j_\lambda \left(\frac{l_\lambda}{l}\right)^{3/2} = j_\lambda \left(\frac{MW_\lambda}{MW}\right)^{3/2} \text{ (ends/ml)}, \tag{2}$$

where $j_\lambda = 3 \cdot 6 \times 10^{11}$ ends/ml and $MW_\lambda = 30 \cdot 8 \times 10^6$ (Davidson & Szybalski, 1971). Note that j is constant for a linear DNA molecule of given length and independent of the DNA concentration. It depends on the ionic strength, however, because b is ionic strength dependent (Hearst, cited in Wang & Davidson, 1968), but this will not be considered here.

If the single-stranded ends of a duplex linear DNA are self-complementary†, then their total concentration/ml, i, is given by

† Two ends are considered self-complementary if they are identical and complementary. Such ends are generated in duplex DNA by the *Eco*RI endonuclease

$$\text{pA-A-T-T} \rule[0.5ex]{2.5em}{0.4pt}\rule[0.2ex]{2.5em}{0.4pt} \text{T-T-A-Ap}$$

(Hedgpeth *et al.*, 1972) or by the *Hind*III endonuclease

$$\text{pA-G-C-T} \rule[0.5ex]{2.5em}{0.4pt}\rule[0.2ex]{2.5em}{0.4pt} \text{T-C-G-Ap}$$

(Old *et al.*, 1975).
The type of ends generated by the *Eco*RII endonuclease

$$\text{pC-C-A-G-G} \rule[0.5ex]{2.5em}{0.4pt}\rule[0.2ex]{2.5em}{0.4pt} \text{G-G-T-C-Cp}$$

(Boyer *et al.*, 1973) or those found in bacteriophage λ DNA

$$\text{pG-G-G-C-G-G-C-G-A-C-C-T} \rule[0.5ex]{2.5em}{0.4pt}\rule[0.2ex]{2.5em}{0.4pt} \text{C-C-C-G-C-C-G-C-T-G-G-Ap}$$

(Wu & Taylor, 1971) are not identical. They are cohesive and complementary, but not "self-complementary". For the latter type of ends, eqn (3) will become:

$$i = N_0 M \times 10^{-3} \text{ (ends/ml)}. \tag{3a}$$

$$i = 2N_0\, M \times 10^{-3}\ (\text{ends/ml}), \tag{3}$$

where N_0 is Avogadro's number, and M is the molar concentration of the DNA molecules. Combining equations (2) and (3), the concentration can be calculated for which $j = i$:

$$M = \frac{j_\lambda}{2N_0 \times 10^{-3}} \left(\frac{MW_\lambda}{MW}\right)^{3/2}. \tag{4}$$

Substituting $N_0 = 6{\cdot}022 \times 10^{23}$ molecules/mol, $j = 3{\cdot}6 \times 10^{11}$ ends/ml, $MW_\lambda = 30{\cdot}8 \times 10^6$, and converting the molar concentration of DNA molecules, M, to DNA concentration, [DNA] (in g/l), gives the simple relation:

$$[\text{DNA}] = \frac{51{\cdot}1}{(MW)^{1/2}}. \tag{5}$$

Above this DNA concentration (i.e. when $i > j$), linear n-mers should be favored, while below this concentration ($i < j$) more circularization should occur.

Equations (2) and (3) can be rearranged to calculate the $j{:}i$ ratio for a DNA solution of given molecular weight and concentration:

$$\frac{j}{i} = \frac{j_\lambda \left(\dfrac{MW_\lambda}{MW}\right)^{3/2}}{2N_0\, M \times 10^{-3}} = \frac{51{\cdot}1}{[\text{DNA}]\,(MW)^{1/2}}, \tag{6}$$

which can be transformed into

$$MW = \left(\frac{51{\cdot}1}{\dfrac{j}{i}\,[\text{DNA}]}\right)^2. \tag{7}$$

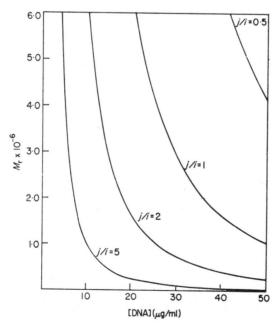

FIG. 1. A graph of eqn (7) showing the relation between molecular weight and DNA concentration for four different $j{:}i$ ratios.

A graph of equation (7) (for four j/i values: see Fig. 1) can be used to predict the type of products to be expected for the ligation of DNAs of different molecular weights at various concentrations.

3. Materials and Methods

(a) *DNA*

Simian virus (SV40) DNA was purified as described by Dugaiczyk *et al.* (1974). Bacteriophage ϕ80 DNA was prepared from purified phage particles produced from lytically infected *E. coli* strain W1485 growing in L broth (C. Yanofsky, personal communication). The crude lysate was centrifuged at low speed to remove bacterial debris and the supernatant (5·5 l; 3×10^{14} plaque-forming units total) was adjusted to 1 mM-magnesium acetate and incubated for 2 h at room temperature with 1·5 mg RNAase A/l. Phage was precipitated overnight at 4°C in large separatory funnels after the addition of 140 g polyethylene glycol/l and 40 g NaCl/l (stirred until dissolved). The phage (white flocculent precipitate) was concentrated by centrifugation for 30 min at 8000 revs/min in a Sorval type GSA rotor. The pellet was suspended in 100 ml of T1 buffer (0·6 mM-MgSO$_4$·7 H$_2$O, 0·5 mM-CaCl$_2$, 6 mM-Tris·HCl, pH 7·3) and stirred overnight (with a few drops of chloroform). The suspension was centrifuged at 12,000 g for 20 min and the supernatant saved. The precipitate was homogenized with 30 ml of T1 buffer in a Teflon homogenizer, stirred for 2 h, then centrifuged at 12,000 g for 20 min. The combined supernatants were centrifuged at 100,000 g for 1 h to pellet the phage. The precipitate was resuspended in the smallest possible volume (5 ml or less) of 50 mM-Tris·HCl (pH 7·5) with 1 drop of antifoam and layered on CsCl step gradients of the following density composition: 1·7 g/cm^3 (2·5 ml), 1·5 g/cm^3 (3 ml), 1·3 g/cm^3 (3 ml), 1·2 g/cm^3 (2·6 ml), and a top layer of 36% sucrose (1·5 ml), all in 50 mM-Tris·HCl (pH 7·5). The gradients were centrifuged for 1·5 h at 33,000 revs/min in a Spinco SW40 rotor. The phage band was aspirated, resuspended in CsCl (density, 1·485 g/cm^3) and centrifuged to equilibrium for 48 h under the conditions described above. Purified phage was dialyzed against 50 mM-Tris·HCl (pH 7·5), 10 mM-MgSO$_4$, then diluted to give an absorbance at 260 nm of about 20 and extracted with phenol that had been equilibrated with 50 mM-Tris·EDTA (pH 7·5). The aqueous phase was extracted with ether, made 0·3 M with respect to sodium acetate, the DNA was precipitated by the addition of 2 vol. ethanol at -20°C and recovered by centrifugation.

Covalently closed, circular plasmid pSC101 DNA was isolated by an adaptation of a sodium chloride/sodium dodecyl sulfate cleared-lysate procedure (Hirt, 1967; Guerry *et al.*, 1973). DNA concentrations were determined spectrophotometrically in 10 mM-Tris, 1 mM-EDTA (pH 7·4); $A_{260\,nm}$ of 1·0 = 50 μg DNA/ml (Padmanabhan & Wu, 1972).

(b) *Enzymes*

E. coli polynucleotide ligase was purified as described by Modrich *et al.* (1973). Its spec. act. was 6000 units/mg protein, as defined by Modrich *et al.* (1973). It had no detectable endonucleolytic (nicking) activity when tested on supercoiled SV40 DNA under the conditions of the ligation experiments (see below), even at a 1000-fold excess of the enzyme compared to the amount used in these experiments. Any endonucleolytic activity would be readily detectable on supercoiled SV40 DNA when analyzed by gel electrophoresis (see Plate I, column a). It also had no detectable phosphatase or exonuclease activity when tested on linear SV40 DNA that was labeled with ^{32}P at the 5'-ends of the EcoRI cleavage (see control in Fig. 1). EcoRI endonuclease was prepared according to Greene *et al.* (1974). Polynucleotide kinase, bacterial alkaline phosphatase, and RNAase A were commercial preparations (Worthington).

(c) *Preparations of* EcoRI *DNA fragments*

DNA was digested using the EcoRI endonuclease in 100 mM-Tris·HCl (pH 7·5), 5 mM-MgCl$_2$, 50 mM-NaCl, as described by Greene *et al.* (1974). After completion of the reaction (assayed by agarose gel electrophoresis), the DNA was purified by extraction with phenol and then precipitation with ethanol at -20°C. In some experiments SV40 DNA cleaved by the EcoRI endonuclease was labeled at the 5′-termini with ^{32}P by the use of poly-nucleotide kinase as described by Dugaiczyk *et al.* (1974). The EcoRI endonuclease digest of ϕ80 DNA was separated by electrophoresis (see below) on a 1·2% agarose slab gel (17 cm × 36 cm) in an apparatus of our construction. Then 1 mg of digested DNA was loaded on an 8·2 cm^2 slot area of the gel and subjected to separation for 9 h at 125 V. Five of the smallest fragments were recovered from the gel by electrophoresis into dialysis bags, extraction with phenol and precipitation with ethanol at −20°C. When agarose was found to contaminate such preparations, the DNA was further purified by absorbing it onto small benzoyl-naphthoyl-DEAE-cellulose (Serva) columns (Pasteur pipet, about 1 cm of the resin), washing the column with 0·1 M-NaCl, 100 mM-Tris·HCl (pH 7·6), and eluting the DNA with 1 M-NaCl, 100 mM-Tris·HCl (pH 7·6).

(d) *Molecular weights*

The molecular weight of pSC101 DNA was taken from Cohen *et al.* (1973), those for the 6 EcoRI endonuclease-generated fragments of λ DNA are electron microscopic determinations (Thomas & Davis, 1975). The 10 EcoRI-generated fragments of ϕ80 DNA (Helling *et al.*, 1974) are numbered, starting with the largest fragment ϕ80(1). The 5 smallest fragments, ϕ80(6) to ϕ80(10), are used in this work. The length of the ϕ80(6) fragment was compared with the length of linear SV40 DNA on the same electron microscope grids. In 7 measurements, the length of the fragment was found to be 54·56% ($\sigma = 0.659$) of the length of SV40 DNA ($M_r = 3.33 \times 10^6$), and therefore its molecular weight is 1.82×10^6. It should be noted that the molecular weight of SV40 DNA is variable as a result of frequent insertions and deletions occurring during lytic growth at high multiplicities of infection (Yoshike & Furuno, 1969); reported values range from 3.25×10^6 (Helling *et al.*, 1974) to 3.6×10^6 (Tooze, 1973). However, a value of $M_r = 3.33 \times 10^6$ for SV40 DNA and a value of 1.85×10^6 for the ϕ80(6) fragment, both as monomers and as ligated n-mers, fit best with the electrophoretic mobility of the EcoRI fragments of λ DNA; for example, it can be seen from Plate I (column d) that the mobility of the ϕ80(6) dimer is the same as that of a λ DNA fragment with $M_r = 3.70 \times 10^6$. In the same way, the molecular weights of the smaller ϕ80 DNA fragments were estimated, i.e. by comparing the electrophoretic mobility of a given n-mer of their ligation products with the mobility of a λ DNA fragment whose molecular weight was known from electron microscopy.

(e) *Ligation of DNA fragments*

The reaction was carried out in volumes ranging from 20 to 200 μl, containing 50 mM-Tris·HCl (pH 8·0), 5 mM-MgCl$_2$, 10 mM-(NH$_4$)$_2$SO$_4$, 25 μM-NAD, 50 μg crystalline bovine albumin/ml, and varying amounts of DNA (the original conditions of Modrich *et al.* (1973) for the ligase reaction call for MgCl$_2$ and EDTA, but the latter was omitted in our work). After preincubation for 15 min at a given temperature, the reaction was started by the addition of the ligase (0·5 to 5 μl, containing 1·4 to 14 units of activity). Progress of ligation was followed by withdrawing samples and either measuring ^{32}P that was rendered phos-phatase resistant, i.e. converted from external 5′-phosphomonoesters into internal phos-phodiesters (Dugaiczyk *et al.*, 1974), or by electrophoretic separation of the ligated products on agarose gels (Helling *et al.*, 1974).

(f) *Agarose gel electrophoresis*

Gels containing 0·7 to 1·2% agarose (SeaKem) were made in a slab type apparatus (E-C Apparatus Corp.). Agarose was melted by refluxing in water, then buffer was added to a final concentration of 50 mM-Tris (pH 8·05), 20 mM-sodium acetate, 18 mM-NaCl and 2 mM-EDTA adjusted with glacial acetic acid originally to pH 8·20 in 10× concentrated buffer. At 2 h or more after the gel was poured, samples were loaded containing 0·02 to 10 μg DNA, 0·01% bromphenol blue, and 20% sucrose in a volume of 15 to 45 μl/0·1 cm^2 of slot area. Electrophoresis was performed at 100 V (17 cm long gels) for about 5 h at 25°C. The gels were soaked for about 15 min in ethidium bromide (4 μg/ml) and photo-graphed in u.v. light through a yellow no. 9 Kodak Wratten gelatin filter (Sharp *et al.*, 1973; Helling *et al.*, 1974). The films were traced using a Joyce–Loebl microdensitometer to quantitate the relative distribution of the products of ligation shown in the Plates.

4. Results

(a) *Temperature optimum*

The effect of temperature on the rate of ligation of SV40 DNA was investigated by determining the conversion of phosphatase-sensitive ^{32}P on terminally labeled *Eco*RI endonuclease-generated linear SV40 DNA into phosphatase-resistant internal phosphodiesters by *E. coli* polynucleotide ligase (Fig. 2). This reaction is mediated *via* the self-complementary, single-stranded *Eco*RI-generated ends. The optimum temperature for ligation of nicked DNA is 37°C, but at this temperature the hydrogen-bonded *Eco*RI ends of DNA are quite unstable, having a t_m of 5 to 6°C for melting a hydrogen-bonded circular monomer to a linear monomer of SV40 DNA (Mertz & Davis, 1972). The experiment was designed to find the optimum temperature for ligation of hydrogen-bonded *Eco*RI termini; it was found that for SV40 DNA the rates of ligation increased by increasing the temperature from 0°C to 5°C to 10°C (Fig. 2). At 15°C the rate of ligation remained the same as that at 10°C (Fig. 2). Although the temperature dependence of the rate of ligation will vary with the concentration and length

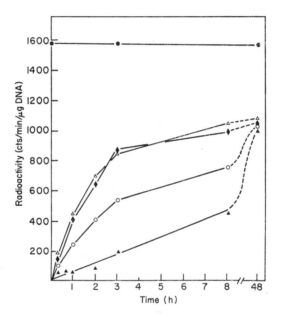

Fig. 2. Ligation of *Eco*RI-cleaved SV40 DNA, labeled with ^{32}P at the 5′-ends of the *Eco*RI endonuclease break. At the indicated temperatures (▲, 0°C; ○, 5°C; ◆, 10°C; △, 15°C; ●, control at 10°C) the reaction was carried out in 100 μl of the ligase buffer (see Materials and Methods), containing 10 μg of DNA ($j:i = 0.28$) and 17 units of activity of *E. coli* polynucleotide ligase, which was added once at zero time. Progress of ligation was followed by measuring ^{32}P that was converted from phosphatase-hydrolyzable, 5′-phosphomonoesters to phosphatase-resistant internal phosphodiesters in the acid-precipitable DNA. This was done by withdrawing 10-μl portions (1580 cts/min) from the ligation reaction into 100 μl of 0.1 M-Tris·HCl (pH 8.0), 10 mM-EDTA, containing 2 units of activity of bacterial alkaline phosphatase, and incubating for 30 min at 37°C or 65°C. DNA was precipitated together with 80 μg of carrier calf thymus DNA by the addition of 2 ml ice-cold 5% trichloroacetic acid. After 15 min at 0°C, the samples were filtered on glass-fiber filters, washed with cold 2 N-HCl, then ethanol, dried, and counted in a scintillation counter. Hydrolysis by the phosphatase at 37°C and at 65°C gave the same values, indicating that ligation occurred always at both DNA strands. In the control, the phosphatase treatment was omitted; it represents total "acid-precipitable" counts in the DNA during the process of ligation.

of the DNA fragments, for subsequent experiments 12·5°C was chosen as an arbitrary optimum for all ligations.

(b) *Cross-ligation of SV40 DNA with a $\phi80(6)$ DNA fragment*

$\phi80(6)$ is a DNA fragment ($M_r = 1·85 \times 10^6$) obtained from the *Eco*RI endonuclease digest of $\phi80$ DNA. At a concentration of 156 μg/ml ($j = 2·45 \times 10^{13}$ ends/ml; $i = 1·01 \times 10^{14}$ ends/ml; $j:i = 0·24$), ligation at 12·5°C gives rise to several products that separate into distinct bands on agarose gels (Plate I, column d). Column b (Plate I) of the same gel shows the separation of the products of ligation of (*Eco*RI-cleaved) SV40 DNA ($M_r = 3·33 \times 10^6$) at a concentration of 26 μg/ml ($j = 1·01 \times 10^{13}$ ends/ml; $i = 9·41 \times 10^{12}$ ends/ml; $j:i = 1·08$) and in column c, the separation of the cross-ligation products between the two DNA species can be seen. New bands among the cross-ligation products that have different mobility in the gel must correspond to cross-ligated SV40 $\times$ $\phi80(6)$ recombinant DNA. In the cross-ligation reaction, the $\phi80(6)$ DNA concentration was 52 μg/ml ($j = 2·45 \times 10^{13}$ ends/ml; $i = 3·38 \times 10^{13}$ ends/ml), that of SV40 DNA was 26 μg/ml ($j = 1·01 \times 10^{13}$ ends/ml; $i = 9·41 \times 10^{12}$ ends/ml). The total concentration of the identical ends of the two DNA species will be the sum $i = 3·38 \times 10^{13} + 9·41 \times 10^{12} = 4·32 \times 10^{13}$ ends/ml, whereas j will remain unaffected by the presence of a second DNA species; therefore, for $\phi80(6)$ DNA, $j:i = 0·57$; for SV40 DNA, $j:i = 0·23$. The molar ratio of the two DNA fragments ($\phi80(6):$SV40) is 3·6, and is equal to the ratio of their i values.

The molecular weights of the ligation products were estimated from a graph relating log molecular weight *versus* electrophoretic mobility of a DNA molecule in the gel (not shown). The molecular weights of the products of ligation of both SV40 DNA and $\phi80(6)$ DNA fragment correspond to integral multiples of the given monomer. In fact, such ligation mixtures provide a very good set of markers that can be used as convenient DNA standards over a wider range of molecular weights than is generally available.

From the molecular weights of some of the SV40 $\times$ $\phi80(6)$ cross-ligation products, the hybrids were identified as one SV40 $\times$ one $\phi80(6)$ ($M_r = 5·18 \times 10^6$), one SV40 $\times$ two $\phi80(6)$ ($M_r = 7·03 \times 10^6$), one SV40 $\times$ three $\phi80(6)$ ($M_r = 8·88 \times 10^6$), and two SV40 $\times$ one $\phi80(6)$ ($M_r = 8·51 \times 10^6$) (Plate I).

The mobility in the gel of covalently closed circular forms of DNA is quite different from that of linear DNA of the same molecular weight. Under the conditions used in the present work, superhelical DNA moves faster than its linear form, whereas the covalently closed "open circle" (with no superhelicity) migrates slower than the linear species. In this way, for example, the three forms of SV40 DNA can be separated (Plate I, column a). Similarly, the separation of the "open circle" and the linear form of pSC101 DNA can be achieved (Plate II, column a). The superhelical form of the latter DNA also moves faster than its linear form (not shown). The sequence of mobility of the open circle, linear and superhelical form of DNA was found to be the same in the voltage range tested of 50 to 200 V and at an agarose concentration of 0·7 to 1·2%. However, this is not so in a Tris·borate buffer (Fisher & Dingman, 1971), in which at 50 V (1% agarose) the sequence of mobility of the three DNA forms is the same as that in the Tris·acetate buffer, i.e. the superhelical form being the fastest, followed by the linear DNA; but by increasing the voltage to 200 V, the linear form becomes the fastest followed by the superhelical. Using agarose/acrylamide gels and the same Tris·borate buffer, similar observations were made by Dingman *et al.* (1972). On the basis of the different electrophoretic behavior in the gel of the circular forms of DNA, the faint but distinct bands that are seen among the main ligation products in the gel (Plate I) and that do not migrate as integral multimers of the starting monomeric DNA fragment, are thought to be circular forms of DNA. Their discrete spacing pattern, running into the position of the fastest, naturally supercoiled DNA (Plate I), seems to indicate that these bands represent circular DNA species with discrete and increasing numbers of superhelical turns.

(c) *Cross-ligation of pSC101 DNA with a φ80(6) DNA fragment*

This experiment is similar to the previous cross-ligation of SV40 DNA with the same φ80(6) DNA fragment, except that it is done with a higher DNA concentration (lower $j:i$ ratio). The gel separation of the products is shown in Plate II. Column b shows the ligation products up to a tetramer of pSC101 DNA, ligated at a concentration of 30 μg/ml ($j = 4 \cdot 41 \times 10^{12}$ ends/ml; $i = 6 \cdot 23 \times 10^{12}$ ends/ml; $j:i = 0 \cdot 71$). Ligation of the φ80(6) DNA fragment at a concentration of 156 μg/ml ($j = 2 \cdot 45 \times 10^{13}$ ends/ml; $i = 1 \cdot 01 \times 10^{14}$ ends/ml; $j:i = 0 \cdot 24$) gives rise to n-mers up to a decamer (column d). In the cross-ligation reaction (column c), the pSC101 DNA concentration was 47·8 μg/ml ($j = 4 \cdot 41 \times 10^{12}$ ends/ml; $i = 9 \cdot 92 \times 10^{12}$ ends/ml; $j:i = 0 \cdot 049$), that of the φ80(6) DNA fragment was 124 μg/ml ($j = 2 \cdot 45 \times 10^{13}$ ends/ml; $i = 8 \cdot 09 \times 10^{13}$ ends/ml; $j:i = 0 \cdot 27$). For the calculation of the $j:i$ ratios in the cross-ligation reaction, the total concentration of the identical ends of the two DNA species is taken, i.e. $i = 9 \cdot 92 \times 10^{12} + 8 \cdot 09 \times 10^{13} = 9 \cdot 08 \times 10^{13}$ ends/ml.

The combination of the molecular weights of the two DNA species in this experiment presents a stringent test for the resolving power of the gels. The molecular weight of the pSC101 DNA monomer ($5 \cdot 80 \times 10^6$) is close to that of the φ80(6) DNA trimer ($5 \cdot 55 \times 10^6$). The φ80(6) DNA tetramer will have a molecular weight ($7 \cdot 40 \times 10^6$) similar to that of the cross-ligation product of one pSC101 × one φ80(6) DNA ($7 \cdot 65 \times 10^6$), and the φ80(6) DNA pentamer will have a molecular weight ($9 \cdot 25 \times 10^6$) again similar to that of the cross-ligation product of one pSC101 × two φ80(6) DNA ($9 \cdot 50 \times 10^6$). An examination of the gel separation of the products of self-ligation and cross-ligation (Plate II) reveals that all of the above species of DNA of similar molecule weights can be discerned. Still higher polymerization products can be identified among the cross-ligation reaction mixture. For example, the band (Plate II, column c) that has an intermediate mobility between the φ80(6) DNA heptamer ($M_r = 12 \cdot 95 \times 10^6$) and the φ80(6) DNA octamer ($M_r = 14 \cdot 80 \times 10^6$; both in column d of Plate II) must be a mixture of two cross-ligation products: two pSC101 × one φ80(6) DNA ($M_r = 13 \cdot 45 \times 10^6$) and one pSC101 × four φ80(6) DNA ($M_r = 13 \cdot 20 \times 10^6$). Even the latter two DNA species might be expected to separate in the gel if the time of electrophoresis was extended.

(d) *Ligation of a φ80(9) DNA fragment*

φ80(9) is a DNA fragment ($M_r = 0 \cdot 38 \times 10^6$) obtained from the *Eco*RI endonuclease digest of φ80 DNA. It can be expected that shorter DNA fragments and decreased DNA concentration should favor the formation of circular forms that can be ligated into covalently closed circles. Furthermore, small DNA rings are expected to have a small number (one, or a few) of superhelical turns.

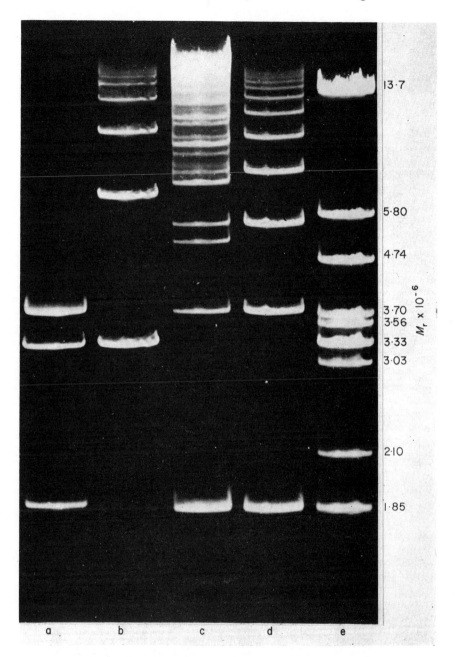

PLATE I. Gel electrophoretic separation of the products of ligation of SV40 DNA (column b), of the φ80(6) DNA fragment (column d), and of the products of cross-ligation of the two DNA species (column c). DNA concentrations of this experiment are given in Table 1 and in the text. The predominant products of ligation are linear concatemers. Column a shows the separation of the three forms of SV40 DNA: circular (top), linear (middle), and supercoiled (bottom). Column e shows the separation of linear DNA standards.

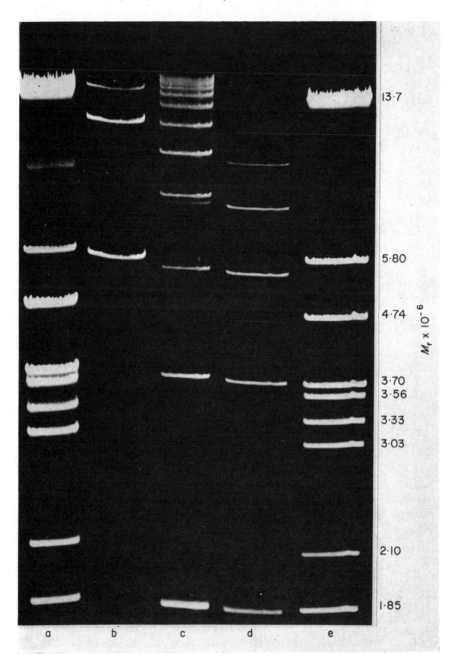

PLATE II. Gel electrophoretic separation of the products of ligation of pSC101 DNA (column b), of the ϕ80(6) DNA fragment (column d), and of the products of cross-ligation of the two DNA species (column c). Columns a and e show the separation of linear DNA standards; column a shows, in addition, the mobility of the circular form of pSC101 DNA. DNA concentrations of this experiment are given in Table 1 and in the text. The predominant products of ligation are linear concatemers.

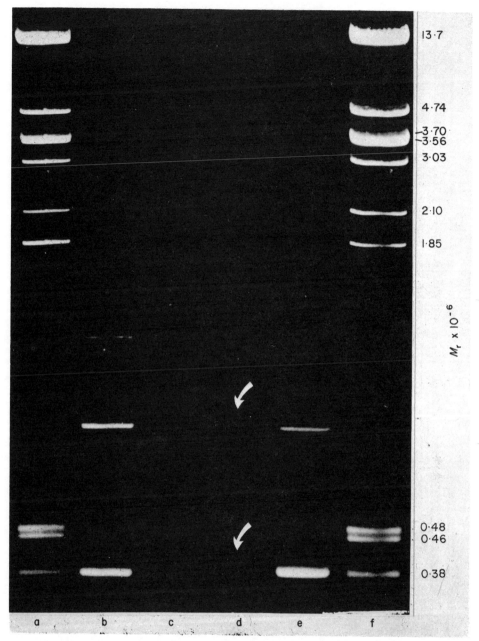

PLATE III. Gel electrophoretic separation of the products of ligation of a $\phi80(9)$ DNA fragment after different times of ligation and at two different DNA concentrations. Columns b and c show the composition of the reaction mixture after 15 min (column b) and after 2 h (column c) of ligation (initial $j:i$ ratio of 1·66 of the starting DNA fragment). Columns d and e show the composition of the reaction after 15 min (column e) and after 2 h (column d) of ligation (initial $j:i$ ratio of 6·64 of the same DNA fragment). DNA recovered from bands shown by arrows is composed of covalently closed circles and it is shown in Plate IV. Columns a and f show linear DNA standards.

A. Dugaiczyk, H. W. Boyer, and H. M. Goodman

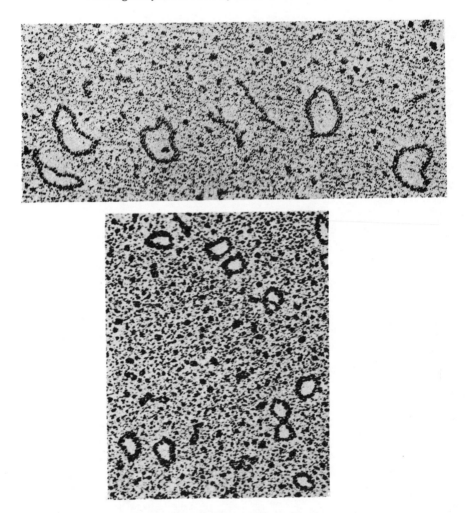

PLATE IV. Electron microscopic photograph of DNA recovered from bands shown by arrows in Plate III. Both photographs were taken under the same magnification and the circles have contour lengths of the monomeric and the dimeric $\phi80(9)$ DNA fragment.

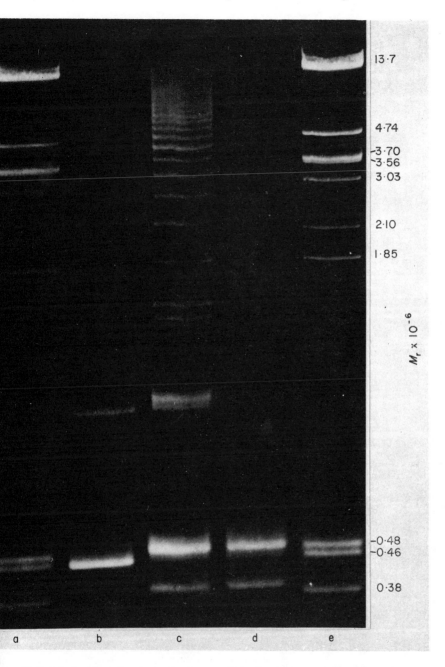

PLATE V. Gel electrophoretic separation of the products of ligation of a $\phi80(8)$ DNA fragment. Columns b and c show the products of ligation of the DNA fragment after 15 min (column b) and after 4 h (column c) of reaction (initial $j:i$ ratio of 1·56 of the starting DNA fragment). Column d shows the products of ligation after 4 h of reaction (initial $j:i$ ratio of 12·45 of the starting fragment). Columns a and e show linear DNA standards.

187

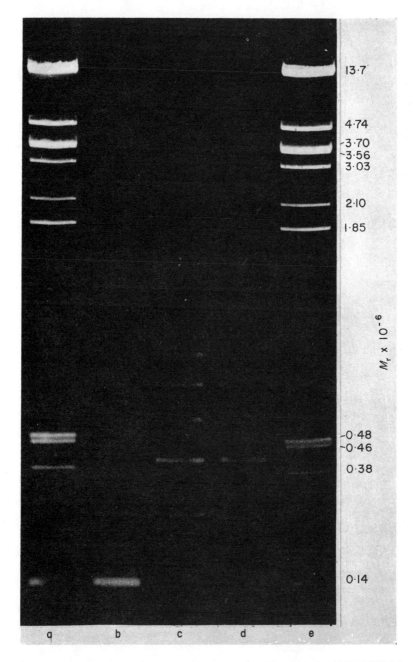

PLATE VI. Gel electrophoretic separation of the products of ligation of a $\phi80(10)$ DNA fragment at an initial $j:i$ ratio of $2\cdot80$ of the starting DNA. Column b shows the composition of the reaction mixture after 5 min, column c after 2 h, and column d after 3 h of reaction. External columns a and e show linear DNA standar

An analysis of the products of ligation of the $\phi80(9)$ DNA fragment is shown by the gel in Plate III. The reaction is compared at two different DNA concentrations and after different incubation times. At the higher DNA concentration (49·6 μg/ml, $j = 2\cdot58\times10^{14}$ ends/ml; $i = 1\cdot55\times10^{14}$ ends/ml; $j{:}i = 1\cdot66$) after 15 minutes of reaction (column b), linear n-mers up to a hexamer are formed. After two hours of ligation of the same DNA mixture (column c), almost all of the starting $\phi80(9)$ DNA monomer disappears giving rise to still higher linear n-mers and to two "satellite" bands (one above, one below) the band of the starting linear monomers and the band of the linear dimers. Similar satellite bands above and below the band of the linear trimer DNA can be faintly recognized in column c of Plate III, and were also observed in other experiments (not shown). At a four times lower DNA concentration (12·4 μg/ml, $j = 2\cdot58\times10^{14}$ ends/ml; $i = 3\cdot88\times10^{13}$ ends/ml; $j{:}i = 6\cdot64$) after 15 minutes of reaction (column e), linear n-mers are formed only up to a trimer but the two satellite bands already appear below and above the band of the linear $\phi80(9)$ DNA. Although after two hours of ligation of the same DNA mixture, linear n-mers up to a hexamer appear (column d), the predominant reaction products are the two satellite bands around the bands of the linear monomer and around that of the linear dimer.

DNA isolated from the "satellite" band (lower arrow in column d) above the band of the linear monomer was examined by electron microscopy and was found to be composed of covalently closed circular molecules having a contour length of the linear $\phi80(9)$ monomer. DNA isolated from the satellite band (upper arrow in column d) above the band of linear dimers was composed of covalently closed circles of a contour length twice as long. Plate IV shows both electron micrographs taken under the same magnification. DNA isolated from the satellite bands below the linear monomers and dimers, when examined in the electron microscope, gave the same length circular structures as DNA from the upper bands that are shown in Plate IV.

(e) *Ligation of a $\phi80(8)$ DNA fragment*

A similar experiment was done with a different, slightly larger DNA fragment, to see if circularization is independent of the source and/or composition of the DNA. Another $\phi80$ DNA fragment, designated $\phi80(8)$ ($M_r = 0\cdot46\times10^6$) was obtained from the *Eco*RI endonuclease digest of $\phi80$ DNA. As can be seen from the gel in Plate V, ligation at a DNA concentration of 48·4 μg/ml ($j = 1\cdot97\times10^{14}$ ends/ml; $i = 1\cdot27\times10^{14}$ ends/ml; $j{:}i = 1\cdot56$) gives rise to higher polymers (column c) as compared to ligation at eight times lower DNA concentration (6·05 μg/ml, $j = 1\cdot97\times10^{14}$ ends/ml; $i = 1\cdot58\times10^{13}$ ends/ml; $j{:}i = 12\cdot45$) for the same period of time (column d). In the first 15 minutes of ligation at the higher DNA concentration, only linear n-mers are formed up to a trimer or tetramer (column b). Later during the reaction the bands of the circular forms of DNA appear. Columns c and d show the molecular distribution of the products of the reaction at the two different DNA concentrations after four hours of ligation, which was about two hours beyond the time when the starting substrate, the linear monomeric $\phi80(8)$ DNA fragment, had disappeared from the reaction. As can be seen from this experiment, the satellite bands below and above the band of the linear monomers do not disappear even after prolonged ligation, which is consistent with the finding that they are covalently closed circular molecules, and hence are not substrates for the ligase. Electron microscopic examination of the

DNA isolated from these bands showed that these are circular molecules of the length of the ϕ80(8) DNA fragment (not shown).

(f) *Ligation of a ϕ80(10) DNA fragment*

Ligation of the smallest available DNA fragment ϕ80(10) ($M_r = 0.14 \times 10^6$) at a concentration of 47·8 μg/ml ($j = 1.21 \times 10^{15}$ ends/ml; $i = 4.21 \times 10^{15}$ ends/ml; $j:i = 2.80$) reveals a difference from the two previous results. The molecular distribution of the products of the reaction after five minutes, two hours and three hours of ligation is shown by the same gel in Plate VI, columns b, c and d, respectively. The characteristic bands of circular forms of DNA that are observed around the band of the starting linear monomer (Plates III and V) are absent here. However, bands do appear above the band of the linear DNA dimer and also around the bands of higher linear n-mers, which are thought to be circular dimers and circular n-mers. The lack of evidence for circular forms of the monomeric ϕ80(10) DNA fragment is particularly conspicuous, since the reaction was carried out under conditions more favorable for their formation than the reaction of ligation of the next larger ϕ80(9) DNA fragment, where monomeric circles were obtained. It would appear that the length of the ϕ80(10) DNA fragment is not sufficient to permit the formation of circular molecules.

5. Discussion

(a) *Temperature optimum*

The substrates used in the present work for the polynucleotide ligase were DNA fragments held together by weak forces of hydrogen bonding and stacking interaction of their single-stranded complementary ends. Although the optimum temperature for the polynucleotide ligase is expected to be 37°C, short hydrogen-bonded DNA termini like the *Eco*RI termini are not very stable at that temperature. Consequently, ligation must be carried out at a temperature that is a compromise between the optimum for the ligase and a temperature sufficiently close to the melting temperature, t_m, of cohered DNA fragments to provide a large number of hydrogen-bonded molecules. Any increase in the t_m (up to 37°C) of the cohesive ends permits a more efficient ligation at a higher temperature, closer to the optimum for the enzyme.

All of our present DNA fragments were terminated by the single-stranded (5') pA-A-T-T sequence (Hedgpeth *et al.*, 1972) and thus held together by forces of hydrogen bonding of four A·T pairs and a stacking interaction of six base-pairs, resulting in a t_m of 5 to 6°C for melting such ends in an SV40 circular monomer to a linear monomer (Mertz & Davis, 1972). Stronger cohesive ends, like those generated by the *Hind*III endonuclease (pA-G-C-T) (Old *et al.*, 1975), or by the *Eco*RII endonuclease (pC-C-A-G-G + pC-C-T-G-G) (Bigger *et al.*, 1973; Boyer *et al.*, 1973), or those of λ or ϕ80 DNA, have a higher t_m, and hence can be ligated at a higher temperature with better efficiency.

The value of t_m depends further on the concentration and the length of the DNA fragments. For linear condensation, t_m (the ratio of dimers to monomers) increases with the total concentration of the monomeric DNA fragments. Circular condensation, on the other hand, is governed by j, the effective concentration of one-end in the neighbourhood of the other-end on the same DNA fragment, and this is independent of the concentration of the fragments. But since j varies as $l^{-3/2}$ (eqn (1)), the t_m

for the circular forms will increase with the decreasing contour length, l, of the DNA fragments. Therefore, the higher the concentration and/or the smaller the DNA fragments, the higher the temperature that can be used for their ligation. The t_m of cohered ends depends also on the ionic strength of the medium, but any alteration of the ionic composition may have an adverse effect on the ligase.

At a DNA concentration of 100 μg/ml, the temperature for the optimum rate of ligation of SV40 DNA terminated by the *Eco*RI (pA-A-T-T) ends was found to be 12·5°C. As discussed above, this optimum will vary with the concentration and the length of the DNA fragments.

(b) *Ligation into linear* versus *circular structures*

As discussed above, the length and base composition of the cohesive ends and the ionic composition of the medium will affect the stability of the hydrogen-bonded ends (t_m) and hence their ligation. The circular and linear mode of condensation will be affected equally. To preferentially promote circularization, one has to increase the $j{:}i$ ratio of the DNA fragments in question. This can be achieved either by decreasing i *via* a decrease of the concentration of the fragments (j being concentration independent) or by increasing j. The latter can be increased only by decreasing the contour length of the fragment, l, or by decreasing the random coil segment value b (see eqn (1)). If a given DNA fragment is to be ligated, l is constant, but its b value can be decreased by increasing the ionic strength of the medium (see On the Probability of Ring Closure of Linear DNA). Thus, an increase in [Na$^+$] will have a twofold effect: (1) it will increase the t_m due to the electrostatic effect on the cohesive ends, which will influence equally the circular and linear modes of condensation, and (2) it will decrease the random coil segment b and thus increase j of the DNA fragment and preferentially promote its circularization. However, since changing the ionic strength may have an adverse effect on the activity of the ligase, the only way to change the ratio of linear to circular products is to change the concentration of the DNA fragments themselves.

Table 1 summarizes the parameters for the DNA fragments used in this work and the main type of products obtained in the progress of their ligation. As can be seen from these results, ligation of DNA fragments under conditions where the calculated concentration of one-end in the neighborhood of the other-end on the same molecule (j) is equal to the total concentration of ends (i) leads predominantly to the production of linear concatemers of the monomeric DNA. This is true for all the DNA fragments examined, starting with the largest, pSC101 DNA ($M_r = 5{\cdot}80 \times 10^6$), to the smallest, ϕ80(10) DNA fragment ($M_r = 0{\cdot}14 \times 10^6$). Under the above conditions ($j = i$), any desired integral multimer up to a pentamer or decamer can be obtained from the starting DNA monomer. When two molecules of different lengths are combined, all of the expected combinations are observed and can be separated and recovered in the less complex situations. It should be pointed out that, due to the vectorial character of the DNA fragments and the identity of their cohesive ends, linear dimers will be represented by three types of molecules ($\rightarrow \rightarrow$; $\rightarrow \leftarrow$; $\leftarrow \rightarrow$) (Mertz & Davis, 1972), expected to be biologically distinct. The situation will become more complex in cases of higher concatemers, but all vectorial combinations will be represented.

Covalently closed circular structures of DNA could be demonstrated among the products of ligation only when the $j{:}i$ ratio was 2 or higher; high $j{:}i$ ratios can be

TABLE 1

Molecular distribution of linear and circular products of ligation of DNA fragments terminated by the pA-A-T-T $\rightleftharpoons$ T-T-A-Ap sequence

Fragment	$M_r \times 10^{-6}$	l/b	Concn (μg/ml)	j/i	Reaction time	Monomer	Dimer	Trimer	Tetramer	Pentamer	Hexamer	Plate
pSC101	5·80	34·7	30	0·71	1 h	40l	32l	20l	8l			II(b)
SV40	3·33	19·9	26	1·08	1 h	30l	25l	11l	7l	1l		I(b)
φ80(6)	1·85	11·1	156	0·24	1 h	30l	21l	16l	13l	10l	6l	I(d)
φ80(8)	0·46	2·75	48·4	1·56	15 min	53l	26l	14l	4l			V(b)
φ80(8)	0·46	2·75	48·4	1·56	4 h	35c	20c 5l	5c 5l	⟵————	30 ————⟶		V(c)
φ80(8)	0·46	2·75	6·05	12·4	4 h	64c	12c 8l	4c 4l	8l			V(d)
φ80(9)	0·385	2·30	49·6	1·66	15 min	35l	25l	18l	13l	5l	1l	III(b)
φ80(9)	0·385	2·30	49·6	1·66	2 h	8c 4l	10c 10l	13l 3c	9l	8l	6l	III(c)
φ80(9)	0·385	2·30	12·4	6·64	15 min	60l 5c	20l	10l	5l			III(e)
φ80(9)	0·385	2·30	12·4	6·64	2 h	28c 5l	15c 15l	15l 3c	10l	5l	2l	III(d)
φ80(10)	0·14	0·84	47·8	2·80	5 min	65l	34l					VI(b)
φ80(10)	0·14	0·84	47·8	2·80	2 h	6l	13c 13l	20l	⟵————	48	————⟶	VI(c)
φ80(10)	0·14	0·84	47·8	2·80	3 h	<1l	12c 12l	21l	⟵————	54	————⟶	VI(d)
λ	30·8	184·1										

Linear (l) and circular (c) n-mers, given as percentage of total.

more readily obtained with smaller DNA fragments without having to work at inconveniently low DNA concentrations (see Fig. 1). The relative amount of circular products appearing in the process of ligation also increases with the time of the reaction. This can be predicted from Figure 1. For example, DNA fragments of molecular weight 10^6 and at a DNA concentration of 50 μg/ml ($j:i = 1$) will be initially ligated into linear concatemers. But after converting 80% of the monomers into higher polymers (contributing increasingly less to the total concentration of ends, i), the remaining 20% (10 μg/ml) of the monomers will have a $j:i$ ratio of 5. If ligation is continued beyond that point, these remaining 20% will be converted predominantly into circular monomers. Such predictions were confirmed experimentally. Table 1 shows that the relative amount of circular forms of DNA indeed increases with the time of the reaction. Eventually, all of the DNA should be converted into circular structures, which have no ends and cease to be substrate for the ligase.

In order to obtain significant amounts of circular molecules, it was necessary to carry out the ligations at high $j:i$ ratios, while the theory (see On the Probability of Ring Closure of Linear DNA) predicts equal amounts of circular and linear products at $j:i$ ratio of only 1. This might be the result of the short DNA fragments used in the present work. The random coil model will give correct values of j from equation (1) for large DNA molecules, like λ, which is almost 200 times the segment length b. Table 1 gives the $l:b$ ratios for the DNA fragments we have tested, and it can be seen that even the largest DNA used in this work, pSC101, has only 35 segment lengths, SV40 DNA has 20, while all the remaining fragments have even fewer. In addition, ligase interacting with DNA may increase b, leading consequently to a decrease of j (eqn (1)), and hence a decreased circularization of the DNA fragment; linear polymerization would remain unaffected by such an increase in b.

Our failure to find circular forms of the monomeric $\phi80(10)$ DNA fragment is to be expected. Its contour length l, calculated from its molecular weight, is 6×10^{-2} μm, which is already smaller than b ($7 \cdot 17 \times 10^{-2}$ μm), the random-coil segment of DNA (Hearst & Stockmayer, 1962). The failure of the latter DNA fragment to circularize was not caused by any damage of its cohesive ends, because it was readily ligated into a circular dimer and into high linear polymers. It must be concluded that the ridigity of double-stranded DNA is such as to impede the circularization of this approximately 225 base-pairs long DNA fragment.

We thank Dr Ned Mantei for exonuclease III and ^{3}H-labeled d(A-T)$_n$ copolymer. This investigation was supported by a U.S. Public Health Service grant (CA14026, GM14378).

REFERENCES

Bigger, D. H., Murray, K. & Murray, N. E. (1973). *Nature New Biol.* **244**, 7–10.

Boyer, H. W. (1974). *Fed. Proc. Fed. Amer. Soc. Exp. Biol.* **33**, 1125–1127.

Boyer, H. W., Chow, L. T., Dugaiczyk, A., Hedgpeth, J. & Goodman, H. M. (1973). *Nature New Biol.* **244**, 40–43.

Chang, A. C. Y. & Cohen, S. N. (1974). *Proc. Nat. Acad. Sci., U.S.A.* **71**, 1030–1034.

Cohen, S. N., Chang, A. C. Y., Boyer, H. W. & Helling, R. B. (1973). *Proc. Nat. Acad. Sci., U.S.A.* **70**, 3240–3244.

Davidson, N. & Szybalski, W. (1971). In *Cold Spring Harbor Laboratory*, pp. 45–82, Cold Spring Harbor, New York.

Dingman, C. W., Fisher, M. P. & Kakefuda, T. (1972). *Biochemistry*, **11**, 1242–1250.

Dugaiczyk, A., Hedgpeth, J., Boyer, H. W. & Goodman, H. M. (1974). *Biochemistry*, **13**, 503–512.

Fisher, M. P. & Dingman, C. W. (1971). *Biochemistry*, **10**, 1895–1899.

Greene, P. M., Betlach, M., Goodman, H. M. & Boyer, H. W. (1974). *Methods in Molecular Biology* (Wickner, R., ed.), **7**, 87–111.

Guerry, P., LeBlanc, D. J. & Falkow, S. (1973). *J. Bacteriol.* **116**, 1064–1066.

Hearst, J. & Stockmayer, W. H. (1962). *J. Chem. Phys.* **37**, 1425–1433.

Hedgpeth, J., Goodman, H. M. & Boyer, H. W. (1972). *Proc. Nat. Acad. Sci., U.S.A.* **69**, 3448–3452.

Helling, R. B., Goodman, H. M. & Boyer, H. W. (1974). *J. Virol.* **14**, 1235–1244.

Hershfield, V., Boyer, H. W., Yanofsky, C., Lovett, M. A. & Helinski, D. R. (1974). *Proc. Nat. Acad. Sci., U.S.A.* **71**, 3455–3459.

Hirt, B. (1967). *J. Mol. Biol.* **26**, 365–369.

Jacobson, H. & Stockmayer, W. H. (1950). *J. Chem. Phys.* **18**, 1600–1606.

MacHattie, L & Thomas, C. (1964). *Science.* **144**, 1142–1144.

Mertz, J. E. & Davis, R. W. (1972). *Proc. Nat. Acad. Sci., U.S.A.* **69**, 3370–3374.

Modrich, P., Anraku, Y. & Lehman, I. R. (1973). *J. Biol. Chem.* **248**, 7495–7501.

Morrow, J. F., Cohen, S. N., Chang, A. C. Y., Boyer, H. W., Goodman, H. M. & Helling, R. B. (1974). *Proc. Nat. Acad. Sci., U.S.A.* **71**, 1743–1747.

Nathans, D., Adler, S. P., Brockman, W. W., Danna, K. J., Lee, N. H. & Sack, G. H., Jr (1974). *Fed. Proc. Fed. Amer. Soc. Exp. Biol.* **33**, 1135–1138.

Old, R., Murray, K. & Roizes, G. (1975). *J. Mol. Biol.* **92**, 331–339.

Padmanabhan, R. & Wu, R. (1972). *J. Mol. Biol.* **65**, 447–464.

Sharp, P. A., Sugden, B. & Sambrook, J. (1973). *Biochemistry*, **12**, 3055–3063.

Thomas, M. & Davis, R. W. (1975). *J. Mol. Biol.* **91**, 315–328.

Tooze, J. (1973). Editor of *Cold Spring Harbor Laboratory*, p. 275, Cold Spring Harbor, New York.

Wang, J. C. & Davidson, N. (1966). *J. Mol. Biol.* **19**, 469–482.

Wang, J. C. & Davidson, N. (1968). *Cold Spring Harbor Symp. Quant. Biol.* **33**, 409–415.

Wu, R. & Taylor, E. (1971). *J. Mol. Biol.* **57**, 491–511.

Yoshike, K. & Furuno, A. (1969). *Fed. Proc. Fed. Amer. Soc. Exp. Biol.* **28**, 1899.

21

Reprinted from *Gene* 1:81–92 (1976). Originally published by Elsevier/North-Holland Biomedical Press B.V.

A GENERAL METHOD FOR INSERTING SPECIFIC DNA SEQUENCES INTO CLONING VEHICLES

(Synthetic decadeoxyribonucleotides; linker sequences; *lac* DNA; plasmids; restriction endonucleases; cloned DNA)

CHANDER P. BAHL, KENNETH J. MARIANS and RAY WU*

Section of Biochemistry, Molecular and Cell Biology, Cornell University, Ithaca, NY 14853 (U.S.A.)

and

JACEK STAWINSKY and SARAN A. NARANG

Division of Biological Sciences, National Research Council of Canada, Ottawa, Ontario (Canada)

(Received September 30th, 1976)
(Revision received and accepted October 5th, 1976)

SUMMARY

A general method has been developed to introduce any double-stranded DNA molecule into cloning vehicles at different restriction endonuclease sites. In this method a chemically synthesized decadeoxyribonucleotide duplex, containing a specific restriction endonuclease sequence, is joined by DNA ligase to both ends of the DNA to be cloned. The resulting new duplex DNA is cut by the same restriction endonuclease to generate the cohesive ends. It is then inserted into the restriction endonuclease cleavage site of the cloning vehicle. To demonstrate the feasibility of this new method, we have inserted separately the synthetic *lac* operator DNA at the *Bam*I and *Hind*III cleavage sites of the plasmid pMB9 DNA.

INTRODUCTION

Molecular cloning has become a powerful tool for the amplification of

---

*To whom correspondence should be addressed.
This is paper XXIX in a series on "Nucleotide Sequence Analysis of DNA".
Paper XXVIII is by Jay and Wu, 1976. This work was supported by Grants GM 18887 and CA 14989 from the National Institutes of Health, and No. 15576 from the National Research Council of Canada.
Abbreviations: IPTG, isopropyl thiogalactoside; ONPG, *o*-nitrophenyl-β-D-galactoside; X-gal, 5-bromo-5-chloro-3-indoyl-β-D-galactoside.

specific DNA fragments and their subsequent isolation in high yields. Two basic steps are involved in molecular cloning. First the DNA fragments to be cloned are joined in vitro to an autonomously replicating vehicle molecule, e.g. plasmid DNA (Cohen et al., 1973; Tanaka and Weisblum, 1975) or λ phage DNA (Thomas et al., 1974; Murray and Murray, 1974). The hybrid DNA formed is then introduced into *E. coli* by transformation and then cloned by single colony isolation or plaque formation.

In one cloning method, two different DNA molecules are cut by the same restriction endonuclease to produce identical cohesive ends. The DNA molecules are annealed to one another and then covalently joined by DNA ligase. This method limits the size and kind of DNA fragments that can be cloned since it often requires cloning of a much larger DNA fragment than one is interested in. For example, if one wants to clone a small DNA fragment such as a promoter (e.g. an RNA polymerase protected fragment), the nearest restriction endonuclease recognition sites may be relatively distant, and thus extraneous DNA sequences must be included in the cloned DNA. Furthermore, many DNA fragments cannot be cloned because of the lack of a suitable res-ₜriction enzyme to produce molecules with cohesive ends. In this communica-tion we report a general procedure to overcome these limitations. The proced-ure utilizes a chemically synthesized decadeoxynucleotide such as d(C-C-G-G-A-T-C-C-G-G) or d(A-C-A-A-G-C-T-T-G-T) to generate restriction enzyme recog-nition sites at the ends of any natural or synthetic DNA molecule to be cloned, thereby making the cloning procedure much more selective and ver-satile. By this method, we have inserted the synthetic *lac* operator DNA at the *Bam*I and *Hind*III sites, respectively, of pMB9 plasmid. The insertion of the *lac* operator into the *Eco*RI site of pMB9 has recently been accomplished for the large scale production of *lac* operator (Marians et al. and Heyneker et al., personal commun.).

MATERIALS AND METHODS

The lactose operator duplex DNA was chemically synthesized as described by Bahl et al. (1976). Lactose repressor (Platt et al., 1973) and T4 DNA ligase (Weiss et al., 1968 and Planet et al., 1973) were purified by published methods. All the procedures for labeling and purifying DNA have been described (Wu et al., 1976).

Chemical synthesis of the decadeoxyribonucleotides

The two decadeoxyribonucleotides d(C-C-G-G-A-T-C-C-G-G) (*Bam*I linker sequence) and d(A-C-A-A-G-C-T-T-G-T) (*Hind*III linker sequence) were syn-thesized by the improved phosphotriester method developed in our laboratory (Itakura et al., 1973, Itakura et al., 1975a, Katagiri et al., 1975, Itakura et al., 1975b, and Stawinsky et al., 1976). The scheme for synthesis and the reaction conditions are summarized in Fig. 1 and Table I, respectively. Two new steps were introduced in these syntheses: the dimethoxytrityl group was removed

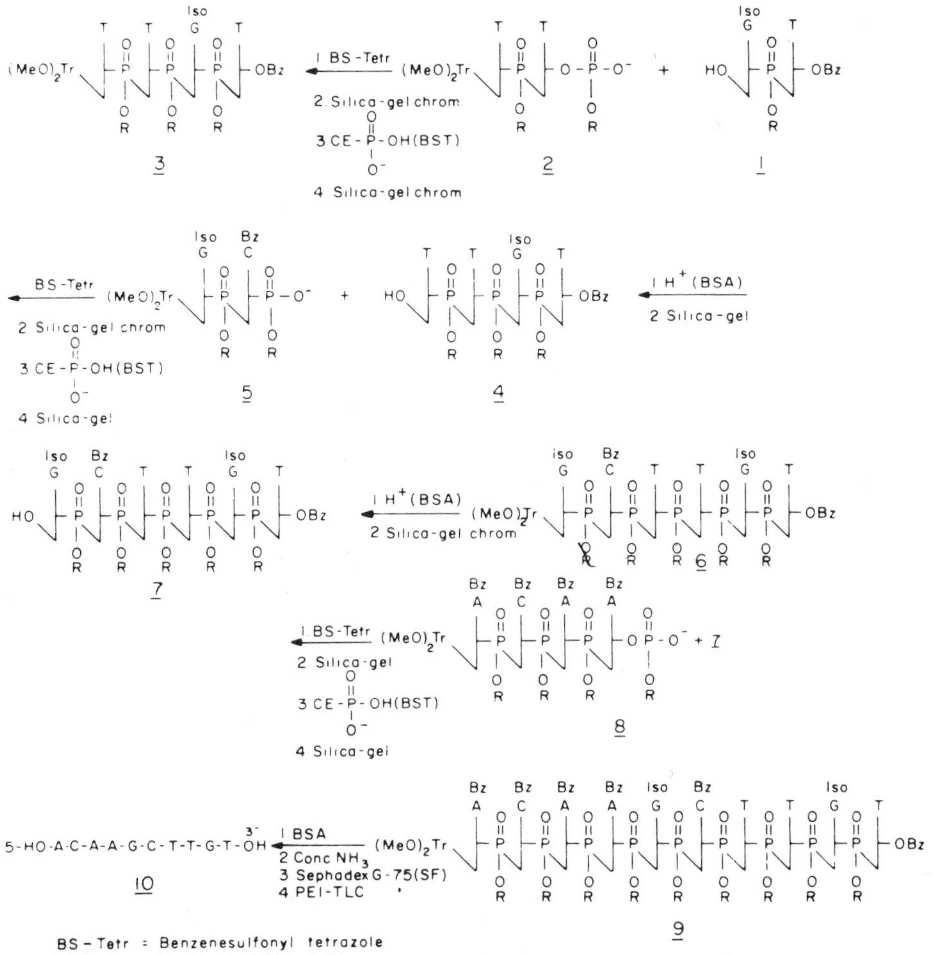

BS – Tetr = Benzenesulfonyl tetrazole
BSA = Benzenesulfonic Acid
R = p - Chorophenyl

Fig. 1. The modified phosphotriester approach for the synthesis of the decadeoxyribonucleotide d(A-C-A-A-G-C-T-T-G-T).

by a 2% solution of benzene-sulfonic acid in chloroform and the chlorophenyl phosphate protecting group was removed by treatment with concentrated ammonium hydroxide for 4—6 h (Stawinsky et al., 1976). The oligonucleotides were characterized by 2-D electrophoresis-homochromatography of their partial venom phosphodiesterase digestion products (Jay et al., 1974 and Tu et et al., 1976). The two-dimensional maps of the two oligonucleotides and the assignment of various spots are shown in Fig. 2, which verify the sequences of the two synthetic decadeoxynucleotides.

TABLE I

REACTION CONDITIONS OF VARIOUS CONDENSATION STEPS AND THE YIELDS OF THE PRODUCTS

5'-Protected component (mmole)	5'-Hydroxyl component (mmole)	Benzene-sulfonyl tetrazole (mmole)	Reaction time (h)	Product yield (%)
	I. 5'HO-A-C-A-A-G-C-T-T-G-T-OH3'			
[(MeO)₂Tr]dIsoG∓bzC-ClPh (0.4)	dT∓T∓IsoG∓T(OBz) (0.35)	1.2	1	[(MeO)₂Tr]dIsoG∓bzC∓T∓T∓IsoG∓T(Obz) (75)
[(MeO)₂Tr]dbzA∓bzC∓bzA∓bzA-ClPh (0.15)	dIsoG∓bzC∓T∓T∓IsoG∓T(OBz) (0.1)	0.45	12	[(MeO)₂Tr]dbzA∓bzC∓bzA∓bzA∓IsoG∓bzC∓T∓T∓OspG∓T(OBz) (60)
	II. 5'HO-C-C-G-G-A-T-C-C-G-G-OH3'			
[(MeO)₂Tr]dbzA∓T-ClPh (0.25)	dbzC∓bzC∓IsoG∓IsoG(OBz) (0.21)	0.75	1	[((MeO)₂Tr]dbzA∓T∓bzC∓bzC∓IsoG∓IsoG(OBz) (71)
[(MeO)₂Tr]dbzC∓bzC∓IsoG∓IsoG-ClPh (0.09)	dbzA∓T∓bzC∓bzC∓IsoG∓IsoG(OBz) (0.06)	0.27	12	[(MeO)₂Tr]dbzC∓bzC∓IsoG∓IsoG∓bzA∓T∓bzC∓bzC∓IsoG∓IsoG(OBz) (58)

Abbreviations are as suggested by the IUPAC-IUB, Biochemistry 9, 4022 (1970). A phosphodiester linkage is represented by hyphen and phosphotriester linkage is represented by (∓) symbol. Each internal internucleotidic phosphate is protected with p-chlorophenyl group (ClPh).

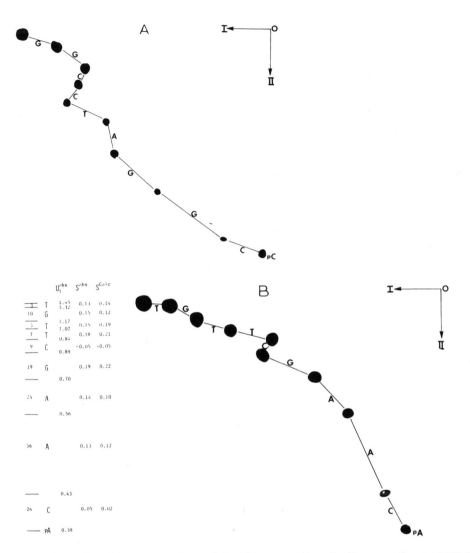

Fig. 2. Two-dimensional map of a partial snake venom phosphodiesterase digest of (A) the decanucleotide d(C-C-G-G-A-T-C-C-G-G) and (B) the decanucleotide d(A-C-A-A-G-C-T-T-G-T). Dimension I, electrophoresis on cellulose acetate strip at pH 3.5, dimension II, homochromatography on DEAE-cellulose thin-layer plate run in Homomixture VI (Jay et al., 1974). The δ values and mobility shift values (S^{calc} and S^{obs}) are as described by Tu et al. (1976).

Synthesis of duplex DNA linker molecules and their digestion by restriction enzymes

Chemically synthesized decadeoxynucleotides (400 pmoles) were phosphorylated at the 5' end using [γ-^{32}P] ATP and polynucleotide kinase (Wu et al., 1976). The labeled decadeoxynucleotides were dissolved in 100 μl of 100 mM Tris—HCl (pH 7.5), heated to 90°C for 1 min, quickly chilled to 0°C and then incubated at 70°C for 30 min. The duplex of the decadeoxynucleotide was formed by slowly cooling the sample to room temperature and then to 4°C. *Bam*I linker when digested with *Bam*I restriction endonuclease (Wilson and Young 1975) gave the expected trinucleotide ^{32}pC-C-G as the labeled product. The duplex *Hin*dIII linker molecule, when digested with *Alu*I (Roberts et al., 1976 and Jay and Wu, 1976) gave the expected pentanucleotide ^{32}pA-C-A-A-G as the labeled product.

Insertion of lac operator sequence into pMB9 plasmid DNA by the use of chemically synthesized linker molecules and blunt-end ligation

As shown in Fig. 3, the synthetic duplex linker (B) (10 pmole) and the synthetic *lac* operator (A) (0.6 pmole) were joined end-to-end (Sgaramella et al., 1970) by incubation with 3 units of T4 DNA ligase in 50 μl of a solution containing 20 mM Tris—HCl (pH 7.5), 10 mM dithiothreitol, 10 mM MgCl$_2$ and 35 μM ATP at 20°C for 6 h to produce molecule (C). The solution was heated at 70°C for 5 min to inactivate the ligase and cooled slowly to room temperature. 2 vol. of ethanol were added, and after 12 h at −20°C the DNA was pelleted at 10 000 g for 1 h. The pellet was dissolved in 50 μl of a solution containing 6.6 mM Tris—HCl (pH 7.5), 6.6 mM MgCl$_2$ and 1 mM dithiothreitol. To this solution was added 1 μg of pMB9 DNA (Rodriguez et al., 1976) and 2 units of the restriction endonuclease (*Bam*I or *Hin*dIII for the corresponding linker). The sample was incubated at 37°C for 12 h to produce molecule (D) and linear pMB9 DNA. The samples were then heated at 70°C for 5 min, cooled slowly to room temperature, and the DNA was precipitated with 2 vol. of ethanol as before. The DNA pellet was dissolved in 50 μl of a solution containing 20 mM Tris—HCl (pH 7.5), 10 mM MgCl$_2$, 10 mM dithiothreitol and 35 μM ATP. Three units of DNA ligase were added and the samples were incubated at 12.5°C for 24 h to produce the hybrid *lac*-pMB9 DNA. After heating at 70°C for 5 min, and slow cooling to room temperature, the hybrid *lac*-pMB9 DNA was directly used for the transformation step.

Transformation and selection of lac-pMB9 hybrid plasmids

The hybrid *lac*-pMB9 DNA was used to transform competent *E. coli* HB129 cells (Rodriguez et al., 1976). The procedure for transformation was exactly as that described in an earlier paper by Marians et al. (1976). The clones were screened for β-galactosidase escape synthesis (Marians et al., 1976) by using the reaction with ONPG or X-gal (Miller, 1972). The DNA from the clones showing positive color reactions due to over-production of β-galactosidase was isolated and tested for *lac* repressor binding by the millipore filter assay (Riggs et al., 1970).

RESULTS AND DISCUSSION

Chemically synthesized oligonucleotides with defined sequences have a great potential for the study of various problems in molecular biology. In this communication we describe a method that exploits the usefulness of synthetic oligonucleotides to create cohesive ends at the termini of any duplex DNA molecule so that the latter can be inserted into cloning vehicles. We have synthesized two decadeoxynucleotide fragments, d(C-C-G-G-A-T-C-C-G-G) and d(A-C-A-A-G-C-T-T-G-T). Each synthetic fragment is self-complementary and can form a duplex molecule which contains the recognition sequence of the restriction endonuclease *Bam*I (Wilson and Young, 1975) and *Hind*III (Old et al., 1975), respectively. The synthetic fragments can serve as linker molecules* for the insertion of any duplex DNA molecules into cloning vehicles. The principle of this general method is illustrated in Fig. 3. Its feasibility has been tested

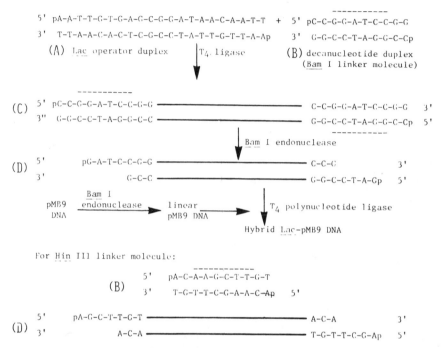

Fig. 3. A general scheme for introducing cohesive ends at the termini of any duplex DNA molecule for molecular cloning. Short dotted lines represent the recognition site of a restriction endonuclease; long double lines represent the 21 nucleotide-long duplex of *lac* operator DNA.

---

*These synthetic oligonucleotides are also useful model substrates for studying the mechanism of cleavage of DNA by a number of restriction endonucleases. For example, the synthetic decanucleotide d(C-C-G-G-A-T-C-C-G-G) in the duplex form also contains the recognition sites for *Hpa*II (Garfin and Goodman, 1974) and *Mbo*I (Roberts, R., personal commun.) restriction enzymes. Similarly, the oligonucleotide d(A-C-A-A-G-C-T-T-G-T) contains the recognition sequence for *Alu*I (Roberts et al., 1976; Jay and Wu, 1976). Moreover, the sequence d(C-C-G-G-A-T-C-C-G-G) as a duplex can be polymerized by T4 ligase and all the fragments, dimer onwards will contain restriction endonuclease recognition sites for *Hae*III (Middleton et al., 1972) and *Bal*I (Roberts, R., personal commun.) endonucleases. A study on the cleavage of a synthetic octanucleotide, d(T-G-A-A-T-T-C-A), by *Eco*RI endonuclease has been reported by Green et al. (1975).

by introducing a synthetic *lac* operator duplex into pMB9 plasmid. In the first step, a duplex linker molecule (B) is enzymatically ligated to both ends of the *lac* operator DNA duplex (A) to produce molecule (C). The latter is cleaved by the proper restriction enzyme to produce molecule (D) which contains the desired cohesive ends at both ends. The DNA molecule (D) is then hybridized to pMB9 DNA which has been made linear with the same restriction enzyme and thus contains the same cohesive ends. The hydrogen-bonded molecules are enzymatically ligated to form hybrid *lac*-pMB9 DNA. The latter is used to transform *E. coli* cells to tetracycline resistance. The resulting colonies are screened for the over-production of β-galactosidase, which indicates the insertion of the *lac* operator DNA. The cells which contain the hybrid *lac*-pMB9 DNA are purified by streaking on agar plates which contain tetracycline and X-gal (Miller, 1972; Marians et al., 1976).

The hybrid *lac*-pMB9 DNA was isolated, after amplification by the addition of chloroamphenicol to the bacterial culture, labeled by nick translation (Maniatis et al., 1975), and studied for *lac* repressor binding properties. Typical binding curves are shown in Fig. 4. The inhibitory effect of IPTG shows that the binding is specific, and thus confirms that a *lac* operator has been inserted into the plasmid.

The hybrid plasmid DNA was further characterized by digestion with the appropriate restriction endonuclease (*Bam*I or *Hind*III corresponding to the linker used) and then labeled at the 3'-ends by repair synthesis in the presence of [α-^{32}P] dNTP (Wu et al., 1976). It gave two fragments on polyacrylamide gel electrophoresis as shown in Fig. 5. The large fragment corresponds to the linear pMB9 DNA. The small fragment corresponds to the *lac* operator since it binds specifically to the *lac* repressor.

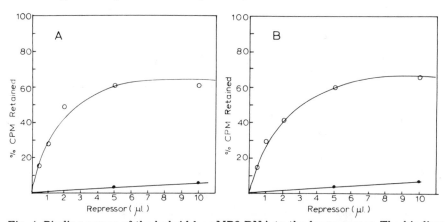

Fig. 4. Binding curves of the hybrid *lac*-pMB9 DNA to the *lac* repressor. The binding of ^{32}P-labeled *lac*-pMB9 DNA (0.1 μg per point) to *lac* repressor (15 ng/μl) on millipore membrane filters was carried out as described by Bahl et al. (1976). (A) Hybrid plasmid constructed with *Bam*I linker to introduce *lac* operator into pMB9 plasmid. (B) Hybrid plasmid constructed with the use of *Hind*III linker. ○, no isopropyl thiogalactoside (IPTG); ●, with 1 mM IPTG.

Fig. 5. An autoradiograph of gel electrophoresis of hybrid *lac*-pMB9 plasmid. (A) pMB9 plasmid (as control) digested by *Bam*I restriction endonuclease. (B) Hybrid plasmid constructed with *Bam*I linker and digested by *Bam*I restriction endonuclease. (C) Hybrid plasmid constructed with *Hind*III linker and digested by *Hind*III restriction endonuclease. Experimental details are the same as described by Marians et al., 1976.

In the insertion of the 21 nucleotide-long duplex *lac* operator into pMB9 plasmid, the length of the linker molecules was so designed that after linking, three extra nucleotides were introduced on either side of the *lac* operator. This leaves the reading frome for pMB9-directed protein synthesis unchanged. Rodriguez et al. (1976) reported that *Hind*III and *Bam*I restriction endonuclease sites in pMB9 are located in the gene which confers tetracycline resistance. We found that the insertion of 27 nucleotides to produce the hybrid *lac*-pMB9 DNA does not change the tetracycline resistance of cells harboring this hybrid plasmid. This observation can have two interpretations. First, the *Bam*I site and *Hind*III sites are not in the tetracycline resistance gene. Alternatively, insertion of 27 nucleotides does not alter the activity of the protein for which it codes.

Chemically synthesized linker molecules are extremely useful tools in molecular cloning since the same oligonucleotide linker can serve to introduce any double-stranded DNA molecule into cloning vehicles at specific sites. The double-stranded DNA may be obtained by cleavage with any restriction enzyme or by other means. If the ends of the DNA are even, they can be joined directly to the linker molecule. If the ends are uneven, they can be digested with Aspergillus nuclease S_1 (Ando et al., 1966; Ghangas and Wu, 1975) to produce even-end DNA molecules.

Another advantage of this method is the facility with which two different

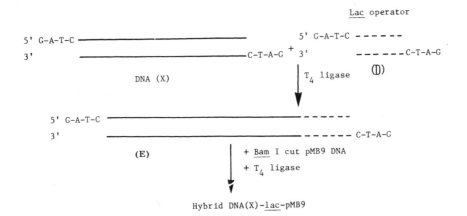

Fig. 6. A general scheme for linking any DNA molecule with *lac* operator and pMB9 plasmid to make a hybrid DNA(X)-*lac*-pMB9 which can be easily screened for the overproduction of β-galactosidase activity (Marians et al., 1976).

DNA molecules can be cloned together. One DNA (e.g. *lac* operator) is chosen to allow for an easy screening for cells harboring hybrid plasmids, as shown by the following example. The *lac* operator sequence with the *Bam*I cohesive ends (such as molecule D in Fig. 3) can be ligated in vitro to another DNA molecule (X) containing similar cohesive ends (Fig. 6). After separation of the unreacted *lac* operator from the hybrid DNA (X)-*lac* operator molecule (molecule E), the latter can then be joined to a cloning vehicle. The lactose operator sequence in the hybrid DNA (X)-*lac*-pMB9 should make the cells constitutive for β-galactosidase (Marians et al., 1976) which are easy to screen. In this way, the presence of DNA (X) in the hybrid plasmid can be easily detected. The purification of hybrid DNA (X)-*lac*-pMB9 can also be carried out by binding to *lac* repressor on a millipore filter and eluting with IPTG (Riggs et al., 1970; Marians et al., 1976). Once the hybrid DNA (X)-*lac*-pMB9 is characterized, it can be produced in large amounts by the usual method for the amplification of pMB9. Thus, large quantities of DNA (X) can be obtained, after digestion with the same restriction enzyme used for its insertion, for DNA sequence analysis and other types of studies.

In conclusion, the cloning procedure described in this paper has a general application for the cloning of a variety of DNA molecules, and will facilitate the solving of different problems in molecular biology.

REFERENCES

Ando, T., A nuclease specific for heat-denatured DNA isolated from a product of *Aspergillus oryzae*, Biochim. Biophys. Acta, 114 (1966) 158—168.
Bahl, C.P., Wu, R., Itakura, K., Katagiri, N. and Narang, S.A., Chemical and enzymatic synthesis of lactose operator of *Escherichia coli* and its binding to lactose repressor, Proc. Natl. Acad. Sci. USA, 73 (1976) 91—94.

Cohen, S.N., Chang, A.C.Y., Boyer, H.W. and Helling, R.B., Construction of biologically functional bacterial plasmids in vitro, Proc. Natl. Acad. Sci. USA, 70 (1973) 3240–3244.

Garfin, D.E. and Goodman, H.M., Nucleotide sequences at the cleavage sites of two restriction endonucleases from *Haemophilus parainfluenzae*, Biochem. Biophys. Res. Commun. 59 (1974) 108–116.

Ghangas, G.S. and Wu, R., Specific hydrolysis of the cohesive ends of bacteriophage λDNA by three single strand-specific nucleases, J. Biol. Chem., (1975) 4601–4606.

Green, P.J., Poonian, M.S., Nussbaum, A.L., Tobias, L., Garfin, D.E., Boyer, H.W. and Goodman, H.M., Restriction and modification of a self-complementary octanucleotide containing the *Eco*RI substrate, J. Mol. Biol., 99 (1975) 237–261.

Heyneker, H.L., Shine, J., Goodman, H.M., Boyer, H., Rosenberg, J., Dickerson, R.E., Narang, S.A., Itakura, K., Lin, S. and Riggs, A.D. Chemically synthesized *lac* operator DNA is functional in vivo, Nature (in press).

Itakura, K., Bahl, C.P., Katagiri, N., Michniewicz, J.J., Wightman, R.H. and Narang, S.A., A modified triester method for the synthesis of deoxyribopolynucleotides, Canad. J. Chem., 51 (1973) 3649–3651.

Itakura, K., Katagiri, N., Bahl, C.P., Wightman, R.H. and Narang, S.A., Improved triester approach for the synthesis of pentadecathymidylic acid, J. Am. Chem. Soc., 97 (1975a) 7327–7332.

Itakura, K., Katagiri, N., Narang, S.A., Bahl, C.P., Marians, K.J. and Wu, R., Chemical synthesis and sequence studies of deoxyribo-oligonucleotides which constitute the duplex sequence of the lactose operator of *Escherichia coli*, J. Biol. Chem., 250 (1975b) 4592–46

Jay, E., Bambara, R., Padmanabhan, R. and Wu, R., DNA sequence analysis; a general, simple and rapid method for sequencing large oligodeoxyribonucleotide fragments by mapping, Nucleic Acid Res., 1 (1974) 331–353.

Jay, E. and Wu, R., *Arthrobacter luteus* restriction endonuclease recognition sequence and its cleavage map of SV40 DNA, Biochemistry, 15 (1976) 3612–3620.

Katagiri, N., Itakura, K. and Narang, S.A., The use of arylsulfonyltetrazoles for the synthesis of oligonucleotides by the triester approach, J. Am. Chem. Soc., 97 (1975) 7332–7337.

Maniatis, T., Jeffrey, A. and Kleid, D.G., Nucleotide sequence of the rightward operator of phage λ, Proc. Natl. Acad. Sci. USA, 72 (1975) 1184–1188.

Marians, K.J., Wu, R., Stawinsky, J., Hozumi, T. and Narang, S.A., Cloned synthetic *lac* operator DNA is biologically active, Nature, (in press).

Middleton, J.H., Edgell, M.H. and Hutchison, C.A. III, Specific fragments of φX174 deoxyribonucleic acid produced by a restriction enzyme from *Haemophilus aegyptius*, endonuclease Z, J. Virol., 10 (1972) 42–50.

Miller, J.H., Experiments in molecular genetics, Cold Spring Harbor Lab., (1972) 48.

Murray, N.E. and Murray, K., Manipulation of restriction targets in phage λ to form receptor chromosomes for DNA fragments, Nature, 251 (1974) 476–481.

Old, R., Murray, K. and Roizes, G., Recognition sequence of restriction endonuclease III from *Haemophilus influenzae*, J. Mol. Biol., 92 (1975) 331–339.

Panet, A., van de Sande, J.H., Loewen, P.C., Khorana, H.G., Raae, A.J., Lillehauge, J.R. and Kleppe, K., Physical characterization and simultaneous purification of bacteriophage T4 induced polynucleotide kinase, polynucleotide ligase and deoxyribonucleic acid polymerase, Biochemistry, 12 (1973) 5045–5050.

Platt, T., Files, J.G. and Weber, K., Specific proteolytic destruction of the NH$_2$-terminal region and loss of the deoxyribonucleic acid-binding activity, J. Biol. Chem., 248 (1973) 110–121.

Riggs, A.D., Suzuki, H. and Bourgeois, S., Lac-repressor—operator interaction, I. Equilibrium studies, J. Mol. Biol., 48 (1970) 67–83.

Roberts, R.J., Myers, P.A., Morrison, A. and Murray, K., A specific endonuclease from *Arthrobacter luteus*, J. Mol. Biol., 102 (1976) 157–165.

Rodriguez, R.H., Boliver, F., Goodman, H.M., Boyer, H.W. and Betloch, M., Construction and characterization of cloning vehicles, Proceedings of the 1976 ICN-UCLA Symposium on Molecular Biology (in press).

Sgaramella, V., van de Sande, J.H. and Khorana, H.G., A novel joining reaction catalyzed by
    T4 polynucleotide ligase, Proc. Natl. Acad. Sci. USA, 67 (1970) 1468—1475.
Stawinsky, J., Hozumi, T., Narang, S.A., Bahl, C.P. and Wu, R., Arylsulfonyltetrazoles, new
    coupling reagents and further improvements in the triester method for the synthesis
    of deoxyribo-oligonucleotides (communicated).
Tanaka, T. and Weisblum, B., Construction of a colicin E1-R factor composite plasmid in
    vitro, means for amplification of deoxyribonucleic acid, J. Bacteriol., 121 (1975) 354—
    362.
Thomas, M., Cameron, J.R. and Davis, R.W., Viable molecular hybrids of bacteriophage
    lambda and eukaryotic DNA, Proc. Natl. Acad. Sci. USA, 71 (1974) 4579—4583.
Tu, C.D., Jay, E., Bahl, C.P. and Wu, R., A reliable mapping method for sequence deter-
    mination of oligodeoxyribonucleotides by mobility shift analysis, Anal. Biochem., 74
    (1976) 73—93.
Weiss, B., Jacquemin-Sablon, A., Live, T.R., Fareed, G. and Richardson, C.C., Enzymatic
    breakage and joining of deoxyribonucleic acid, VI, J. Biol. Chem., 243 (1968) 4543—
    4555.
Wilson, G.A. and Young, F.E., Isolation of a sequence specific endonuclease (*Bam*I) from
    Bacillus amyloliquefaciens H., J. Mol. Biol., 97 (1975) 123—125.
Wu, R., Jay, E. and Roychoudhury, R., Nucleotide sequence analysis of DNA, Methods
    in Cancer Res., 12 (1976) 88—176.

Reprinted from *Gene* 1:255–280 (1977). Originally published by Elsevier/North-Holland Biomedical Press B.V.

IN VITRO PACKAGING OF A λ D*am* VECTOR CONTAINING *Eco*RI DNA FRAGMENTS OF *Escherichia coli* AND PHAGE P1*

(bacteriophage P1; cloned DNA; λD⁻ vector)

NAT STERNBERG**, DAVID TIEMEIER and LYNN ENQUIST

Laboratory of Molecular Genetics, NICHD, NIH Bethesda, Maryland 20014 (U.S.A.)

(Received November 4th, 1976)
(Accepted January 5th, 1977)

SUMMARY

In this report we describe a coliphage λ vector system for cloning endo R. *Eco*RI DNA fragments. This system differs significantly from those previously described in two ways. First, restricted and ligated DNA is encapsidated in vitro. Second, with increasing λ DNA size in the range 78 to 100% that of wild-type, the efficiency of DNA encapsidation into infectious phage particles markedly increases. For λ wild-type DNA the efficiency of in vitro packaging (10^6 to 10^7 plaques produced per µg of added DNA) is equal to, or better than, the standard $CaCl_2$ transfection method. The use of a D*am* mutation to facilitate recognition of size classes of inserted fragments is described. Using this vector and in vitro packaging, several *E. coli* and phage P1 endo R.*Eco*RI fragments were cloned.

INTRODUCTION

The in vitro construction of coliphage λ carrying DNA fragments from diverse sources is a relatively simple process (Murray and Murray, 1974; Thomas et al., 1974; Borck et al., 1976). The technique typically involves a specialized receptor (or vector) phage capable of taking up DNA fragments generated by restriction endonucleases. The DNA fragments are joined to the receptor phage DNA in vitro using DNA ligase. These joined molecules are

---

*These experiments were approved by the N.I.H. Biohazard Safety Committee and were performed under EK1, P1 conditions.
**Present address: Basic Research Program, NCI Frederick Cancer Research Center, P.O. Box B, Frederick, Maryland 21701 (U.S.A).
Abbreviations: *am*, amber; kb, kilobase pairs; *pD*, the product of gene *D*; *ts*, temperature-sensitive; TEMED, *N,N,N',N'*-tetramethylethylenediamine.

ten added to CaCl$_2$ treated cells (transfection) and viable phage particles are produced.

In this report , we describe the in vitro packaging of recombinant DNA into viable λ phage particles, an alternative procedure more efficient than transfection. Briefly, recombinant DNA is added to an extract containing partially assembled λ virion proteins, the added DNA is incorporated (or packaged) into the phage head, and tails are joined to the filled heads to give an infectious particle. In addition we describe the use of an *am*[*] mutation in gene D of the vector to allow recognition of different size classes of hybrid phage. Finally, we describe the use of such a D^- vector in combination with in vitro packaging and transfection to clone specific endo R. *Eco*RI DNA fragments of *E. coli* and bacteriophage P1.

MATERIALS AND METHODS

(a) Phages

The phage strains used are listed in Table I. The nomenclature of targets for restriction enzymes uses the abbreviation for the enzyme (Smith and Nathans, 1973) followed by the genome concerned and the number of the site within the genome. Thus, sites for endo R.*Eco*RI are described by *sr*Iλ1, *sr*Iλ2, etc. and the sites for endo R.*Bam*HI would be *sbam*HIλ1, *sbam*HIλ2, etc. For convenience, we have shortened the latter to *sbam*λ1, *sbam*λ2, etc.

All of the experiments described in this report use a specifically constructed λ receptor phage for fragments of DNA generated by endo R.*Eco*RI. The phage has a single target for this enzyme at *sr*1λ3, a site located in the phage *red*α gene (Murray and Murray, 1974). Since the remaining targets are either deleted or mutated, fragment insertion always inactivates this phage gene. The phage has a net 22% deletion with most of the nonessential regions removed allowing insertion of fragments up to 14kb. In addition, the phage carries a D*am* mutation that confers several novel properties to the vector, which will be described below. The complete genotype of the phage is λ*Dam*15*b*538*sr*Iλ 3*c*Its857*sr*Iλ4°*nin*5*sr*Iλ5°. For convenience we refer to the vector as λD^- *sr*Iλ3. It was constructed by crossing phage λ*Dam*15*b*538*imm*21 and λgt·λC (Thomas et al., 1974). The desired recombinant was obtained from the cross diagrammed in Fig.1 by isopycnic CsCl centrifugation of the lysate and selection for phage with the *nin*5 deletion. These phage form plaques on strain NS377 (Sternberg, 1976). Such phage from a density region corresponding to a DNA content of 78% that of λ wild-type were purified and tested for the *b*538 deletion, the presence of only a single endo R.*Eco*RI site and the D*am*15 mutation as follows: The *b*538 deletion removes the entire *att-int* region and part of the *xis* gene (Enquist and Weisberg, in press). As expected, the isolate did not complement λ*int*$^-$ or *xis*$^-$ phage for red plaque formation (Enquist and Weisberg, 1976). That the isolate contained a single endo R.*Eco*RI site was determined by measurement of restriction ratios as described by Murray and

TABLE I

BACTERIAL AND PHAGE STRAINS

Bacterial strains	Relevant characteristics and uses	References
Bacteria		
YMC	sup^+ (supF); used to assay λam mutants	Dennert and Henning, 1968
NS62	YMC(λcI$^+$);used to assay λimm434 phage	This report
NS617	YMC(λimm434); used to assay λimmλ phage	This report
C600	sup^+ (supE) used to assay λam mutants	Appleyard, 1954
594	sup^-; used to assay λam^+	Campbell, 1965
N99	sup^-; source of *E. coli* DNA	Shimada et al., 1972
N205	sup^-; recA	Sternberg and Weisberg, 1975
NS377	N99; nusA-1, rif^R-1	Sternberg, 1976b
NS428	N205(λAam11b2red3cIts857Sam7)	This report
NS433	N205(λEam4b2red3cIts857Sam7)	This report
NS309	AB1157 (λcI$^+$); *his, arg, leu, pro, thr*	Sternberg and Weisberg, 1975
NS686	C600(P1cIr100$^-$ m$^-$)	This report
N3098	YMC; *lig*ts7	Gottesman et al., 1973
NS698	N3098(λimm434); used to assay λimmλ phage in the presence of λimm434 and λimmλ red^- phage	This report
K175	sup^+ (supD); used to assay phage P1	Scott, 1973
K175(P1cry)	Permissive for P1, not permissive for P1r^-m$^-$	Scott, 1973
HN404	N99-R1 (ϕ80psupF); used to assay restriction ratios	This report from Howard Nash
HN405	N99 (ϕ80psupF)	This report from Howard Nash
LE289	YMC[λ(int-FII)$^\Delta_{gal}$T]; used to assay λint and xis function	Enquist and Weisberg, 1976
JC5183	sup^+ (supD), endo1$^-$,recB21, recC22, sbcA5	Barbour et al., 1970
Phage Strains		
Phage		
λcIts857	λ wild-type (*wt*)	DNA size references
λimm21cIts	0.95 λwt DNA size	
λcIts857nin5	0.945 λwt DNA size	
λ$pgal$49cIts857	0.90 λwt DNA size	
λb2cIts857	0.87 λwt DNA size	
λgt·λC	λcIts857srIλ4°srIλ5°nin5, lacking the λB fragment; 0.85 λwt DNA size	
λb221cIts857	0.78 λwt DNA size	
λDam15b538cIts857		Sternberg and Weisberg, in preparation
λD^-sr1λ3	λDam15b538srIλ3cIts857srIλ4°nin5srIλ5°	This report
λimm434 cI$^+$	Density marker; DNA size 0.97 that of λwt	
λb538imm434cSam7	Density marker; DNA size 0.805 that of λwt	
P1cI r100r$^-$ m$^-$		This report

[a]The sizes of phage λ DNAs other than those of λgal49 and λgt·λC are given in Davidson and Szybalski, 1971 and Fiandt et al. (1971). The size of λgal49 DNA is given in Nash (1974) and λgt·λC in Thomas et al. (1974).

Murray (1974). The D*am*15 mutation in λD^- *sr*Iλ3 cannot be detected in the conventional manner by testing ability to form plaques on *sup*⁺ but not *sup*⁻ hosts because D*am* phage with deletions exceeding 18% the λ wild-type DNA content form plaques equally well on either *sup*⁺ or *sup*⁻ hosts (Sternberg and Weisberg, in press). However, such *D*-deficient phage do have a recognizable property: deletion phages with D*am* mutations are extremely sensitive to chelating agents (e.g., EDTA) after growth in a *sup*⁻ host but remain resistant to such agents after growth in a *sup*⁺ host. The λD^- *sr* Iλ3 vector exhibits this property (data not shown). Further confirmation that this phage carries a cryptic D*am* mutation comes from experiments described in this report where endo R.*Eco*RI fragments greater than 1.9 kb were inserted into the *sr*Iλ3 site of the vector. Hybrids of this type exhibit conventional behavior for phage carrying a D*am* mutation: They form plaques only on a *sup*⁺ host and in a *sup*⁻ host complement phage with *am* mutations in all phage head genes except gene *D*. Analysis of fragments produced after restriction of vector DNA with endo R.*Eco* RI and endo R.*Bam*HI verified the presence of a single endo R.*Eco*RI site at *sr*Iλ3 and two endo R.*Bam*HI sites at *sbam*λ1 and *sbam*λ4. A summary of these analyses is presented in Fig.1.

For the purposes of this report, we denote λD^- *sr*1λ3 carrying heterologous DNA fragments in shorthand by the symbol "λ·" followed by the fragment carried. For example, λ·*E. coli* or λ·P1 are λD^- *sr*Iλ3 carrying *E. coli* or P1 fragments at the *sr*I λ3 site. In the case of λ·P1 hybrids, specific P1 fragments are given number designations corresponding to their position in an agarose gel. For example, λ·P1-1 is λD^- *sr* Iλ3 carrying the largest (slowest migrating) fragment, λ·P1-2 carries the next largest fragment etc. When the function specified by a fragment is known, the hybrids are noted as follows: λ·*pro* or λ·*thr* are λD^- *sr* Iλ3 carrying specific fragments expressing functions involved in proline and threonine biosynthesis.

(b) Bacterial strains

The bacterial strains used are listed in Table I. Strains with NS or LE numbers were constructed in this laboratory.

(c) Media

LB broth contained per liter Difco Tryptone, 10 g, Difco yeast extract, 5 g, and NaCl, 5 g, and was adjusted to pH 7.2. LB broth was routinely made 0.02 M in $MgSO_4$ by the addition of 20 ml 1 M $MgSO_4$ to 1 liter sterilized medium. Tryptone broth (TB) is LB broth without the yeast extract. TB was also made 0.02 M in $MgSO_4$ as described above. The presence of Mg^{2+} is essential for stability of D^- phages grown in the absence of *p*D (Sternberg and Weisberg, in preparation). TB and LB were solidified by the addition of 1.2% Difco agar.

Phage stocks were prepared in liquid LB plus 0.02 M $MgSO_4$ for λ and in LB plus 0.002 M $CaCl_2$, 0.002 M $MgSO_4$ for P1. Lambda assays were made on

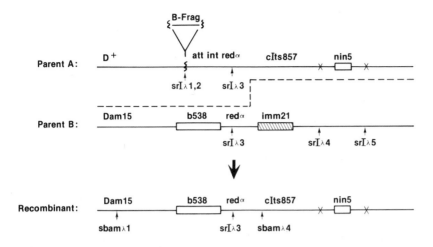

Fig. 1. Construction of λ Dam15b538sr Iλ3cIts857srIλ4°nin5srIλ5°.

Phage parent A was λgt·λC obtained from Dr. R. Davis (Thomas et al., 1974). Phage parent B was λDam15b538srl λ3imm21srl λ4srl λ5 from our collection. Strain YMC was infected with both parents at a multiplicity of 4 of each phage. After 10 min adsorption at 32° C, the infected cells were diluted 1000-fold into tryptone broth containing 20 mM MgSO$_4$ and incubated at 37° C for 60 min. Chloroform was added to complete lysis and debris was removed by low speed centrifugation. A solution of CsCl in 10 mM Tris—HCl pH 7.4 and 10 mM MgSO$_4$ was prepared giving a density of 1.485 g/ml. 0.2 ml of the lysate was added to 4.0 ml of the CsCl solution. About 10^6 λb2cIts857 and λb221cIts857 were added as marker phages. The solution was mixed and run at 35 000 rpm in the SW56 rotor for 24 h at 4° C. The rotor was brought to an unbraked stop, the bottom of the tube was punctured and 6 drop fractions were collected in 0.5 ml TMG. The position of the marker phages was determined and phage plating on NS377 from the region of λb221 cIts857 were purified. The recombinants were then tested as described in MATERIALS AND METHODS, section (a). One recombinant was chosen and purified. The position of restriction endonuclease cleavage sites was determined after enzyme digestion and agarose gel electrophoresis (see MATERIALS AND METHODS, section i). Digestion with endo R.*Eco* RI yielded two fragments; a large 23.5 kb piece from the left portion of the vector and a 13.5 kb fragment from the right. A similar analysis with endo R.*Bam* HI gave 3 fragments: a 5.3 kb piece containing the left end, a 19.9 kb central fragment and a 11.2 kb fragment from the right end.

solidified TB plus 0.02 M MgSO$_4$ and P1 assays were made on solidified LB plus 0.002 M CaCl$_2$.

LB chloramphenicol agar is LB broth, 1.2% Difco agar and 12.5 μg/ml chloramphenicol.

M9 medium is M9 salts (Shimada et al., 1972) supplemented with 2% casamino acids, 0.2% glucose and 0.002 M MgSO$_4$. TMG buffer is 0.01 M Tris— HCl pH 7.4, 0.01 M MgSO$_4$ and 0.1% gelatin.

(d) General techniques

The methods for preparing cells and phage stocks are described in Shimada et al. (1972) and Schrenk and Weisberg (1975). Procedures for phage plaque assays, spot test complementation for λ *am* mutations and genetic crosses are given in Sternberg (1976a). Procedures for determination of *int* and *xis* function using the red plaque test are described in Enquist and Weisberg (1976). P1 was assayed as described in Rosner (1972). Determination of restriction ratios is described in Murray et al. (1974). Integration proficiency was assayed as described by Gottesman and Yarmolinsky (1968). Tests for the *red⁻* phenotype were done by plating on strain N3098 (*ligts*7) at 32° C. Phages defective in *red* function do not plate on this host (Zissler et al., 1971; Gottesman et al., 1973).

(e) Preparation of phage lysates

(i) Phage λ. Plate lysates were prepared as described in Schrenk and Weisberg (1975). Liquid lysates were prepared as follows. A phage plaque containing 10^5 to 10^6 phage was added to 1 ml of a fresh culture of either strain C600 or 594 grown overnight in TB. After 5 to 10 min for phage adsorption at 38° C, the infected cells were diluted 200-fold into LB broth containing 0.02 M $MgSO_4$. The culture was vigorously aerated at 38° C until lysis occurred, usually 4 to 6 h. Lysis was completed by shaking the culture an additional 5 to 10 min with several drops of chloroform.

(ii) Phage P1. A lysogenic culture carrying a P1 prophage with a thermosensitive repressor mutation (NS686) was grown overnight at 32° C and then diluted 100-fold into fresh LB broth with $CaCl_2$ and $MgSO_4$ and grown to about $2 \cdot 10^8$ cells/ml. The culture was then shifted to 42° C for 30 min with vigorous aeration and then back to 38° C with continuing shaking until lysis occurred (usually 30 min after the shift to 38° C). Lysis was completed by the addition of several drops of chloroform.

(f) Purification of phage

Phage were purified from liquid lysates as follows. Solid NaCl was added to 0.5 M, the lysate chilled to 4° C and kept at that temperature for at least 60 min (lysates were often kept overnight at 4° C). Bacterial debris was removed by low speed centrifugation (8000 rpm for 10 min in the Sorval G50 rotor). Powdered polyethylene glycol (PEG 6000, Baker) was added to give a 10% solution (w/v), the PEG dissolved at room temperature and the solution was kept at 4° C for at least 60 min. The PEG precipitate was collected by centrifugation (8000 rpm for 15 min in the Sorval G50 rotor). This precipitate, containing better than 90% of the phage, was resuspended in 1/50 the original lysate volume in 0.01 M Tris—HCl, 0.01 M $MgSO_4$ pH 7.4. For each 3.5 ml of resuspended PEG pellet either 2.7 g CsCl (phage λ) or 2.4 g CsCl (phage P1) was added and the resulting solution was centrifuged to equilibrium in an SW56 rotor at 30 000 rpm for 24 h. After centrifugation a visible phage λ band was present in the center of the tube while the PEG remained at the top. For P1 typically 10—20% of the phage band at lighter positions in the gradient than that of the major visible phage band. The one phage λ band and all of the phage P1 bands were removed from the top using a Pasteur pipette and typically rebanded in a second equilibrium density gradient. The phage were then used for DNA extraction.

(g) DNA preparation

   (i) Phage λ and P1 DNA. Phage suspensions ($2 \cdot 10^{11}$ to $2 \cdot 10^{12}$ plaque-forming units/ml) were adjusted to 1.0 mg/ml pronase (Calbiochem, nuclease free) and dialyzed against 100 volumes 20 mM Tris—HCl (pH 7.5), 0.1 M NaCl, 1.0 mM EDTA, and 0.002% Triton X-100 at 37°C for 60 min. The suspensions were extracted twice by gentle rolling with an equal volume of phenol saturated with 20 mM Tris—HCl (pH 7.5), 0.1 M NaCl, 1.0 mM EDTA. The aqueous phases were dialyzed against 1000 volumes of 20 mM Tris—HCl (pH 7.4), 0.1 mM EDTA for 20 h with one change of buffer. The resulting DNA solutions (10—100 μg/ml) were stored at 4°C.

   (ii) Bacterial DNA. DNA was extracted from strain N99 by the method of Marmur (1961) and stored in 10 mM Tris—HCl pH 7.4, 10 mM NaCl, 10 mM EDTA.

(h) Enzymes and chemicals

   Pronase (nuclease free) was purchased from Calbiochem, San Diego, CA. Restriction endonuclease *Eco* RI was isolated from *E. coli* BRY 13.3/1100.5 kindly provided by Dr. Julian Davies. Its purification will be described in detail elsewhere. Some of the endo R.*Eco* RI preparations used were generously provided by Dr. S. Tilghman and Dr. F. Polsky. Restriction endonuclease *Bam* HI was purchased from Bethesda Research Laboratories, Rockville, MD. T4 polynucleotide ligase was purchased from Miles Laboratories, Inc., Elkhart, IN. Cesium chloride was obtained from Schwartz/Mann, Orangeburg, NY. Phenol was obtained from Mallinckrodt, St. Louis, MO. Agarose was obtained from SeaKem, MCI Biomedical, Rockland, ME. Acrylamide; *N,N'*-methylene bisacrylamide and TEMED were obtained from Bio-Rad Laboratories, Richmond, CA. Ethidium bromide was obtained from Calbiochem, San Diego, CA. Whenever possible, chemicals used were AR grade.

(i) Restriction endonuclease digestion and gel electrophoresis

   Restriction endonucleases were assayed by digestion of 0.5 μg λcIts857 DNA in 50 μl reactions for 30 min at 37°C. Endo R.*Eco*RI reactions were performed in 100 mM Tris—HCl, pH 7.9, 50 mM NaCl, 12 mM $MgCl_2$, 0.1 mM EDTA. Endo R.*Bam* HI reactions were performed in 20 mM Tris—HCl, pH 7.5, 60 mM NaCl, 7 mM $MgCl_2$ and 2 mM β-mercaptoethanol. Reactions were stopped by incubation for 5 min at 70°C and chilled on ice. The extent of reaction was determined by electrophoresis in 1% (w/v) agarose gels. For electrophoresis, 1/10 volume of 50% sucrose, 0.15% bromphenol blue was added to each reaction. The reaction mixtures were applied to 20 cm × 20 cm × 0.3 cm agarose slab gels (apparatus from Blaircraft, Cold Spring Harbor, NY) and electrophoresed as described in Helling et al. (1974) with 50 mM Tris-acetate, pH 8.05, 20 mM sodium acetate, 18 mM NaCl, 1.0 mM EDTA. The DNA products were visualized and photographed under a UV lamp after immersion for 3 min in 0.01% ethidium bromide.

The same reaction conditions were employed to prepare endo R.*Eco* RI fragments for subsequent ligation. However, the extent of $\lambda D^- sr1\lambda 3$ DNA restriction was additionally monitored by transfection of JC5183 (see "k" below) with the endonuclease reaction products. Sufficient endo R.*Eco* RI was used to reduce the transfecting efficiency of $\lambda D^- sr$ I$\lambda 3$ to 0.1% to 0.3% that of the intact DNA.

(j) Ligase reactions

Solutions of 25 μl to 250 μl containing 20–40 μg/ml restricted $\lambda D^- sr$I$\lambda 3$ DNA and 5-15 μg/ml restricted foreign DNA (to be inserted) in 100 mM Tris—HCl, pH 7.4, 50 mM NaCl, 12 mM $MgCl_2$, and 0.1 mM EDTA (1 $\times$ *Eco* RI buffer) were incubated for 10 min at 37°C to separate hydrogen-bonded fragments. The mixtures were chilled on ice and adjusted to 10 mM dithiothreitol, 50 μg/ml BSA, and 0.1 mM ATP by addition of 1/5 volume of a ligase buffer containing the 1 $\times$ *Eco* RI buffer plus the concentrated components at 60 mM, 0.3 mg/ml, and 0.6 mM, respectively. The reactions were initiated by addition of T4 polynucleotide ligase to 0.5 units/ml and incubated at 9°C for 2 to 5 days. The yield of recombinant phage by either transfection or in vitro packaging reached a plateau after 36 to 48 h (data not shown).

(k) Transfection of E. coli to recover recombinant phages

Cells competent for transfection were prepared essentially as described by Mandel and Higa (1970). All steps except where indicated were performed at 0°C, and cells were collected at each step by centrifugation for 5 min at 2000 g. A fresh overnight culture of JC5183 was diluted 1/25 in LB broth plus 50 μg/ml thymidine and grown at 37°C to an $A_{630} = 0.60$. The cells were washed with an equal volume of 10 mM NaCl and suspended with gentle swirling in 1/2 volume 50 mM $CaCl_2$ for 20 min. The competent cells were finally suspended in 1/10 volume 50 mM $CaCl_2$ and used within 30 min. For each transfection reaction, 0.2 ml competent cells were added to 0.1 ml 10 mM Tris—HCl, pH 7.5, 0.1 mM EDTA containing as much as 0.1 μg DNA. DNA samples from ligase reactions were added directly to the Tris—HCl/EDTA buffer to give a final concentration of 1.0 μg/ml DNA. The suspensions were incubated 30 sec at 37°C, 60 min at 0°C, and finally 2 min at 42°C. The reactions were chilled and immediately plated at 37°C with 2 drops of an overnight culture of YMC plus 2.5 ml tryptone top agar. The efficiency of transfection ranged from $6 \cdot 10^5$ to $4 \cdot 10^6$ plaque-forming units per μg DNA ($3 \cdot 10^{-5}$ to $2 \cdot 10^{-4}$ /molecule).

(l) In vitro packaging procedure

(i) Preparation of packaging extracts. Cultures of strains NS428 and NS433 were prepared by adding a loopful of a frozen glycerol stock to M9 medium and growing them overnight at 32° C. For 10 packaging reactions we routinely prepared 1.5 ml of the NS428 overnight culture and 7.5 ml of the NS433 overnight culture. These overnight cultures were then diluted 100-fold with the same medium and grown at 32° C to $2 \cdot 10^8$ cells/ml, at which point they were induced by swirling them in a 90° C water bath until the temperature of the cultures had risen to 42° C. Incubation was continued with vigorous aeration at 42° C for 20 min. The cultures were then shifted to 38° C for 70 min and then chilled to 4° C. The two packaging extracts were prepared as follows (see also Becker and Gold, 1975). *Extract A.* 150 ml of each of the two induced cultures were mixed and centrifuged at 8000 rpm for 10 min at 4° C in a G50 Sorval rotor. The pellet was resuspended in 1/500 volume (0.6 ml) buffer A (20 mM Tris—HCl pH 8.0, 1 mM EDTA, 3 mM MgCl$_2$, 5 mM β-mercaptoethanol) and disrupted by sonic vibration (12 2-second bursts) in a chilled tube (Falcon tube No. 2052) using the fine tip of a Bronwill Biosonik sonicator at the lowest setting. Sonic disruption for a more extended period of time produced a less active extract. *Extract B.* Cells from the remainder of the NS433 culture (600 ml) were concentrated by centrifugation as described above and then resuspended in 1/500 volume (1.2 ml) of a solution consisting of 10% sucrose and 50 mM Tris—HCl pH 7.4. The resuspended cell pellet was then transferred to a 25 ml erlenmeyer flask and twice subjected to freezing (in liquid nitrogen) and thawing (at 32° C). To complete the formation of this extract we added per 100 μl of cell suspension, first 5 μl of lysozyme (egg white, Sigma at 1 mg/ml in 0.25 M Tris—HCl pH 7.4) and, after 30 min at 4° C, 10 μl of buffer B (6 mM Tris—HCl pH 7.4, 15 mM ATP, 16 mM MgCl$_2$, 60 mM spermidine, 30 mM β-mercaptoethanol). The two extracts could be stored separately without loss of activity for at least one month by adding glycerol to 20% and then adding 100 μl droplets directly into a plastic tube filled with liquid nitrogen. The frozen droplets were stored immersed in liquid nitrogen.

(ii) Packaging reaction. The packaging reaction was divided into two stages. The stage I reaction contains 30 μl buffer A, 5 μl DNA (usually at 5 μg/ml), 2 μl buffer B and 20 μl extract A and is incubated for 10 min at 22° C. Following this incubation period the stage II reaction is initiated by adding 100 μl of extract B and incubating the reaction for 60 min at 35° C. The latter reaction is terminated by adding 150 μl TMG containing 10 μg/ml DNAase I (RNAase free, Worthington) and incubating an additional 10 min at 35° C. For a typical packaging reaction with wild-type λ DNA, 5 μl of the DNAase-treated mixture gave about 3000 plaques.

Exogenous phage DNA (either directly from the ligase reaction or dialyzed with 10 mM Tris— HCl pH 8.0, 0.1 mM EDTA) for the stage I reaction was heated for 3 min at 70° C and quick cooled just prior to use. The efficiency of packaging (plaques per μg of added DNA) was found to be constant in the range 0.002 to 0.2 μg of DNA. In the absence of exogenous phage DNA or either of the two extracts the packaging efficiency was reduced to 0.001% that observed when wild-type λ DNA was added to the reaction. The reason for the inefficient packaging of endogenous phage DNA will be discussed below (see RESULTS).

(m) Harvesting of plaques

Following transfection or in vitro packaging, the resultant plaques were harvested in the same way as a plate lysate. The phage from approximately 1000 plaques were pooled into one tube. Generally 2 ml of TMG was added and the pooled plaques were mixed well. The bacterial debris and agar were removed by low speed centrifugation and the supernatant fluid used directly for analysis.

(n) CsCl density gradient analysis

CsCl gradients were prepared by mixing 2.7 g solid CsCl with 3.5 ml 0.1 M Tris—HCl pH 7.4, 0.01 M $MgSO_4$ containing the phages of interest. Generally, marker phage of known density were added at about 10^6 phage per gradient. The CsCl solutions were centrifuged to equilibrium in the SW56 rotor at 30 000 rpm (24 h, 5° C). The distribution of phage in such gradients was determined by puncturing the bottom of the centrifuge tube and collecting 7 drop fractions into 0.1 ml TMG buffer.

(o) EDTA inactivation

Phage were diluted 100-fold into either TMG buffer or 0.02 M Tris—HCl containing 0.02 M EDTA pH 7.4 and incubated at either 32° C or 41° C for 15 min. Phage inactivation by EDTA was stopped by adding 1 M $MgSO_4$ to 0.05 M and samples were removed to assay phage survival.

(p) The isolation of λ·pro and λ·thr transducing phage

Recombinant phages of λD^- sr $I\lambda 3$ and *E.coli* originally obtained from 400 plaques matured by in vitro packaging were adsorbed for 10 min at 32° C to 10^8 cells of strain AB1157 lysogenic for wild-type λ (NS309). The multiplicity of infection was 4. Infected cells were then spread on minimal agar-B1-biotin plates (Sternberg and Weisberg, 1975) supplemented with all combinations of the 4 amino acids required by this strain but not the fifth. The plates were incubated at 32° C for 48 h and the number of colonies was scored. Plates lacking proline or threonine contained 100 to 500 more colonies than control plates with uninfected cells. Several *pro*+ and *thr*+ transductants were purified and the λ·pro and λ·thr phage isolated from them as follows. The transductants were grown to about 10^8 cells/ml, washed at the same cell density with 0.01 M $MgSO_4$, exposed to 440 erg/mm² UV irradiation and then spread on solidified TB plates containing a lawn of strain YMC. The plates were incubated overnight at 41° C and phage plaques isolated. Lysates were prepared from plaques containing *red*⁻ phage and the phage then tested for their ability to transduce strain NS309 to *pro*+ or *thr*+.

RESULTS

Modification of the system of Becker and Gold (1975) for in vitro packaging of recombinant DNA

The procedure developed by Becker and Gold (1975) for in vitro packaging of λ was designed to assay the activity of the λ *A*-gene protein and encapsidates both endogenous and exogenous DNA. We have modified the procedure so that only exogenous DNA is efficiently incorporated into infectious λ particles. The two λ lysogens used to prepare complementing extracts for in vitro packaging carry the following prophage and host mutations.

(1) *Am* mutations in either capsid genes *A* or *E*. Mixed extracts prepared from induced lysogens containing either *A* or *E* defective prophages should contain all phage functions necessary for phage capsid formation (Hohn and Hohn, 1974; Kaiser et al., 1975; Becker and Gold, 1975). The necessity of using two types of complementing extracts is a peculiarity of the system and is discussed by Becker and Gold (1975).

(2) The λ*Sam*7 mutation (Goldberg and Howe, 1969). In the absence of the phage gene *S* product, induced λ lysogens fail to lyse, continue to replicate and produce intracellular phage products for 2 to 3 h beyond the time of normal lysis (Adhya et al., 1971; Reader and Siminovitch, 1971). This property permitted us to harvest cells 90 min after prophage induction, concentrate the unlysed cells and prepare active packaging extracts.

(3) The λ*b*2 mutation. This phage deletion damages the λ attachment site and prevents prophage excision after induction (Campbell, 1965; Gottesman and Yarmolinsky, 1968). The induced λ*b*2 prophage is effectively trapped in the *E. coli* chromosome even though many rounds of prophage replication occur. Since trapped prophage DNA cannot be packaged into plaque forming virions, the packaging extracts contain no endogenous source of packageable DNA (Sternberg and Weisberg, 1975). Under these conditions, only the exogenous DNA is available for in vitro encapsidation into plaque forming phage.

(4) The *rec*A mutation in the host and the *red*3 mutation in phage λ. These two mutations effectively inactivate the major general recombination systems present in the packaging extracts. It was necessary to prevent recombination because the addition of exogenous DNA to *rec⁺ red⁺* extracts resulted in the production of phage that carried markers present in both endogenous and exogenous DNA (data not shown). For the experiments reported here, this recombination was undesirable. In the absence of both recombination systems, none of the plaques obtained by in vitro packaging carry markers present in the endogenous prophage DNA.

Efficiency of packaging is a function of DNA size

An initial comparison of the packaging efficiency of λ wild-type DNA and λ*D⁻ sr* Iλ3 DNA (78% the size of λ wt DNA) indicated that the former DNA was packaged at least 200 times more efficiently than the latter. Therefore, we determined packaging efficiency for λDNA ranging in size from 78% to 100% that of wild-type (Fig.2) and found that the in vitro packaging efficiency decreased markedly as the size of the DNA decreased. In contrast to this size dependence for efficient in vitro packaging, $CaCl_2$ transfection efficiency was independent of DNA size (Fig.2). In the experiments described in this report, in vitro packaging of λ wild-type DNA routinely gave 2- to 10-fold more plaques per μg of DNA added than did transfection.

λD⁻ sr Iλ3 with inserted endo R. EcoRI fragments is preferentially packaged in vitro

DNA from λD^- sr Iλ3 was digested with endo R.*Eco*RI, mixed with similar fragments of *E. coli* or phage P1 DNA and treated with T4 ligase as described in MATERIALS AND METHODS. Aliquots of the ligated mixtures were added to $CaCl_2$ treated cells and to in vitro packaging extracts. The number of plaques per μg of added DNA was determined as well as the number of hybrid phage produced (Table II).

If a mixture containing equal amounts of endo R.*Eco* RI cleaved λD^- srIλ3 DNA and other endo R.*Eco* RI fragments is treated with ligase, the predominant plaque forming DNA molecule produced should be rejoined λD^- sr1λ3 vector with no inserted DNA. We expect this because the insertion of a fragment requires two independent ligase joining events, whereas the reformation of the parental vector requires only one such event. We expect therefore, that $CaCl_2$ transfection will detect mainly these reformed vector DNA molecules, while in vitro packaging will be biased toward detection of the molecules with inserted fragments because large molecules are packaged more efficiently than small molecules. In agreement with this, we found no stimulation of vector transfection efficiency by addition of *E. coli* or P1 endo R. *Eco*R1 fragments, but did find at least a 10-fold stimulation in vector packaging efficiency with addition of heterologous DNA fragments (Table II, lines 4, 5, 6). The fraction of plaques obtained by transfection or packaging containing inserted fragments (determined by scoring their *red⁻* phenotype) is also given in this table. As expected, phages carrying inserted fragments are a minority fraction (0.05—0.10) of plaques obtained by transfection. On the other hand, the vast majority of phages (0.84—0.91) arising from in vitro packaging of an identical DNA mixture contained inserted fragments.

We expected that on the average packaged hybrid phage would contain larger inserts than hybrids obtained by transfection. To confirm this, we assumed that phage with inserted DNA fragments would have increased density In CsCl gradients. Using CsCl gradient analysis, we determined the DNA content of phage obtained from about 400 to 6000 plaques derived from either transfection or in vitro packaging of a ligase reaction mixture containing endo R.*Eco*RI fragments of λD^- srIλ3 vector DNA and *E.coli* DNA (Fig.3). The results are consistent with the data given in Table II. The majority of phage obtained by transfection have the density expected of λD^- sr Iλ3. The small peaks (about 5% of the total phage) with increased density are *red* , as ex-

TABLE II

THE ENCAPSIDATION OF $\lambda D^- sr$Iλ3 HYBRID DNA EITHER BY IN VITRO PACKAGING
OR BY TRANSFECTION

DNA from λcIts857 (λ wild-type) or $\lambda D^- sr$Iλ3 as well as DNA from *E. coli* N99 and
phage P1cl r100r⁻ m⁻ was purified as described in MATERIALS AND METHODS. The
DNA was either untreated or treated with endo R.*Eco*RI (abbreviated here as RI). For
some experiments T4 ligase was added after endo R.*Eco*RI digestion. All enzyme treatments
are described in MATERIALS AND METHODS. The DNA was then packaged into phage
particles in vitro or mixed with $CaCl_2$ treated cells for transfection as described in
MATERIALS AND METHODS. The numbers in the table are the normalized efficiencies of
plaque production per μg of added DNA with respect to λ wild-type DNA.

DNA source	Enzyme treatment	Packaging efficiency		Transfection efficiency	
		Exp.	Exp. 2	Exp. 1	Exp. 2
λ wild-type	none	1[a]	1[b]	1[c]	1[d]
$\lambda D^- sr$Iλ3	none	0.004	0.006	1.5	1.1
$\lambda D^- sr$Iλ3	RI	0.0001	0.0001	0.004	0.005
$\lambda D^- sr$Iλ3	RI, ligase	0.0008	0.0007	0.07	0.11
$\lambda D^- sr$Iλ3 + E. coli	RI, ligase	0.01 (0.84)[e]	---	0.08 (0.05)[e]	---
$\lambda D^- sr$Iλ3 + P1	RI, ligase	---	0.03 (0.91)[e]	---	0.12 (0.10)[e]

[a] 3.9·10⁶ plaques/μg.
[b] 1.1·10⁷ plaques/μg.
[c] 2.2·10⁶ plaques/μg.
[d] 1.2·10⁶ plaques/μg.
[e] The number in parentheses is the fraction of total plaques that contained an inserted fragment
as determined by their *red* phenotype (see MATERIALS AND METHODS).

pected for hybrid phage. A similar analysis for phage obtained by in vitro
packaging shows a strikingly different profile. Here most of the phages have a
density greater than the $\lambda D^- sr$Iλ3 vector. Subsequent analysis verified that all
of these heavy phages were *red*⁻ and therefore contained DNA inserted in the
*sr*Iλ3 site.

The size of the peaks in CsCl gradients is a function of the number of DNA
fragments of a particular size class, the packaging efficiency of the $\lambda D^- sr$Iλ3
vector carrying that size fragment and selective factors operating during growth.
To show the size discrimination of in vitro packaging, we used phage P1, which
upon endo R.*Eco* RI digestion, yields about 25 fragments ranging from a few
hundred base pairs in length to about 17.9 kb. With few exceptions, each
fragment has a unique size. By comparing the density (Fig.4) of phages made
by transfection or by in vitro packaging of ligase treated endo R.*Eco* RI digested $\lambda D^- sr$ Iλ3 and P1 DNA, we could demonstrate that packaging produced
phages carrying a broader range of fragments. Again, the absolute number of
hybrids obtained was far greater for in vitro packaging than for transfection.

The agarose gel analysis of an endo R.*Eco* RI digested mixture of λ·P1
hybrid DNA obtained from phage packaged in vitro revealed that many P1

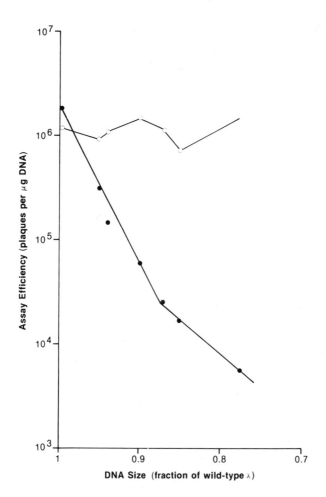

Fig. 2. In vitro packaging and transfection with λ DNA of different sizes. Phage (λcIts857, λimm21cIts, λcIts857nin5, λpgal49cIts857, λb2cIts857, λgt·λB εχλ λb221cIts857) with DNA contents ranging from 1.0 to 0.78 that of λ wild-type (see Table I) were grown and purified as described in METHODS, sections (e) and (f). DNA was extracted from $2-5\cdot10^{11}$ phage, diluted to 5 μg/ml in 10 mM Tris—HCl pH 7.5, 0.1 mM EDTA and either encapsidated in an in vitro packaging reaction or used in a transfection reaction. The closed and open circles represent, respectively, the yield of phage produced per μg of added DNA in the in vitro packaging reaction and the infectious centers produced per μg of added DNA in the transfection reaction.

fragments could be inserted into the vector (Fig. 5, Slot *b*). One notable exception was the large 17.9 kb fragment (fragment 1) which, if added to λD⁻ *sr*Iλ3 would exceed the packaging limit of the λ head (about 109% of the wild-type DNA content) (Weil et al., 1973; Feiss et al., in press; Sternberg and Weisberg, in preparation). Several other fragments, including P1 fragment 2 and many of the smaller P1 fragments were not found in the hybrid pool. We

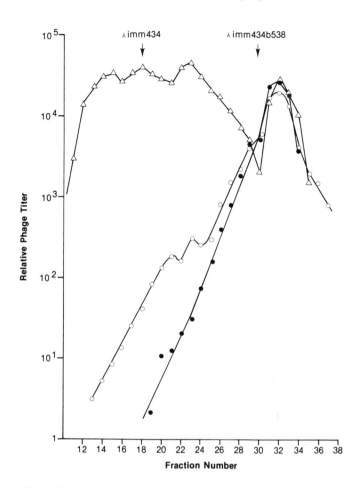

Fig. 3. The density distribution of λD^- sr Iλ3·$E.$ $coli$ hybrid phage obtained either by in vitro packaging or transfection. This figure is a composite of two CsCl equilibrium gradients prepared and assayed as described in METHODS, section (n) and aligned so that the marker phages (λimm434 (0.97 λ wild-type DNA content) and λb538imm434cSam7 (0.805 λ wild-type DNA content)) overlap. A total of 60 to 62 fractions were collected. One gradient contains 10^6 plaque-forming phage obtained by harvesting 400 plaques (84% of which were hybrids) derived from an in vitro packaging reaction containing λD^- sr Iλ3 and $E.$ $coli$ restricted and ligated DNA. The second gradient contains 10^6 phage obtained by harvesting 6000 plaques (5% of which were hybrids) derived from a transfection reaction with the same DNA as that used in the in vitro packaging reaction. Following addition of the two λimm434 marker phages, gradients were centrifuged and collected as described in MATERIALS AND METHODS. The total immλ phage obtained by in vitro packaging (△——△) or transfection (○——○) were assayed on strain NS617. The immλ phage obtained by transfection that were not hybrid (●——●) were assayed by virtue of their ability to plaque on strain NS698 (λD^- sr Iλ3 is red^+ and so can plaque on this host). The imm434 marker phage were assayed as clear and turbid plaque formers on strain NS62. The fractions not shown in the figure contain no more than the background level of phage.

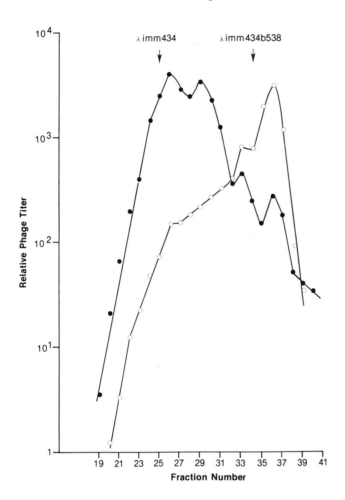

Fig.4. The density distribution of λD^- sr Iλ3·P1 hybrids obtained either by in vitro packaging or transfection. The figure is a composite of two CsCl equilibrium gradients prepared and assayed as described in METHODS, section (n), and aligned so that the marker phages, λimm434 and λb538imm434cSam 7 overlap. One gradient (•——•) contains 10^6 phage obtained from 1000 plaques (91% of which are hybrids) derived from an in vitro packaging reaction containing λD^- sr Iλ3 and P1clr100r^- m^- restricted and ligated DNA. The second gradient (○——○) contains 10^6 phage obtained by harvesting 4000 plaques (10% of which were hybrids) derived from a transfection reaction containing the same DNA as that used in the in vitro packaging reaction. Phage with immλ were assayed on strain NS617 and phage with imm434 on strain NS62.

are uncertain if the smaller fragments were incompatible with λ during lytic growth or if the hybrids were in small quantity and thus would not show up on the gel at the DNA concentration used. We have cloned several small fragments not visible in the hybrid pool using a genetic procedure described below. The failure to detect P1 fragment 2 (12.7 kb) is surprising since it should still be possible to insert this fragment into λD^- sr I λ 3 and not exceed the packaging limit of the λ head. The presence of this fragment might be incompatible with normal λ lytic growth. We note here that an agarose gel analysis of an endo R.*Eco* RI digested mixture of an equivalent amount of a λ·P1 hybrid DNA pool obtained from phage matured by transfection revealed no detectable P1 fragments (Fig.5, Slot *c*). This result is consistent with the small amount of hybrid DNA (less than 10%) present in such a pool.

The cloning of specific E. coli and P1 fragments in λD^- srIλ3

We isolated hybrid phages carrying specific endo R.*Eco* RI fragments by two general techniques: density selection and genetic selection. The density selection has been discussed in the previous section. Briefly, several hundred plaques obtained by plating aliquots of transfection or in vitro packaging mixtures are pooled, debris removed by low speed centrifugation and the supernatant fluid centrifuged to equilibrium in CsCl. The gradient fractions; are collected, titered and stored in the cold. If the desired fragment size is known, then it is straightforward to calculate the density of λD^- sr I λ 3 carry-

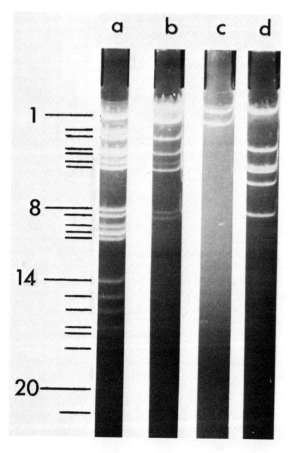

ing this fragment. By selecting plaques from this region of the gradient, one can greatly enrich for the fragment of interest. To evaluate the validity of this approach, we arbitrarily selected λ·*E. coli* hybrids from several fractions of the CsCl gradient shown in Fig.3 expected to contain phages with inserted fragments of 12.1 kb (104% wild-type λ DNA content) and 8.4 kb (96% wild-type λ DNA content). Single plaques were purified and rebanded in CsCl gradients (Fig.6), their DNA isolated, restricted by endo R.*Eco* RI and analyzed by agarose gel electrophoresis (Fig.7). The phages retained the density of the region in the gradient from which they were selected and contained, in addition to DNA of λ*D⁻sr*Iλ3, inserted fragments of 12.6 kb and 8.4 kb.

Several λ*D⁻sr*Iλ3 hybrid phages carrying specific P1 endo R.*Eco*RI fragments were isolated using the density selection as follows. To isolate hybrids carrying the largest fragments, we picked single plaques from fractions 21 to 32 of the CsCl gradient shown in Fig.4. The DNA from several hybrids was extracted, restricted using endo R.*Eco* RI and analyzed by agarose gel electrophoresis (Fig.8, Slots *d* and *e*). One hybrid carries P1 fragment 6 (5.9 kb) and the other P1 fragment 7 (5.5 kb). Fragment 6 is cleaved by endo R.*Bam* HI and fragment 7 is not.

Using the density selection, we have cloned and are analyzing several other

---

Fig. 5. Restriction enzyme analysis of λ*D⁻ sr* Iλ·P1 recombinants obtained either by in vitro packaging or transfection. Endo R.Eco RI fragments of λ*D⁻ sr*Iλ3 DNA (20 μg/ml) and P1*clr*100r⁻ m⁻ DNA (15 μg/ml) were recombined in vitro as described in METHODS, section (j) in a 50 μl reaction. After 48 h, 5μl aliquots were either used for transfection of JC5183 (see METHODS (k)) or packaged and plated on strain YMC. A pool of 4000 plaques (10% of which were hybrids) obtained after transfection and a pool of 1000 plaques (91% of which were hybrids) obtained after in vitro packaging were amplified in strain YMC by liquid infection (see METHODS (e)). DNA isolated from the amplified stocks was digested with endo R.*Eco* RI and subjected to electrophoresis in 1% agarose slab gels (see METHODS (g) and (i)).

(a) Endo R.*Eco* RI DNA fragments (1.5 μg) derived from phage P1*clr*100r⁻ m⁻ . 21 fragments are detectable in this gel system and are indicated by the hatched marks. From the data of Schultz and Stodolsky (1976) and from ethidium bromide staining of these gels we have assigned bands 4 and 5 and 11 and 12 distinct fragments. Although ethidium bromide staining of gels suggests that band 5 may be composed of two fragments, we are less certain of this than of our other assignments. We are presently attempting to clarify this point. Fragments 17 to 21 are not easily visualized in the contact prints of these gels. We have observed three additional fragments after electrophoresis in 0.5% agarose: 3.0% polyacrylamide gels (data not shown). The light bands observed above fragment 1 here and in Fig.8, lane 6, are due to incomplete digestion of the batch of P1 DNA used for these particular analyses. Band assignments have been confirmed by experiments in which digestion was complete.

(b) Endo R.*Eco* RI DNA fragments (1.0 μg) derived from the hybrid pool obtained by in vitro packaging. The two largest fragments in this panel are the arms of the λ*D⁻ sr*Iλ3 vector.

(c) Endo R.*Eco* RI DNA fragments (0.5 μg) derived from the hybrid pool obtained by transfection. The light band observed above the two major fragments is a complex of left and right vector arms joined by annealing of the λ cohesive termini.

(d) Endo R.*Eco* RI DNA fragments (0.5 μg) derived from λ*c*I*ts*857.

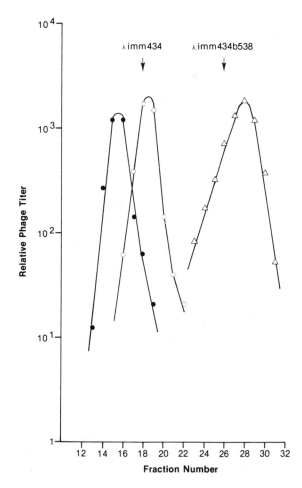

Fig.6. The density distribution of λD^- $sr\mathrm{I}\lambda 3$ and two λD^- $sr\mathrm{I}\lambda 3 \cdot E.$ *coli* hybrids. This figure is a composite of three CsCl equilibrium gradients prepared and assayed in exactly the same way (see METHODS, section (n)) and aligned by the marker phage added to each gradient ($\lambda imm434$ and $\lambda b538imm434cSam7$). One gradient contains the λD^- $sr\mathrm{I}\lambda 3$ vector (△————△) and the other two gradients contain phage isolated from a plaque obtained either from fraction 19 (○——○ ; designated λD^- sr $\mathrm{I}\lambda 3 \cdot E.$ *coli*-96) or fraction 13 (●——● ; designated λD^- sr $\mathrm{I}\lambda 3 \cdot E.$ *coli*-104) of the in vitro packaging gradient shown in Fig.3. The fractions not shown in the figure contain no more than the background level of phage.

P1 fragments. We note here that fragment 7 is unusual because it confers a clear plaque phenotype to the hybrid phage. Analysis of this phenomenon is in progress.

   In the density selection described above, we studied plaques produced by the initial plating of transfection or in vitro packaging mixtures. We found that if the identical plaque mixtures were regrown for several cycles to give much higher phage titers, certain classes of hybrids disappeared. The density distri-

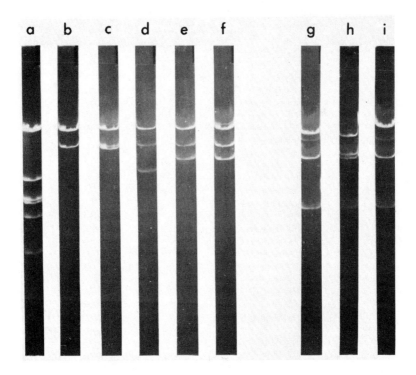

Fig. 7. Restriction enzyme analysis of specific λD^- sr Iλ3·E. coli hybrids. (a) λcIts857; (b) λD^- srIλ3; (c) λD^- srIλ3·E. coli-104; (d) λD^- srIλ3·E. coli-96; (e) λD^- srIλ3·pro; (f) λD^- srIλ3·thr; (g) λD^- srIλ3; (h) λD^- srIλ3·pro; (i) λD^- srIλ3·thr. Slots (a) through (f) represent 0.5 μg DNA digested with endo R.EcoR1 and slots (g), (h) and (i) represent 0.5 μg DNA digested with endo R.Bam HI. λD^- srIλ3·E. coli recombinants were constructed as described in MATERIALS AND METHODS, sections (i) and (j), using N99 DNA. Phage and DNA were obtained as described in MATERIALS AND METHODS, sections (e) through (g). Restriction and gel electrophoresis are described in MATERIALS AND METHODS, section (i). We used the following size estimates of λcIts857 endo R.EcoRI fragments as standards: λA = 20.6 kb; λB = 4.52 kb; λC = 5.16 kb; λD = 7.11 kb; λE = 5.58 kb and λF = 3.21 kb (Helling et al., 1974). The nomenclature for λ fragments is that given by Thomas et al. (1974). As observed in lane g, digestion of λD^- sr Iλ3 with endo R.BamHI yields three fragments: a 5.3 kb piece containing the left end, a 19.9 kb central fragment, and an 11.2 kb fragment from the right end. The light band observed below the 19.9 kb fragment corresponds to the complex of the left and right arms joined by the λ cohesive termini.

bution in CsCl gradients of the in vitro packaged phage population shown in Fig. 3, but now regrown for several cycles, is given in Fig. 9. There appears to be a selection for hybrid phages with a density corresponding to a DNA content of 92—96% that of λ wild-type (phage with inserted fragments of 6.5—8.4 kb). We are uncertain why this should be, but it is obvious that it can be advantageous to select the desired hybrid phage from the initial packaging or transfection mixture rather than after growing these mixtures for many cycles of growth.

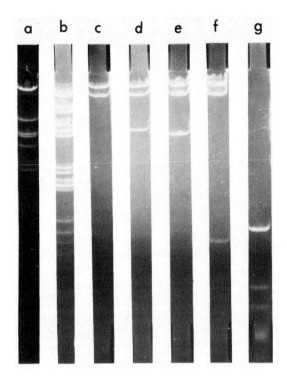

Fig. 8. Restriction enzyme analysis of specific λD^- sr Iλ3·P1 hybrids. (a) λcIts857; (b) P1clr100r$^-$ m$^-$; (c) λD^- sr Iλ3; (d) λD^- srIλ3·P1-6; (e) λD^- srIλ3·P1-7; (f) λD^- srIλ3·P1-16; (g) SV40. Slots (a) through (f) depict 0.5 µg of DNA digested with endo R.EcoRI. Slot (g) represents SV40 DNA digested with endo R.HaeIII. The construction of λD^- sr Iλ3·P1 hybrids is described in MATERIALS AND METHODS, section (j) and Fig. 4. λD^- sr Iλ3·P1-6 and λD^- srIλ3·P1-7 were obtained respectively by in vitro packaging and transfection as described in MATERIALS AND METHODS, section (k) and (l). Restriction and gel analysis are described in MATERIALS AND METHODS, section (i). Size estimates of the small P1clr100r$^-$ m$^-$ endo R.EcoRI fragments were made using SV40 endo R.HaeIII fragments. (Endo R.HaeIII was kindly provided by J. Seidman and SV40 DNA by Norman Salzman.) The SV40 DNA (0.5 µg) was digested with endo R.HaeIII at 37° C for 60 min in 20 mM Tris—HCl, pH 7.5; 60 mM NaCl and 7 mM $MgCl_2$. The reaction was stopped by incubation for 5 min at 70° C. The three visible fragments in this gel system are 1500, 945, and 600 base pairs, respectively.

Several genetic methods can be used to detect λD^- sr Iλ3 carrying inserted endo R.EcoR1 fragments. Because the only site for insertion of fragments is the srIλ3 site located in the phage gene encoding λ exonuclease, all hybrids must be defective in this enzyme (red^-). This defect can be easily recognized because red^- phage do not form plaques on hosts with *pol* A or *ligts* mutations (the *feb* phenotype Zissler et al., 1971). A second method, used for the identification of hybrids as well as the detection of different size classes of hybrids, relies on the observations by Sternberg and Weisberg (in preparation) of the

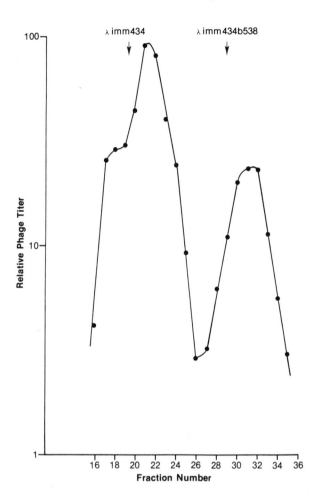

Fig.9. The effect of phage amplification on the distribution of λD^- *sr* Iλ3 *E. coli* hybrids. The hybrid phages shown in Fig.3 obtained after in vitro packaging were amplified by the plate lysate method. Approximately 10^6 amplified phage were centrifuged to equilibrium in a CsCl gradient along with λimm434 and λb538imm434cSam7 marker phage. The *imm*λ phage were assayed on strain NS617 and the *imm*434 phage on strain NS62.

dispensibility of the λ capsid protein pD. Phage λ will grow normally in the absence of pD if the phage contains no more than 82% (38 kb) the wild-type DNA content (46.5 kb). Since the λD^- *sr* Iλ3 vector has 78% the wild-type content (36 kb), it can grow equally well in a *sup*⁺ or *sup*⁻ host. Hybrid phages adding endo R.*Eco*RI fragments of more than 4% the wild-type genome (1.9 kb) should now require pD for growth and will only form plaques on a *sup*⁺ host. In contrast, λD^- *sr* Iλ3 hybrids with inserted fragments of less than 1.9 kb will grow on *sup*⁻ as well as *sup*⁺ hosts.

The validity of these predictions was tested using the hybrid phages isolated by the density selection described above. We found that all hybrids were *red⁻* and those with inserted fragments larger than 1.9 kb did not plate on *sup⁻* hosts. In reconstruction experiments we found that these hybrid phages could easily be detected as turbid plaques on mixed indicator lawns containing both *sup⁻* and *sup⁺* bacteria. The vector λ*D⁻ sr* Iλ3 forms clear plaques on such a lawn. This method can therefore be used to positively select hybrid phages.

We searched for small inserted fragments using λ·P1 recombinants obtained by transfection. About 20% of the hybrids recognized as *red⁻* were able to form plaques on *sup⁻* hosts. This property is expected of hybrids with inserted fragments less than 1.9 kb. Several such hybrids were purified and analysed. The agarose gel pattern after endo R.*Eco*RI restriction of one of these hybrids is shown in Fig.8, Slot *f*. This hybrid carries P1 fragment 16, the size of which is 1.4 kb, confirming the above expectation.

We selected several *E. coli* hybrids by looking for transduction of specific markers. A mixed pool of λ·*E. coli* hybrids obtained by in vitro packaging was used to transduce each of the five individual amino acid markers of strain NS309. We were successful in finding hybrid phages capable of transducing only *pro* and *thr* (see MATERIALS AND METHODS). The presence of wild-type λ in NS309 insured greater homology for integration of the prospective hybrid phage. Recall that the λ*D⁻ sr* Iλ3 vector has a deletion of the entire region required for integration and excision. As a result, the vector cannot integrate by site specific recombination, and the formation of lysogens by the vector is a rare event (1.8·10⁻⁵ lysogens per infected cells). When *E. coli* fragments are inserted into the vector or when λ is integrated into the host chromosome, the frequency of stable lysogens increases more than 20-fold. Because the hybrids are capable of expressing proline and threonine biosynthetic functions when the phage promoters are repressed, the inserted fragments probably contain promoters for expression of these genes. Phage carrying the endo R.EcoR1 fragments of either *pro* or *thr* were isolated from these *pro⁺* and *thr⁺* transductants as described in MATERIALS AND METHODS. Endo R.*Eco* RI and endo R.*Bam*HI restriction analysis of the purified DNA is shown in Fig.7. The endo R.*Eco* RI fragments from both hybrids were identical in mobility in agarose gels (10.2 kb), however, endo R.*Bam*H1 cleaves the *pro* fragment in two at a position approximately 8.8 kb from the rightmost *Eco* RI site of the fragment. The *thr* endo R.*Eco* RI fragment is not cleaved by endo R.*Bam*HI.

DISCUSSION

The CaCl₂ method of transfection is relatively inefficient and strain dependent (Mandel and Higa, 1970). Typically the efficiency of transfection ranges from 10^5 to 10^6 plaques per μg of λ DNA where theoretically one should obtain about 10^{10} plaques per μg. Only a few *E. coli* K12 strains are commonly used: C600 derivatives (see Thomas et al., 1974), 803 derivatives (Borck et al., 1976) and JC5183 derivatives (Tiemeier et al., 1976). In our experience many *E. coli* strains are inefficient or virtually impossible to transfect using the standard $CaCl_2$ transfection methods. We find that the in vitro packaging of recombinant DNA molecules is an attractive alternative to $CaCl_2$ transfection for several reasons. First, for λ DNA of near wild-type size packaging is routinely 10-fold more efficient than transfection. Second, the phage are produced in standard extracts and as such are independent of special strain requirements for phage growth. Third, the marked preference of the packaging system for large DNA provides a strong selection for recombinant phage if the receptor phage chromosome is sufficiently small. Previously this selection was available only with special receptor phages such as λgt·λC (Thomas et al., 1974). With the vector used in this report, yields of 80—90% hybrids were routine. We find that in vitro packaging is particularly useful for cloning large DNA fragments. We note here that any of the λ vectors described to date could be used in the packaging system.

We suggest that in vitro packaging offers certain safety features that may be advantageous in some recombinant DNA experiments, where one might be required to transfect a special host. In our experience most acceptable "safe" hosts are not easily made competent for transfection. The in vitro packaging technique, however, requires only that the bacterium contain λ phage receptors. The packaging extracts involve small volumes of material (less than 0.5 ml), require no aerosol producing manipulation during packaging, and can be sterilized by chloroform. Moreover unreacted DNA is destroyed by DNAse prior to plating for phage particles.

The receptor phage used in this report carried the D*am*15 mutation. We were able to use the special properties of D*am* phages to recognize recombinants carrying inserted fragments of less than or more than 1.9 kb by inspection of plaques. The observations based on the properties of D*am* phages can also be used with most λ receptor phages currently available. We suggest that for cloning small DNA fragments, phages carrying a D*am* mutation offer a novel biological containment property. Phages with less than 82% the wild-type λ DNA content do not require *p*D for growth. However, when such deletion phages are grown in the absence of *p*D, they are extremely unstable. Most chelating agents tested reduce the viability of these phages at least six orders of magnitude within seconds at 4° C (Sternberg and Weisberg, in preparation). In the presence of 20 mM $MgSO_4$, however, they are quite stable. For additional safety, other *ts* λ mutations could be added.

The use of a D*am* phage as a vector offers an advantage also in the subsequent analysis of cloned DNA fragments. DNA deletions can be easily selected in D*am* phages with a DNA content in excess of 82% that of wild-type because such deletions allow these phages to plaque on a *sup⁻* host (Sternberg and Weisberg, in preparation). Since many of these deletions should extend into the inserted fragment of a hybrid phage they can be used to dissect and analyze the composition of that fragment.

We have shown that the packaging system and the λ*D⁻ sr* Iλ3 vector can be used for cloning of specific endo R.*Eco* RI fragments of *E. coli* and bacteriophage P1. A more detailed description of these new hybrids will be presented

in a later communication. We note here that with no modification, the
λD^- srIλ3 receptor could also be used as a receptor for endo R.*Bam*HI and
endo R.*Sal* fragments (unpublished observations).

ACKNOWLEDGEMENTS

We would like to thank Drs. Robert Weisberg and Philip Leder for their
encouragement and constructive criticism during the course of this work. We
also want to thank Linda Fauvie for patiently typing the manuscript.

REFERENCES

Adhya, S., Sen, A. and Mitra, S., The role of gene *S*, in A.D. Hershey (Ed.), The Bacteriophage Lambda, Cold Spring Harbor Laboratory, 1971, pp. 743—746.
Appleyard, R.K., Segregation of new lysogenic types during growth of a doubly lysogenic
   strain derived from *E. coli* K12, Genetics, 39 (1954) 440—452.
Barbour, S., Nagaishi, H., Templin, A. and Clark, A.J., Biochemical and genetic studies of
   recombination proficiency in *E. coli*, II. *rec*⁺ revertant caused by indirect suppression of
   rec⁻ mutations, Proc. Natl. Acad. Sci. USA, 67 (1970) 128—135.
Becker, A., and Gold, M., Isolation of bacteriophage lambda A-gene protein, Proc. Natl.
   Acad. Sci. USA, 72 (1975) 581—585.
Borck, K., Beggs, J.D., Brammar, W.J., Hopkins, A.S. and Murray, N.E., The construction
   in vitro of transducing derivatives of phage lambda, Mol. Gen. Genet., 146 (1976) 199—
   207.
Campbell, A., The steric affect in lysogenization by bacteriophage λ, II. Chromosomal at-
   tachment of a b2 mutant, Virology, 27 (1965) 340—345.
Davidson, N. and Szybalski, W., Physical and chemical characteristics of lambda DNA, in
   A.D. Hershey (Ed.), The Bacteriophage Lambda, Cold Spring Harbor Laboratory, 1971,
   pp. 45—82.
Dennert, G. and Henning, V., Tyrosine-incorporating amber suppressors in *Escherichia coli*
   K12, J. Mol. Biol., 33 (1968) 322—329.
Enquist, L. and Weisberg, R.A., The red plaque test: A rapid method for identification of
   excision defective variants of bacteriophage lambda, Virology, 12 (1976) 147—15.
Enquist, L. and Weisberg, R.A., A genetic analysis of the *att-int-xis* region of coliphage λ,
   J. Mol. Biol., in press.
Feiss, M., Fisher, R.A., Crayton, M.A. and Enger, C., Packaging of the bacteriophage
   lambda chromosome: Effect of chromosome length, Virology, in press.
Fiandt, M., Zdenka, H., Lozeron, H.A. and Szybalski, W., Electron micrographic mapping
   of deletions, insertions, inversions and homologies in the DNA's of coliphages lambda
   and Phi 80, in A.D. Hershey (Ed.), The Bacteriophage Lambda, Cold Spring Harbor
   Laboratory, 1971, pp. 329—354.
Goldberg, A.R. and Howe, M., New Mutations in the *S* cistron of bacteriophage λ affecting
   host cell lysis, Virology, 38 (1969) 200—204.
Gottesman, M.E. and Yarmolinsky, M.B., Integration-negative mutants of bacteriophage λ,
   J. Mol. Biol., 31 (1968) 487—505.
Gottesman, M.M., Hicks, M.L., and Gellert, M., Genetics and function of DNA ligase in
   E. coli, J. Mol. Biol., 77 (1973) 531—547.
Helling, R.B., Goodman, H.M. and Boyer, H.W., Analysis of endonuclease R.*Eco* RI
   fragments of DNA from lambdoid bacteriophages and other viruses by agarose-gel electro-
   phoresis, Virology, 14 (1974) 1235—1244.

Hohn, B. and Hohn, T., Activity of empty, headlike particles for packaging of DNA of bacteriophage λ in vitro, Proc. Natl. Acad. Sci. USA, 71 (1974) 2372—2376.

Kaiser, D., Syvanen, M. and Masuda, T., DNA packaging step in bacteriophage λ head assembly, J. Mol. Biol., 91 (1975) 175—186.

Mandel, M. and Higa, A., Calcium-dependent bacteriophage DNA infection, J. Mol. Biol., 53 (1970) 159—162.

Marmur, J., A procedure for the isolation of DNA from microorganisms, J. Mol. Biol., 3 (1961) 208—218.

Murray, N. and Murray, K., Manipulation of restriction targets in phage λ to form receptor chromosomes for DNA fragments, Nature, 251 (1974) 476—481.

Nash, H., Lambda *att*B-*att*P, a derivative containing both sites involved in integrative recombination, Virology, 57 (1974) 207—216.

Reader, R.W. and Siminovitch, L., Lysis defective mutants of bacteriophage lambda: Genetics and physiology of *S* cistron mutants, Virology, 43 (1971) 607—622.

Rosner, J.L., Formation, induction and curing of bacteriophage P1 lysogens, Virology, 48 (1972) 678—689.

Schrenk, W.J. and Weisberg, R.A., A simple method for making new transducing lines of coliphage λ, Mol. Gen. Genet., 137 (1975) 101—107.

Schulz, G. and Stodolsky, M., Integration sites of foreign genes in the chromosome of coliphage P1: A finer resolution, Virology, 73 (1976) 299—302.

Scott, J.R., Phage P1 cryptic, II. Location and regulation of prophage genes, Virology, 53 (1973) 327—336.

Shimada, K., Weisberg, R.A. and Gottesman, M.E., Prophage lambda at unusual chromosomal locations, I. Location of the secondary attachment sites and the properties of the lysogens, J. Mol. Biol., 93 (1972) 483—503.

Smith, H.D. and Nathans, D., Nomenclature for restriction enzymes, J. Mol. Biol., 81 (1973) 1072—1078.

Sternberg, N., A genetic analysis of bacteriophage λ head assembly, Virology, 71 (1976a) 568—582.

Sternberg, N., A class of rif^R RNA polymerase mutations that interferes with the expression of coliphage λ late genes, Virology, 73 (1976b) 139—154.

Sternberg, N. and Weisberg, R.A., Packaging of prophage and host DNA by coliphage λ. Nature, 256 (1975) 97—103.

Thomas, M., Cameron, J.R. and Davis, R.W., Viable molecular hybrids of bacteriophage λ and eucaryotic DNA, Proc. Natl. Acad. Sci. USA, 71 (1974) 4579—4583.

Tiemeier, D., Enquist, L.W. and Leder, P., An improved derivative of a bacteriophage λ EK2 vector useful in the cloning of recombinant DNA molecules: λgtWES·λB, Nature, 263 (1976) 526—527.

Weil, J., Cunningham, R., Martin, R., Mitchell, E. and Bolling, B., Characteristics of λ*p*4, a λ derivative containing 9% excess DNA, Virology, 50 (1973) 373—380.

Zissler, J., Signer, E. and Schaefer, F., The role of recombination in the growth of bacteriophage λ, I. The gamma gene, in A.D. Hershey (Ed.), The Bacteriophage Lambda, Cold Spring Harbor Laboratory, 1971 pp. 454—468.

23

Reprinted from *Natl. Acad. Sci. (USA) Proc.* **74**:3259–3263 (1977)

Packaging recombinant DNA molecules into bacteriophage particles *in vitro*

(bacteriophage λ morphogenesis/infectious DNA)

BARBARA HOHN* AND KENNETH MURRAY†

* Department of Microbiology, Biozentrum der Universität Basel, Klingelbergstrasse 70, CH-4056 Basel, Switzerland; and † Department of Molecular Biology, University of Edinburgh, Edinburgh, Scotland, United Kingdom

ABSTRACT   Recombinant phage genomes made in reactions with purified enzymes may be recovered directly by packaging into phage heads *in vitro*. The process is efficient and nonselective and offers containment in initial stages of handling recombinant DNA. Ligase [poly(deoxyribonucleotide):poly-(deoxyribonucleotide) ligase (AMP-forming), EC 6.5.1.1] reaction products can recombine with endogenous phage DNA during packaging, but UV-irradiation eliminates the biological activity of the endogenous DNA.

Hybrid molecules may now be constructed *in vitro* between DNA fragments from various sources and bacterial plasmids or bacteriophage λ DNA (1–8). Since the plasmids and phage DNA are capable of autonomous replication, the recombinant molecules enable DNA fragments to be cloned in bacterial cells. In addition to autonomous replication, the principal requirements of the receptor DNA molecules are that they must be able to infect competent bacteria, that they contain one or a very limited number of targets for the restriction enzyme, that insertion of the new DNA fragment does not destroy a gene essential for the function of the plasmid or phage, and that cells carrying the plasmid or phage can be readily selected and the recombinant plasmids or phage distinguished easily from their parents (9–11). For phage λ DNA there is a further constraint in that only molecules in a certain size range can be packaged into a viable virion; some of the DNA must therefore be deleted from inessential regions of the phage genome in order to provide space for the new fragments and the recombinant molecule must have a length between about 75 and 105% of wild-type λ DNA (12). The joining of fragments of vector and donor DNA in reactions with polynucleotide ligase [poly(deoxyribonu-cleotide):poly(deoxyribonucleotide) ligase (AMP-forming), EC 6.5.1.1] usually yields a complex range of products which in practice are seldom fractionated. Instead, the mixtures are used directly to transform (13), or transfect (14), cultures of competent bacterial cells, thus allowing the cell to selectively propagate only those DNA molecules that are viable in this specific host strain.

Yields of transformed cells, or phage particles, from these procedures are variable. With intact DNA from lambdoid phages the yield of phage particles per μg in transfection experiments with *E. coli* is usually between 10^4 and 10^6; the average represents a recovery of less than 1 in 10^5 of the DNA molecules. With DNA preparations that have been restricted and the fragments rejoined by treatment with polynucleotide ligase, the yield of phage per μg of DNA in the reaction mixture is usually at least an order of magnitude lower than this.

Other procedures are available for conversion of DNA molecules into phage particles. Of these, the packaging of DNA

into viable phage particles *in vitro* with extracts of appropriately lysogenized and induced cells gives recoveries as high as 1 in 10^3 of the DNA molecules.

The principle of *in vitro* packaging is the following: in the presence of high concentrations of the phage head precursor, or prehead, and the packaging proteins, which are the products of genes *A*, *Nu1*, *D*, and *FI* (15), λ DNA is packaged (16, 17). The full heads are then matured (*in vitro*) into plaque-forming units (PFU) in the presence of proteins from genes *W* and *FII* and phage tails (18). Practically, *in vitro* packaging is most efficiently performed in a very concentrated mixture of two induced lysogens, one of which is genetically blocked at the prehead stage (by an *amber* mutation in, e.g., gene *D*) and therefore accumulates these precursors, while the other is inhibited from forming capsids by an *amber* mutation in the gene, *E*, for the main capsid protein. These two lysates provide all the necessary components to convert λ DNA into a plaque-forming particle. Endogenous concatemeric phage DNA present in these lysates is packaged and cleaved (17, 19), but if mature linear DNA from another lambdoid phage is added to the reaction mixture, this also becomes packaged (16). It has been reported that circular, monomeric λ DNA, however, cannot be made infective by either transfection (20) or *in vitro* packaging (16, 19). In this communication we describe the application of this packaging system for making *in vitro* recombinant DNA infectious.

MATERIALS AND METHODS

Description of Vectors. The phage vector 540 is described in ref. 21; vectors 590, 598, 607, 641, 705, and 728 are described in ref. 11. The genotypes of the remainder are as follows: 459, λ (*srI*λ1–2) ∇*cI*857 *nin*5; 461, λ*b*538 *srI*λ3 *nin*5; 596, λ [(*srI*λ1–3)∇ (*supE*⁺)] *sus*γ *cI*857 *nin*5; 616, λ*plac*5 *att*⁺ *imm*21 *nin*5; 646, λ*plac*5 *srI-lac* *srI*λ3 *Nam cI*ts *nin*5; 711, λ(*srI*λ1–2)∇ *shn*λ3 (*trpE*⁺) (*att-red*)∇ *sus*γ *c*V (KH54) *nin*5; and 722, λ(*srI*λ1–2)∇ *shn*λ3 (*supF*⁺) (*att-xis*)∇ *red*⁺ *imm*21 *nin*5. (The abbreviations used for restriction enzymes and their targets in phage genomes are given in refs. 11, 21, 22, and 23.)

Restriction and Ligation Reactions. Restriction enzyme digests were normally carried out with about 2 μg of DNA in 50 μl of a solution containing 10 mM Tris·HCl (pH 7.5), 10 mM 2-mercaptoethanol, 10 mM MgCl₂, and 50 mM NaCl for the *Hin*dIII enzyme, but 100 mM NaCl for the *Eco*RI enzyme; solutions were incubated at 37° for 1–2 hr, heated at 70° for 5 min, and kept at 0°. The necessary quantity of the restriction enzyme was determined in trial digests of λ⁺ DNA that were analyzed by electrophoresis in 1% agarose gels (7). The restriction of the vector DNA samples was assessed from the reduction in transfection yield before proceeding with the addition of donor DNA fragments (restriction enzyme digests of

Abbreviation: PFU, plaque-forming units.

Table 1.   Yields of phage (per μg of vector DNA) in transfection and packaging experiments *in vitro*

Exp. no.	Phage DNA (vector)	Restriction targets	Before restriction		After restriction + donor DNA + ligation	
			Transfection	Packaging	Transfection	Packaging
1	461	1 *Eco*RI	2.5×10^4	8×10^6 11×10^6 *		
			2.9×10^4	6.5×10^5 2.5×10^6 *		
	607	1 *Eco*RI	4.6×10^4	1×10^7 3.3×10^6 *	4×10^3	1.5×10^6 1.0×10^6
			5.1×10^4	1.7×10^6 5×10^6 *	4×10^2	7.6×10^4 7.2×10^3 †
	λc I857Sam7	5 *Eco*RI	6.2×10^4	1.3×10^7 1.0×10^7 *		
2	461	1 *Eco*RI	2.9×10^4			
	641	1 *Eco*RI	1.2×10^5		3.7×10^3	8.2×10^4
	646	2 *Eco*RI	1.2×10^5		6×10^4	6.6×10^5
	540	1 *Hind*III	1.1×10^5		1.9×10^4	1.3×10^6
	590	1 *Hind*III	1.5×10^5		5×10^4	1.5×10^6
	598	1 *Hind*III	3.4×10^4		4.2×10^4	8.0×10^5
	705	2 *Hind*III	5.3×10^4		5.6×10^3	1.6×10^5
	711	2 *Hind*III	7.8×10^4		6.8×10^3	3.2×10^4
3‡	596	2 *Eco*RI	$0.15\text{–}5 \times 10^5$	$1.7\text{–}8.0 \times 10^7$	0.8×10^3	$0.1\text{–}0.5 \times 10^5$
	722	2 *Hind*III	$0.09\text{–}1.8 \times 10^5$	$0.1\text{–}6.0 \times 10^7$	$0.1\text{–}9.0 \times 10^3$	$0.1\text{–}7.6 \times 10^5$
	598	1 *Hind*III	1×10^5	4.4×10^5	3×10^3	1.7×10^5

The packaging in Exps. 2 and 3 was performed with UV-irradiated cells (see legend to Fig. 1 for procedure). Phage were plated on *E. coli* C600. The yields of phage shown for the restriction and ligation experiments are total phage; the recombinants were recognized in the various experiments by their clear plaque morphology [vectors 607, 641, 590, and 598 (9, 11)], by complementation of an auxotrophic host strain, or by their ability to plate on a *pel⁻* host strain since this is dependent upon the length of the phage genome (ref. 31; and W. Arber and H. Jütte, personal communication).
* Duplicated experiments were done simultaneously with independently prepared batches of cells for the packaging mixtures.
† An experiment done simultaneously with yeast DNA as donor gave 500 PFU/μg of DNA in transfection, but 4.7×10^4 and 1.4×10^4 PFU/μg of DNA in duplicate packaging experiments *in vitro*.
‡ This group of experiments was carried out by the participants in the EMBO course referred to in the *text*. The combination of donor and receptor DNA was different in each experiment: the donors were DNA from different *E. coli* strains, a λ*trp* transducing phage, or plasmids carrying all or part of the *E. coli* trp operon. The latter were kindly provided by V. Hershfield and D. R. Helinski. The results are given as the range of yields from three independent experiments with the 596 vector and four independent experiments with the 722 vector.

E. coli DNA) and reaction with T4 polynucleotide ligase (23, 24). The solutions were diluted to a DNA concentration (vector molecules) of 5–10 μg/ml and incubated for 3 hr at 10° followed by at least 24 hr at 0° with polynucleotide ligase from *E. coli* infected with phage T4 (Miles Laboratories Ltd., Stoke Poges, Bucks, U.K.), (0.1 unit/ml or 2 μl of the concentrated enzyme preparation per ml of reaction mixture; the optimum concentration was determined experimentally for each batch of enzyme by measuring the restoration of transfection yield of a restriction enzyme digest of a phage DNA with a single target for *Eco*RI) in a mixture containing 66 mM Tris·HCl (pH 7.5), 1 mM EDTA, 10 mM MgCl₂, 5 mM dithiothreitol or 10 mM 2-mercaptoethanol, 40 mM NaCl, and 0.1 mM ATP.

Transfection. *E. coli* 803 *supE⁺ supF⁺* r⁻ₖm⁻ₖ or r⁻ₖm⁺ₖ was grown in a poor medium and starved in 0.1 M CaCl₂ solution (14) or, most often, grown in L broth, washed in 0.01 M MgSO₄ solution, and starved in 0.1 M CaCl₂ solution (13). DNA was usually added to the competent cells to give a final concentration of about 1 μg/ml and diluted to ¹⁄₁₀ its original concentration before plating in the case of intact phage DNA; it was added at a concentration of about 0.5 μg/ml and plated undiluted in the case of restricted DNA solutions. Mixtures from ligase reactions were diluted to a DNA concentration of about 0.5 μg/ml with 150 mM NaCl/15 mM sodium citrate before they were mixed with the competent cells (two volumes) and these also were plated undiluted.

Packaging. Genomes were packaged *in vitro* as described (17) with some slight modifications: single colonies of the strains W3101 (λ*imm*434c*Its*E*am*4S*am*7) and W3101 (λ*imm*434c-*Its*D*am*15S*am*7) were streaked out on an LA plate (17) and

grown overnight at 32°. The cells were suspended in warmed LB medium (17) at an OD₆₀₀ of not more than 0.15 and incubated with shaking until an OD₆₀₀ of 0.3 was reached. Induction was effected by incubation of the cultures at 45° for 15 min, standing. Thereafter they were transferred to 37° and incubated for 3 additional hours with vigorous aeration. (A small sample of each culture, which is lysis-inhibited as a result of the mutation in gene S, should be checked for induction; upon addition of a drop of CHCl₃, the culture should clear.) The two cultures were then mixed, centrifuged at 5000 rpm for 10 min, and resuspended in M9A medium (25), but without casein hydrolysate, at an OD₆₀₀ of 0.3. For experiments without UV treatment (Tables 3 and 4) the cells were concentrated about 500-fold in complementation buffer (40 mM Tris·HCl, pH 8.0/10 mM NaN₃/10 mM spermidine/0.1% 2-mercaptoethanol/7% dimethyl sulfoxide). UV-irradiated cells (Fig. 1) were concentrated in the same buffer. [At this stage the cells can be prepared for storage or used directly. Samples of 10–20 μl are distributed in polypropylene or nylon centrifuge tubes (1 ml), frozen rapidly in liquid N₂, and stored at −57° or below. When needed, a sample is then transferred in liquid N₂ and put on ice; ATP is added to a final concentration of 1.5 mM. If cells are to be used directly, ATP can be added before distribution of samples.] After freezing in liquid N₂ and thawing in ice (which takes 2–4 min), the DNA to be packaged (0.01–0.2 μg) was added and carefully mixed with the concentrate *immediately after thawing*. The mixture was centrifuged to the bottom of the tube by spinning for a few seconds to avoid unnecessary evaporation during the subsequent 30- to 60-min incubation at 37°. At the end of this incubation period 2 μl of a DNase solution (100 μg/ml) was added; after further incubation at 37° for 10 min, 0.5 ml of SMC (17) and a drop of CHCl$_3$ were added. Phage obtained by this procedure can, when freed from debris, be stored as any phage lysate.

Containment Conditions. EK1 bacterial and bacteriophage strains were used. Physical containment was P1 except when yeast DNA was involved, where it was P2.

RESULTS AND DISCUSSION

Efficiency of Packaging In Vitro. Table 1 summarizes a series of experiments in which DNAs from a selection of EcoRI and HindIII vectors as well as recombinants made with E. coli DNA by restriction and ligation reactions were compared for their transfection and packaging efficiencies in vitro. Some of the data were collected by the participants of an EMBO (European Molecular Biology Organization) course on "DNA restriction endonucleases: Reactions and applications" (Basel, 1976), and therefore should be fairly representative of the efficiencies, and the variations thereof, obtainable. Generally, packaging of vector DNA in vitro yielded plaques with an efficiency about a 100-fold higher than transfection. Restricted and ligated DNa was packaged with an efficiency 10- to 100-fold higher than that from transfection.

Storage of Cells for Packaging. Since gentle lysis of the packaging cells is accomplished by freezing in liquid nitrogen and thawing, the possibility of storing the induced lysogens as a concentrated frozen mixture was investigated because it would obviously be a great convenience to make a large batch of cells and remove small samples as required. The competence of such a mixture for packaging in vitro remained at the original level for at least 14 months, the longest time tested, on storage at −57°. Survival of packaging activity upon storage at −24°, however, was less satisfactory.

Packaging Is Nonselective. Lambdoid phages with different genotypes vary appreciably in their growth characteristics on a given host and further differences in efficiency of phage growth are frequently encountered when a given phage is grown on different host strains. Both host and phage recombination systems markedly affect the ease of handling a particular phage; those with chromosomes of abnormal size or with reiterated sequences are relatively unstable and therefore become replaced in a population by more stable derivatives. One would thus not expect all of the members of a population of recombinant DNA molecules made in vitro to be equally viable, so that in transfection experiments with a large mixed population of DNA molecules, there may well be selection against some recombinants and a variable yield of each member of the family of recombinants, some of which may be rapidly lost altogether (26). If single plaques are isolated after transfection and propagated separately, these arguments obviously do not apply, but some recombinants may have been lost already because they were unable to grow on the host strain used for transfection.

Packaging of the DNA molecules in vitro does not offer the same opportunities for selection of some of the recombinants at the expense of others or for selection of the parental molecules if these have higher viability than the recombinants. Since the DNA is merely a passive packaging substrate and only DNA sequences very close to the ends have to be specific (16), the characteristics of the DNA molecule being packaged should be of little consequence and the composition of the population of phages produced in the "lysate" should reflect the relative abundance of the various DNA molecules present.

Table 2 shows that within the normal range of genome length for viable lambdoid phages, there is little difference in yield of viable phages in the packaging of genomes in vitro with deletions of various size. Table 2 also shows that, as expected,

Table 2. Efficiency of packaging in vitro of λ DNA preparations having different genome lengths and red genotype

Exp. no.	Phage DNA (vector)	% λ$^+$ DNA deleted	red genotype	Packaging efficiency, PFU × 10^{-7}/ μg DNA
1	λcI857Sam7	0	+	1.3
	616	10	+	1.7
	459	16	+	1.2
	607	18	+	1.5
	461	20	−	0.78
2	λcI857Sam7	0	+	0.68
	596	6	−	0.32
	598	18	−	1.7
	607	18	+	0.44
	641	20	−	0.94
	728	20	−	1.0

Packaging in vitro was performed with UV-irradiated cells (see legend to Fig. 1 for procedure) and phage were plated on E. coli C600.

red$^-$ phages do not appear to be at a disadvantage compared to red$^+$ phages. More generally, the results of Tables 1 and 2 show little variation in yield attributable to phage genotype. In contrast with these results, Sternberg et al. (27) recently found that partially purified extracts from a different combination of cells preferentially packaged normal sized phage genomes and could thus be made selective for particular classes of recombinants. Also, in several experiments where the system described here was used with a mixture of a receptor genome (having a deletion of about 18%) and its in vitro recombinant, the recombinant was obtained in a noticeably higher yield (about 3-fold) whereas this was not the case with transfections (B. Klein, B. Hohn, and K. Murray, unpublished work). In any event in vitro packaging of a ligation reaction mixture gives a population of phages that has not been exposed to any of the selective pressures that accompany replication and growth of the phage within host cells. Moreover, since packaging is independent of replication, the recovery of independent DNA molecules out of a mixture is possible. The in vitro packaged phage can be kept indefinitely and scored by any selection method on various host strains whenever needed, but when the phage resulting from the packaging reactions are subsequently propagated they are, of course, subject to the normal factors encountered during growth within host cells. At the stage of recovery of the in vitro recombinant DNA molecules, however, the whole population is probably obtained without appreciable selection.

Recombination during Packaging. Several genetic markers were used to investigate recombination in vitro between the endogenous DNA and the added DNA at three different intervals along the λ genome. Table 3 shows that a low level of recombination in the packaging mixture was indeed detected and was more frequent in the central region of the chromosome, possibly due to the action of the phage integration enzyme, which functions in vitro (19, 28, 29).

One of the more useful λ vectors has the immunity region of phage 434 substituted for its own; its only EcoRI restriction target is located within the 434 cI gene (i.e., the gene for repressor) (9, 11). Insertion of a DNA fragment at this site inactivates the cI gene so that recombinant phages are distinguished from the vector because they form clear plaques. When this vector and donor (E. coli) DNA were restricted with EcoRI and the fragments were joined, recombination with endogenous DNA occurred during packaging. This recombination occurred at a variable efficiency that was sometimes high, but that de-

Table 3.   Recombination of unrestricted exogenous DNA with λ DNA endogenous in the packaging mix

Indicator strain (E. coli)	amber mutations of λ suppressed	DNA packaged in vitro, PFU/μg	Frequency of recombination ($\times 10^4$) in interval 1	2	3
C600 (supE+)	A, D, E, but not S	1.4×10^7, turbid plaques			
		3.2×10^3, clear plaques (~10 × higher than S^+ revertants of packaged endogenous DNA)			2.3
		5.6×10^4 Aam+ Dam or Aam+ Eam, turbid plaques		40	
W3101 (sup)⁰	None	3.5×10^3, turbid plaques (7 × higher than Aam+ revertants)	2.5		

Aam   *imm*434
  1   2   3   exogenous DNA
  Eam   *imm*434   Sam   endogenous DNAs
  or Dam   *c*Its

Packaging *in vitro* was performed with nonirradiated cells, as described in the legend to Table 1. λ*imm*434 Aam32 DNA was used as exogenous DNA. When packaged, it produces turbid plaques on *E. coli* C600. Recombinants in interval 3 make a clear plaque on the same indicator at 37° due to the temperature-sensitive repressor. Recombinants in interval 2 plate on a *sup*⁰ (λAam32) lysogen but not on a *sup*⁰ (λEam4) or *sup*⁰ (λDam 15) lysogen. Recombinants in interval 1 form turbid plaques of the *sup*⁰ indicator *E. coli* W3101.

Table 4.   Effect of UV irradiation upon recombination of restricted and ligated vector DNA with λ DNA endogenous in the packaging mix

UV	Efficiency*	Interval†	% recombinants‡	Recombination in interval§ 1	2	1+2	2+3	3
			Exp. no. 1					
−	2.5×10^6	4	89					16/16
			Exp. no. 2					
−	3.3×10^4	1	27	3/30	5/30	4/30	3/30	15/30
−	1.3×10^4	5	7				4/35	31/35
−	1.7×10^5	18	<1					
			Exp. no. 3					
−	2.1×10^4	1	49	1/172			17/172	154/172
+	2.1×10^4	1	<0.0091					
−	3.4×10^5	3	5.2	1/23			1/23	21/23

  b538   *imm*434
                                     607 or 607-hybrid DNA
  1   2   3
                                     endogenous DNA
Dam or Eam *imm*434   Sam
            *c*Its

Packaging *in vitro* was performed as described in the legend to Table 1. For Exps. 1 and 2 only nonirradiated cells were used; but Exp. 3 both unirradiated cells and cells irradiated for 30 min were used. Restriction of vector and donor (*E. coli*) DNA, and the ligation reactions were as described in the legend to Table 1. The ligated samples were kept at 4° for the number of days indicated, without inactivation of the ligase. The proportion of the recombinants between exogenous and endogenous DNA was calculated by comparing the plating properties of individual, purified plaques on *E. coli* C600 (*supE*+) and a *sup*⁰ strain. Unrecombined 607 vector DNA or a ligation product thereof plates on both indicators. The majority of the recombinants (as detected by this plating procedure) had picked up an *amber E* or *D* mutation from the endogenous DNA. *S amber* mutations carried by a recombinant were not screened because their detection requires a *supF*+ suppressor in the indicator strain, and this would have enabled phage containing the endogenous DNA to plate as well. Individual plaques, purified on *E. coli* C600, were tested for: *imm*434 (turbid plaques at both 32° and 40°); insertion of DNA fragment in *imm*434 [clear plaques at both temperatures (9, 11)]; *imm*434 *c*Its (turbid plaques at 32°, but clear ones at 40°); b538 deletion [inability to plate efficiently on GL1 *pel*21 W3101*supE*+, provided by S. Emmons (25)]; and *D*am or *E*am (by their inability to plate on a *sup*⁰ host).

* PFU/μg of vector DNA in ligation reaction mixture.
† Days between ligation and packaging.
‡ % of total PFU plating on C600.
§ Fraction of total recombinants tested.

creased with the period of incubation (at about 40°) of the ligation reaction mixture (Table 4). Moreover, there was a hot spot of recombination at an interval between immunity and gene *S* which gave *imm*434 *c*Its S^+ Eam4 and *imm*434 *c*Its S^+ Dam15 recombinants at a frequency two to three orders of magnitude above the Sam7 to S^+ reversion rate observed in control complementation mixtures from which exogenous DNA was omitted. This recombination was not dependent upon the ligase (which normally is retained, but not inactivated, in the ligated sample) since heat inactivation prior to packaging *in vitro* did not reduce recombination with the endogenous phage DNA. The mechanism of this recombination is unknown, but possibly the ligated DNA molecules, or more probably large DNA fragments remaining after incomplete ligation or perhaps resulting from the action of a contaminating exonuclease, may initiate or enhance recombination with the endogenous phage DNA. The *Eco*RI target within *imm*434 is located 100 base pairs to the left of the O_R operator and therefore most likely to the right of the *ts* marker (V. Pirrotta, personal communication), which is consistent with this view.

Packaging and Containment. Packaging *in vitro* avoids the necessity of cycling phage DNA molecules through living cells in order to recover them as phage particles. When used as the initial step in recovery of DNA molecules made in recombination research *in vitro*, it removes one stage at which viable phage could escape from the laboratory into the wider environment.

In addition to this degree of containment of the recombinant DNA, the final step in the packaging reactions *in vitro* includes treatment with chloroform, a procedure that has been recom-

mended as a chemical means of containment (sterilization in this case) since it kills bacterial cells very effectively (30). However, any procedure that converts DNA or an *in vitro* recombinant thereof into plaque-forming phage should avoid or minimize the possibility for dissemination of a new and conceivably hazardous sequence via exchange with endogenous DNA in the system being used, be this *in vitro* packaging or transfection. The recombination observed between exogenous, unrestricted DNA, as well as the products of restriction and ligation reactions, and DNA endogenous in the *in vitro* packaging mixtures is therefore important in this context.

There are, in principle, two ways to eliminate recombination between the λ DNA added to the packaging extract and the *E. coli* or phage DNA that is already present in it: genetic elimination of all the known recombination systems from the lysogens used for the packaging mixtures, or elimination of the biological activity of the endogenous DNA. The latter course

Biochemistry: Hohn and Murray

Proc. Natl. Acad. Sci. USA 74 (1977) 3263

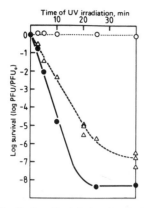

FIG. 1. Inactivation of bacteriophage λ and an *in vitro* packaging mixture by irradiation with UV light. In order to assess the required UV dose, λc1857 Sam7 phage and the packaging mixture prepared as described in the legend to Table 1 were each resuspended in M9 medium at an OD_{600} of 0.3 and distributed in empty sterile petri dishes in 10-ml portions. UV irradiation and all subsequent handling was performed in a darkroom, with the UV lamp (Westinghouse Sterilamp) as the only light source. At various time intervals after irradiation at a 50-cm distance, the phage titer was determined and the packaging cells were concentrated by centrifugation. Packaging *in vitro* was carried out as described in the *text,* using λc1857 Sam7 DNA as the exogenous DNA. Plating was on Ymel(λ) for phage that had packaged endogenous DNA and on Ymel (λimm434) for phage containing the DNA provided exogenously (25). ●, Free phage; △, endogenous DNA packaged; ○, exogenous DNA packaged.

proved to be preferable because introduction of recombination deficiencies into host and prophage of the strains used for packaging reduced the efficiency of *in vitro* packaging by a factor of 2 to 5 (B. Hohn, unpublished work). Extensive irradiation of recombination-proficient packaging cells with UV light prior to use in the *in vitro* packaging reactions did not reduce their efficiency for packaging exogenous DNA, but greatly reduced the yield of endogenous DNA, either packaged or as free phage (Fig. 1). With packaging cells treated in this way no recombinants were detected during packaging of exogenous DNA (Table 4, Exp. 3).

These experiments show that the recombination during *in vitro* packaging can be controlled according to the requirements of a given experiment. With recombinant phage DNA made with a prokaryotic donor DNA, it may sometimes be of interest, or use, to retain this possibility; with donor DNA from higher eukaryotes, it would appear advisable to eliminate the opportunity for recombination as a precaution against possible incorporation of donor DNA sequences into the endogenous DNA from which they could eventually disseminate further. With this added precaution, *in vitro* packaging offers an efficient means for recovery of phage DNA molecules made by the methods currently used to construct heterologous recombinants *in vitro,* and at the same time provides for containment of the recombinants during the process.

We thank Drs. T. Hohn and S. Brenner for suggesting the use of *in vitro* packaging for recombinant DNA research, Dr. M. Birnstiel for recommending the use of UV light to inactivate unwanted DNA, Dr. Ch. Hänni for some of the restriction and ligation reactions, and Dr.

N. E. Murray for many phage strains. The participants of the EMBO course on "DNA restriction endonucleases: Reactions and applications" performed, with interest and enthusiasm, several restriction, ligation, transfection, and *in vitro* packaging experiments and we thank them for allowing their results to be included here. Financial support was provided by grants from the Swiss National Fonds (to T. Hohn) and the Science Research Council, U.K.

1. Chang, A. C. Y. & Cohen, S. N. (1974) *Proc. Natl. Acad. Sci. USA* **71,** 1030–1034.
2. Morrow, J., Cohen, S. N., Chang, A. C. Y., Boyer, H. W., Goodman, H. M. & Helling, R. (1974) *Proc. Natl. Acad. Sci. USA* **71,** 1743–1747.
3. Keddes, C. H., Cohen, S. N., Houseman, D. & Chang, A. C. Y. (1975) *Nature* **255,** 533–538.
4. Chang, A. C. Y., Lansman, R. A., Clayton, D. A. & Cohen, S. N. (1975) *Cell* **6,** 231–244.
5. Murray, N. E. & Murray, K. (1974) *Nature* **251,** 476–481.
6. Borck, K., Beggs, J. D., Brammar, W. J., Hopkins, A. S. & Murray, N. E. (1976) *Mol. Gen. Genet.* **146,** 199–207.
7. Struhl, K., Cameron, J. R. & Davis, R. W. (1976) *Proc. Natl. Acad. Sci. USA* **73,** 1471–1475.
8. Ratzkin, B. & Carbon, J. (1977) *Proc. Natl. Acad. Sci. USA* **74,** 487–491.
9. Murray, K., Murray, N. E. & Brammar, W. J. (1975) *Proceedings of the Xth FEBS Meeting* (North Holland-American Elsevier, Amsterdam), Vol. 38, pp. 193–207.
10. Murray, K. (1977) in "Recombinant DNA," *Xth Miles Symposium* (Raven Press Inc., New York), in press.
11. Murray, N. E., Brammar, W. J. & Murray, K. (1976) *Mol. Gen. Genet.* **150,** 53–61.
12. Bellet, A. J. D., Busse, H. G. & Baldwin, R. L. (1971) in *The Bacteriophage λ,* ed. Hershey, A. D. (Cold Spring Harbor Laboratory, Cold Spring Harbor, NY), pp. 501–513.
13. Lederberg, E. M. & Cohen, S. N. (1974) *J. Bacteriol.* **119,** 1072–1074.
14. Mandel, M. & Higa, A. (1970) *J. Mol. Biol.* **53,** 159–162.
15. Hohn, T., Wurtz, M. & Hohn, B. (1976) *Phil. Trans. R. Soc. London* **276,** 51–61.
16. Hohn, B. (1975) *J. Mol. Biol.* **98,** 93–106.
17. Hohn, B. & Hohn, T. (1974) *Proc. Natl. Acad. Sci. USA* **71,** 2372–2376.
18. Casjens, S., Hohn, T. & Kaiser, A. D. (1972) *J. Mol. Biol.* **64,** 551–563.
19. Syvanen, M. (1974) *Proc. Natl. Acad. Sci. USA* **71,** 2496–2499.
20. Drabkina, L. E., Konevega, L. V., Legina, O. K. & Mosevitsky, M. I. (1976) *Mol. Gen. Genet.* **144,** 83–86.
21. Murray, K. & Murray, N. E. (1975) *J. Mol. Biol.* **98,** 551–564.
22. Nathans, D. & Smith, H. O. (1975) *Annu. Rev. Biochem.* **44,** 273–293.
23. Sharp, P. A., Sugden, B. & Sambrook, J. (1973) *Biochem.* **12,** 3055–3063.
24. Weiss, B. & Richardson, C. C. (1967) *Proc. Natl. Acad. Sci. USA* **57,** 1021–1028.
25. Hohn, T., Flick, H. & Hohn, B. (1975) *J. Mol. Biol.* **98,** 107–120.
26. Cameron, J. H., Panasenko, S. M., Lehman, I. R. & Davis, R. W. (1975) *Proc. Natl. Acad. Sci. USA* **72,** 3416–3420.
27. Sternberg, N., Tiemeier, D. & Enquist, L. (1977) *Gene* **1,** 255–280.
28. Nash, H. A. (1975) *Proc. Natl. Acad. Sci. USA* **72,** 1072–1076.
29. Gottesman, S. & Gottesman, M. E. (1975) *Proc. Natl. Acad. Sci. USA* **72,** 2188–2192.
30. Weissmann, C. & Boll, W. (1976) *Nature* **261,** 428–429.
31. Emmons, S. W., MacCasham, V. & Baldwin, R. L. (1975) *J. Mol. Biol.* **91,** 133–146.

24

Reprinted from *Science* **196**:180–182 (1977)

SCREENING λgt RECOMBINANT CLONES BY HYBRIDIZATION TO SINGLE PLAQUES IN SITU

W. David Benton and Ronald W. Davis

Recombinant DNA technology has made it possible, in principle, to isolate large quantities of DNA corresponding to any single gene. This isolation can be accomplished in any of three ways: (i) purification of the gene or its transcript before insertion into a vector for cloning (*1*); (ii) genetic selection of the desired gene after "shotgun" (*2*) cloning the genome of the donor organism (*3, 4*); or (iii) the use of complementary RNA or DNA probes to screen clones produced through a shotgun experiment. The first method is limited by the practical difficulty of physically purifying the complete nucleic acid sequence of any gene. Application of the second is limited to genes for which appropriate mutations are available in hosts which can express foreign genetic information. The third method is limited only by the availability of complementary nucleic acid probes and has been the most widely applied.

Screening methods have been developed for both recombinant plasmid (*5*) and bacteriophage (*6, 7*) cloning systems. The most rapid of these, the methods of Grunstein and Kramer, are based on a technology developed by Olivera and Bonhoeffer (*8*) for growing and lysing cells on membrane filters. The speed and simplicity of these methods is limited by the need to pick and spot plaques or colonies from plates onto filters for further growth. We have found that a single plaque of a λgt recombinant phage contains enough phage DNA for detectable hybridization to complementary labeled nucleic acid, and that this DNA can be fixed to a nitrocellulose filter by making direct contact between the plaque and the filter. These findings have allowed us to develop a method for screening recombinant phage which has significant advantages over previous methods: (i) physical containment of recombinant molecules (and organisms carrying them) is made simpler because this method requires fewer manipulations of viable phage, requires the growth and handling of much smaller

quantities of recombinant DNA-containing phage and cells, and produces no contaminated materials other than the petri dishes on which the plaques are grown. (ii) At least 2×10^4 plaques can be screened per hour (*9*), which compares with about 100 per hour by the fastest previous method (*7*). (iii) Because up to 2×10^4 plaques can be screened on a single petri dish and nitrocellulose filter, the method requires smaller amounts of materials, including radioactively labeled probe. (iv) Because screening can be done directly from the transfection plate, all the independently constructed phages containing similar or identical inserts can be isolated and individually characterized. This can give information on the heterogeneity of repeated genes or of genes from different somatic tissues.

We believe that this method will be widely applicable and, therefore, present it here in detail.

Cells (*10*) are infected with phage, plated in L-soft agar on L plates (*11*), and grown for at least 12 hours at 37°C. Large plaques are usually desirable, and 0.1 ml of late log phase cells per plate has been found to give large plaques of many λgt phages. The number of phages which can be plated on a single petri dish depends on the relative growth rates of desired recombinant and the background phages. After growth, the plates are placed at 4°C for about 15 minutes to harden the agar. Phage and DNA are transferred to a nitrocellulose filter by placing the dry filter (88 mm, Millipore HA or Schleicher and Schuell, BA 85) on the lawn of cells, no air bubbles being allowed to form between the soft agar and the filter. Phages are allowed to adsorb to the filters for 1 to 20 minutes. If duplicate filters are desired, several filters can be adsorbed sequentially for 30 to 60 seconds each, or filters can be stacked and adsorbed at once. If filters are stacked, they must be allowed to adsorb for at least 20 minutes. During adsorption, it is convenient to mark the filters and plates

for orientation. After adsorption, the filter is carefully lifted from the plate to avoid removing the soft agar layer which occasionally adheres to the filter. Should this occur, the agar can be removed by gently shaking the filter during the denaturation and neutralization steps (see Fig. 1, plate 2a and replicas b to d, for examples). Both the DNA and phage are denatured and fixed in situ by dipping the filters in 0.1*N* NaOH and 1.5*M* NaCl for 20 seconds; the filters are then neutralized by dipping in 0.2*M* tris, *p*H 7.5, and $2 \times$ SSCP(*12*) for 20 seconds. Filters are blotted and baked at 80°C in a vacuum for 1.5 to 2 hours.

Hybridization to the nitrocellulose replicas is carried out in a siliconized glass petri dish. The ^{32}P-labeled probe (*13*) (10^5 to 10^6 counts per minute per filter) is placed in enough $5 \times$ SSCP and 50 percent formamide to cover all filters. The hybridization mixtures were usually incubated for 12 to 18 hours. Buffer and probe can be recovered and reused for at least 2 weeks. After hybridization, the filters are washed in a large (10 to 15 ml per filter) volume of $5 \times$ SSCP and 50 percent formamide at 42°C for 30 minutes and then for 20 to 30 minutes in $2 \times$ SSCP at room temperature. The filters are then covered with plastic wrapping film and placed against x-ray film for autoradiography. From 24 to 48 hours of exposure is usually sufficient with a homogeneous probe of specific activity of 10^5 to 10^6 count min^{-1} μg^{-1}. X-ray intensifying screens (Kodak X-Omatic Regular) have been used with flash-activated film at −70°C to shorten exposure time.

The method described above should be generally applicable to all recombinants of λgt phages. In the experiments described below we used the phage λgt-Sc1109 (*7, 14*) in model systems to optimize the method. Phage grown on *E. coli* C600 for less than 9 hours give negligible autoradiographic signals; 12 to 14 hours of growth is optimal. Although T plates generally yield larger plaques, these ap-

W. D. Benton and R. W. Davis

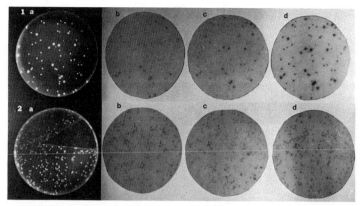

Fig. 1. Plaques 1a and 2a of λgt-Sc1109 phage and autoradiographic replicas (1b to 1d and 2b to 2d). Replicas were produced (as described in the text) by sequential adsorption to filters, beginning with 1d and 2d. The probe was ^{32}P-labeled ribosomal RNA and Chronex 4 x-ray film was exposed for 20 hours for autoradiography. A sector of the top agar from plate 2a was removed with the last filter adsorbed (2b). Note that plaques are visible on the bottom agar and that the autoradiographic replica is not changed in the sector. Plaques 1a and 2a are visualized by the fluorescence of ethidium bromide which is sprayed onto the plates.

parently contain fewer phage particles and less DNA than comparably sized plaques grown on L plates because, at all growth times, nitrocellulose replicas of T plates gave lower signals than those of L plates (11).

To determine the maximum number of plaques per plate which can be screened under various relative growth conditions, two experiments were performed. In the first, to model a situation in which the background grows much better than the desired recombinant phage, a small number of λgt-Sc1109 phage were grown together with increasing numbers of λgt2 (15) phage. No effect on the signals was apparent up to about 2000 plaque-forming units (pfu) per plate, but the signals were weak and variable with more than 4000 pfu per plate. In the second competition experiment, λgt-Sc1109 was plated with increasing numbers of λgt-λB, which has approximately the same plaque size as the recombinant phage. In this case, sufficient signal was obtained with more than 2×10^4 pfu per plate. A practical rule seems to be that the method will accommodate the greatest number of plaques that does not produce confluent lysis.

Time of phage and DNA adsorption to the nitrocellulose replica filter does not seem to be critical; adsorption times of 1 to 60 minutes have been used and little difference noticed. Very long adsorptions resulted in slightly more diffuse plaque spots on the nitrocellulose replicas and the autoradiograms. Stacks of at least five filters can be adsorbed at one time to produce replicates (for hybridization to several different probes, for ex-

ample). For most applications, better replicates are produced by placing the filters sequentially on the plate for a brief adsorption to each (Fig. 1).

Dipping the nitrocellulose replica directly into a beaker of denaturing or neutralizing solution does not result in loss of signal or in distortion of the shapes or relative positions of the plaque spots. Denaturation and neutralization times of 0 to 60 seconds all give acceptable results; 15 to 20 seconds of each is optimal, giving an approximately twofold increase in signal over either no denaturation or no neutralization. Baking the filters before hybridization is not necessary, but in this model system the signal was increased approximately twofold by baking for 90 to 120 minutes.

After autoradiography, it is necessary to align the autoradiogram of the nitrocellulose replica with the original plaque plate to allow the plaques corresponding to positive signals to be picked. The most reliable system we have used for this alignment involves labeling the filters and plates during adsorption by puncturing the filter at three points around its circumference with a hypodermic needle and injecting 0.1 to 0.2 μl of India ink into the agar at each point. The ink diffuses very little during the 2 days normally required for hybridization and autoradiography, and it allows absolute orientation of the filters, autoradiograms, and plates. From a plate with more than 1000 plaques, we normally pick about 20 plaques from the area of each positive signal for rescreening and plaque purification. Once pure clones are obtained, the sizes and restriction

spectra of the inserted DNA's can be rapidly determined by the method of Cameron *et al.* (16).

The large capacity and speed of this method make practical, we believe, the isolation of eukaryotic unique genes by screening all the recombinants produced through shotgun cloning the entire genome. This should make possible isolation of a large number of genes for which physical or genetic purification schemes are not available.

References and Notes

1. M. Thomas, R. L. White, R. W. Davis, *Proc. Natl. Acad. Sci. U.S.A.* **73**, 2294 (1976); B. E. Noyes and G. R. Stark, *Cell* **5**, 301 (1975); S. G. Clarkson, H. O. Smith, W. Schaffer, K. W. Gross, M. L. Birnstiel, *Nucleic Acids Res.* **3**, 2617 (1976); G. N. Yenikolopov, A. P. Ryskov, T. Nitta, G. P. Georgiev, *ibid.*, p. 2645; S. C. Hardies and R. D. Wells, *Proc. Natl. Acad. Sci. U.S.A.* **73**, 3117 (1976); R. Higuchi, G. V. Paddock, R. Wall, W. Salser, *ibid.*, p. 3146.
2. A shotgun experiment is defined (NIH guidelines) as one which involves the production of recombinant DNA's between the vector and total or partially purified DNA from a cellular source.
3. K. Struhl, J. R. Cameron, R. W. Davis, *Proc. Natl. Acad. Sci. U.S.A.* **73**, 1471 (1976).
4. L. Clarke and J. Carbon, *Cell* **9**, 91 (1976).
5. L. Kedes, A. Chang, D. Houseman, S. Cohen, *Nature (London)* **255**, 533 (1975); M. Grunstein and D. Hogness, *Proc. Natl. Acad. Sci. U.S.A.* **72**, 3961 (1975).
6. K. Jones and K. Murray, *J. Mol. Biol.* **96**, 455 (1975); Yu. V. Ilyin, N. A. Tchurikov, G. P. Georgiev, *Nucleic Acids Res.* **3**, 2115 (1976); B. Sanzey, O. Mercereau, T. Temynck, P. Kourilsky, *Proc. Natl. Acad. Sci. U.S.A.* **73**, 3394 (1976).
7. R. A. Kramer, J. R. Cameron, R. W. Davis, *Cell* **8**, 227 (1976).
8. B. M. Olivera and F. Bonhoeffer, *Nature (London)* **250**, 513 (1974).
9. This calculation is based on the worst case in which only 2000 plaques can be grown on each of 50 plates, the screening of which would require about five working hours over a 3-day period.
10. The usual host was *E. coli* C600 rk⁻ mk⁻. Cells and phage are described by J. R. Cameron, S. M. Panasenko, I. R. Lehman, R. W. Davis, *Proc. Natl. Acad. Sci. U.S.A.* **72**, 3416 (1975).
11. The T plates contained 1 percent Bacto-tryptone, 0.5 percent NaCl, 1 percent agar, made to pH 7.5 with NaOH. The L plates contained 1 percent Bacto-tryptone, 0.5 percent Bacto yeast extract, 0.5 percent NaCl, 1.25 percent agar, made to pH 7.5 with NaOH. Seven percent agar was used for the T-soft and the L-soft agar.
12. Standard saline citrate phosphate buffer (SSCP) is (1×) 120 mM NaCl, 15 mM sodium citrate, 13 mM KH₂PO₄, 1 mM EDTA, titrated to pH 7.2 with NaOH.
13. In the model experiments described, the probe was total yeast ribosomal RNA (labeled with ³²P) prepared by the method of G. M. Rubin, *Methods Cell Biol.* **12**, 45 (1975).
14. λgt-Sc1109 (the gift of R. Kramer) is a recombinant λgt phage which carries two Eco RI fragments of *Saccharomyces cerevisiae* DNA, one of which carries the 5 · 8S and 26S rRNA genes. Approximately 5 percent of this phage's DNA is complementary to rRNA.
15. The λgt2 [λgti in Struhl *et al.* (3)] and λgt-λB were the gifts of M. Thomas.
16. J. R. Cameron, P. Philippsen, R. W. Davis, *Nucleic Acids Res.*, in press.
17. We thank T. St. John for helpful discussions and L. Horn for help in manuscript preparation. This work was supported by NSF grant PCM 76-02600 and NIH grant GM 021819-03. One of us (W.D.B.) was supported by NIH training grant 5T01-GM 01156 and by a fellowship from the Bush Foundation, St. Paul, Minn.
* Present address: Department of Genetics and Cell Biology, University of Minnesota, St. Paul 55108.

25

Reprinted from *Natl. Acad. Sci. (USA) Proc.* **72**:3961–3965 (1975)

Colony hybridization: A method for the isolation of cloned DNAs that contain a specific gene

(*Drosophila melanogaster* DNA/recombinant DNA molecules/plasmids/18–28S rRNA genes/autoradiography)

Michael Grunstein* and David S. Hogness†

Department of Biochemistry, Stanford University School of Medicine, Stanford, California 94305

ABSTRACT   A method has been developed whereby a very large number of colonies of *Escherichia coli* carrying different hybrid plasmids can be rapidly screened to determine which hybrid plasmids contain a specified DNA sequence or genes. The colonies to be screened are formed on nitrocellulose filters, and, after a reference set of these colonies has been prepared by replica plating, are lysed and their DNA is denatured and fixed to the filter *in situ*. The resulting DNA-prints of the colonies are then hybridized to a radioactive RNA that defines the sequence or gene of interest, and the result of this hybridization is assayed by autoradiography. Colonies whose DNA-prints exhibit hybridization can then be picked from the reference plate. We have used this method to isolate clones of ColE1 hybrid plasmids that contain *Drosophila melanogaster* genes for 18 and 28S rRNAs. In principle, the method can be used to isolate any gene whose base sequence is represented in an available RNA.

Segments of DNA from *Drosophila melanogaster* chromosomes (Dm segments) can be isolated by cloning hybrid DNA molecules that consist of a Dm segment inserted into the circular DNA of an *Escherichia coli* plasmid. We have previously reported on the use of such cloned segments in the analysis of DNA sequence arrangements in the *D. melanogaster* genome (1–3). However, that analysis has been limited by our inability to isolate cloned Dm segments that contain a specified DNA sequence or gene. In this article we describe a procedure that permits the isolation of such specific Dm segments, and which can be extended to DNA segments from any organism.

Experimental Plan. Consider an experiment in which the Dm segments in a random set are individually inserted into a given *E. coli* plasmid. Transformation of *E. coli* by these hybrid plasmids to a phenotype conferred by genes in the parental plasmid will yield colonies that individually contain a single cloned Dm segment (1–3). If these segments are randomly distributed and exhibit a mean length of 10,000 base pairs, or 10 kb, then we expect that about one colony in 16,000 will contain a particular nonrepetitive *D. melanogaster* DNA sequence the length of a typical structural gene, i.e., 1–2 kb. Hence, the goal is to devise a screening procedure whereby one can rapidly determine which colony in thousands contains such a sequence.

The screening procedure that we have developed is designed to detect sequences that can hybridize with a given

radioactive RNA. In this procedure the colonies to be screened are first grown on nitrocellulose filters that have been placed on the surface of agar petri plates prior to inoculation. A reference set of these colonies is then obtained by replica plating (4) to additional agar plates that are stored at 2–4°C. The colonies on the filter are lysed and their DNAs are denatured and fixed to the filter *in situ* to form a "DNA-print" of each colony. The defining, labeled RNA is hybridized to this DNA and the result of the hybridization is monitored by autoradiography on x-ray film. The colony whose DNA-print exhibits hybridization with the defining RNA can then be picked from the reference set.

The characteristics of this procedure and its application to the isolation of hybrid plasmids containing the *D. melanogaster* genes for '18' and '28'S rRNAs are described in this paper.

MATERIALS AND METHODS

Bacteria. *E. coli* K12 strains HB101, HB101 [pDm103], and C600 [pSC101] are those used previously (plasmids are indicated in brackets) (3). Strain W3110 has been described (5), and W3110 [ColE1] was obtained from D. R. Helinski.

DNAs, Complementary RNAs (cRNAs), and Enzymes. pDm103 (3) and ColE1 (6) DNAs were generously provided by D. M. Glover and D. J. Finnegan, respectively, and were prepared from HB101 [pDm103] and W3110 [ColE1] according to the indicated references, except that the ColE1 was amplified by overnight incubation of W3110 [ColE1] in the presence of chloramphenicol (7) prior to lysis. ^{32}P- and ^{3}H-labeled cRNAs were transcribed *in vitro* from these DNAs with *E. coli* RNA polymerase (8), as described by Wensink *et al.* (1). The RNA polymerase was prepared according to the indicated reference, and was the generous gift of W. Wickner. Pancreatic ribonuclease and proteinase K were obtained from Worthington Biochemical Corp. and E. Merck Laboratories, respectively.

Colony hybridization

Formation of the Filter and Reference Sets of Colonies. Colonies are formed on Millipore HA filters (0.45 μm pores) that have been washed three times in boiling H_2O (1 min per wash), placed between sheets of absorbant paper, autoclaved at 120° for 10 min, and dried for 10 min in the autoclave. The filter is then placed on an L-agar petri plate (1) and the desired bacteria are transferred to the filter surface either by spreading or using sterile toothpicks to obtain ≤7 colonies per cm² after incubation of the filter-plate at 37°. The reference set is produced by replica plating of the colonies that develop on the filter to L-agar plates and is stored at 2–4°.

Abbreviations: kb (kilobases), 1000 bases or base pairs in single- or double-stranded nucleic acids, respectively; Dm, a segment of *Drosophila melanogaster* DNA; cDm and pDm, hybrid plasmids consisting of a Dm segment inserted into ColE1 and pSC101 DNAs, respectively; SSC = 0.15 M NaCl, 0.015 M sodium citrate; cRNA, RNA complementary to DNA; rDNA, DNA coding for ribosomal RNA.

* Present address: Molecular Biology Institute and Department of Biology, University of California, Los Angeles, Calif. 90024.

† To whom reprint requests should be sent.

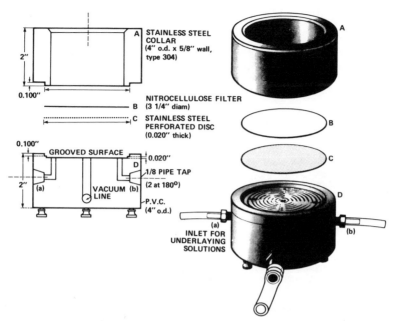

FIG. 1.   Apparatus for treatment of colonies on filters. To wet the underside of the filter, solutions are introduced through ports (a) or (b), while the tube connected to the vacuum port is clamped off. Solutions are removed through the vacuum port which is connected to a water aspirator. Other procedures are described in the text. ", inches (2.54 cm); o.d., outside diameter; P.V.C., polyvinyl chloride.

Lysis, DNA Denaturation, and Fixation. To prevent movement of the bacteria or DNA from their colonial sites during lysis, denaturation and fixation, the solutions used to effect these reactions are applied to the underside of the filter and allowed to diffuse into the colony. The apparatus shown in Fig. 1 has been designed for this purpose. The filter is lifted from the agar plate and placed on the perforated disc that is set in a plastic cylinder which has ports cut into it to introduce solutions sequentially to the underside of the filter and to apply vacuum. Unless otherwise indicated, all operations are carried out at room temperature (20–25°).

Lysis and DNA denaturation are effected by introducing 0.5 N NaOH beneath the filter until it barely floats. After 7 min the NaOH is slowly removed with a minimum of vacuum, and replaced by 1.0 M Tris·HCl (pH 7.4) for 1 min. This solution is replaced with the same buffer, after which the pH of the solution in contact with the filter should be approximately neutral. The last wash is replaced by 1.5 M NaCl, 0.5 M Tris·HCl (pH 7.4), which is removed after 5 min. The stainless steel collar is then placed over the filter, and full vacuum is applied for approximately 2 min until the colonial residues assume a dry appearance. At this point there is less danger of movement from the colonial site and the remaining solutions can be layered on the upper side of the filter.

A 2 mg/ml solution of proteinase K in 1 × SSC (0.15 M NaCl, 0.015 M sodium citrate) is added to just cover the filter. After 15 min, it is removed by vacuum filtration, and 95% ethanol (1 ml/cm² of filter) is similarly passed through the filter. After five washes effected by passing chloroform through the filter (2 ml/cm² per wash), the filter is removed from the apparatus, dipped into 0.3 M NaCl to remove loose cellular debris, and baked at 80° *in vacuo* for 2 hr.

Hybridization and ³²P-Autoradiography or ³H-Fluorography. The dry filter is moistened with a 5 × SSC, 50% formamide solution containing the labeled RNA, using 10–15 μl/cm² of filter. The filter is covered with mineral oil, incubated for 16 hr at 37° to allow hybridization, and then washed for 10 min in a beaker containing chloroform that is gently agitated on a shaking platform. Two more identical chloroform washes are followed by 10 min washes in 6 × SSC, 2 × SSC, and 2 × SSC containing 20 μg/ml of pancreatic ribonuclease. If the RNA is ³²P-labeled, the filter is blotted to remove excess liquid, covered with Saran Wrap, and placed under Kodak RPS/54 x-ray film for autoradiography. If the RNA is ³H-labeled, the filter is dried for 30 min at 80° *in vacuo*, and 40 μl of 7% 2,5-diphenyloxazole (PPO) in ether is applied per cm² of filter. The dry filter is then placed under x-ray film for fluorography at −82° (9).

RESULTS

Colony hybridization distinguishes between [ColE1]⁺ and [ColE1]⁻ bacteria

We have turned increasingly toward the use of the colicinogenic plasmid, ColE1, as a cloning vector because one can obtain much higher cellular concentrations of its hybrids (7) than is the case for the tetracycline resistance plasmid, pSC101, which we used previously (1–3). The first test system for colony hybridization therefore consisted of ³²P-labeled cRNA made by transcription of ColE1 DNA *in vitro* with *E. coli* RNA polymerase, and *E. coli* containing or not containing ColE1, i.e., [ColE1]⁺ or [ColE1]⁻ bacteria.

Fig. 2A shows the autoradiographic response obtained after hybridization of [³²P]cRNA to the DNA-prints of [ColE1]⁺ and [ColE1]⁻ colonies formed on nitrocellulose fil-

Biochemistry: Grunstein and Hogness

Proc. Nat. Acad. Sci. USA 72 (1975) 3963

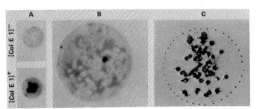

FIG. 2. Hybridization of ColE1 cRNA to [ColE1]⁻ and [ColE1]⁺ colonies. The procedures for colony hybridization, autoradiography, and fluorography are described in *Materials and Methods*, as are the W3310 and W3110 [ColE1] *E. coli* strains used to form the [ColE1]⁻ and [ColE1]⁺ colonies, respectively. (A) 1 × 10⁵ cpm of [³²P]cRNA (5 × 10⁷ cpm/μg) were applied to each 13-mm filter (area = 1.3 cm²) in a 20 μl volume. After hybridization, the DNA-prints of [ColE1]⁺ colonies contained an average of 1.8 × 10² cpm per colony, which is 30-fold greater than the background radiation from an equivalent area on the filter. Exposure time = 45 min. (B) A mixture of [ColE1]⁺ and [ColE1]⁻ bacteria in a 1:100 ratio was spread on a 47-mm filter (area = 17.3 cm²) to obtain a total of 1 to 2 × 10² colonies per filter; 5 × 10⁵ cpm of [³²P]cRNA (3 × 10⁷ cpm/μg) in 250 μl were applied to the filter. Exposure time = 4 hr. (C) A 1:1 mixture of [ColE1]⁺ and [ColE1]⁻ bacteria was spread on a 47-mm filter to obtain a total of 93 colonies, of which 52 gave the A⁺ response seen in the figure; 1 × 10⁶ cpm of [³H]cRNA (2 × 10⁷ cpm/μg) in 200 μl were applied to the filter. Exposure time = 24 hr.

ters. The positive response given by the [ColE1]⁺ colonies is abbreviated by A⁺ and the negative response of [ColE1]⁻ colonies by A⁻. Colonies obtained by spreading mixtures of [ColE1]⁺ and [ColE1]⁻ bacteria in different ratios gave the expected frequencies of A⁺ and A⁻ responses. Fig. 2B shows the result obtained when [ColE1]⁺/[ColE1]⁻ = 1/100.

A more precise measure of the specificity of colony hybridization of mixtures is given by the following experiment in which a 1:1 mixture of [ColE1]⁺ and [ColE1]⁻ bacteria was spread on a filter to yield 31 colonies. Hybridization and autoradiography revealed that 16 were A⁺ and 15 A⁻. Bac-

TOTAL cRNA		[pDm 103]⁺	[pDm 103]⁻
cpm	ng		
750	30.0		
750	0.038		
1500	0.075		
3750	0.19		
7500	0.38		
15,000	75.0		
30,000	1.5		

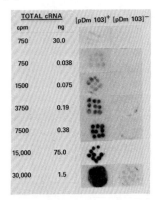

FIG. 3. Hybridization of different amounts of pDm103 [³²P]cRNA to [pDm103]⁺ and [pDm103]⁻ colonies. Colonies were obtained by transferring HB101 [pDm103] or HB101 bacteria, respectively, to 13-mm filters with toothpicks. In the experiments where ≤1.5 ng of cRNA were applied per filter, the specific activity = 2 × 10⁷ cpm/μg. The lower specific activities used for the other two experiments were obtained by mixing this cRNA with unlabeled pDm103 cRNA. The weak response observed for [pDm103]⁻ colonies could result either from *E. coli* DNA impurities in the pDm103 DNA preparations used to prepare the [³²P]cRNA, or from some similarity of sequence in pDm103 and *E. coli* DNAs.

teria from each of the corresponding colonies on the agar replica plate were then tested for colicin production according to an overlay technique described by Finnegan and Willets (10). All 16 A⁺ colonies were colicin-positive (i.e., [ColE1]⁺); all 15 A⁻ colonies were colicin-negative and therefore presumed to be [ColE1]⁻.

Fig. 2A and B show that the position of A⁻ colonies can be detected on the autoradiograph because of the higher background radiation from the filter itself. While this background radiation is convenient for the direct visualization of A⁻ colonies and is not critical to the observation of the A⁺ response obtained with cRNAs, it may become an important factor with other RNAs if they give a weaker A⁺ response. Our observations indicate that the level of this background varies with the preparation of labeled RNA and, possibly, with the batch of filters, but we have not examined such factors in detail.

Fig. 2C shows that the colony hybridization procedure can be adapted to ³H-labeled cRNA by impregnating the filter with 2,5-diphenyloxazole after hybridization and prior to placement on the x-ray film (*Materials and Methods*). Of the 93 colonies obtained by spreading a 1:1 mixture of [ColE1]⁺ and [ColE1]⁻ bacteria, 52 were A⁺ and 41 A⁻. We estimate from the extent of the A⁺ response that this ³H-fluorography is about one-twentieth as efficient as the ³²P-autoradiography.

The autoradiographic response is proportional to the total radioactivity of the applied cRNA and insensitive to its specific activity

We next examined the dependence of the A⁺ response on the total and the specific radioactivity of the applied cRNA. In this case, the ³²P-labeled cRNA was transcribed *in vitro* from a hybrid plasmid called pDm103, and hybridized to DNA-prints of colonies that either contained this hybrid, [pDm103]⁺, or did not, [pDm103]⁻. The pDm103 hybrid was formed between pSC101 plasmid DNA (9 kb) and a segment of *D. melanogaster* DNA (Dm103; 17 kb) that contains the gene for '18' and '28'S rRNAs (3).

Fig. 3 shows that the autoradiographic response obtained when pDm103 [³²P]cDNA was hybridized to 13-mm filters containing [pDm103]⁺ colonies is roughly proportional to the total radioactivity. It is clearly insensitive to the mass of cRNA containing that radioactivity, i.e., to its specific activity. For example, the response to 750 cpm of [³²P]cRNA is approximately the same whether contained in 0.038 ng or in 30 ng. Similarly the response to 15,000 cpm contained in 75 ng is intermediate between that to 7,500 cpm and 30,000 cpm, although the last two samples contained only 0.38 and 1.5 ng, respectively. This would suggest that the RNA·DNA hybridization is occurring under conditions of DNA excess even when 75 ng of pDm103 cRNA are applied per 13 mm filter. However, we have calculated that there is only some 2 ng of pDm103 DNA per colony [i.e., (2 × 10⁷ cells per colony) × (4 pDm103 per cell) 2.9 × 10⁻⁸ ng DNA per pDm103]. This value is based on our observation of 2 × 10⁷ cells per 1 mm colony and the presence of 4 pDm103 per cell in liquid culture (3). Evidently only a small fraction of the applied cRNA can react with the DNA-prints on the filter even though the reaction is occurring ostensibly in DNA excess. A similar result was observed when ColE1 cRNA was hybridized to [ColE1]⁺ colonies (legend, Fig 2A). Of 2 ng cRNA applied to each filter only 0.004 ng (i.e., 0.2%) hybridized per [ColE1]⁺ colony. A 1 mm [ColE1]⁺ colony is estimated to contain 3–4 ng of ColE1 DNA.

Proc. Nat. Acad. Sci. USA 72 (1975)

A simple explanation of these results is obtained if one assumes that most or all of the cRNA in the small fraction of the RNA solution which wets a DNA-print will hybridize, and that the remainder of the cRNA will not hybridize at a significant rate, due perhaps to its slow diffusion through the nitrocellulose, or because of other barriers. Thus a DNA-print from a 1-mm colony, which occupies 0.6% of the area of a 13-mm filter, would be expected to hybridize ≤0.6% of the applied RNA, an expectation that is compatible with the 0.2% observed. For a given ratio of colony to filter area, the fraction of applied cRNA that hybridizes to a DNA-print, in conditions of local DNA excess, would therefore be constant and independent of the total applied cRNA over a wide range of values.

Colony hybridization with cRNA to pDm103 provides a screen for cDm plasmids containing *D. melanogaster* rDNA

Hybrid plasmids consisting of a Dm segment inserted into ColE1 DNA are called cDm plasmids, as distinguished from pDm plasmids where the Dm segment has been inserted into pSC101. In this section we describe two applications of colony hybridization that result in the isolation of cDm plasmids that contain DNA from the repeating gene-spacer units for '18–28'S rRNAs (i.e., rDNA) in *D. melanogaster* (3). In the first application, [^{32}P]cRNA to pDm103 was used to isolate clones of cDm103 plasmids; i.e., plasmids in which the Dm103 segment is inserted into ColE1 DNA at its single *Eco*RI endonuclease cleavage site (7). In the second application, the same [^{32}P]cRNA was used to screen a large set of random cDm clones for rDNA. cRNA formed by transcription of the entire pDm103 DNA can be used for these purposes since we have demonstrated that pSC101 and ColE1 sequences do not interact to give a significant A$^+$ response (data not shown).

Cleavage of circular pDm103 DNA with the *Eco*RI restriction endonuclease yields intact Dm103 segments and linear pSC101 DNA (3). In cooperation with D. M. Glover, we treated a mixture of *Eco*RI-cleaved pDm103 and ColE1 DNAs with *E. coli* ligase under previously described conditions (3), and then transformed colicin-sensitive *E. coli* to colicin E1 immunity with this mixture of ligated DNAs (11). Since the *Eco*RI termini of the linear ColE1, pSC101, and Dm103 molecules can be randomly joined by the ligase, any of the following circular products of this ligation may be present in the colonies of transformants: (*i*) recycled ColE1 (monomers, dimers, etc), (*ii*) molecules containing one ColE1 and one pSC101 segment [abbreviated by (c)$_1$(p)$_1$], (*iii*) (c)$_1$(Dm103)$_1$ molecules, i.e., the desired cDm103 plasmids, or (*iv*) rarer more complex combinations, such as (c)$_1$(p)$_1$(Dm103)$_1$, which contain one or more copies of ColE1.

Forty-eight of the transformants were screened for the presence of either pSC101 or Dm103 segments by colony hybridization with [^{32}P]cRNA to pDm103 (Fig. 4A), and for the presence of the pSC101 segment by testing for resistance to tetracycline. Of the eight A$^+$ transformants shown in Fig. 4A, six were tetracycline resistant and probably contain (c)$_1$(p)$_1$ plasmids. They were not examined further. The remaining two (indicated by 1 and 2 in Fig. 4A) were tetracycline sensitive, and were assumed to contain cDm103 plasmids; they were designated cDm103/1 and cDm103/2, respectively.

Proof of this assumption was obtained by electron microscopic examination of the plasmids isolated from the two

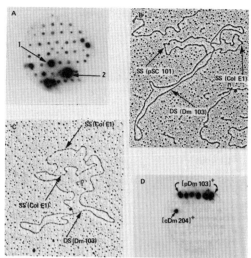

FIG. 4. (A) The screen for cDm103 hybrids. 5 μg of pDm103 DNA and 0.25 μg of ColE1 DNA were cleaved to completion with *Eco*RI endonuclease (in 0.120 ml of 0.1 M Tris·HCl, pH 7.5, 0.01 M MgSO$_4$), heated for 5 min at 65° to inactivate the enzyme and brought to 4°. The DNAs were then incubated at 14° with DNA ligase (14 μg/ml) in 0.1 M Tris·HCl, pH 7.5, as well as a reaction buffer consisting of 0.1 mM DPN, 1 mM EDTA, 10 mM (NH$_4$)$_2$SO$_4$, 10 mM MgSO$_4$ with 100 μg/ml of bovine serum albumin for 120 min in a total volume of 0.140 ml. The solution was then diluted 3-fold with the same reaction buffer and incubated for 36 hr at 14° in the presence of ligase (10 μg/ml). The ligated mixture of *Eco*RI-cleaved pDm103 and ColE1 DNAs (see *text*) was used to transform HB101 to colicin E1 immunity as described previously (11). Each of 48 transformants were transferred by toothpick to a 47-mm filter for colony hybridization (*Materials and Methods*), and to L-agar plates containing 15 μg of tetracycline per ml. 5 × 10^5 cpm of pDm103 [^{32}P]cRNA (2 × 10^7 cpm/μg) were used for the colony hybridization, which after a 6-hr exposure yielded the above autoradiograph. The colonies marked 1 and 2 contain cDm103/1 and cDm103/2, respectively. (B) Electron micrograph of a pDm103·cDm103/2 heteroduplex. pDm103 and cDm103/2 circular DNAs were randomly nicked (broken in one strand) by x-rays. The procedures for denaturation and renaturation of these DNAs to form heteroduplexes, for spreading in 40% formamide prior to electron microscopy, and for measuring contour lengths have been described (1). pSC101 (9.2 kb; ref. 1) was used as an internal reference for double-stranded lengths (DS in the figure); no reference was used for single-stranded lengths (SS), as only the ratio of two SS-lengths is used in the analysis (see *text*). (C) Electron micrograph of a cDm103/1·cDm103/2 heteroduplex. The procedures are given in (B) above. See *text* for explanation. (D) The screen for cDm hybrids containing *D. melanogaster* rDNA. Hybrids between *Eco*RI-cut ColE1 and randomly broken Dm segments were formed as indicated in the *text*, and then used to transform HB101 to colicin E1 immunity as in (A) above. 300 independent transformants were transferred to six 47-mm filters, each of which contained six control colonies of HB101 [pDm103] at the top of the pattern. 5 × 10^5 cpm of pDm103 [^{32}P]cRNA (2 × 10^7 cpm/μg) was applied per filter for the colony hybridization. The autoradiograph in the figure resulted from one of the six filters after a 5-hr exposure, and shows one of the five rDNA hybrids (cDm204) identified by this screening procedure.

transformants, and of heteroduplexes formed between pDm103 and cDm103/2, and between cDm103/1 and cDm103/2. The mean lengths ±SD (*n* = 18) of cDm103/1 and cDm103/2 are 23.0 (±1.2) kb and 21.7 (±1.5) kb, respectively. The sum of the lengths of Dm103 (17 kb) and

ColE1 (6 kb; ref. 7) is 23 kb, in reasonable agreement with these values.

A heteroduplex formed between pDm103 and cDm103/2 is shown in Fig. 4B. It consists of a 17 kb double-stranded element whose ends are connected by each of two single-stranded elements that exhibit a length ratio of 1.5. This is the structure expected if cDm103/2 consists of a Dm103 segment inserted at the *Eco*RI cleavage site of ColE1; i.e., the double-stranded element represents the paired Dm103 segments of the two plasmid strands, and the larger and smaller single-stranded elements represent the pSC101 and ColE1 segments respectively (expected length ratio = 9 kb/6 kb = 1.5).

The heteroduplex formed between cDm103/1 and cDm103/2 consists of a 17 kb duplex whose ends are connected by two single-stranded elements of equal length (Fig. 4C). The simplest explanation of this structure is that the Dm103 segments were oppositely inserted into ColE1 during formation of cDm103/1 and cDm103/2. If the Dm103 segments in the single strands of two such oppositely oriented plasmids pair to create a 17 kb duplex element, then the two single-stranded ColE1 segments would contain identical rather than complementary base sequences, and could not pair.

The last experiment consists in screening hundreds of different [cDm]$^+$ colonies for rDNA. The [cDm]$^+$ colonies were obtained by transformation of colicin-sensitive *E. coli* to immunity with a heterogeneous population of cDm molecules constructed from *Eco*RI-cleaved ColE1 and random Dm segments (obtained by shear breakage) by the poly(dA)·poly(dT) joining method (1). These transformants were provided by D. J. Finnegan and G. Rubin. They were individually transferred by toothpick to six 47-mm nitrocellulose filters, each filter containing about 50 independent transformants. Colony hybridization with pDm103 [^{32}P]cRNA indicated no A$^+$ colonies on three filters, 1 A$^+$ colony on two filters, and 3 A$^+$ colonies on one filter. The autoradiograph of one of the two filters containing a single A$^+$ colony, cDm204, is given in Fig. 4D (the top row of A$^+$ colonies on the filter are [pDm103]$^+$ controls). When each of the 5 A$^+$ colonies was retested by repeating this colony hybridization on subclones, such subclones were consistently A$^+$.

Since pSC101 and ColE1 sequences do not interact to give an A$^+$ response, we presume that the cDm plasmids in these 5 A$^+$ colonies contain sequences present in Dm103; i.e., they contain rDNA from *D. melanogaster*. Indeed, D. M. Glover and R. L. White (personal communication) have shown recently that the 28 kb Dm segment in cDm204 contains the same arrangement of '18'–'28'S and spacer sequences as is found in Dm103.

DISCUSSION

In principle, colony hybridization of cloned hybrid plasmids can be used to isolate any gene, or other DNA segment, whose base sequence is represented in an available RNA. We used cRNA to pDm103 for the isolation of cDm plasmids containing rDNA. However, as we have observed that [pDm103]$^+$ colonies give an adequate A$^+$ response with ^{3}H-labeled '18' plus '28'S rRNAs isolated from *D. melanogaster* cell cultures (3), the isolation could have been accomplished with these rRNAs. For rRNA the genes are repeated hundreds of times per genome, and this is the reason that we were able to isolate several hybrids containing rDNA by screening only a few hundred colonies.

By contrast, we calculate that it would be necessary to screen approximately 50,000 hybrid clones to have a 95% chance of finding a hybrid containing a nonrepeated structural gene of typical length from *D. melanogaster*. From the data given in Fig. 3 and assuming 24-hr exposures, we estimate that this would require a total of approximately 4 × 10^6 cpm of [^{32}P]mRNA (specific activity ≥ 4 × 10^5 cpm/μg) applied to about one hundred thirty-five 82-mm filters. Thus a screen of this size is quite feasible. The isolation of nonrepeated genes from larger genomes would, of course, proportionately increase the number of colonies to be screened and hence the total required radioactivity.

An important advantage of colony hybridization is that it facilitates containment of any potentially hazardous hybrid plasmids that may be cloned in such large screening operations. By confining the reproductive state of the hybrid-clones to colonies, the probability of escape is reduced over that for liquid cultures because the number of bacteria per clone is generally smaller and aerosols or accidental spills are less likely. Furthermore the screening operation can be confined to small, controllable areas.

M.G. is grateful to R. T. Schimke for his support and encouragement. We thank D. J. Finnegan, D. M. Glover, G. M. Rubin, and R. L. White for helpful discussions and for providing materials. We are especially thankful to D. M. Glover for his help and advice in isolating and characterizing the cDm103 hybrids. This work was supported by grants from the National Science Foundation (BMS74-21774) and the National Institutes of Health (GM20158, GM14931).

1. Wensink, P. C., Finnegan, D. J., Donelson, J. E. & Hogness, D. S. (1974) *Cell* 3, 315–325.
2. Hogness, D. S., Wensink, P. C., Glover, D. M., White, R. L., Finnegan, D. J. & Donelson, J. E. (1975) in *The Eukaryote Chromosome*, eds. Peacock, W. J. & Brock, R. D. (Australian National University Press, Canberra), in press.
3. Glover, D. M., White, R. L., Finnegan, D. J. & Hogness, D. S. (1975) *Cell* 5, 149–157.
4. Hayes, W. (1965) in *The Genetics of Bacteria and Their Viruses* (John Wiley & Sons, New York), pp. 185–188.
5. Lederberg, E. M. (1960) *Symp. Soc. Gen. Microbiol.* 10, 115–131.
6. Katz, L., Kingsbury, D. K. & Helinski, D. R. (1973) *J. Bacteriol.* 114, 577–591.
7. Hershfield, V., Boyer, H. W., Yanofsky, C., Levett, M. A. & Helinski, D. R. (1974) *Proc. Nat. Acad. Sci. USA* 71, 3455–3459.
8. Berg, D., Barrett, K. & Chamberlin, M. (1971) in *Methods in Enzymology*, eds. Grossman, L. & Moldave, K. (Academic Press, New York), Vol XXI, pp. 506–519.
9. Randerath, k. (1970) *Anal. Biochem.* 34, 188–205.
10. Finnegan, D. J. & Willetts (1972) *Mol. Gen. Genet.* 119, 57–66.
11. Glover, D. (1975) in *New Techniques in Biophysics and Cell Biology*, eds.   , R. & Smith, B. (John Wiley & Sons, New York), in press.

26

Reprinted from *J. Mol. Biol.* **98**:503–517 (1975)

Detection of Specific Sequences Among DNA Fragments Separated by Gel Electrophoresis

E. M. SOUTHERN

Medical Research Council Mammalian Genome Unit
Department of Zoology
University of Edinburgh
West Mains Road, Edinburgh, Scotland

This paper describes a method of transferring fragments of DNA from agarose gels to cellulose nitrate filters. The fragments can then be hybridized to radioactive RNA and hybrids detected by radioautography or fluorography. The method is illustrated by analyses of restriction fragments complementary to ribosomal RNAs from *Escherichia coli* and *Xenopus laevis*, and from several mammals.

1. Introduction

Since Smith and his colleagues (Smith & Wilcox, 1970; Kelly & Smith, 1970) showed that a restriction endonuclease from *Haemophilus influenzae* makes double-stranded breaks at specific sequences in DNA, this enzyme and others with similar properties have been used increasingly for studying the structure of DNA. Fragments produced by the enzymes can be separated with high resolution by electrophoresis in agarose or polyacrylamide gels. For studies of sequences in the DNA that are transcribed into RNA, it would clearly be helpful to have a method of detecting fragments in the gel that are complementary to a given RNA. This can be done by slicing the gel, eluting the DNA and hybridizing to RNA either in solution, or after binding the DNA to filters. The method is time consuming and inevitably leads to some loss in the resolving power of gel electrophoresis. This paper describes a method for transferring fragments of DNA from strips of agarose gel to strips of cellulose nitrate. After hybridization to radioactive RNA, the fragments in the DNA that contain transcribed sequences can be detected as sharp bands by radioautography or fluorography of the cellulose nitrate strip. The method has the advantages that it retains the high resolving power of the gel, it is economical of RNA and cellulose nitrate filters, and several electrophoretograms can be hybridized in one day. The main disadvantage is that fragments of 500 nucleotide pairs or less give low yields of hybrid and such fragments will be under-represented or even missing from the analysis.

2. Materials, Methods and Results

(a) Restriction endonucleases

EcoRI prepared according to the method of Yoshimuri (1971) was a gift of K. Murray. HaeIII prepared by a modification of the method of Roberts (unpublished data) was a gift of H. J. Cooke.

(b) *Gel electrophoresis*

Gels were cast between glass plates (de Wachter & Fiers, 1971). The plates were separated by Perspex side pieces 3 mm thick and along one edge was placed a "comb" of Perspex, which moulded the sample wells in the gel. The Perspex pieces were sealed to the glass plates with silicone grease and the plates clamped together with Bulldog clips. The assembly was stood with the comb along the lower edge. Agarose solution (Sigma electrophoresis grade agarose) was prepared by dissolving the appropriate ·weight in boiling electrophoresis buffer (E buffer of Loening, 1969). The solution was cooled to 60 to 70°C and poured into the assembly, where it was allowed to·set for at least an hour. The assembly was then inverted, the comb removed and the wells filled with electrophoresis buffer. Samples made 5% with glycerol were loaded from a drawn-out capillary by inserting the tip below the surface and blowing gently. Electrophoresis buffer was layered carefully to fill the remaining space and a filter-paper wick inserted between the glass plates along the top edge. The lower end of the assembly was immersed in a tray of electrophoresis buffer containing the platinum anode, and the paper wick dipped into a similar cathode compartment. Electrophoresis was at 1·0 to 1·5 mA/cm width of gel for a period of about 18 h. Bromophenol blue marker travels about 3/4 the length of the gel under these conditions, but it should be noted that small DNA fragments move ahead of the bromophenol blue, especially in dilute gels. Cylindrical gels were cast in Perspex tubes 9 mm i.d. and either 12 or 24 cm long. These were run at 3 to 5 mA/tube in standard gel electrophoresis equipment.

Dr J. Spiers donated ribosomal DNA that had been purified on actinomycin/caesium chloride gradients from DNA made from the pooled blood of several animals, and also ^{3}H-labelled 18 S and 28 S RNAs prepared from cultured *Xenopus laevis* kidney cells. *Escherichia coli* DNA was prepared by Marmur's (1961) procedure from strain MRE600. ^{32}P-labelled *E. coli* RNA was prepared from cells grown in low phosphate medium with ^{32}Pi at a concentration of 50 μCi/ml and fractionated by electrophoresis on 10% acrylamide gels. ^{32}P-labelled rat DNA was a gift of M. S. Campo. DNA from human placenta was a gift of H. J. Cooke, DNA from rat liver was a gift of A. R. Mitchell, DNA from mouse and rabbit livers were gifts of M. White. Calf thymus DNA was purchased from Sigma Biochemicals. For digestion with restriction endonucleases, the DNAs were dissolved in water to a concentration of approximately 1 mg/ml. One-tenth volume of the appropriate buffer was added and sufficient enzyme to give a complete digestion overnight at 37°C. Enzyme activity was checked on phage λ DNA and digests of this DNA were also used as size markers in gel electrophoresis, using the values given by Thomas & Davis (1975).

(c) *Method of transfer*

This section describes the method finally adopted: preliminary experiments and controls are described in later sections.

After electrophoresis, the gel is immersed for 1 to 2 h in electrophoresis buffer containing ethidium bromide (0·5 μg/ml), and photographed in ultraviolet light (254 nm) with a red filter on the camera. A rule laid alongside the gel aids in matching the photograph of the fluorescence of the DNA to the final radioautograph of the hybrids. Strips to be used for transfer from flat gels are cut from the gel using a flamed blade. The strips should be 0·5 cm to 1 cm wide and normally extend from the origin to the anode end of the gel. The gels used in this laboratory are 3 mm thick, and the length from the origin to the anode end is 18 cm but the method can be adapted to gels with different dimensions and to cylindrical gels. Strips of gel are then transferred to measuring cylinders containing 1·5 M-NaCl, 0·5 M-NaOH for 15 min and this solution is then replaced by 3 M-NaCl, 0·5 M-Tris·HCl (pH 7) and the gel is left for a further 15 min. The depth of liquid in the cylinders should be greater than the length of the gel strips and the cylinders should be inverted from time to time. For cylindrical gels (9 mm diam.), the times required for denaturation and neutralization are 30 and 90 min. Each gel transfer requires:

One piece of thick filter paper 20 cm × 18 cm, soaked in 20 × SSC (SSC is 0·15 M-NaCl, 0·015 M-sodium citrate).

Two pieces of thick filter paper 2 cm × 18 cm soaked in 2 × SSC.

One strip of cellulose nitrate filter (e.g. Millipore 25 HAWP), 2·2 cm × 18 cm, soaked in 2 × SSC. These strips are immersed first by floating them on the surface of the solution; otherwise air is trapped in patches, which leads to uneven transfer.

Three pieces of glass or Perspex, 5 cm × 20 cm and the same thickness as the gel.

Four or five pieces of thick, dry filter paper, 10 cm × 18 cm.

Transfer of the denatured DNA fragments is carried out as follows.

The large filter paper soaked in 20 × SSC is laid on a glass or plastic surface, care being taken to avoid trapping air bubbles below the paper. 20 × SSC is poured on so that the surface is glistening wet. One of the glass or Perspex sheets is laid on top of the wet paper. The gel strip is taken from the neutralizing solution and laid parallel to the glass or Perspex sheet, 2 to 3 mm away from it. The second glass or Perspex sheet is laid 2 to 3 mm away from the other side of the gel (Fig. 1(a)). The cellulose nitrate strip is then laid on top of the gel with its edges resting on the sheets of Perspex or glass, so that it bridges the two air spaces (Fig. 1(b)). The two narrow pieces of filter paper, moistened with 2 × SSC are laid with their edges overlapping the cellulose nitrate strip by about 5 mm (Fig. 1(c)) and the dry filter paper is then placed on top of these (Fig. 1(d)).

For cylindrical gels, the arrangement is similar, but in this case, the Perspex that supports the Millipore filter may be in contact with the gel because an air space is retained over the top of the gel. Several cylindrical gels can be transferred at the same time using the apparatus shown in Fig. 2 and similar arrangements can be used for flat gels.

20 × SSC passes through the gel drawn by the dry filter paper and carries the DNA, which becomes trapped in the cellulose nitrate. The minimum time required for complete transfer has not been measured: it depends on the size of the fragments and probably also depends on the gel concentration. A period of 3 h is enough to transfer completely all HaeIII fragments of E. coli DNA from 2% agarose gels 3 mm thick. But even after 20 h, transfer of large EcoRI fragments of mouse DNA from 9 mm diam. cylindrical gels is not complete. DNA remaining in the gel can be seen by the fluorescence of the ethidium bromide, which is not completely removed during treatment of the gel. During the period of the transfer, it is necessary occasionally to add more 20 × SSC to the bottom sheet of filter paper. If the paper dries too much, the gel shrinks against the cellulose nitrate strip and liquid contact is broken. The paper may be flooded, but care must be taken that liquid does not fill the air spaces between the gel and the side-pieces and soak the paper, bypassing the gel. It may be found convenient to leave the cellulose nitrate in position overnight: if the supply of

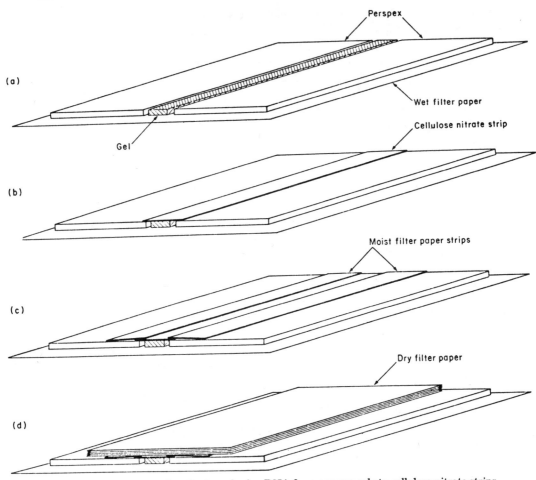

FIG. 1. Steps in the procedure for transferring DNA from agarose gels to cellulose nitrate strips.

20 × SSC has dried up it will be found that the gel has shrunk against the cellulose nitrate, but this does not impair the transfer. At the end of the transfer period the cellulose nitrate strip is lifted carefully so that the gel remains attached to its under-side. It is turned over and the outline of the gel marked in pencil by a series of dots. The gel is peeled off the cellulose nitrate, the area of contact cut out with a flamed blade, and immersed in 2 × SSC for 10 to 20 min. The strip is then baked in a vacuum oven at 80°C for 2 h.

(d) *Hybridization*

Radioactive RNAs are usually available in small quantities only and it is important to keep the volume of the solution used for hybridization as small as possible so that the RNA has a reasonable concentration. Two procedures can be used for hybridizing the cellulose nitrate strips after transferring the restriction fragments.

The procedure that uses the smallest volume is carried out by moistening the strip in hybridization mixture and then immersing it in paraffin oil. A drop of RNA solution (0·3 ml for a strip 1 cm × 18 cm) is placed on a plastic sheet. One end of the

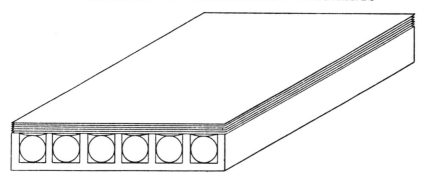

FIG. 2. Apparatus for transferring DNA from a number of cylindrical gels.

The apparatus is constructed of Perspex. The uprights which separate the gels and support the sheet of cellulose nitrate should be about 0·5 mm higher than the diameter of the gels, so that the cellulose nitrate sheet dips down to touch the gel. Thus an air gap is left between the cellulose nitrate sheet and the filter paper, above the line of contact between the gel and cellulose nitrate sheet. The apparatus is laid in a shallow tray containing 20 × SSC and the gels are then inserted into the troughs, care being taken to avoid trapping air bubbles beneath the gel. The cellulose nitrate sheet, wet with 2 × SSC, is laid over the gels and one piece of wet filter paper is laid over this. A stack of dry filter paper is then placed over the whole assembly. If necessary, a glass plate can be used to weigh down the filter papers. The depth of 20 × SSC in the tray should be enough to cover the lower part of the gels, but not so much that the air space between the Perspex and the cellulose nitrate becomes flooded.

cellulose nitrate strip is floated on the drop and when liquid is seen to soak through, the strip is drawn slowly over the surface of the drop. When it is completely wetted from one side, it is turned over and any remaining liquid is used to wet the other side. The strip is then immersed in paraffin oil saturated with the hybridization solution at the hybridization temperature. It should be borne in mind that baking the strip in 2 × SSC introduces salt, which must be taken into account when deciding on a solvent for the RNA if this method of hybridization is used. For example, if hybridization is to be carried out in 6 × SSC the RNA should be dissolved in 4 × SSC. Though this method can give good results (see Plate I) it often leads to high and uneven background. Kourilsky et al. (1974) found that this problem is removed if the hybridization is carried out in 2 × SSC, 40% formamide at 40°C. I have not tried this method, because this solvent removed DNA from the filters (see later section). It may well be the best method for hybridization to large fragments. I have found it convenient to carry out the hybridization in a vessel designed to hold the strip in a small volume of liquid.

The vessel (Fig. 3), which is easily made from Perspex, has internal dimensions of 0·8 mm deep by 2 cm high and about 1 cm longer than the strip to be hybridized. The vessel is filled with the solvent to be used for hybridization and the strip is fed in through the narrow opening in the top. The solvent is then drained off and the RNA solution introduced. Around 1 ml of solution is needed for a strip 1 cm × 18 cm. The wide sheets of cellulose nitrate used for transferring several gels (e.g. using the apparatus shown in Fig. 2) are too wide to be hybridized in this type of vessel. They can be hybridized in a small volume by wrapping them around a cylinder of Perspex, which is then inserted into a close-fitting tube. In this way, it is possible to hybridize a sheet 24 cm × 8 cm with about 4 ml of solution. If hybridization is carried out in a water-bath, it is not necessary to seal the top of the vessel provided the water-bath

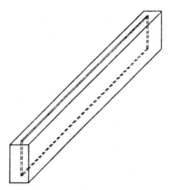

FIG. 3. Vessel used for hybridization of narrow strips.

itself is covered. The liquid in the vessel evaporates very slowly and can be replenished by small additions of water. A further advantage of this method of hybridization is that the RNA can be recovered and used again.

The period allowed for hybridization depends on the RNA concentration, its sequence complexity, its purity, and on the conditions of hybridization (see for example Bishop, 1972). After the appropriate period, strips are removed from the solution or paraffin oil, blotted between sheets of filter paper and washed, with stirring, for 20 to 30 min in a large volume of the hybridization solvent at the hybridization temperature. If the background is high, they may then be treated with a solution of RNAase A (20 μg/ml in 2 × SSC for 30 min at 20°C). After a final rinse in 2 × SSC they are dried in air.

So far the method has been tested with ^{32}P, ^{3}H, ^{35}S and ^{125}I-labelled RNAs. [^{32}P]RNAs have been detected by radioautography. For this the cellulose nitrate strips are laid on X-ray film and flattened against it with light pressure. ^{3}H, ^{125}I, ^{35}S and ^{14}C may be detected by fluorography. The cellulose nitrate strip is dipped through a solution of PPO in toluene (20%, w/v) dried in air, laid against X-ray film (Kodak RP-Royal Xomat) and kept at −70°C.

(e) Completeness of transfer and retention of DNA

Preliminary experiments showed that loading of DNA on to cellulose nitrate filters in 6 × SSC, conditions widely used in hybridization work, did not give complete retention of small fragments and a systematic study was made of the effect of salt concentration on retention. ^{3}H-labelled X. laevis DNA was sonicated to a single-strand molecular weight of 10^4 and denatured by boiling in 0·1 × SSC. Samples were made up to various salt concentrations and 0·1-ml portions of these solutions were pipetted on to cellulose nitrate filters, previously moistened with 2 × SSC, which were resting on glass-fibre filters. The solution that passed through the cellulose nitrate filter was thus collected in the glass-fibre filter. Both filters were then immersed in 5% trichloroacetic acid for 10 min, dried for 30 min in a vacuum oven at 80°C, and counted. It can be seen (Fig. 4) that the fraction of DNA retained by the cellulose nitrate increases with the salt concentration, and at concentrations above 10 × SSC the DNA is almost completely retained.

Losses of DNA at various stages of the transfer procedure were measured using ^{32}P-labelled E. coli DNA. The DNA was digested with EcoRI to give fragments in

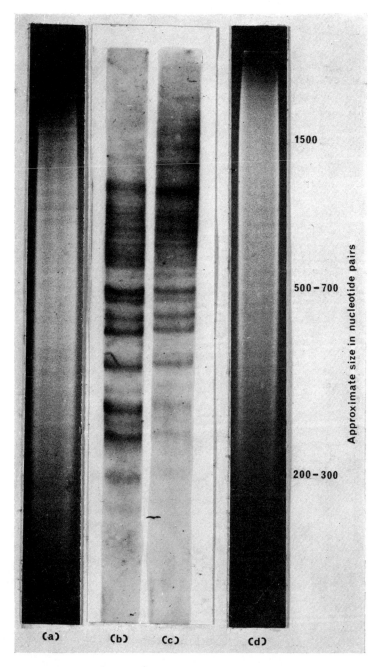

PLATE I. HaeIII digest of *E. coli* MRE600 DNA analyzed by electrophoresis on 2% agarose gel. DNA was then transferred to cellulose nitrate and hybridized with ^{32}P-labelled, high molecular weight RNA. (a) and (d) Photographs of ethidium bromide fluorescence. (b) and (c) Radioautographs of hybrids.

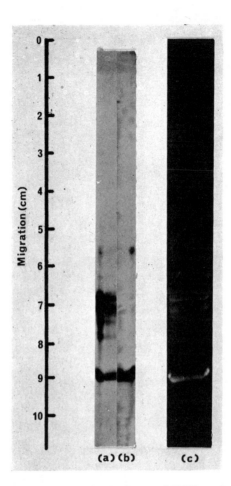

PLATE II. EcoRI digest of purified *X. laevis* ribosomal DNA analyzed by electrophoresis on 1% agarose gel. DNA was transferred to a cellulose nitrate strip, which was then cut longitudinally in two. The left-hand side was hybridized to 18 S RNA and the right-hand side to 28 S RNA (spec. act. of RNAs, $1\cdot5 \times 10^6$ c.p.m. per μg). Hybridization was done in $1 \times$ SSC at 65°C using the vessel shown in Fig. 3. A large excess of cold 28 S RNA was added to the labelled 18 S RNA to compete out any 28 S contamination. After hybridization, the strips were washed in $1 \times$ SSC at 65°C for $1\cdot5$ h, and dried. They were then dipped through a solution of PPO in toluene (20%, w/v) dried in air and placed against Kodak RP Royal X-ray film at -70°C for 2 months. Photograph of ethidium bromide fluorescence (c). Fluorograph of 18 S hybrids (a). Fluorograph of 28 S hybrids (b).

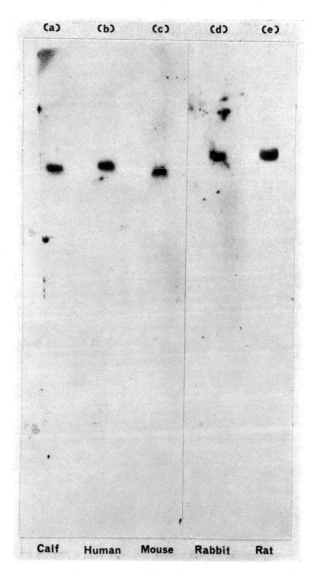

PLATE III. EcoRI digests of five mammalian DNAs, hybridized to 28 S RNA. Calf (a), human (b), mouse (c), rabbit (d) and rat (e) DNAs were digested to completion with EcoRI and separated by electrophoresis on 1% agarose gels (9mm × 12 cm, approx. 40 μg DNA per tube, 3 mA/tube for 16 h). The gels were pretreated as usual and the DNA fragments transferred to a single sheet of cellulose nitrate filter (12 cm × 8 cm) using the apparatus shown in Fig. 2. The top end of each gel was carefully aligned with one edge of the cellulose nitrate sheet. After 20 h, traces of DNA could still be seen, by ethidium bromide fluorescence, in the high molecular weight region of the gel. The filter was hybridized with 28 S RNA and radioautographed as described in the legend to Fig. 8.

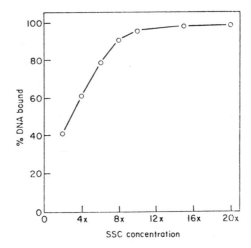

FIG. 4. Effect of salt concentration on efficiency of binding sonicated DNA to cellulose nitrate filters.

the large size range and with HaeIII to give small fragments. The fragments were then separated on a flat 1% agarose gel and transferred in the usual way. The solutions, the gel and the cellulose nitrate strip were counted. It can be seen (Table 1) that, whereas a small proportion of the DNA is leached out into the solutions during denaturation and neutralization, only traces remain in the gel after transfer.

TABLE 1

Losses of DNA at stages of the procedure

	EcoRI fragments	HaeIII fragments
	DNA lost (%)	
Denaturing solution	2·1	4·8
Neutralizing solution	1·3	4·4
Remaining in gel after transfer	0·21	0·31

Two samples of *E. coli* DNA (0·1 μg; spec. act. approx. 10^6 c.p.m. per μg) were digested with EcoRI and HaeIII. The fragments were separated by electrophoresis on 1% gels in 1-cm wide slots, and then transferred to cellulose nitrate strips as described in Materials and Methods. The transfer was left overnight. The radioactivity leached out of the gel by the denaturing and neutralizing solutions, that remaining in the gel, and that which had been trapped on the cellulose nitrate filter were measured in a liquid scintillation counter (Cerenkov radiation).

(f) *Effect of DNA size on yield of hybrid*

Melli & Bishop (1970) have shown that hybridization by the filter method gives low yields with low molecular weight DNA. Their results were obtained using a single set of hybridization conditions and it seemed possible that losses might be reduced by using high salt concentrations. The effect of salt concentration on loss of

DNA from the filters was examined by loading filters with radioactive *X. laevis* DNA, single-strand molecular weight about 10^4, and incubating them in various salt solutions at different temperatures. Increasing the salt concentration does improve the retention of the DNA at any given temperature (Table 2) but the gain does not appear to be useful, because with increasing salt concentration it is necessary to use higher temperatures for hybridization, and this cancels the advantage of the high salt concentration. For example, the loss in $2 \times$ SSC at 65°C is the same as that in $6 \times$ SSC at 80°C and these are both typical hybridization conditions. Further experiments showed that it is disadvantageous to perform hybridization at high salt concentrations, below the optimum temperature. The optimum temperature for rate of hybridization of *X. laevis* 28 S RNA is around 80°C in $6 \times$ SSC but the rate at 70°C is still appreciable (Fig. 5). Below 70°C the rate falls rapidly. 28 S RNA was hybridized

TABLE 2

Effects of temperature and solvent on retention of sonicated DNA on cellulose nitrate filters

Solvent	50°C	Temperature		90°C
		65°C	80°C	
		DNA retained (%)		
$2 \times$ SSC		77	62	48
$6 \times$ SSC		97	76	56
$10 \times$ SSC		95	83	73
$20 \times$ SSC		97	88	81
$6 \times$ SSC in 50% formamide	58	50		

[3]H-labelled *X. laevis* DNA (spec. act. approx. 5×10^5 c.p.m. per μg) was dissolved in ice-cold $0.1 \times$ SSC and sonicated in six 15-s bursts. Between each treatment the solution was cooled in ice for 1 min. The solution was boiled for 5 min, made to $20 \times$ SSC and cooled. Samples of this solution were pipetted on to 13-mm circles of cellulose nitrate, which were then washed in $2 \times$ SSC at room temperature. Approximately 650 c.p.m. were loaded on each filter, and there was no loss caused by washing in $2 \times$ SSC. The filters were dried, baked at 80°C for 2 h in a vacuum oven and immersed in 10 ml of the solvent equilibrated at the temperature used for incubation. After 90 min, the filters were removed, washed in $2 \times$ SSC at room temperature, dried under vacuum and counted in a liquid scintillation counter.

to high molecular weight and sonicated DNA in $6 \times$ SSC at 70 and 80°C (Fig. 6). As expected, the rate of hybridization at 70°C was lower than the rate at 80°C, but against expectation, both the rate and the final extent of hybridization were lower at the lower temperature, for the sonicated but not for the high molecular weight DNA. This result was unexpected because Melli & Bishop did not find an effect of DNA size on the rate of hybridization. They suggested that the decrease in yield for low molecular weight DNA is due to a loss of hybrid from the filter and it would be expected that such losses would increase with temperature. The lower yield for low molecular weight DNA at low temperature remains unexplained, but shows that there is no advantage to be gained in using high salt concentrations and low temperatures to retain small fragments of DNA during hybridization reactions. The advantage of using $6 \times$ SSC at optimum temperature is that the rate is greatly increased over the rate with, say, $2 \times$ SSC. A disadvantage is that the background of RNA that sticks to filters that have no DNA, increases with increasing salt concentration.

(g) *Methods of detecting and measuring hybrids: advantages of film detection*

Radioactive RNA may be detected and measured either by radioautography (or fluorography for weak β-emitters) or by cutting the strip into pieces, which can be counted in a scintillation counter. Film detection methods have the advantages over

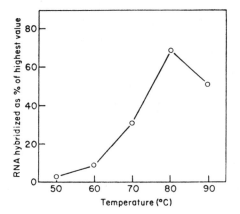

FIG. 5. Temperature dependence of hybridization of 28 S rRNA to *X. laevis* DNA.

X. laevis DNA was loaded on cellulose nitrate filters (17 μg DNA/13-mm diameter disc), which were pretreated as usual for hybridization. ³H-labelled 28 S RNA from *X. laevis* kidney cells (spec. act. 1.5×10^6 c.p.m./μg) was dissolved in $6 \times SSC$ (0·28 μg/ml) and warmed to the temperature used for hybridization. Two filters loaded with DNA and 2 blank filters were introduced into the solutions and left for 30 min. They were washed in 2 l of $2 \times SSC$ at room temperature, treated with 200 ml of RNAase A (20 μg/ml in $2 \times SSC$) at room temperature for 20 min, washed in 200 ml of $2 \times SSC$ for 10 min, dried under vacuum and counted. Hybridization is expressed as a percentage of that obtained after 5 h at 80°C.

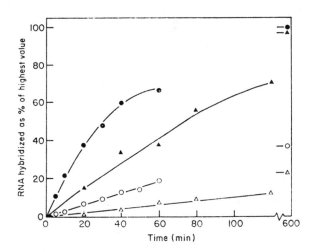

FIG. 6. Time course of hybridization of 28 S RNA to sonicated and high molecular weight DNA at 70 and 80°C.

Filters were loaded as described in the legend to Fig. 5. Two sets were loaded: one with high molecular weight DNA and one with DNA sonicated as described in the legend to Table 2. Hybridization and subsequent treatment of the filters was carried out as described in the legend to Fig. 6 and filters removed at the times indicated. $6 \times SSC$ at 80°C, high molecular weight DNA (●); $6 \times SSC$ at 70°C, high molecular weight DNA (▲): $6 \times SSC$, 80°C sonicated DNA (○): $6 \times SSC$ at 70°C, sonicated DNA (△).

counting that they are more sensitive, give higher resolution, and can reveal artifacts not seen by counting.

The high sensitivity is illustrated by the analysis of *E. coli* rDNA (Plate I(b)). None of the bands that is clearly visible in the radioautograph contained more than 10 c.p.m. The strip of cellulose nitrate was cut into 150, 1-mm pieces and the pieces counted in a liquid scintillation counter. None of the pieces gave counts more than twice background and none of the features visible in the radioautograph was discernible from the counts. Around 100 c.p.m. of ^{32}P in a single band 1 cm wide can be detected with an overnight exposure. The radioautograph shown in Plate I was exposed for 1 week. Fluorography of ^{3}H is not so sensitive; about 3000 d.p.m. in a 1-cm band are needed to give a visible exposure overnight. The fluorograph shown in Plate II was exposed for 2 months.

The greater resolution of film detection is illustrated by a comparison of Plate II with Figure 7(c). Plate II is a fluorograph of the strip and Figure 7(c) shows the pattern of counts obtained by cutting the strip into 1-mm pieces. Many of the bands seen in the fluorograph are not discernible in the pattern of counts (compare also the tracing of the fluorograph (Fig. 7(b)) with (c)).

For ionizing radiation, blackening of the X-ray film is proportional to the amount of incident radiation, up to the limit where a high proportion of silver grains are exposed. The relative amount of radioactivity in bands can therefore be compared by tracing radioautographs in a densitometer and comparing peak areas. However, like all other photosensitivie materials, X-ray films suffer from "reciprocity failure" at low intensities of illumination by non-ionizing radiation and it is likely that bands which contain only a few counts of ^{3}H will not be detected by fluorography even after long exposures. I have not determined the lower limit of detection. Bonner & Laskey (1974) found that 500 d.p.m. of ^{3}H in a band 1 cm × 1 mm could be detected in one week and in my own experience, less than 20 d.p.m. can be detected with longer exposure. Reciprocity failure could affect quantitation of fluorographs by densitometry but comparison of Figure 7(b) and (c) suggests that the response of the film is linear within the limits of this experiment. Clearly, quantitation of ^{32}P by densitometry can be accurate and more sensitive than counting, but film response to ^{3}H may not be linear for low amounts.

An additional advantage of film detection is that non-specific binding of RNA to the cellulose nitrate is more easily distinguished from bands of hybrid. Plate III illustrates this point. In this radioautograph, non-specific binding can be seen as dots and streaks with an appearance clearly different from that of a band. Had this strip been analysed by counting, non-specific binding would not have been distinguishable from the hybrids.

(h) *Analysis of ribosomal DNA in* X. laevis

A total of 0·6 μg of purified *X. laevis* rDNA was digested with EcoRI and the fragments separated by electrophoresis in 1% agarose gels (Plate II(c)). The pattern of fragments is similar to that described by Wellauer *et al.* (1974). They compared the secondary structures of the denatured DNA fragments with those of the ribosomal RNAs and showed that the fastest running fragment (M_r approx. 3×10^6) contained most of the DNA coding for 28 S RNA, all of the transcribed spacer, and a small portion of the DNA coding for 18 S RNA. The larger fragments (M_r 4 to 6×10^6) contained most of the DNA coding for 18 S RNA, all of the non-transcribed

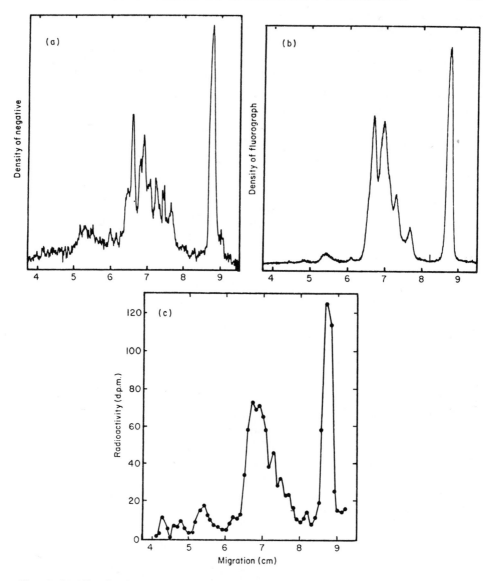

Fig. 7. (a) Microdensitometer tracing of the negative of Plate II(c). (b) Microdensitometer tracing of Plate II(a). (c) Distribution of counts in the Millipore strip which on fluorography gave Plate II(a). The strip was cut into 1 mm pieces, which were counted in a liquid scintillation counter at an efficiency of 40%.

spacer, and a small portion of the DNA coding for 28 S RNA. Different lengths of non-transcribed spacer DNA accounted for the variation in size of the longer fragments. The digest shown in Plate II(c) was transferred to cellulose nitrate as described previously. The strip was cut longitudinally into 2 parts and 1 part was hybridized with 18 S RNA and the other with 28 S RNA. Hybrids were detected by fluorography of the ^{3}H-labelled RNA (Plate II(a) and (b)). Comparison of Plate II(a) and (c) shows that the resolution of the fine bands containing the 18 S coding sequence is not as high in the fluorograph as it is in the photograph of the gel. Whereas 9 bands can be distinguished in the photograph, only 7 can be distinguished with confidence in the fluorograph. From this analysis it is possible to locate the EcoRI site within the DNA coding for 18 S RNA. As Wellauer et al. (1974) showed, 1 of the 2 breaks in the rDNA occurs towards one end of the 18 S region and the other is close to the distal end of the 28 S region. The 3×10^6 mol. wt fragment accounts for virtually all of the hybridization to 28 S RNA and for about 30% of the hybridization to the 18 S RNA (27% measured from the tracing of the fluorograph (Fig. 7(b)) and 31% from the counts). Only traces of 28 S RNA hybridize to the heterogeneous collection of fragments with molecular weights between 4 and 6×10^6, whereas about 70% of the 18 S hybridization is accounted for in these fragments. Thus the break in the 28 S region of the DNA is very close to the end of the coding sequence and the break in the 18 S region is about one-third of the way into the coding sequence.

(i) Analysis of mouse and rabbit ribosomal DNAs: evidence for long, non-transcribed spacer DNA

An EcoRI digest of total mouse DNA was separated by electrophoresis on cylindrical 1% agarose gels and transferred to strips of cellulose nitrate paper. One strip was hybridized to 18 S RNA and another to 28 S RNA prepared from rat myoblasts labelled with ^{32}P. The 28 S hybrids showed a strong, sharp band at the position of about 5.2×10^6 daltons and a very faint, broad band in the region around 14×10^6 daltons (Fig. 8(b)). The 18 S hybrids showed corresponding bands but in this case the slower moving, broad band was relatively more intense (Fig. 8(a)). From this information, a partial structure can be derived for the ribosomal DNA in mouse. Assuming that the ribosomal genes are tandemly linked, it is clear that EcoRI makes at least 2 breaks in the sequence; one in the 18 S and one in the 28 S region. Transcription of ribosomal genes in mammals produces a precursor RNA corresponding to a DNA mol. wt of about 6×10^6, and it follows that the EcoRI fragment of about 5.2×10^6, which contains both 28 S and 18 S sequences, must also encompass much of the transcribed spacer. The heterogeneous fragments with a mol. wt of 14×10^6 must contain a long stretch of non-transcribed spacer, and may contain some of the transcribed spacer too.

A similar analysis was carried out with rabbit DNA and gave similar results, although the size of the fragments was different from the corresponding fragments from mouse DNA. The band containing most of the 28 S sequence was larger (M_r approx. 6×10^6), whereas that containing most of the 18 S sequence was smaller (M_r approx. 12×10^6) and more homogeneous than the corresponding fragment in the mouse. The structures of mouse and rabbit ribosomal DNAs are thus rather similar to that of X. laevis but with longer spacer regions. The overall length of the unit in mouse is at least twice as long as that in X. laevis.

(j) EcoRI sites in the rDNA of five mammals

The analyses described above, taken with those of Wellauer et al. (1974) suggest that the two EcoRI sites in the ribosomal genes have been conserved since the amphibians and mammals diverged. In this case it would be expected that all

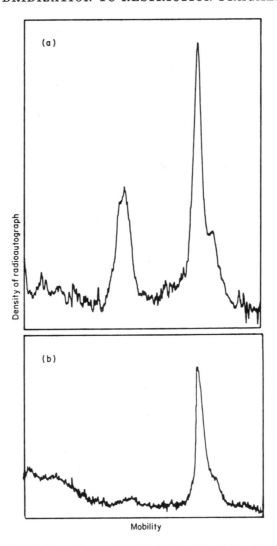

Fig. 8. EcoRI digest of mouse DNA hybridized to 18 S and 28 S RNA.

Total mouse DNA was digested to completion with EcoRI. The digest was separated by electrophoresis on 1% cylindrical agarose gels (9 mm × 24 cm, 5 mA/tube for 20 h, 40 µg of DNA/gel).

The gels were stained, photographed, and the DNA transferred to cellulose nitrate as described in Materials and Methods. One gel was hybridized to ^{32}P-labelled 18 S RNA and another to 28 S RNA. The RNA concentration was 0·1 µg/ml in 6 × SSC and hybridization was carried out at 80°C for 4 h. The filters were then washed in 2 × SSC (4 l) at 60°C for 30 min, dried and radioautographed using Kodak Blue Brand X-ray film.

(a) Densitometer tracing of the 18 S hybrids. (b) Densitometer tracing of the 28 S hybrids.

mammalian rDNAs would have equivalent EcoRI sites. Total DNAs from calf thymus, human placenta, and from livers of mouse, rabbit and rat were digested with EcoRI and the fragments separated by electrophoresis on cylindrical gels. The fragments were then transferred to a single sheet of cellulose nitrate filter and hybridized with ^{32}P-labelled rat 28 S RNA. All 5 DNAs showed a strong band in the radioautograph of the sheet. Each band was in the mol. wt region of 5 to 6×10^6, but there were small differences in their mobilities (Table 3). This result suggests that the two EcoRI sites have indeed been conserved in the rDNA of the mammals. The different fragment size

TABLE 3

Size of EcoRI fragments that hybridize to ribosomal RNAs

Species	Size of RI fragment bearing 28 S sequences $(\times 10^{-6})$	Size of fragments bearing 18 S sequences $(\times 10^{-6})$
Calf	5·7	
Human	5·7	
Mouse	5·2	5·2 and approx. 14
Rabbit	6·0	6·0 and approx. 12
Rat	6·0	
X. laevis	3·0	3·0 and 4 to 6

Sizes were estimated from mobilities in 1% agarose gels by comparison with EcoRI fragments of λ-phage DNA. The sizes of the large fragments from mouse and rabbit DNAs hybridizing to 18 S RNA are approximate estimates because there was only one marker in this region of the gel and in this region large differences in size result in small mobility differences.

can readily be accounted for by differences in the size of the transcribed spacer between 28 S and 18 S regions. Different sizes for the ribosomal RNA precursor have been reported for HeLa cells and mouse L-cells (Grierson *et al.*, 1970).

3. Conclusion

The method described here provides a simple way of detecting DNA fragments that are complementary to RNAs, after the DNA frqgments have been separated by gel electrophoresis. Transfer of the DNA from the gel to the cellulose nitrate filter is almost complete for a wide range of fragment sizes. However, large fragments $(M_r > 10^7)$ diffuse rather slowly and small fragments hybridize inefficiently. These factors should be taken into account when the method is used for quantitative work.

Much of this work was carried out when I was on leave of absence in the Institut fur Molekularbiologie II, Zurich University, supported in part by the Swiss Science Foundation (grant no. 3.8630.725R) and I am grateful to Professor M. L. Birnstiel for hospitality during this period.

REFERENCES

Bishop, J. O. (1972). In *Karolinska Symposia on Research Methods in Reproductive Endocrinology* (DiczFalnsy, E. & DiczFalnsy, A., eds), pp. 247–273, Karolinska Institute, Stockholm.
Bonner, M. & Laskey, R. A. (1974). *Eur. J. Biochem.* 46, 83–88.
Grierson, D., Rogers, M. E., Sartirana, M. L. & Loening, U. E. (1970). *Cold Spring Harbor Symp. Quant. Biol.* 35, 589–598.
Kelly, T. J. & Smith, H. O. (1970). *J. Mol. Biol.* 51, 393–409.
Kourilsky, Ph., Mercereau, O. & Tremblay, G. (1974). *Biochimie,* 56, 1215–1221.
Loening, U. E. (1969). *Biochem. J.* 113, 131–138.

Marmur, J. (1961). *J. Mol. Biol.* **3**, 208–218.
Melli, M. & Bishop, J. O. (1970). *Biochem. J.* **120**, 225–235.
Smith, H. O. & Wilcox, K. (1970). *J. Mol. Biol.* **51**, 379–391.
Thomas, M. & Davis, R. W. (1975). *J. Mol. Biol.* **91**, 315–328.
de Wachter, R. & Fiers, W. (1971). In *Methods in Enzymology* (Grossman, L. & Moldave, K., eds), vol. 21D, pp. 167–178, Academic Press Inc., New York and London.
Wellauer, P. K., Reeder, R. H., Carroll, D., Brown, D. D., Deutch, A., Higashinakagawa, T. & Dawid, I. B. (1974). *Proc. Nat. Acad. Sci., U.S.A.* **71**, 2823–2827.
Yoshimuri, R. N. (1971). Doctoral Thesis, University of California at San Francisco.

27

Reprinted from *Natl. Acad. Sci. (USA) Proc.* **73**:2294–2298 (1976)

Hybridization of RNA to double-stranded DNA: Formation of R-loops

(gene isolation/gene mapping/electron microscopy/restriction endonucleases)

MARJORIE THOMAS, RAYMOND L. WHITE, AND RONALD W. DAVIS

Department of Biochemistry, Stanford University School of Medicine, Stanford, California 94305

ABSTRACT   RNA can hybridize to double-stranded DNA in the presence of 70% formamide by displacing the identical DNA strand. The resulting structure, called an R-loop, is formed in formamide probably because of the greater thermodynamic stability of the RNA·DNA hybrid when it is near the denaturation temperature of duplex DNA. The rate of R-loop formation is maximal at the temperature at which half of the duplex DNA is irreversibly converted to single-stranded DNA (the strand separation temperature or t_{ss}) of the duplex DNA and falls precipitously a few degrees above or below that temperature. This maximal rate is similar to the rate of hybridization of RNA to single-stranded DNA under the same conditions. At temperatures above the t_{ss} the rate is proportional to the RNA concentration. However, at temperatures below t_{ss} the rate of R-loop formation is less dependent upon the RNA concentration. Once formed, the R-loops display considerable stability; the formamide can be removed and the DNA can be cleaved with restriction endonucleases without loss of R-loop structures.

It has recently been observed by R. L. White and D. Hogness (manuscript in preparation) that rRNA can hybridize to duplex DNA that codes for rRNA (rDNA) of *Drosophila melanogaster*. As shown in the electron microscope, these R-loop structures appear similar to the D-loop structures reported by Robberson *et al.* (1). However, the duplex portion of the loop is an RNA·DNA hybrid rather than a DNA·DNA hybrid and has thus been termed an R-loop. In order to optimize the formation and utilization of R-loops observed by White and Hogness, we have studied their rate of formation and kinetic stability under a variety of conditions. The ability to form R-loops at high efficiency under controlled conditions may facilitate the mapping and isolation of DNA sequences complementary to specific RNAs.

MATERIALS AND METHODS

Reagents. λgt-Scl109 DNA and yeast total rRNA were generously supplied by R. Kramer. Formamide was purified by crystallization at −3° in a salt-ice bath. The mixture was stirred with a motor-driven, stainless steel paddle to the consistency of soft ice cream. A 250-ml polycarbonate centrifuge tube was cut in half and a nylon screen cemented between the halves. The formamide crystals were recovered by centrifugation onto the nylon screen. A hole, drilled just below the screen, was used to remove the sedimented liquid. The formamide was stored at −20°. For hybridization reactions, small glass test tubes (5 × 60 mm) were siliconized by treatment with a 1% solution of dimethyl dichlorosilane dissolved in benzene. The tubes were then dried in an oven of 200° and extensively washed with water.

Abbreviations: Pipes, piperazine-*N*,*N*′-bis(2-ethanesulfonic acid)-Na$_{1.4}$; t_{ss}, temperature at which half of the duplex DNA is irreversibly converted to single-stranded DNA; t_m, temperature at which 50% of the base pairs are no longer paired; t_{max}, temperature at which maximum reaction rate occurs; rDNA, DNA that codes for rRNA.

R-loop Formation Buffer. A buffered formamide solution was first prepared by mixing 0.42 ml of formamide, 50 μl of 1 M Pipes [piperazine–*N*,*N*′-bis(2-ethanesulfonic acid) Na$_{1.4}$] at pH 7.8, 12 μl of 0.5 M Na$_3$EDTA, and 18 μl of H$_2$O. Fifty microliters of this solution were delivered with Teflon or polyethylene tubing to the bottom of a 5- × 60-mm siliconized glass test tube. Ten microliters of a solution containing approximately 30 μg/ml each of RNA and DNA, 0.1 M NaCl, and 0.05 M Tris·HCl at pH 7.5, was then added. The final cation concentration is approximately 0.17 M. The solution was covered with paraffin oil and the tube sealed with Parafilm. The test tubes were incubated in a Haake constant temperature bath filled with ethylene glycol.

Determination of Strand Separation Temperature (t_{ss}). DNA cleaved with *Eco*RI endonuclease was placed in the R-loop formation buffer as described above. Samples of 4 μl each were taken at 1° intervals starting at 45°. Five minutes were allowed for equilibration between intervals. Each sample was diluted 20-fold into 0.35 M ammonium acetate at pH 8, 0.01 M Na$_3$EDTA, and 75 μg/ml of cytochrome c at 0°. Samples were mounted for electron microscopy by the aqueous drop method. Drops of 25 μl of the above DNA plus cytochrome solutions were placed on a polished teflon bar. A parlodion-coated microscope grid was touched to the side of each drop and was stained with uranyl acetate (2). At the strand separation temperature (t_{ss}) of a specific *Eco*RI DNA fragment, the duplex strands are converted into collapsed single-strand bushes. The t_{ss} of the RNA·DNA hybrid was determined in an identical manner, except that RNA was first hybridized to its *Eco*RI-cleaved complementary DNA strand at 47° in the R-loop formation buffer.

Our initial work on R-loop formation was complicated by DNA degradation and nonreproducible reaction rates. We believe this was the consequence of changes in the reaction constituents and their concentrations during the course of the reaction. The following revisions have resulted in our ability to obtain reproducible R-loop formation rates with no detectable DNA degradation when assayed by electron microscopy: (*1*) recrystallized formamide was used; (*2*) the reactions were conducted under oil in sealed siliconized glass test tubes, which prevented evaporation and solvent condensation on the walls; (*3*) the temperature was maintained to within ±0.2°; and (*4*) the decomposition of the RNA and the pH change of the formamide solution was minimized by using Pipes buffer at pH 7.8.

RESULTS

Model system for R-loop formation

The study of R-loop formation has been greatly facilitated by the use of a simple model system. A viable hybrid DNA molecule (λgt-Scl109) containing the 2.6 kb (kilobase pair) repetitive

Table 1.   Strand separation temperatures (t_{ss})
of *Eco*RI segments of λgt-Sc1109 DNA

*Eco*RI segment	t_{ss} °C	Approximate % G+C (5, 6)
λB	45	37
rDNA	49	47
λgt right	50	47
λgt left	55	57
rDNA·rRNA	58	47

*Eco*RI DNA fragment from *Saccharomyces cerevisiae* (yeast) covalently inserted into the middle of bacteriophage λ DNA was used. The inserted DNA fragment hybridizes to yeast 26S rRNA, which can easily be prepared from yeast cells in large quantities. This λ-yeast hybrid is one of several isolated from large pools of λ-yeast hybrids by *in situ* hybridization of RNA to individual plaques (Kramer, Cameron, and Davis, manuscript in preparation). As shown below, the yeast DNA fragment coding for rRNA (rDNA) is about ¾ of the length of the 26S rRNA and hybridizes at one end of the 26S rRNA.

Strand separation temperatures of the λgt-Sc *Eco*RI DNA fragments

The strand separation temperatures of the *Eco*RI DNA fragments from λgt-Sc1109 were determined under the R-loop reaction conditions (70% formamide, 0.1 M Pipes, 0.01 M-Na_3EDTA). Samples were taken at 1° intervals starting at 45° and examined in the electron microscope as described in *Materials and Methods*. The temperature at which half of the duplex DNA is irreversibly converted to single-stranded DNA is defined as the strand separation temperature (t_{ss}) of this fragment. The strand separation of the rDNA fragment occurred within a 1° interval, while the strand separation of the left and right end fragments of bacteriophage λ occurred with partial denaturation within some of the molecules over a 4° range. Table 1 shows the t_{ss} of the *Eco*RI fragments rDNA, λB, λgt left arm, and λgt right arm (3), and their approximate base compositions (4, 5). The t_{ss} is conceptually different from the melting temperature (t_m) where ½ of the base pairs are no longer paired.

R-loop formation to saturation

An initial attempt to obtain a uniform population of DNA

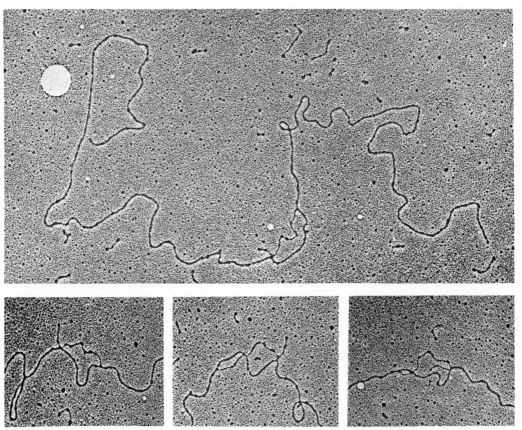

FIG. 1.   R-loops were made by heating 5 μg/ml of λgt-Sc1109 DNA and 5 μg/ml total rRNA in 70% vol/vol formamide, 0.1 M Pipes at pH 7.8, and 0.01 M Na_3EDTA at 47° for 20 hr. The reaction was performed under oil in a sealed, siliconized glass tube. All 500 RNA molecules examined contained an R-loop similar to those shown. The sample was mounted for electron microscopy by the formamide technique (2). Grids were stained with uranyl acetate and shadowed with Pt/Pd.

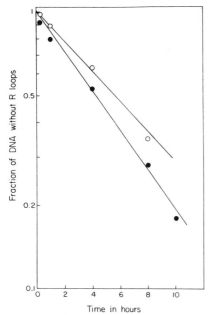

FIG. 2. R-loop formation was performed by heating 5 μg/ml of DNA and 5 μg/ml (●) or 0.5 μg/ml (○) of total rRNA in 70% vol/vol formamide, 0.1 M Pipes at pH 7.8, and 0.01 M Na₃EDTA at 43°. The reaction was carried out under oil in a sealed siliconized glass tube. Samples were taken with 26-gauge thin-walled Teflon tubing. The tubing was wiped with a Kimwipe soaked in ethanol to remove adhering oil. The sample was mounted for electron microscopy as described in Fig. 1.

molecules containing R-loops was made by heating a mixture of λgt-Sc1109 DNA at 5 μg/ml and total rRNA at 5 μg/ml at 47° for 20 hr. As shown in Fig. 1, clear R-loops were formed. Each of 500 DNA molecules examined contained an R-loop. Therefore, it is possible to saturate DNA for the formation of R-loops. Measurements of the length of the DNA·RNA portion of the R-loops indicate that the rDNA fragment is homologous to about ¾ of the length of the 26S rRNA. Single-stranded tails of RNA were only observed on one end of each R-loop, indicating that the rDNA fragment is homologous to one end of the 26S rRNA molecule.

Determination of the rate of R-loop formation

The reaction conditions used here were selected to give a reaction rate which could be readily measured. Much higher reaction rates can be achieved by increasing the RNA and salt concentrations. The course of R-loop formation under various conditions was followed by examination of samples in the electron microscope. The fraction of DNA molecules containing an R-loop was scored. The size of the R-loop was not considered. A total of 100–200 molecules were scored at each time point and 2–4 time points were taken for each condition. The accuracy of scoring in the electron microscope is greatest when ½ of the molecules have reacted. Therefore, time points near the ½ reaction time were scored and, in this work, the rate will be expressed as the time for ½ reaction since this value is directly measured.

For the reaction DNA + RNA → DNA·RNA in which RNA is in molar excess so that its concentration does not change during the reaction, a plot of the log of the fraction of unreacted

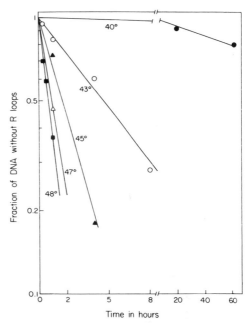

FIG. 3. R-loop formation was followed as described in the legend to Fig. 2; an RNA concentration of 5 μg/ml was used. A number of samples were taken during each reaction. Those samples containing R-loops in approximately half of the DNA molecules were scored. The repeat experiment at 43° shows a ½ reaction time similar to that shown in Fig. 2.

DNA as a function of time should give a straight line. This is supported by the data in Fig. 2 at two different RNA concentrations.

Rate of R-loop formation as a function of temperature

The time course of R-loop formation was followed at a series of temperatures using a 20-fold molar excess of RNA. Some of these time courses are shown in Fig. 3. The ½ reaction time was determined from such plots for each temperature. A plot of the log of the ½ reaction time as a function of temperature is shown in Fig. 4. The rate of R-loop formation is very dependent upon the temperature of the reaction. This plot also demonstrates that the temperature (t_{max}) at which maximum reaction rate occurs is very close to the t_{ss} (49°) of the DNA. The temperature at which the largest and most uniform-appearing R-loops were formed was from t_{ss} −2° to t_{ss} +1°. At t_{ss} −9°, most of the R-loops were quite small and were formed from small RNA fragments. At t_{ss} +2°, denatured fragments of λgt-Sc1109 DNA were observed.

Rate of R-loop formation as a function of RNA concentration and temperature

For a simple bimolecular reaction, the rate of R-loop formation should be directly proportional to the RNA concentration. This prediction is verified for the reaction at t_{ss} +1° as shown in Table 2. However, the rate at t_{ss} −6°, as shown in Fig. 2 and Table 2, is increased by a factor of only 1.5 when the RNA concentration is increased by a factor of 10. This behavior is also found at t_{ss} −9°. Between t_{ss} −6° and t_{ss} +1°, the reaction rate converts from one virtually independent of RNA concentration to one directly proportional to RNA concentration.

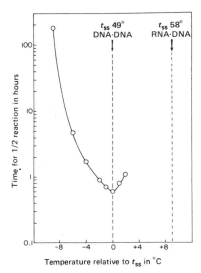

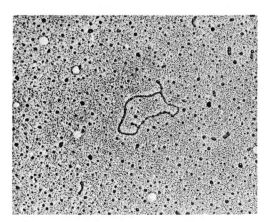

FIG. 5. A preparation of DNA containing 100% R-loops was dialyzed into 0.1 M NaCl, 0.05 M Tris·HCl at pH 7.5, 10^{-4} M EDTA. The cleavage reaction was started by the addition of $MgSO_4$ to 0.01 M and of *Eco*RI endonuclease. The mixture was digested at 37° for 10 min. The resulting sample was mounted for electron microscopy as indicated in the legend to Fig. 1. The R-loop shown was formed with a fragment of 26S rRNA and therefore forms a smaller R-loop.

FIG. 4. The ½ reaction time was determined from the plots shown in Fig. 3. The t_{ss} for duplex DNA (DNA·DNA) and for the RNA·DNA hybrid was determined by electron microscopy as described in *Materials and Methods.*

Rate of R-loop formation as a function of ionic strength

The rate of DNA renaturation increases about 10-fold when the sodium ion concentration is increased from 0.17 M to 0.67 M (6). The rate of R-loop formation was determined for this same increase in salt concentration. The addition of 0.5 M NaCl to the reaction mixture should increase the t_m of DNA by 9° (6). The ½ reaction time at 56° (t_{ss} −2°) in 0.5 M NaCl, 0.1 M Pipes at pH 7.8, 0.01 M Na_3EDTA with 5 μg/ml of DNA and 5 μg/ml of RNA is about 3.5 min. This represents approximately a 15-fold increase in reaction rate over the same reaction in the absence of NaCl.

Stability of R-loops

The large R-loops used in this work, once formed, are quite stable. They are quantitatively retained after storage for several days at 5° in the formamide reaction mixture. Two samples, stored for 2 wk after dialysis into 1 M NaCl–0.01 M Tris·HCl at pH 7.5–0.01 M Na_3EDTA or 0.1 M NaCl 0.01 M Tris·HCl at pH 7.5–0.01 M Na_3EDTA, that originally contained 100% R-loops, were reduced to 99% and 98%, respectively. These dialyzed R-loop samples were also partially resistant to heat. About 25% of the R-loops remained in the sample containing 1 M NaCl after heating to 70° (t_m − 30°) for 5 min, and about

Table 2. Ratio of R-loop formation rate at 5 μg/ml and 0.5 μg/ml of total rRNA as a function of temperature

Temperature relative to t_{ss} (°C)	$\dfrac{\text{Rate at 5 μg/ml}}{\text{Rate at 0.5 μg/ml}}$
−9	1.5
−6	1.5
−1	5.6
+1	10

The rate of R-loop formation was determined as given in Fig. 2.

75% of the R-loops remained in the sample containing 0.1 M NaCl after heating to 50° (t_m − 30°) for the same length of time.

Since such stability of the R-loops was observed, their stability during and after restriction endonuclease cleavage of R-loops containing DNA at 37° was tested. Fig. 5 shows one example of R-loops which were easily observed after cleavage of the DNA with *Eco*RI endonuclease.

Removal of R-loops

Upon removal of formamide, R-loops are thermodynamically unstable at t_m −30° and are displaced by branch migration (7). However, this displacement of large R-loops by identical DNA strands is slow. Therefore, destruction of the RNA seems necessary for the elimination of R-loops. Two methods were successfully used. (1) The RNA was hydrolyzed by addition of 0.2 M NaOH and incubated at 37° for 10 min. Denaturation of the DNA was prevented by addition of $MgSO_4$ to a final concentration of 0.01 M prior to addition of alkali. After incubation, the solution was neutralized with Tris·HCl and the Mg ion was removed by increasing the EDTA concentrations to 0.02 M. Some DNA aggregation was observed in the electron microscope with this method. (2) The R-loops are quite sensitive to RNase. Incubation for 5 min at 37° in 0.1 M NaCl, 0.05 M Tris·HCl at pH 7.5, 10^{-4} M EDTA with 10 μg/ml of RNase A resulted in the total loss of R-loops.

DISCUSSION

The temperature for maximum rate (t_{max}) of R-loop formation is within one degree of the t_{ss} of the DNA. From the values for t_{ss} of three λDNA *Eco*RI fragments with different known base compositions (Table 1), the following equation relating t_{max} to base composition can be derived for the conditions used in these experiments:

$$t_{max} = 26.5 + 0.50 \ (\%G+C).$$

If one assumes that the effect of salt concentration on the denaturation of DNA in formamide is the same as in aqueous solution, then from Schildkraut's equation (8) in 70% formamide:

$$t_{max} = 39 + 0.50 \, (\%G{+}C) + 16.6 \log \text{Na}^+.$$

Also from Schildkraut's equation, the effect of formamide can be expressed as:

$$t_{max} = 81.5 + 0.50 \, (\%G{+}C) + 16.6 \log \text{Na}^+$$
$$-0.60 \, (\% \text{ formamide}).$$

These equations have not been rigorously tested and may not apply if the conditions are significantly varied from those used in this work.

R-loops are not thermodynamically stable at $t_m -30°$ after the formamide is removed since the RNA is slowly displaced by the identical DNA strand through branch migration (7). At low temperatures ($t_m -70°$), in the absence of formamide, the stability may result from random base pairing in the single-stranded regions which block branch migration. At higher temperatures ($t_m -30°$), branch migration commences and the RNA is found to be partially displaced. However, if displacement occurs at both ends, the RNA must thread through the R-loop once for each turn of duplex RNA·DNA hybrid removed. As the RNA is displaced, the loop becomes progressively smaller, and, concurrently, the displacement may become progressively slower. Consistent with this notion are the observations that, after heating in the absence of formamide, most R-loops are very small and that the rate of displacement at $t_m -30°$ is slower in low ionic strength buffer where electrostatic repulsion between the strands is greater. As evidenced by the RNase sensitivity of the R-loops, if the RNA is destroyed as it is displaced, these blocks to the branch migration are removed.

The formation of R-loops is probably equally complicated. The most likely mechanism for their formation is the generation of denatured regions in the duplex DNA and nucleation of RNA within these denatured regions. The size and frequency of these denatured regions would depend upon the temperature, being small and infrequent below the t_{ss} of the DNA. If the size of a denatured region in the DNA is small relative to the RNA, then it seems unlikely that irreversible nucleation can occur easily in the center of the RNA because the RNA would have to thread through a very small loop. Nucleation in this case probably occurs most readily at the end of the RNA. Thus, if there are nonhomologous sequences on the end of the RNA, as is the case in this study, the rate of R-loop formation may be reduced.

The observation that the rate of R-loop formation is virtually independent of the RNA concentration at temperatures below $t_{ss} -5°$ is difficult to understand. One possible explanation is that the rate-limiting step under these conditions is the formation of denatured regions. Another possible explanation is that the rate-limiting step under these conditions is subsequent to the nucleation event. The addition of base pairs to the nucleation complex may be slow due to steric interference in winding. The nucleation complexes that do not proceed to full R-loop formation dissociate before or during mounting for electron microscopy. The rate limiting step above t_{ss} should be the rate of RNA nucleation. Therefore, above t_{ss} the rate of R-loop formation should be directly proportional to the RNA concentration as is found in this study. Also, as is found in this study, the rate of R-loop formation above the t_{ss} of the DNA should decrease because of the decreased stability of the nucleation events. With low RNA concentrations at temperatures below t_{ss}, the rate-limiting step could be the RNA nucleation. Consistent with this suggestion is the observation that, at low RNA concentration between $t_{ss} -6°$ and t_{ss}, the rate of R-loop formation is fairly independent of temperature.

It is not clear if very small R-loops can be generated since our initial attempts to form R-loops with yeast 5S RNA have not been successful. Small R-loops have an additional complication in that the RNA is probably rapidly displaced.

The experiments presented here indicate that the R-loop formation reaction is quite complex. There are additional variables that were not quantitatively investigated. For example, the secondary structure in the RNA which is not totally removed under these conditions could affect the R-loop formation rate. Also, the size of the RNA may well affect the rate since it was observed that the ratio of R-loops derived from small RNA to those derived from large RNA was greater at temperatures below t_{ss} than at temperatures above t_{ss}.

The base sequence and composition may also affect the rate of formation and stability of R-loops. R-loops are sufficiently stable to withstand dialysis into a buffer of low ionic strength (0.05 M Tris·HCl at pH 7.5, 0.1 M NaCl) and cleavage with an *Eco*RI restriction endonuclease. This should greatly facilitate the mapping of a number of rRNA and mRNA sequences in DNA. The end of the R-loop can be rather precisely mapped relative to nearby restriction sites.

The use of R-loops should facilitate the physical isolation of specific genes. For example, R-loops might be formed on unfractionated DNA with use of a specific mRNA. The resulting preparation containing duplex DNA and R-loops on the specific gene sequence could be sedimented to equilibrium in a CsCl density gradient. The R-loop should give the DNA containing the specific gene sequence a density shift; this would result in considerable purification.

1. Robberson, D. L., Kasamatsu, H. & Vinograd, S. (1972) *Proc. Natl. Acad. Sci. USA* **69**, 737–741.
2. Davis, R. W., Simon, M. N. & Davidson, N. (1971) in *Methods in Enzymology*, eds. Grossman, L. & Moldave, K. (Academic Press, New York), Vol. 21, pp. 413–428.
3. Thomas, M., Cameron, J. R. & Davis, R. W. (1974) *Proc. Natl. Acad. Sci. USA* **71**, 4579–4583.
4. Davidson, N. & Szybalski, W. (1971) in *The Bacteriophage Lambda*, ed. Hershey, A. D. (Cold Spring Harbor Laboratories, Cold Spring Harbor, N.Y.), p. 65.
5. Retel, J. & Van Keulen, H. (1975) *Eur. J. Biochem.* **58**, 51–57.
6. Wetmur, J. G. & Davidson, N. (1968) *J. Mol. Biol.* **31**, 349–370.
7. Lee, C. S., Davis, R. W. & Davidson, N. (1970) *J. Mol. Biol.* **48**, 1–22.
8. Schildkraut, C. & Lifson, S. (1965) *Biopolymers* **3**, 195.

28

RESTRICTION AND MODIFICATION ENZYMES AND THEIR RECOGNITION SEQUENCES

RICHARD J. ROBERTS

Cold Spring Harbor Laboratory, P.O. Box 100, Cold Spring Harbor, NY 11724 (U.S.A.)

INTRODUCTION

During the last few years many bacterial strains have been examined for the presence of Type II restriction endonucleases and a large number of these enzymes have now been characterized. Much of the information available has never been formally published. While this reflects the lengthy time which can elapse between discovery and publication, increasingly it results from the fact that a newly discovered endonuclease is an isoschizomer of a more familiar one. Thus, unless the new source offers some advantage, there is a natural trend to avoid formal publication. To some extent, review articles [e.g. 86, 86a, 127] fill this gap; however, they quickly become outdated. The present compilation is an attempt to extend current awareness of the enzymes now available.

In forming this list, all endonucleases cleaving DNA at a specific sequence have been considered to be restriction enzymes, although in most cases, there is no direct genetic evidence for the presence of a restriction modification system. In addition to the examples listed, other strains are known to contain specific endonucleases. For example, many different *Staphylococcus aureus* isolates contain an isoschizomer of *Sau*3A [102], while a large-scale screening [63] of a wide range of gliding bacteria (orders: *Myxobacterales* and *Cytophagales*) has shown that 27 out of 120 strains examined contained restriction enzymes. These are not included in this list because, at the moment, they are insufficiently characterized.

Within the table, the source of each microorganism is given either as an individual or a National Culture Collection. If further information is required, it is available either in the first reference shown, which in each case refers to the purification procedure for the restriction enzyme, or from the individuals who provided their unpublished results. Where more than one reference appears, the second concerns the recognition sequence for the restriction enzyme: the third contains the purification procedure for the methylase, and the fourth describes the recognition sequence for the methylase. In some cases, two references appear in one of these categories when two independent groups have reached similar conclusions.

For those who may have seen earlier, informal versions of this list, or who come across literature references to enzymes not included in the present list, it is worthwhile noting that several strains which produced useful enzymes, have been lost over the years. Perhaps the most notable is *Brevibacterium umbra* (*Bum*I) which originally arose "out of darkness" as a contaminant of *Brevibacterium albidum* and disappeared just as mysteriously. *Apo*I, from *Arthrobacter polychromogenes*, and *Zra*I from *Zoogloea ramigera* were almost certainly the products of contaminants and have not since been isolated from these sources. *Uba*I, from an Unidentified *ba*cterium, is now *Acc*I following the identification of this organism as *Acinetobacter calcoaceticus*. *Bsp*I, from a *Bacillus* sp., is now *Bme*I following the identification of the organism as *Bacillus megaterium*. *Tac*I, from *Thermoplasma acidophilus*, was changed to *Tha*I to avoid confusion with *Taq*I from *Thermus aquaticus*.

Finally, let me make a plea for a uniform nomenclature! The original proposals of Smith and Nathans [100] have proved valuable and have been followed for the most part. However, many cases have since arisen which fall outside their scope. Different strains of *Moraxella nonliquefaciens* often produce endonucleases of differing specificity and in naming these enzymes, I have slightly extended the original idea by retaining the first initial of both the genus and the species but have distinguished the various enzymes by using a third letter from elsewhere in the species name. (Only these first three letters should be italicized, with the first letter being capitalized.) Perhaps this idea, or some other *simple* system, could be employed in other cases which give similar difficulties.

Microorganism	Source	Enzyme[a]	Sequence[b]	λ	Ad2	SV40	φX174	References
Achromobacter immobilis	ATCC 15934	*AimI*	?	?	?	?	?	24
Acinetobacter calcoaceticus	R.J. Roberts	*AccI*	GT↓$\binom{A}{C}\binom{G}{T}$AC	7	8	1	3	125
Agrobacterium tumefaciens	ATCC 15955	*AccII* (*FnuDII*)	CGCG	>50	>50	0	14	125
Agrobacterium tumefaciens	E. Nester	*AtuAI*	?	>30	>30	?	?	92
Agrobacterium tumefaciens B6806	E. Nester	*AtuBI* (*EcoRII*)	CC$\binom{A}{T}$GG	>35[d]	>35	16	2	88
Agrobacterium tumefaciens ID135	C. Kado	*AtuII* (*EcoRII*)	CC$\binom{A}{T}$GG	>35[d]	>35	16	2	56
Agrobacterium tumefaciens C58	E. Nester	*AtuCI* (*BclI*)	TGATCA	7[d]	5	1	0	92
Anabaena catanula	CCAP 1403/1	*AcaI*	?	?	?	?	?	45
Anabaena subcylindrica	K. Murray	*AsuI*	G↓GNCC	>30	>30	11	2	45
Anabaena variabilis	K. Murray	*AvaI*	C↓PyCGPuG	8	?	0	1	72
		AvaII	G↓G$\binom{A}{T}$CC	>17	>30	6	1	72, 27, 107
Anabaena variabilis UW	E.C. Rosenvold	*AvaX*	ATGCAT	?	?	3	0	87, 20a, 98
		AvrI (*AvaI*)	CPyCGPuG	8	?	0	1	88a
		AvrII	CCTAGG	1	2	2	0	88a
Arthrobacter luteus	ATCC 21606	*AluI*	AG↓CT	>50	>50	35	24	83
Arthrobacter pyridinolis	R. DiLauro	*ApyI*	CC$\binom{A}{T}$GG	>35[d]	>35	16	2	21
Bacillus amyloliquefaciens F	ATCC 23350	*BamFI* (*BamHI*)	GGATCC	5	3	1	0	97
Bacillus amyloliquefaciens H	F.E. Young	*BamHI*	G↓GATCC	5	3	1	0	120, 85
Bacillus amyloliquefaciens K	T. Kaneko	*BamKI* (*BamHI*)	GGATCC	5	3	1	0	97
Bacillus amyloliquefaciens N	T. Ando	*BamNI* (*BamHI*)	GGATCC	5	3	1	0	96
		BamN$_x$	?	?	?	?	?	95, 96
Bacillus brevis S	A.P. Zarubina	*BbvSI*	G$\overset{*}{C}\binom{T}{A}$GC	specific methylase				118
Bacillus brevis	ATCC 9999	*BbvI*	GC$\binom{T}{A}$GC	>30	>30	23	14	33
Bacillus caldolyticus	A. Atkinson	*BclI*	T↓GATCA	7[a]	5	1	0	6, 92
Bacillus cereus	ATCC 14579	*Bce*14579	?	>10	?	?	?	97
Bacillus cereus	IAM 1229	*Bce*1229	?	>10	?	?	?	97
Bacillus cereus	T. Ando	*Bce*170 (*PstI*)	CTGCAG	18	25	2	1	97

Microorganism	Source	Enzyme[a]	Sequence[b]	Number of cleavage sites[c]				Reference
				λ	Ad2	SV40	φX174	
Bacillus cereus Rf sm st	T. Ando	BceR (FnuDII)	CGCG	>50	>50	0	14	97
Bacillus globigii	G.A. Wilson	BglI	?	22	12	1	0	22a, 121
		BglII	A↓GATCT	6	12	0	0	22a, 121, 77
Bacillus megaterium 899	B899	Bme899	?	>5	?	?	?	97
Bacillus megaterium B205-3	T. Kaneko	Bme205	?	>10	?	?	?	97
Bacillus megaterium	J. Upcroft	BmeI	?	>10	4	?	?	31
Bacillus pumilus AHU1387	T. Ando	BpuI	?	6	>30	2	?	47
Bacillus sphaericus	IAM 1286	Bsp1286	?	?	?	?	?	97
Bacillus sphaericus R	P. Venetianer	BspRI (HaeIII)	GGCC	>50	>50	19	11	49
Bacillus stearothermophilus 1503-4R	N. Welker	BstI (BamHI)	GGATCC	5	3	1	0	17
Bacillus stearothermophilus 240	A. Atkinson	BstAI	?	?	?	?	?	6
Bacillus stearothermophilus ET	N. Welker	BstEI	?	?	?	?	?	65
		BstEII	?	11	8	0	?	65
		BstEIII	?	>7	?	?	?	65
Bacillus subtilis strain X5	T. Trautner	BsuRI (HaeIII)	GG↓*CC	>50	>50	19	11	11, 12, 43a, 38
Bacillus subtilis Marburg 168	T. Ando	BsuM	?	>10	?	?	?	97
Bacillus subtilis	ATCC 6633	Bsu6633	?	>20	?	?	?	97
Bacillus subtilis	IAM 1076	Bsu1076 (HaeIII)	GGCC	>50	>50	19	11	97
Bacillus subtilis	IAM 1114	Bsu1114 (HaeIII)	GGCC	>50	>50	19	11	97
Bacillus subtilis	IAM 1247	Bsu1247 (PstI)	CTGCAG	18	25	2	1	97, 44
Bacillus subtilis	ATCC 14593	Bsu1145	?	>20	?	?	?	97
Bacillus subtilis	IAM 1192	Bsu1192	?	>10	?	?	?	97
Bacillus subtilis	IAM 1193	Bsu1193	?	>30	?	?	?	97
Bacillus subtilis	IAM 1231	Bsu1231	?	>20	?	?	?	97
Bacillus subtilis	IAM 1259	Bsu1259	?	>8	?	?	?	97
Bordetella bronchiseptica	ATCC 19395	BbrI (HindIII)	AAGCTT	6	11	6	0	82
Brevibacterium albidum	ATCC 15831	BalI	TGG↓CCA	15	17	0	0	32
Brevibacterium luteum	ATCC 15830	BluI (XhoI)	C↓TCGAG	1	5	0	1	34
		BluII (HaeIII)	GGCC	>50	>50	19	11	115
Caryophanon latum L	H. Mayer	ClaI	AT↓CGAT	12	?	0	0	61
Chloroflexus aurantiacus	A. Bingham	CauI	?	>30	>30	15	?	5
		CauII	?	>30	>30	0	?	5

Microorganism	Source	Enzyme	Sequence					References
Chromobacterium violaceum	ATCC 12472	CviI	?	?	?	?	?	24
Corynebacterium humiferum	ATCC 21108	ChuI (*Hind*III)	AAGCTT	6	11	6	0	24
		ChuII (*Hind*II)	GTPyPuAC	34	>20	7	13	24
Corynebacterium petrophilum	ATCC 19080	CpeI (*Bcl*I)	TGATCA	7[d]	5	1	0	26a
Diplococcus pneumoniae	S. Lacks	DpnI	GÄTC	?	?	?	0	51, 52
Diplococcus pneumoniae	S. Lacks	DpnII (*Mbo*I)	GATC	>50[d]	>50	7	0	51, 52
Enterobacter cloacae	H. Hartmann	EclI	?	15	?	?	?	42
Enterobacter cloacae	DSM 30060	EcaI	G↓GTNACC	12	?	0	0	62
Escherichia coli RY13	R.N. Yoshimori	EcoRI	G↓AATTC	5	5	1	0	37, 43, 37, 22
		EcoRI'	PuPuA↓TPyPy	>10	>10	24	16	71
Escherichia coli R245	R.N. Yoshimori	EcoRII	↓CC(A/T)GG	>35[d]	>35	16	2	124, 3 and 7, 124
Escherichia coli B	W. Arber	EcoB	TGA(N)8TGCT[e]	?	?	?	?	26, 54 and 79, 55
Escherichia coli K	M. Meselson	EcoK	?	?	?	?	?	66, -, 40
Escherichia coli (PI)	K. Murray	EcoPI	?	?	?	?	?	39, -, 8 and 9, 10
Escherichia coli P15	W. Arber	EcoP15	?	?	?	?	?	80
Fusobacterium nucleatum A	M. Smith	FnuAI (*Hinf*I)	G↓ANTC	>50	>50	10	21	57
		FnuAII (*Mbo*I)	GATC	>50	>50	7	0	82
Fusobacterium nucleatum C	M. Smith	FnuCI (*Mbo*I)	↓GATC	>50[d]	>50	7	0	57
Fusobacterium nucleatum D	M. Smith	FnuDI (*Hae*III)	GG↓CC	>50	>50	19	11	57
		FnuDII	CG↓CG	>50	>50	0	14	57
		FnuDIII (*Hha*I)	GCG↓C	>50	>50	2	18	57
Fusobacterium nucleatum E	M. Smith	FnuEI (*Mbo*I)	↓GATC	>50	>50	7	0	57
Fusobacterium nucleatum 84	M. Smith	Fnu84I	?	?	?	?	?	57
Haemophilus aegyptius	ATCC 11116	HaeI	(A/T)GG↓CC(T/A)	?	?	11	6	74
		HaeII	PuGCGC↓Py	>30	>30	1	8	81, 114
		HaeIII	GG↓CC	>50	>50	19	11	67, 12, 59
Haemophilus aphrophilus	ATCC 19415	HapI	?	>30	?	?	?	82
		HapII (*Hpa*II)	C↓CGG	>50	>50	1	5	110, 104
Haemophilus gallinarum	ATCC 14385	Hgal	GACGC[f]	>50	>50	0	14	110, 14 and 103
Haemophilus haemoglobinophilus	ATCC 19416	HhgI (*Hae*III)	GGCC	>50	>50	19	11	82
Haemophilus haemolyticus	ATCC 10014	HhaI	GCG↓C	>50	>50	2	18	84
		HhaII (*Hinf*I)	GANTC	>50	>50	10	21	58

Microorganism	Source	Enzyme[a]	Sequence[b]	Number of cleavage sites[c]				References
				λ	Ad2	SV40	ϕX174	
Haemophilus influenzae 1056	J. Stuy	Hin1056I(FnuDII)	CGCG	>50	>50	0	14	76
		Hin1056II	?	>30	>30	?	5	76
Haemophilus influenzae serotype b, 1076	J. Stuy	HinbIII(HindIII)	AAGCTT	6	11	6	0	76
Haemophilus influenzae R$_b$	C.A. Hutchison	HinbIII(HindIII)	AAGCTT	6	11	6	0	68 and 82
Haemophilus influenzae serotype c, 1160	J. Stuy	HincII(HindII)	GTPyPuAC	34	>20	7	13	76
Haemophilus influenzae serotype c, 1161	J. Stuy	HincII(HindII)	GTPyPuAC	34	>20	7	13	76
Haemophilus influenzae R$_c$	A. Landy, G. Leidy	HincII(HindII)	GTPyPuAC	34	>20	7	13	53
Haemophilus influenzae R$_d$ (exo$^-$ mutant)	S.H. Goodgal	HindI	CÅC	specific methylase				89, 90
		HindII	GTPy↓PuÅC	34	>20	7	13	101, 48, 89, 90
		HindIII	A↓AGCTT	6	11	6	0	75, 75, 89, 90
		HindIV	GÅC	specific methylase				89, 90
Haemophilus influenzae R$_d$ 123	V. Tanyashin	HindGLU	?	?	?	?	?	111
Haemophilus influenzae R$_f$	C.A. Hutchison	HinfI	G↓ANTC	>50	>50	10	21	68, 46 and 73
		HinfII(HindIII)	AAGCTT	6	11	6	0	60
Haemophilus influenzae H-1	M. Takanami	HinHI(HaeII)	PuGCGCPy	>30	>30	1	8	110
Haemophilus parahaemolyticus	C.A. Hutchison	HphI	GGTGA[g]	>50	>50	4	9	68, 50
Haemophilus parainfluenzae	J. Setlow	HpaI	GTT↓AAC	11	6	4	3	93, 29
		HpaII	C↓CGG	>50	>50	1	5	93, 29, 59
Haemophilus suis	ATCC 19417	HsuI(HindIII)	AAGCTT	6	11	6	0	82
Herpetosiphon giganteus HP1023	J.H. Parish	HgiAI	G(^T_A)GC(^T_A)↓C	20	?	0	3	16
Klebsiella pneumoniae OK8	J. Davies	KpnI	GGTAC↓C	2	8	1	0	99, 112
Microcoleus sp.	D. Comb	MstI	TGCGCA	>10	>15	0	1	19, 33
Moraxella bovis	ATCC 10900	MboI	↓GATC	>50[d]	>50	8	0	30
		MboII	GAAGA[h]	>50	>50	16	11	30, 13 and 23
Moraxella glueidi LG1	J. Davies	MgiI	?	?	?	?	?	99
Moraxella glueidi LG2	J. Davies	MgiII	?	?	?	?	?	99
Moraxella nonliquefaciens	ATCC 19975	MnoI(HpaII)	C↓CGG	>50	>50	1	5	82, 78
		MnoII	?	>10	>6	2	?	82

Organism	Source/Strain	Enzyme	Sequence					Ref.
Moraxella nonliquefaciens	ATCC 17953	MnlI	CCTC[i]	>50	>50	51	35	126
Moraxella nonliquefaciens	ATCC 17954	MnnI(HindII)	GTPyPuAC	34	>20	7	13	41
		MnnII(HaeIII)	GGCC	>50	>50	19	11	41
		MnnIII	?	>50	>50	?	?	41
		MnnIV(HhaI)	GCGC	>50[d]	>50	2	18	41
Moraxella osloensis	ATCC 19976	MosI(MboI)	GATC	>50	>50	8	0	30
Moraxella sp.	R.J. Roberts	MspI(HpaII)	CCGG	>50	>50	1	5	116
Myxococcus virescens	H. Reichenbach	MviI	?	1	?	?	?	70
		MviII	?	?	?	?	?	70
Neisseria gonorrhoeae	G. Wilson	NgoI(HaeII)	PuGCGCPy	>30	>30	1	8	122
Neisseria gonorrhoeae	CDC 66	NgoII(HaeIII)	GGCC	>50	>50	19	11	20
Proteus vulgaris	ATCC 13315	PvuI	CGATCG	4	7	0	0	33
		PvuII	CAG↓CTG	15	22	3	0	33
Providencia alcalifaciens	ATCC 9886	PalI(HaeIII)	GGCC	>50	>50	19	11	31
Providencia stuartii 164	J. Davies	PstI	CTGCA↓G	18	25	2	1	99, 15
Pseudomonas facilis	M. Van Montagu	PfaI	?	>30	>30	?	?	76 and 115
Rhodopseudomonas sphaeroides	R. Lascelles	RspI(PvuI)	?	3	12	?	?	4
Rhodopseudomonas sphaeroides	S. Kaplan	RshI(PvuI)	CGATCG	4	7	0	0	28
Serratia marcescens S$_b$	C. Mulder	SmaI	CCC↓GGG	3	12	0	0	36, 25
Serratia sp. SAI	B. Torheim	SspI	?	?	?	?	?	113
Staphylococcus aureus 3A[1]	E.E. Stobberingh	Sau3A(MboI)	GATC	>50	>50	8	0	105
Staphylococcus aureus PS96	E.E. Stobberingh	Sau96I(AsuI)	G↓GNCC	>30	>30	11	2	106
Streptococcus faecalis subsp. zymogenes	R. Wu	SfaI(HaeIII)	GG↓CC	>50	>50	19	11	123
Streptococcus faecalis ND547	D. Clewell	SfaNI	GATGC	>50	>30	6	12	92
Streptomyces achromogenes	ATCC 12767	SacI	GAGCT↓C	2	>10	0	0	1
		SacII	CCGC↓GG	3	>25	0	1	1
		SacIII	?	>30	>30	?	?	1
Streptomyces albus	CMI 52766	SalPI(PstI)	CTGCAG	18	25	2	1	18
Streptomyces albus subsp. pathocidicus	KCC S0166	SpaI(XhoI)	CTCGAG	1	5	0	1	109
Streptomyces albus G	J.M. Ghuysen	SalI	G↓TCGAC	2	3	0	0	2
		SalII	?	>30	?	?	?	2
Streptomyces bobiliae	ATCC 3310	SboI	?	?	?	?	?	108
Streptomyces bradiae	ATCC 3535	SbrI	?	?	?	?	?	108
Streptomyces cuspidorus	KCC S0316	ScuI(XhoI)	CTCGAG	1	5	0	1	109

Microorganism	Source	Enzyme[a]	Sequence[b]	λ	Ad2	SV40	φX174	References
					Number of cleavage sites[c]			
Streptomyces exfoliatus	KCC S0030	*SexI (XhoI)*	CTCGAG	1	5	0	1	109
Streptomyces goshikiensis	KCC S0294	*SgoI (XhoI)*	CTCGAG	1	5	0	1	109
Streptomyces griseus	ATCC 23345	*SgrI*	?	0	7	0	?	1
Streptomyces hygroscopicus	?	*ShyI*	?	2	?	?	?	118
Streptomyces lavendulae	ATCC 8644	*SlaI (XhoI)*	C↓TCGAG	1	5	0	1	108
Streptomyces luteoreticuli	KCC S0788	*SluI (XhoI)*	CTCGAG	1	5	0	1	109
Streptomyces stanford	S. Goff,	*SstI(SacI)*	GAGCT↓C	2	>10	0	0	35, 69
		SstII(SacII)	CCGC↓GG	3	>25	0	1	35
	A. Rambach	*SstIII(SacIII)*	?	>30	>30	?	?	35
Thermoplasma acidophilum	D. Searcy	*ThaI (FnuDII)*	CG↓CG	>50	>50	0	14	64
Thermopolyspora glauca	ATCC 15345	*TglI (SacII)*	CCGCGG	3	>25	0	1	33
Thermus aquaticus YTI	J.I. Harris	*TaqI*	T↓CGA	>50	>50	1	10	91
		TaqII	?	>30	>30	4	6	82
Xanthomonas amaranthicola	ATCC 11645	*XamI(SalI)*	GTCGAC	2	3	0	0	2
Xanthomonas badrii	ATCC 11672	*XbaI*	T↓CTAGA	1[d]	4	0	0	128
Xanthomonas holcicola	ATCC 13461	*XhoI*	C↓TCGAG	1	5	0	1	34
		XhoII	?	>20	>20	4	?	76
Xanthomonas malvacearum	ATCC 9924	*XmaI*	C↓CCGGG	3	12	0	0	25
		XmaII(PstI)	CTGCAG	18	25	2	1	25
Xanthomonas nigromaculans	ATCC 23390	*XniI (PvuI)*	CGATCG	4	7	0	0	41
Xanthomonas oryzae	M. Ehrlich	*XorI (PstI)*	CTGCAG	18	25	2	1	94
		XorII (PvuI)	CGATCG	4	7	0	0	94
Xanthomonas papavericola	ATCC 14180	*XpaI(XhoI)*	C↓TCGAG	1	5	0	1	34

[a] When two enzymes recognize the same sequence, i.e., are isoschizomers, the prototype (i.e., the first example isolated) is indicated in parentheses.

[b] Recognition sequences are written from 5'→3', only one strand being given, and the point of cleavage is indicated by an arrow (↓). When no arrow appears, the precise cleavage site has not been determined. For example, G↓GATCC is an abbreviation for

5' G↓GATCC 3'
3' CCTAG↑G 5'

Bases appearing in parentheses signify that either base may occupy that position in the recognition sequence. Thus, *AccI* cleaves the sequences GTAGAC, GTATAC, GTCGAC, and GTCTAC. Where known, the base modified by the corresponding specific methylase is indicated by an asterisk. A is N^6-methyladenosine. C is 5-methylcytosine.

[c] These columns indicate the frequency of cleavage by the various specific endonucleases on bacteriophage lambda DNA (λ), adenovirus-2 DNA (Ad2), simian virus 40 DNA (SV40), and φX174 Rf DNA (φX174). In the latter two cases, the sites were checked by computer search of the published sequences.

[d] In most *E. coli* strains, bacteriophage lambda DNA is partially modified against the action of *Atu*BI, *Atu*II, *Atu*CI, *Apy*I, *Bcl*I, *Cpe*I, *Dpn*II, *Eco*RII, *Fnu*CI, *Mbo*I, *Mos*I and *Xba*I.

[e] The following methylated oligonucleotides have been isolated from *Eco*B modified DNA [117]:

$$\text{TG}\overset{*}{\text{A}},\ \text{C}\overset{*}{\text{A}}\text{C},\ \text{AG}\overset{*}{\text{A}}\text{C},\ \overset{*}{\text{A}}\text{AT},\ \text{(A,G,C)}\overset{*}{\text{A}}$$

[f] *Hga*I cleaves as indicated:

```
5'-G A C G C N N N N N      -3'
3'-C T G C G N N N N N N N N N N↑-5'
                       ↓
```

[g] *Hph*I cleaves as indicated:

```
5'-G G T G A N N N N N N N ↓  -3'
3'-C C A C T N N N N N N N ↑   -5'
```

[h] *Mbo*II cleaves as indicated:

```
5'-G A A G A N N N N N N N ↓  -3'
3'-C T T C T N N N N N N N ↑  -5'
```

[i] *Mnl*I cleaves 5 to 10 bases in 3' direction from the recognition sequence.

REFERENCES

1 Arrand, J.R., Myers, P.A. and Roberts, R.J., unpublished observations.
2 Arrand, J.R., Myers, P.A. and Roberts, R.J., J. Mol. Biol., 118 (1978) 127—135.
3 Bigger, C.H., Murray, K. and Murray, N.E., Nature New Biol., 244 (1973) 7—10.
4 Bingham, A.H.A., Atkinson, A. and Darbyshire, J., unpublished observations.
5 Bingham, A.H.A. and Darbyshire, J., unpublished observations.
6 Bingham, A.H.A., Sharp, R.J. and Atkinson, A., unpublished observations.
7 Boyer, H.W., Chow, L.T., Dugaiczyk, A., Hedgpeth, J. and Goodman, H.M., Nature New Biol., 244 (1973) 40—43.
8 Brockes, J.P., Biochem. J., 133 (1973) 629—633.
9 Brockes, J.P., Brown, P.R. and Murray, K., Biochem. J., 127 (1972) 1—10.
10 Brockes, J.P., Brown, P.R. and Murray, K., J. Mol. Biol., 88 (1974) 437—443.
11 Bron, S., Murray, K. and Trautner, T.A., Mol. Gen. Genet., 143 (1975) 13—23.
12 Bron, S. and Murray, K., Mol. Gen. Genet., 143 (1975) 25—33.
13 Brown, N.L., Hutchison, C.A. III and Smith, M., J. Mol. Biol., in press.
14 Brown, N.L. and Smith, M., Proc. Natl. Acad. Sci. USA, 74 (1977) 3213—3216.
15 Brown, N.L. and Smith, M., FEBS Lett., 65 (1976) 284—287.
16 Brown, N.L., McClelland, M. and Whitehead, P.R., unpublished observations.
17 Catterall, J. and Welker, N., J. Bacteriol., 129 (1977) 1110—1120.
18 Chater, K., Nucl. Acids Res., 4 (1977) 1989—1998.
19 Comb, D., Schildkraut, I. and Roberts, R.J., unpublished observations.
20 Clanton, D.J., Woodward, J.M. and Miller, R.V., J. Bacteriol., 135 (1978) 270—273.
20a Denniston-Thompson, K., Moore, D.D., Kruger, K.E., Furth, M.E. and Blattner, F.R., Science, 198 (1978) 1051—1056.
21 DiLauro, R., unpublished observations.
22 Dugaiczyk, A., Hedgpeth, J., Boyer, H.W. and Goodman, H.M., Biochemistry, 13 (1974) 503—512.
22a Duncan, C.H., Wilson, G.A. and Young, F.E., J. Bacteriol., 134 (1978) 338—344.
23 Endow, S.A., J. Mol. Biol., 114 (1977) 441—450.
24 Endow, S.A. and Roberts, R.J., unpublished observations.
25 Endow, S.A. and Roberts, R.J., J. Mol. Biol., 112 (1977) 521—529.
26 Eskin, B. and Linn, S., J. Biol. Chem., 247 (1972) 6183—6191.
26a Fisherman, J., Gingeras, T.R. and Roberts, R.J., unpublished observations.
27 Fuchs, C., Rosenvold, E.C., Honigman, A. and Szybalski, W., Gene, 4 (1978) 1—23.
28 Gardner, J. and Kaplan, S., unpublished observations.
29 Garfin, D.E. and Goodman, H.M., Biochem. Biophys. Res. Commun., 59 (1974) 108—116.
30 Gelinas, R.E., Myers, P.A. and Roberts, R.J., J. Mol. Biol., 114 (1977) 169—180.
31 Gelinas, R.E., Myers, P.A. and Roberts, R.J., unpublished observations.
32 Gelinas, R.E., Myers, P.A., Weiss, G.A., Roberts, R.J. and Murray, K., J. Mol. Biol., 114 (1977) 433—440.
33 Gingeras, T.R. and Roberts, R.J., unpublished results.
34 Gingeras, T.R., Myers, P.A., Olson, J.A., Hanberg, F.A. and Roberts, R.J., J. Mol. Biol., 118 (1978) 113—122.
35 Goff, S.P. and Rambach, A., Gene, 3 (1978) 347—352 and unpublished observations.
36 Greene, R. and Mulder, C., unpublished observations.
37 Greene, P.J., Betlach, M.C., Goodman, H.M. and Boyer, H.W., Methods Mol. Biol., 7 (1974) 87—111.
38 Gunthert, U., Freund, M. and Trautner, T.A., Abstracts of 12th FEBS Symposium, Dresden (1978).
39 Haberman, A., J. Mol. Biol., 89 (1974) 545—563.
40 Haberman, A., Heywood, J. and Meselson, M., Proc. Natl. Acad. Sci. USA, 69 (1972) 3138—3141.
41 Hanberg, F., Myers, P.A. and Roberts, R.J., unpublished observations.
42 Hartmann, H. and Goebel, W., FEBS Lett., 80 (1977) 285—287.

43 Hedgpeth, J., Goodman, H.M. and Boyer, H.W., Proc. Natl. Acad. Sci. USA, 69 (1972) 3448—3452.

43a Heininger, K., Horz, W. and Zachau, H.G., Gene, 1 (1977) 291—303.

44 Hoshino, T., Uozumi, T., Horinouchi, S., Ozaki, A., Beppu, T. and Arima, K., Biochim. Biophys. Acta, 479 (1977) 367—369.

45 Hughes, S.G., Bruce, T. and Murray, K., unpublished observations.

46 Hutchison, C.A. and Barrell, B.G., unpublished observations.

47 Ikawa, S., Shibata, T. and Ando, T., J. Biochem., 80 (1976) 1457—1460.

48 Kelly, T.J., Jr. and Smith, H.O., J. Mol. Biol., 51 (1970) 393—409.

49 Kiss, A., Sain, B., Csordas-Toth, E. and Venetianer, P., Gene, 1 (1977) 323—329.

50 Kleid, D., Humayun, Z., Jeffrey, A. and Ptashne, A., Proc. Natl. Acad. Sci. USA, 73 (1976) 293—297.

51 Lacks, S. and Greenberg, B., J. Biol. Chem., 250 (1975) 4060—4072.

52 Lacks, S. and Greenberg, B., J. Mol. Biol., 114 (1977) 153—168.

53 Landy, A., Ruedisueli, E., Robinson, L., Foeller, C. and Ross, W., Biochemistry, 13 (1974) 2134—2142.

54 Lautenberger, J.A., Kan, N.C., Lackey, D., Linn, S., Edgell, M.H. and Hutchison, C.A. III., Proc. Natl. Acad. Sci. USA, 75 (1978) 2271—2275.

55 Lautenberger, J.A. and Linn, S., J. Biol. Chem., 247 (1972) 6176—6182.

56 LeBon, J.M., Kado, C., Rosenthal, L.J. and Chirikjian, J., submitted for publication.

57 Lui, A., McBride, B.C. and Smith, M., unpublished results.

58 Mann, M.B., Rao, R.N. and Smith, H.O., Gene, 3 (1978) 97—112.

59 Mann, M.B. and Smith, H.O., Nucl. Acids Res., 4 (1977) 4211—4221.

60 Mann, M.B. and Smith, H.O., unpublished observations.

61 Mayer, H., Grosschedl, R., Schutte, H. and Hobom, G., unpublished observations.

62 Mayer, H., Schwarz, E., Melzer, M. and Hobom, G., unpublished observations.

63 Mayer, H. and Reichenbach, H., J. Bacteriol., in press.

64 McConnell, D., Searcy, D. and Sutcliffe, G., Nucl. Acids Res., 5 (1978) 1729—1739.

65 Meagher, R.B., unpublished observations.

66 Meselson, M. and Yuan, R., Nature, 217 (1968) 1110—1114.

67 Middleton, J.H., Edgell, M.H. and Hutchison, C.A. III, J. Virol., 10 (1972) 42—50.

68 Middleton, J.H., Stankus, P.V., Edgell, M.H. and Hutchison, C.A. III, unpublished observations.

69 Muller, F., Stoffel, S. and Clarkson, S.G., unpublished observations.

70 Morris, D.W. and Parish, J.H., Arch. Microbiol., 108 (1976) 227—230.

71 Murray, K., Brown, J.S. and Bruce, S.A., unpublished observations.

72 Murray, K., Hughes, S.G., Brown, J.S. and Bruce, S., Biochem. J., 159 (1976) 317—322.

73 Murray, K. and Morrison, A., unpublished observations.

74 Murray, K., Morrison, A., Cooke, H.W. and Roberts, R.J., unpublished observations.

75 Old, R., Murray, K. and Roizes, G., J. Mol. Biol., 92 (1975) 331—339.

76 Olson, J.A., Myers, P.A. and Roberts, R.J., unpublished observations.

77 Pirrotta, V., Nucl. Acids Res., 3 (1976) 1747—1760.

78 RajBhandary, U.L. and Baumstark, B., unpublished observations.

79 Ravetch, J.V., Horiuchi, K. and Zinder, N.D., Proc. Natl. Acad. Sci. USA, 75 (1978) 2266—2270.

80 Reiser, J. and Yuan, R., J. Biol. Chem., 252 (1977) 451—456.

81 Roberts, R.J., Breitmayer, J.B., Tabachnik, N.F. and Myers, P.A., J. Mol. Biol., 91 (1975) 121—123.

82 Roberts, R.J. and Myers, P.A., unpublished observations.

83 Roberts, R.J., Myers, P.A., Morrison, A. and Murray, K., J. Mol. Biol., 102 (1976) 157—165.

84 Roberts, R.J., Myers, P.A., Morrison, A. and Murray, K., J. Mol. Biol., 103 (1976) 199—208.

85 Roberts, R.J., Wilson, G.A. and Young, F.E., Nature, 265 (1977) 82—84.

86 Roberts, R.J., CRC Crit. Rev. Biochem., 4 (1976) 123—164.

86a Roberts, R.J., in DNA Insertion Elements, Plasmid, and Episomes (Bukhari, A.I., Shapiro, J.A. and Adhya, S.L., eds.) Cold Spring Harbor Laboratory, N.Y., 1977, pp. 757—768.
87   Roizes, G., unpublished observations.
88   Roizes, G., Patillon, M. and Kovoor, A., FEBS Lett., 82 (1977) 69—70.
88a Rosenvold, E.C. and Szybalski, W., unpublished observations.
89   Roy, P.H. and Smith, H.O., J. Mol. Biol., 81 (1973) 427—444.
90   Roy, P.H. and Smith, H.O., J. Mol. Biol., 81 (1973) 445—459.
91   Sato, S., Hutchison, C.A. and Harris, J.I., Proc. Natl. Acad. Sci. USA, 74 (1977) 542—546.
92   Sciaky, D. and Roberts, R.J., unpublished observations.
93   Sharp, P.A., Sugden, B. and Sambrook, J., Biochemistry, 12 (1973) 3055—3063.
94   Shedlarski, J., Farber, M. and Ehrlich, M., unpublished observations.
95   Shibata, T. and Ando, T., Mol. Gen. Genet., 138 (1975) 269—380.
96   Shibata, T. and Ando, T., Biochim. Biophys Acta, 442 (1976) 184—196.
97   Shibata, T., Ikawa, S., Kim, C. and Ando, T., J. Bacteriol., 128 (1976) 473—476.
98   Shimatake, H. and Rosenberg, M., unpublished observations.
99   Smith, D.L., Blattner, F.R. and Davies, J., Nucl. Acid Res., 3 (1976) 343—353.
100 Smith, H.O. and Nathans, D., J. Mol. Biol., 81 (1973) 419—423.
101 Smith, H.O. and Wilcox, K.W., J. Mol. Biol., 51 (1970) 379—391.
102 Stobberingh, E.E., Schiphof, R. and Sussenbach, J.S., J. Bacteriol., 131 (1977) 645—649.
103 Sugisaki, H., Gene, 3 (1978) 17—28.
104 Sugisaki, H. and Takanami, K., Nature New Biol., 246 (1973) 138—140
105 Sussenbach, J.S., Monfoort, C.H. Schiphof, R. and Stobberingh, E.E., Nucl. Acids Res., 3 (1976) 3193—3202.
106 Sussenbach, J.S., Steenbergh, P.H., Rost, J.A., Van Leeuwen, W.J. and van Embden, J.D.A., Nucl. Acids Res., 5 (1978) 1153—1163.
107 Sutcliffe, J.G. and Church, G.M., Nucl. Acids Res., 5 (1978) 2313—2319.
108 Takahashi, H., Saito, H. and Ikeda, Y. and Sugisaki, H., Gene, 5 (1979) in press.
109 Takahashi, H., Shimotsu, H. and Saito, H., unpublished observations.
110 Takanami, M., Meth. Mol. Biol., 7 (1974) 113—133.
111 Tanyashin, V.I., Li, L.I., Muizhnieks, I.O. and Bayev, A.A., Dokl. Akad. Nauk. SSSR, 231 (1976) 226—228.
112 Tomassini, J., Roychoudhury, R., Wu, R. and Roberts, R.J., unpublished observations.
113 Torheim, B., personal communication.
114 Tu, C-P.D., Roychoudhury, R. and Wu, R., Biochem. Biophys. Res. Commun., 72 (1976) 355—362.
115 Van Montagu, M., unpublished observations.
116 Van Montagu, M., Myers, P.A. and Roberts, R.J., unpublished observations.
117 Van Ormondt, H., Lautenberger, J.A., Linn, S. and DeWaard, A., FEBS Lett., 33 (1973) 177—180.
118 Vanyushin, B.F. and Dobritsa, A.P., Biochim. Biophys. Acta, 407 (1975) 61—72.
119 Walter, F., Hartmann, M. and Roth, M., Abstracts of 12th FEBS Symp., Dresden, 1978
120 Wilson, G.A. and Young, F.E., J. Mol. Biol., 97 (1975) 123—126.
121 Wilson, G.A. and Young, F.E., in Microbiology 1976 (D. Schlessinger, ed.), Am. Soc. Microbiol., Washington, 1976, pp. 350—357.
122 Wilson, G.A. and Young, F.E., unpublished observations.
123 Wu, R., King, C.T. and Jay, E., Gene, 4 (1978) 329—336.
124 Yoshimori, R.N., Ph.D. Thesis, 1971.
125 Zabeau, M. and Roberts, R.J., unpublished observations.
126 Zabeau, M., Greene, R., Myers, P.A. and Roberts, R.J., unpublished observations.
127 Zabeau, M. and Roberts, R.J., in Mol. Genet., (J.H. Taylor, ed.), in press.
128 Zain, B.S. and Roberts, R.J., J. Mol. Biol., 115 (1977) 249—255.

29

Reprinted from *Nucl. Acids Res.* **3**:2387–2398 (1976)

A SIMPLE METHOD FOR DNA RESTRICTION SITE MAPPING

Hamilton O. Smith and Max L. Birnstiel

*Institut für Molekularbiologie II
der Universität Zürich*

ABSTRACT

When a DNA molecule, enzymatically labelled with ^{32}P at one end, is partially digested with a restriction enzyme labelled DNA fragments are obtained which form an overlapping series of molecules, all with a common labelled terminus. A restriction map can then be constructed from an analysis of the size distribution of these molecules. This technique has been used for the restriction site mapping of cloned histone DNA (h22) where as many as 35 cleavage sites may be accurately determined in a single experiment.

INTRODUCTION

Development of a restriction enzyme cleavage map for a given DNA molecule is often a prerequisite for the detailed study of its genetic organization and for the determination of its base sequence. Most of the commonly used mapping procedures involve identification of the cleavage fragments in a complete restriction enzyme digest and the subsequent laborious ordering of these fragments by various methods, most commonly by analysis of a subset of fragments contained in each of several overlapping partial digestion products (1). We describe here a simple and rapid procedure for direct mapping of the restriction sites relative to the termini of the DNA molecule. The method is similar in concept to that utilized by Gilbert and Maxam for DNA sequencing (2) and we suggest that our restriction mapping technique will be most useful in connection with their sequencing protocol.

The principle of the site mapping procedure is as follows: The DNA in question is labelled at the 5' termini with ^{32}P-phosphoryl groups using polynucleotide kinase and (γ-^{32}P) ATP. The labelled DNA is then cleaved assymmetrically with a suitable

restriction enzyme into two fragments that are separable by
gel electrophoresis. Each DNA segment, labelled at only one
end, is recovered and digested with the chosen restriction enzyme so as to produce a partial digest. A large spectrum of
partial digestion products may be produced, but the labelled
fragments form a simple overlapping series, all with a common
labelled terminus. These are fractionated according to molecular
weight by gel electrophoresis and detected by autoradiography.
The relative mobility of each labelled fragment allows determination of its molecular weight by comparison with a set of DNA
molecular weight standards and in turn locates the distance, in
base pairs, of the respective restriction sites from the labelled
terminus. The order of the fragments and their lengths thus
correspond directly to the order of restriction sites along the
DNA molecule. Partial DNA digests using several different
restriction enzymes can be analyzed simultaneously in adjacent
slots of a slab gel, allowing the relative positions of all the
restriction sites to be read directly from the autoradiogram.

    To illustrate the general method we describe its appli-
cation to mapping of a number of restriction sites in a cloned
6 kb repeat unit of the sea urchin Psammechinus miliaris histone DNA.

MATERIALS AND METHODS

    Enzymes. EcoRI was a gift from Ch. Weissmann, Zürich,
Hae III enzyme (activity) was obtained from Miles Laboratories.
Hpa II (2.5u/µl) was prepared according to DeFilippes (3)
HindII and HindIII (10u/µl) were prepared by an adaptation of
the DeFilippes procedure (4). The preparation of Alu I (1u/µl)
by a similar protocol will be described elsewhere (4). In all
cases, one unit yields a complete digest of 1µg DNA under
standard reaction conditions in 1 h at 37°C. T4 polynucleotide
kinase (Miles Laboratories) was supplied at 3u/µl in 50% glycerol. Bacterial alkaline phosphatase (Worthington, BAPC 20u/ml)
was supplied as a 70% ammonium sulfate suspension. The enzyme
was pelleted at 12000 x g for 5 min, redissolved in 10mM Tris,
10mM $MgCl_2$, pH 7.6 at 60 u/ml and heated at 95° for 10 min to
inactivate any contaminating nucleases.

DNAs. Calf thymus DNA was dissolved in TE buffer (10mM
Tris-Cl, 1mM EDTA, pH 7.6) at 2.5 mg/ml, sonicated for 1 min at
maximum intensity on a Branson sonicator, extracted with phenol,
ethanol precipitated and redissolved in TE buffer at 10mg/ml.
Wild-type phage λDNA was obtained by phenol extraction of CsCl
purified λ Sam 7 phage (5).

For the recovery of the 6 kb h22 DNA, the recombinant λ
histone DNA was first cleaved with HindIII restriction enzyme.
The reaction mixture contained 1 mg λh22 DNA in 1 ml TE buffer,
10µl 1M $MgCl_2$, 20µl 1M Tris-Cl, pH 7.6, 0.1ml HindIII enzyme
and was incubated 2.5 h at 37°. The reaction was terminated by
addition of 0.1 ml 0.5M EDTA, pH 7.6; 2ml of 0.2M glycine, 15mM
NaOH buffer; 10% (final) glycerol, and 0.1ml ethidium bromide
2mg/ml.

Isolation of the 6 kb histone DNA by preparative gel
electrophoresis. A 5 cm 0.6% agarose gel was prepared in a 6 cm
x 20 cm glass tube supported at the bottom by a dialysis membrane
held in place with a rubber band. A large tube was clamped verti-
cally and suspended in a 2 litre beaker filled with electrophoresis buffer (0.2M glycine; 0.015M NaOH; pH 8.3). The glass column
was filled with buffer to the same height as in the beaker.
Circular platinum electrodes were placed around the outside and
on the inside of the glass tube. The entire DNA digest was loaded
onto the gel and electrophoresis begun at approximately 60V and
50mamps. The DNA was easily visible with a longwave UV lamp. As
soon as it had entered the gel, the buffer compartments were
circulated by pump. After 3 h voltage was decreased to 25V and
continued for 16 h at which time the 6 kb histone DNA fragment
was about 3 cm into the gel and the 7×10^6 and 19×10^6 dalton
(right and left arm of the phage) were approximately 1.5 and
0.5 cm into the gel. Electrophoresis was continued at 80-90V
under surveillance until the histone fragment was just about to
exit. At this point the gel was slipped out of the tube and inverted so that the band was ready to emerge upward. The buffer
was replaced and a 2cm layer of 20% glycerol in electrophoresis
bu-fer was layered over the gel. Electrophoresis was continued
at 90V with reversed polarity until the histone DNA was completely

into the glycerol layer. This was then removed by pipetting.

The DNA was loaded onto a DE52 column of 1.4ml bed volume equilibrated with 50mM Tris-Cl, 10mM EDTA, pH 7.6, washed with 50ml of equilibrium buffer, 25ml 0.15M NaCl, 20mM Tris-Cl, 10mM EDTA, pH 7.6, and then eluted with 4ml 1M NaCl into this buffer. Two volumes of ethanol were added, the tube was stored at -20° overnight, and centrifuged at 15,000 rpm for 30 min. The precipitate was rinsed with 95% ethanol, dried and dissolved in 1 ml of TE buffer. The overall yield was 46μg of histone DNA from 1mg of recombinant DNA.

$5'-^{32}P$ labelling of the 5.1 kb fragment. For labelling, the isolated 6 kb cloned histone DNA h22 was treated with phosphatase to remove 5'-phosphoryl groups. The termini were then labelled using $(\gamma-^{32}P)$ ATP and T4 polynucleotide kinase. The reaction mixture contained 100μl h22 DNA (4.6μg), 1μl of $MgCl_2$ 1M, 0.5μl of 1M dithiothreitol, and 5μl of bacterial alkaline phosphatase. Incubation was for 1 h at 37°. Phosphatase was inactivated at pH 2 by addition of 10μl of 0.1M HCl and incubated at 23° for 10 min. The pH was readjusted by addition of 5μl of 1M Tris-Cl, pH 8. Then 1.2μl of 5M NaCl, 10μl $(\gamma-^{32}P)$ ATP (400 Ci/mmole, 10μM), and 2μl of kinase (3000 u/ml) were added and incubation continued at 37° for 30 min. The labelling reaction was terminated by addition of 2μl of 5 mM ATP and immediately raising the temperature to 75° for 5 min. The terminally labelled h22 DNA was then cleaved into two fragments of 5.1 kb and 0.9 kb (1) by incubation with 5μl of EcoRI enzyme at 37° for 1 h. The reaction was terminated by addition of 30μl of a 0.12% bromphenol blue, 2% sarkosyl, 25% glycerol, 0.125M EDTA (glycerol-dye mixture) and heating at 65° for 5 min.

The reaction mixture was divided into 3-50μl aliquots and electrophoresed on a 1% agarose cylindrical gel (0.9 x 12cm) at 65V until the dye marker had progressed 6 cm. The lower half containing unincorporated radioactivity was cut off and discarded. The upper portion was stained for 10 min in 0.5μg/ml ethidium bromide. The larger and the smaller bands were visualized by long wave UV light and excised. The agar slices containing the large and the small fragments were separately pooled and dissolved in 3ml of 4M KI, 14mM 2-mercaptoethanol at 45°. One µl of calf thymus DNA (sonicated, 10mg/ml) was added to each tube as carrier. The DNA fragments were adsorbed onto small hydroxy-apatite columns (0.3 ml), washed with several volumes of 4M KI, 14mM 2-mercaptoethanol followed by several volumes of 10mM sodium phosphate, pH 6.7, and then eluted with about 1 ml of 0.4M phos-phate buffer. The phosphate was removed by chromatography on a 8-9 ml G50 Sephadex column in TE buffer. The DNAs contained in about 2 ml were removed in the region of the void column. The DNA was precipitated by addition of sodium acetate to 0.3M and magne-sium acetate to 10 mM, and 2 volumes of ethanol. After 2 h at -20° the DNA was pelleted in a SW50 rotor at 40,000 rpm for 30 min. Each labelled fragment was dissolved in 50µl of TE buffer. The large and small fragment preparations contained about 1200 cpm/µl and 4000 cpm/µl respectively.

   Electrophoresis and autoradiography for restriction site mapping.   Labelled digestion products (in 5-12µl) containing 0.2 volume of glycerol-dye mixture (see above) were electro-phoresed on 1-2% agarose slab gels (0.2 cm x 14 cm x 40 cm or 0.3 cm x 16 cm x 18 cm) in a 0.2M glycine, 0.015M NaOH buffer (pH 8.3) at approximately 5V/cm. At completion of the run, the gel was removed onto a sheet of Saranwrap on a glass plate, covered with a sheet of Whatman DE81 paper, 4-6 sheets of 3MM paper and another glass plate. The 3MM paper was changed every 5-10 min as it became wet. 1% gels were reduced to a thin skin-like film in 30 min and 2% gels in about 1 h. The Saranwrap-covered gel still adhering to the DE81 paper was covered with a sheet of Kodak XR-5 X-Omat film, sandwiched between 2 glass plates, and exposed 1-3 days at room temperature. In some cases, to decrease exposure time by a factor of 2-3, a sheet of Fast Tungstate (Ilford) was placed on top of the film sheet and expo-sure was at -70° for 1-3 days. Film was developed using Kodak DX-80 developer and FX-40 fixer.

RESULTS

   Location of Hpa II restriction sites in H22 DNA. A partial digestion in which all of the partial products are adequately

represented is essential for the site mapping procedure. The conditions for this are best determined by analyzing a number of
early time points in a reaction so constituted as to reach completion in 1-2 h. An example of this approach is shown in Fig.1.

Sea urchin histone DNA (h22) for restriction site mapping
was obtained from λred⁻Sam7 Pm hist22, i.e. a phage λred⁻Sam7
hybrid carrying one 6 kb repeat unit of Psammechinus histone DNA
inserted at a single HindIII site in the cI gene (3). The phage
was grown on 803 $r_k^-m_k^-$ rec A⁻supII Met⁻ (a gift of N. Murray,
Edinburgh). A 1.8 litre lysate was purified by two consecutive
CsCl gradients. After dialysis against TE buffer (see Methods)
the phage was extracted twice with phenol and the DNA precipitated with ethanol, washed with 94% ethanol and finally dissolved
in 5ml of TE buffer at a concentration of 1mg/ml.

The 6 kb histone DNA (h22 DNA) was excised from the recombinant DNA with HindIII restriction enzyme. The digest yielded
the 7×10^6 dalton and 19×10^6 dalton right and left arms of λ
in addition to the excised 4×10^6 dalton h22 DNA. The histone
DNA was separated and recovered electrophoretically (see Methods).
The 6 kb h22 DNA was treated with phosphatase, labelled with
^{32}P-ATP by means of T_4 polynucleotide kinase and cleaved with
EcoRI into the 5.1 and the 0.9 kb DNA fragments. Both kinds of
molecules were isolated and subjected to further analysis.

For restriction site mapping analysis the 5.1 kb EcoRI
fragment now labelled at only one of its 5' termini was digested
with Hpa II enzyme and samples removed for electrophoretic analysis at 2, 4, 10, 20, and 40 min. The 2 min sample still contained a large portion of undigested DNA but 4 partials can be
seen at increasing intensities with time of incubation (Figure 1).
The 4 and 10 min samples reveal nearly optimal intensities for
each fragment band while at 20 and 40 min the reaction is nearing
completion. The number of base pairs (bp) of each partial fragment was determined graphically from a semi-logarithmical plot
of standard DNA fragment sized in kb units versus relative
mobility in cm. Reading from the labelled terminus there are
four Hpa II sites at 975, 2400, 3150, and 3920 bp. The RI site

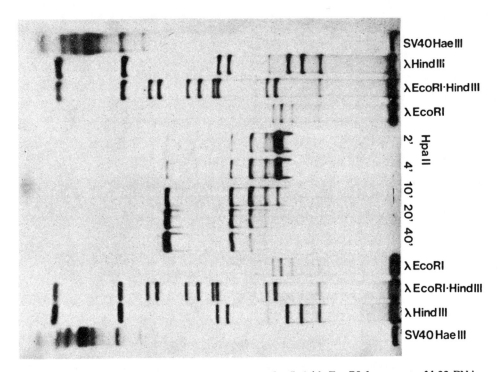

Fig. 1: Mapping of Hpa II restriction sites in the 5.1 kb EcoRI fragment of h22 DNA. For restriction site mapping of the 5.1 kb unit the DNA was treated with Hpa II enzyme as follows: The reaction mixture (50μl) contained 7500 cpm of h22, 5.1 kb EcoRI fragments, 10mM MgCl pH 7.6, 0.5mM EDTA, 14mM 2-mercapto-ethanol, and 1μl of Hpa II enzyme. Incubation was at 37°. Aliquots of 10μl were removed at 2, 4, 10, 20, and 40 min in 2.5μl of dye mixture and heated at 65° for 5 min prior to loading onto a 1% agarose slab gel (0.3 x 16 x 18 cm) and electrophoresis at 120V for 2 h (bromophenol blue marker at 11.3 cm). The gel was dried onto DE81 paper as described in the Methods section and a 3 day autoradiographic exposure was made. Molecular weight markers of terminally labelled total digest fragments were included as indicated in the Figure. These included HindIII, EcoRI and a double digest with EcoRI and HindIII of λDNA. Molecular weights for the fragments were as given by Murray & Murray (12). A labelled Hae III digest of SV40 DNA (2.5μg) was similarly prepared (3400 cpm/μl). Molecular weights of the fragments as given by Yang et al. (13).

marking the extreme end of the 5.1 kb fragment is at a distance of 5050 bp. Four very light bands can be seen above and below the smallest fragment in the 20 and 40 min digests. These might be due to contamination with a trace of an as yet undescribed Hpa enzyme or conceivably to a relaxed specificity of Hpa II, similar to EcoRI* (6). The 0.9 kb EcoRI fragment was analyzed in a similar way. Two restriction sites were discovered at

125 and 250 from the labelled terminus at 930 bp (data not
shown).

<u>Simultaneous mapping of the restriction sites for Hpa II,
HindII, Hae III, and Alu I in h22 DNA</u>. In all cases where sites
for several different enzymes are closely spaced along the DNA
molecule, ordering is most reliably done by electrophoresing
each partial enzyme digest in adjacent, parallel tracts of a
gel. This is illustrated in Fig.2. The 5.1 kb fragment was digested by each of the four restriction enzymes <u>Hpa</u> II, <u>Hind</u>II,
<u>Hae</u> III, and <u>Alu</u> I under predetermined conditions for partial
digestion and then electrophoresed on a 40 cm 1.5% agarose gel
along with standards (see Legend Fig.2). The total number of
visible bands, each representing a restriction site, for <u>Hpa</u> II,
<u>Hind</u>III, <u>Hae</u> III and <u>Alu</u> I are 4, 4, 16, and 11, respectively.
The distance from the terminus in base pairs for these enzymes
are given in Table 1. The 0.9 fragment was also subjected to
restriction analysis with <u>Alu</u> I. Two sites were found at ∿ 100
and 825 bp. From the unpublished work of W. Schaffner (Zürich)
it is known that <u>Hae</u> III cleaves the same DNA segment three
times. These sites have not as yet been mapped.

<u>Distribution of restriction sites in h22 DNA</u>. The histone
DNA of <u>Psammechinus miliaris</u> (h22) comprises AT-rich spacer and
GC-rich cistronic DNA in about equal proportions (5). The
distribution and sites of these distinct kinds of DNA sequences
in the 6 kb DNA unit have been determined by partial denaturation
mapping in the electron-microscope (7), and are depicted schema-
tically in Fig.3. <u>Hpa</u> II and <u>Hae</u> III restriction enzymes recognize and cleave the sequences CCGG (8) and GGCC (9), respectively and would therefore be expected to cut spacer DNA with a GC
content of 36% (1,7), on a random basis, every 950 bp. As anticipated, the number of restriction sites in the AT-rich spacer
DNA is low for these two enzymes. <u>Alu</u> I, with the restriction
site AGCT (10) should cleave every 330 bp, on a random basis,
but only few sites are detected experimentally. Spacer DNA, then,
clearly has a strong base sequence bias which excludes recognition sites for <u>Alu</u> I. However, tight cluster of nine <u>Alu</u> sites
are found in the cistronic region coding for the H1 mRNA, and

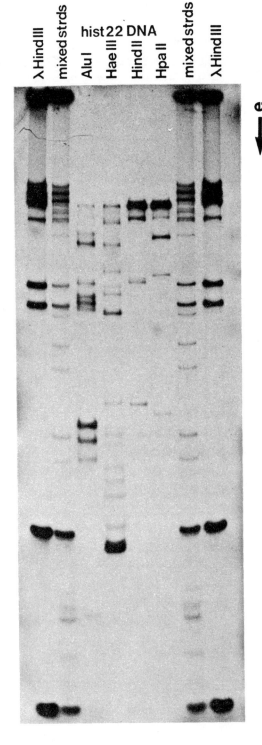

Fig. 2: Simultaneous mapping of the Hpa II, HindII, HindIII, Hae III and Alu I restriction sites in the 5.1 kb EcoRI fragment of the h22 DNA. All reaction mixtures contained 10mM Tris-Cl pH7.6, 10mM MgCl₂, 14mM 2-mercaptoethanol and 500 cpm/μl of the 5.1 kb EcoRI fragment of H22 DNA. The Hpa II reaction mixture (5μl)contained 0.13μl of enzyme and was incubated at 37° for 15 min. The HindII reaction mixture (5μl) contained 0.1μl of enzyme, 50mM NaCl, and was incubated for 10 min at 37°. The Hae III mixture (10μl) contained 1μl enzyme and was incubated for 60 min at 37°. The Alu I mixture (10μl) contained 1.2μl enzyme and was incubated for 3 min at 37°. The reaction was terminated with 0.2 volumes of dye mixture and heating to 65° for 5 min. Each reaction mixture was layered onto a slot of a 1.5% agarose slab gel (0.2 x 14 x 40 cm) and electrophoresed at 200V for 6.5 h (dye marker at 25 cm) at room temperature. The gel was dried onto DE81 paper (See Methods). Autoradiographic exposure was 60 h at - 70° using a Fast Tungstate intensifier screen. Molecular weight standards were included as indicated in the Figure. "Mixed standards" consisted of λ digested with EcoRI, with HindIII, with EcoRI and HindIII together and SV40 cleaved with Hae III (12, 13).

Table 1: Lengths (in base pairs) of partially digested RNA molecules obtained from the

5.1 kb unit				0.9 kb unit			
Hpa II	HindII	Hae III	Alu I	Hpa II	HindII	Hae III	Alu I
4210	4080	(3700)	3620	250	none	three	825
3200	2430	(2980)	3320	~125		n.d.	~100
2950	2390	2480	3020				
1020	1150	2290	2330				
		2170	2130				
		1910	2030				
		1090	1960				
		950	960				
		860	880				
		760	790				
		710	345				
		655					
		555					
		515					
		505					
		490					

an additional four in or near the H2A cistronic DNA segment. The restriction sites for <u>Hae</u> III are similarly clustered in the cistronic regions H2A and H1 of the histone DNA. Therefore, restriction of histone DNA by <u>Alu</u> I and <u>Hae</u> III provides a

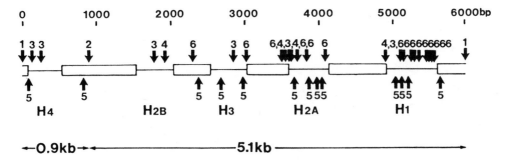

Legend to Figure 3:   Distribution of some restriction cleavage sites in histone DNA (h22).   The positions of the restriction sites along the histone DNA 5.1 and 0.9 kb fragments were combined with the partial denaturation map of the same DNA molecule. The spacer DNAs are represented as double lines, the cistronic DNA as single lines. The Hae III sites have not as yet been mapped in the 0.9 kb unit, nor is it certain that all restriction sites have been discovered in the distal part of the 5.1 kb unit (see Discussion).   There is a relative lack of restriction sites in spacer DNA while the cistronic regions, especially those containing the H2A and the H1 coding sequences, are cleaved many times.   Restriction cleavage sites: 1 <u>Hind</u>III; 2 <u>Eco</u>RI; 3 <u>Hpa</u>II; 4 <u>Hind</u>II; 5 <u>Alu</u> I; 6 <u>Hae</u> III

sufficient number of defined DNA segments to facilitate rapid
sequencing of those two genes by the Maxam and Gilbert procedure (2).

DISCUSSION

     The experiments of Fig.2 illustrate the power of the termi-
nal labelling - partial digestion procedure for the ordering of
multiple restriction sites. All sites in the region of 0-3 kb
have been accurately mapped from this single gel (see Table 1).
This work has led to the discovery of a new HindII site in the
interval 2390 - 2430, yielding a DNA fragment with only 40 nucleotides which had previously escaped our notice (1). Certain partial
fragments band at lesser or greater intensity than others but this
is no point for concern so long as the restriction enzymes are
reasonably pure. This finding suggests that not all sites are
equivalent substrates.

     In the region of 3-5.1 kb the bands are not sufficiently
well spaced out in the gel to distinguish sites that are within
0.1 - 0.2 kb. For this reason we cannot for the moment exclude
that there may be additional Hae III and Alu I sites in this
region of histone DNA. There are several possible methods for
detecting all possible sites. One could resort to polyacrylamide
gradient gels of the sort described by Jeppesen (11). Alternatively, a lower percentage agarose gel could be used to expand
the high molecular weight region while sacrificing the resolution
of the smaller fragments. Perhaps the best method would be to
obtain a different assymmetric fragment labelled from the opposite end to complete the fine structure mapping of the distal
3-5.1 kb region.

     The limitations of the gel electrophoresis for the analysis
of DNA molecules more than approximately 10×10^6 daltons, make
our procedure of most value for relatively small molecules. A
large viral genome would need to be mapped in segments. However, because of the rapidity and ease of the procedure we believe it will be useful in such cases.

     Another major advantage of our procedure is the ability to

detect and accurately order sites that are only 20-50 bases apart; in other procedures, very small fragments often go undetected in electropherograms of complete digests where band intensities by staining or uniform labelling depend on fragment mass. Since partials are identical in sequence except for the terminal regions, mobility artefacts due to sequence composition are minimized.

ACKNOWLEDGEMENT

H.O.S. was a stipendiate of the Roche Foundation. The work was supported in part by the Schweiz. Nationalfonds, grant No. 3.602-0.75 SR and by the State of Zürich.

REFERENCES

1 Schaffner, W., Gross, K., Telford, J. and Birnstiel, M. (1976) Cell 8, 471
2 Gilbert, W., Maxam, A. and Mirzabekov, A. in "Context of Ribosome Synthesis", eds Kjelgaard, N.O. and Maaloe, O. (The Alfred Benzon Symposium IX, Munksgaard, Copenhagen, Denmark), in press
3 DeFilippes, F. (1974) Biochem.Biophys.Res.Comm. 58, 586
4 Clarkson, S.G., Smith, H.O. and Kurer, V. (in preparation)
5 Clarkson, S.G., Smith, H.O., Schaffner, W., Gross, K. and Birnstiel, M., Nucleic Acid Research, submitted
6 Polisky, B., Green, P., Garfin, O.E., McCarthy, B.J., Goodman H.M. and Boyer, H.W.(1975) Proc.Natl.Acad.Sci.US 72, 3310
7 Portmann, R. and Birnstiel, M., in preparation
8 Garfin, D.C. and Goodman, H.M. (1974) Biochem.Biophys.Res. Comm. 59, 109
9 Murray, K. and Morrison, A., unpublished observations
10 Roberts, R.,Myers, Ph., Morrison, A. and Murray, K. (1976) J.Mol.Biol. 102, 157
11 Jeppeson, P.G.N. (1974) Anal.Biochem. 58, 195
12 Murray, K. and Murray, N.E. (1975) J.Mol.Biol. 98, 551
13 Yang,R.C.A., Van de Voorde, A. and Fiers, W. (1976) Europ. J.Biochem. 61, 101

Part IV

GENE ISOLATION
FROM COMPLEX GENOMES

Editors' Comments
on Papers 30 Through 33

30   LEDER et al.
The Cloning of Mouse Globin and Surrounding Gene Sequences in Bacteriophage λ

31   VANDE WOUDE et al.
Cloning of Integrated Moloney Sarcoma Proviral DNA Sequences in Bacteriophage λ

32   CARBON et al.
The Construction and Use of Hybrid Plasmid Gene Banks in Escherichia coli

33   BLATTNER et al.
Cloning Human Fetal γ Globin and Mouse α-Type Globin DNA: Preparation and Screening of Shotgun Collections

In recombinant DNA technology, a single DNA fragment is removed from a population and propagated in a vector for further study. A basic problem existed in identifying the specific DNA fragment of interest, a problem that became more difficult as the complexity of the organism studied increased. Thus while the genome of phage λ gives rise to six Eco R1 fragments, the genome of its host, *E. coli*, yields almost a thousand. Higher organisms (for example, mice, rabbits, and humans) have genomes that yield about a million Eco R1 fragments. The problem of finding one DNA fragment in a million was solved by two general approaches. In the first, DNA fragments were partially purified using a variety of methods including column chromatography or preparative gel electrophoresis. The partially purified fragments were then inserted into a vector. In the second approach, a large collection of independent recombinant DNA molecules was made from unfractionated DNA. This collection was made sufficiently large to ensure the presence of all possible DNA fragments. This latter approach, the so-called shotgun method (see Paper 33) also has been called the "gene bank" method (see Paper 32) or the "genomic library" method (see Maniatis et al. 1978).

Implicit in both approaches was that a method was available to identify the DNA fragment of interest. In most cases, the specific RNA transcript of the desired gene(s) was available, and hybridization techniques such as the Southern method, the Grunstein and Hogness method, or the Benton and Davis method were employed (see Part III). Such hybridization methods often used as radioactive probe, complementary DNA (cDNA) made from RNA species by reverse transcriptase. In several of the specific examples reprinted here, the cDNA copies were inserted into *E. coli* plasmids and then used as hybridization probes. In fact, the specific messenger RNAs (mRNA) for globin were some of the first eukaryotic gene transcripts propagated in *E. coli* (Maniatis et al. 1976; Rabbits 1976; Rougeon et al. 1975).

The partial purification approach was first applied to the mouse genome and the β-globin gene by Leder and his colleagues (Paper 30). A different purification procedure was developed concurrently by Tonegawa and coworkers (1977). The methodology introduced by Leder's group used two consecutive purification steps, each with high capacity and resolution. The first step used RPC-5 column chromatography to fractionate DNA fragments. The second step used preparative gel electrophoresis of specific RPC-5 fractions. This electrophoresis step involved the design of an automated horizontal electrophoresis apparatus controlled by an electronic timer (Polsky et al. 1978). The fragments so purified were enriched some 500- to 1000-fold for the gene of interest. The partially purified fragments were inserted into the λgt*WES* vector (Paper 12). Recombinant phages were screened using the Benton and Davis protocol (Paper 24), and about one in a thousand phages contained the desired fragment. The results of the method gave Leder's group the first direct look at a mammalian gene and its surprising architecture, namely, the presence of intervening sequences within structural genes.

Vande Woude and colleagues (Paper 31) used the partial purification scheme of Leder to isolate specific DNA fragments from the genome of normal mink cells or transformed mink cells containing integrated RNA tumor viruses. An unusual feature of this work was that it represented one of the first experiments to be done under P4 maximum containment conditions. This work demonstrates the advantages of the partial purification method for identification of sequences, given the availability of a specific probe. It also shows that once a genome has been fractionated, one can return to the samples for additional experiments. Vande Woude's group demonstrated that the mink sequences bracketing the integrated RNA tumor virus could be used as hybridization probes to locate and isolate this site of virus integration in normal cells.

The "gene bank" or "library" approach was applied initially to organisms like *E. coli* and yeast whose genomes were less complex than those of mammals (see, for example, Clarke and Carbon 1976; Cramer et al. 1976; Struhl et al. 1976). This method is summarized by Carbon and colleagues in Paper 32. Their approach was to break *E. coli* or yeast DNA by random scission using hydrodynamic shear, rather than by restriction endonuclease action. In this way, they hoped to generate an overlapping collection of fragments in which the desired gene cluster was expected to be intact on at least one fragment of the sheared DNA. The terminal transferase tailing procedure (see Papers 1 and 2) was then used to insert these fragments into a colE1 plasmid vector. Carbon's group derived a simple formula to calculate how many independent recombinant plasmids were needed to give a gene bank representing any genome of known size. For example, to be 95 percent sure of having the entire *E. coli* genome represented, some 940 independent recombinants must be obtained. Similarly, 3,000 and 30,000 independent recombinants were needed for a gene bank of yeast and Drosophila, respectively. Carbon and colleagues discuss some useful methods for maintaining and screening gene banks from simple organisms. In addition, there is discussion on expression of yeast functions in *E. coli* (see Part V).

The use of this approach obviously becomes more laborious as the genome complexity, and consequently the size of the gene bank, increases. Kedes and coworkers (1975) were able to analyze a moderately complex genome (sea urchin) by screening hybrids derived from unfractionated DNA. However, for organisms like mammals, 10^6 to 10^7 individual clones must be made and screened to be sure of finding a single copy gene. Paper 33 by Blattner and colleagues describes their rationale and methodology developed to examine mammalian gene banks (called "shotgun" collections in this paper). Concurrently, Maniatis and coworkers (1978) developed a similar approach and introduced the term "genomic libraries" to indicate a collection of recombinant DNAs encompassing an entire genome. Both groups utilized the increased efficiency accorded by *in vitro* packaging of λ vectors to overcome the inefficient transfection step (see Papers 22 and 23). The strategy used by Maniatis and colleagues is unusual in that they made use of the enzyme Eco R1 methylase to inactivate all Eco R1 sites in the DNA to be inserted into the vector. This methylated DNA was then partially cleaved by restriction enzymes that create blunt or flush ends (Alu I or Hae III) and fragments of about 20,000 base pairs were purified. The combination of incomplete cleavage and the high frequency occurrence of Alu I or Hae III cleavage sites in most DNA meant that a large collection of 20,000 base pair fragments purified from such a partial digest would represent the entire genome. These fragments were then prepared for insertion into a λ Eco R1 vector by addition of synthetic Eco R1 linkers using blunt-end ligation (see Papers 4 and 21). By protecting the internal Eco R1 sites with the specific methylation introduced earlier by Eco R1 methylase, genes having Eco R1 sites within them could be isolated intact. This method was formally similar to random shear except that using the Maniatis approach, the ends of the molecules obtained were of known sequence and structure.

Paper 33 by Blattner's group describes a novel strategy for producing and screening millions of recombinant phage plaques. They developed a high efficiency, *in vitro* packaging system from several concepts described in the earlier systems (Becker and Gold 1975, and Papers 22 and 23). They then devised a "megaplate" variation of the Benton and Davis plaque hybridization procedure (see Paper 24) that used cafeteria trays rather than standard petri dishes. In this way several million plaques could be screened at one time on a few trays. The footnotes of this paper contain many detailed methods and suggestions including the complete *in vitro* packaging protocol.

One advantage of the gene bank or library is that once made, it can be propagated and distributed to others. Indeed, the Clarke and Carbon collection (1976) of colE1-*E. coli* hybrids has been propagated and screened by many laboratories. An important assumption for this method is that no DNA segment will be lost or altered during propagation of the recombinant collection.

The fractionation methodology has the advantage of identifying fragments prior to inserting them into a vector. For example, fragments too small or too large to be inserted into a given vector will be absent from gene banks generated by restriction endonucleases rather than random breakage methods. Similarly, fragments that impose selective disadvantages on vector growth will be underrepresented in the gene bank. Prior fractionation enables one to identify desired fragments and adapt the experimental details accordingly. Another potential problem that can be simplified by fractionation and prior screening is determination of how many different fragments of interest are in the genome. Once the desired recombinant has been obtained, one can determine whether it has been altered in a major way (that is, by a deletion that occurred during growth in *E. coli*) by direct comparison of the sizes of the starting material and of the cloned fragment.

REFERENCES

Becker, A., and M. Gold. 1975. Isolation of the Bacteriophage λ A Gene Protein. *Natl. Acad. Sci. (USA) Proc.* **72**:581–585.

Clarke, L., and J. Carbon. 1976. A Colony Bank Containing Synthetic colE1 Hybrid Plasmids Respresentative of the Entire *E. coli* Genome. *Cell* **9**:91–99.

Cramer, R. A., J. R. Cameron, and R. W. Davis. 1976. Isolation of Bacteriophage λ Containing Yeast Ribosomal Genes: Screening by *in situ* RNA Hybridization to Plaques. *Cell* **8**:227–232.

Kedes, L. H., R. H. Cohn, J. C. Lowry, A. C. Y. Chang, S. N. Cohen. 1975. The Organization of Sea Urchin Histone Genes. *Cell* **6**:359–369.

Maniatis, T., S. G. Kee, A. Efstratiadis, F. C. Kafatos. 1976. Amplification and Characterization of a β-globin Gene Synthesized *in vitro*. *Cell* **8**:163–182.

Maniatis, T., R. C. Hardison, E. Lacy, J. Lauer, C. O'Connell, D. Quon, G. K. Sim, and A. Efstratiadis. 1978. The Isolation of Structural Genes from Libraries of Eukaryotic DNA. *Cell* **15**:687–701.

Polsky, F., M. H. Edgell, J. G. Seidman, P. Leder. 1978. High Capacity Gel Preparative Electrophoresis for Purification of Fragments of Genomic DNA. *Anal. Biochem.* **87**:397–410.

Rabbits, T. H. 1976. Bacterial Cloning of Plasmids Carrying Copies of Rabbit Globin Messenger RNA. *Nature* **260**:221–225.

Rougeon, F., P. Kourilsky, and B. Mach. 1975. Insertion of a Rabbit β-globin Gene Sequence into an *E. coli* Plasmid. *Nucl. Acid Res.* **2**:2365–2378.

Struhl, K., J. R. Cameron, and R. W. Davis. 1976. Functional Genetic Expression of Eukaryotic DNA in *Escherichia coli*. *Natl. Acad. Sci. (USA) Proc.* **73**:1471–1475.

Tonegawa, S., E. Brack, N. Hozumi, R. Schuller. 1977. Cloning of an Immunoglobulin Variable Region Gene from Mouse Embryo. *Natl. Acad. Sci. (USA) Proc.* **74**:3518–3522.

30

Reprinted from *Cold Spring Harbor Symp. Quant. Biol.* **42**:915–920 (1978)

The Cloning of Mouse Globin and Surrounding Gene Sequences in Bacteriophage λ

P. Leder, S. M. Tilghman, D. C. Tiemeier, F. I. Polsky, J. G. Seidman, M. H. Edgell, L. W. Enquist, A. Leder, and B. Norman

Laboratory of Molecular Genetics, National Institute of Child Health and Human Development, National Institutes of Health, Bethesda, Maryland 20014

We have developed a rather straightforward procedure that brings virtually any segment of the mammalian genome within cloning range of a very versatile EK2 bacteriophage λ vector. We illustrate the procedure in terms of cloning a segment of the mouse genome containing a β-like globin gene and its surrounding sequences. Surprisingly, the sequence cloned does not contain a continuous representation of the globin messenger RNA (mRNA) sequence, but instead is interrupted about two-thirds through by an approximately 550-base-long *intervening sequence*. The significance of this structure and its possible generality are discussed together with the notion that single polypeptide chains may be encoded by discrete and separate gene sequences.

The Technical Problem and Its Solution: Purification of Globin Genes

The essential difference between cloning specific segments of a mammalian genome and those of a prokaryotic genome is the 1000-fold greater complexity of the former. When we began this work 3 years ago, we calculated that the available screening techniques would make it possible, though uncomfortable, to screen through several thousand cloned fragments in order to find a sequence of interest. Inasmuch as the mammalian genome is divided into approximately one million fragments by the *Escherichia coli* restriction endonuclease, *Eco*RI, this would require us to purify a given fragment approximately 1000-fold prior to cloning. With this degree of purification in mind, we have adopted two procedures which meet our requirements for high capacity and resolution.

The first dimension is RPC-5, a reverse phase chromatographic system originally developed for the separation of tRNAs by Pearson et al. (1971). Hardies and Wells (1976) and Landy et al. (1976) have shown that this technique can be easily adapted for the purpose of separating restriction fragments of bacteriophage λ DNA. We have applied this medium for the resolution of *Eco*RI fragments of total mouse genomic DNA — derived from the mouse plasmacytoma MOPC-149 — as shown in Figure 1 (Tiemeier et al. 1977). The DNA is eluted over approximately 150 fractions, each having the approximate degree

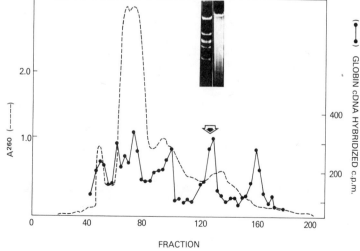

Figure 1. RPC-5 chromatography of *Eco*RI-digested mouse genomic DNA. An *Eco*RI digest of mouse genomic DNA was chromatographed on an RPC-5 column as described in Tilghman et al. (1977). (– – –)A₂₆₀; (●——●) S₁-resistant hybridization of ³²P-labeled globin cDNA. The insert above fractions 124–130 illustrates an ethidium-bromide-stained 1% analytical agarose gel of an *Eco*RI digest of 0.5 μg of wild-type λcI857 DNA (left) (where migration is from top to bottom and the kilobase pair sizes of individual fragments are 21.3, 7.35, 5.79, 5.40, 4.69, and 3.3 kb) and 2 μg of fraction 130 (right).

of complexity shown by the fraction illustrated in the inset in Figure 1. Each fraction is then assayed for the presence of globin gene sequences using hybridization to a mixture of α and β globin ^{32}P-labeled cDNAs. Several peaks were observed and that migrating at approximately fraction 130 was used for further purification.

As shown in the inset in Figure 1, the fraction containing a globin sequence represents a mixture of EcoRI fragments ranging in size from over 20,000 base pairs to less than 1000, which migrate essentially as a smear. Obviously, this size heterogeneity provides a logical basis for a second dimension for purification. To take advantage of it and to achieve the capacity and resolution required, a high-capacity, horizontal slab gel electrophoresis apparatus controlled with an electronic timer was designed (Polsky et al. 1977). With this device, the pooled fractions from the RPC-5 column can be highly enriched according to size, and the globin sequences localized as shown in Figure 2. As can be seen, the single peak fraction from RPC-5 contained two EcoRI fragments which hybridized to globin cDNA, one approximately 7000 bases long and the other approximately 15,000 bases long. A rough estimate of the purity of the lower-molecular-weight fraction suggested that the globin-containing sequence was represented approximately once per 2000 fragments (Tilghman et al. 1977). The larger fraction was of even greater purity.

The Vector System and Identification of Positive Clones

Not only did the degree of purification noted above meet our early requirements, but bacteriophage λ proved to be a far more powerful cloning vector than we had originally expected. The vector system used in these studies represents an EK2 derivative of the λgt system originally constructed by Thomas et al. (1974). We have introduced a phenotypically inert fragment of λ DNA (indicated as the B fragment

in Fig. 3), as well as certain mutations designed to meet the guidelines established by the National Institutes of Health Advisory Committee on Recombinant DNA Research. As presently constituted, this λ vector offers both a positive selection for the insertion of a fragment of DNA and a negative selection against the reformation of the parental-type phage (Tiemeier et al. 1977; Leder et al. 1977). The reason for this is that the left and right arms of the phage contain all the genes necessary for lytic phage growth, but lack sufficient DNA for efficient packaging of viral phage particles. This provides a positive selection in that only those phage which have incorporated a foreign fragment of DNA will grow. On the other hand, parental-type recombinants can be selected against by either purifying the left and right arms by RPC-5 chromatography (Tiemeier et al. 1977) or by utilizing endonuclease SstI sites located only in this fragment (cf. Fig. 3). When the vector is digested with both EcoRI and SstI, the two cloning arms are freed of the central B fragment which is destroyed by SstI, thus providing a negative selection against parental-type recombinants.

We estimated that one in several thousand of the purified fragments used for cloning would contain globin sequences. Although we originally used several methods for detecting positive clones, by far the most efficient was that recently described by Benton and Davis (1977). This technique allows the simple screening of literally tens of thousands of clones, provided a radioactive hybridization probe is available for their detection. Using this technique, we screened several plates in order to detect several positive clones as illustrated in Figure 4. These were then plaque-purified and preparatively grown for further characterization.

Characterization of the Globin-gene-containing Sequence

To determine whether the sequence cloned was related to an α- or β-globin sequence, positive hybrid

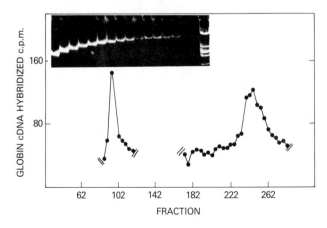

Figure 2. Detection of globin-containing fragments obtained from a preparative discontinuous electroelution gel. Approximately 2 mg of DNA from fractions 124–130 of the RPC-5 chromatography run (Fig. 1) were electrophoresed through a preparative 1% agarose discontinuous electroelution gel (Polsky et al. 1977). (●——●) S₁-resistant hybridization of pooled fractions to ³²P-labeled globin cDNA. The insert contains an ethidium bromide stain of an analytical 1% agarose gel containing 50 μl of individual fractions which correspond to the values at the bottom of the diagram. Size markers are provided by an EcoRI digest of λcI857 as in Fig. 1.

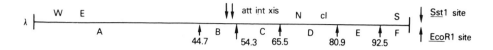

Figure 3. Modified λ phage suitable for cloning DNA from higher organisms. The lines represent the genome of phage λ. The length of the line drawn represents the full length of the genome of wild-type bacteriophage λ. Letters over each line refer to specific λ genes. Letters under each line refer to *Eco*RI restriction fragments of λ, with each arrow indicating an *Eco*RI site. The number under each arrow represents the position of the site as a percentage of the λ genome. Arrows over each line indicate the position of an *Sst*I site (note inversion of the *Eco*RI B fragment as compared to wild type). Scored boxes represent deleted portions of the λ genome. X represents the point at which an *Eco*RI site has been eliminated by mutation. Details of the construction of the two vectors have been described by Enquist et al. (1976) and Tiemeier et al. (1976).

phage were tested for their ability to anneal to plasmids containing the reverse transcript of either α or β mouse and rabbit globin sequences (Rougeon and Mach 1977; Efstratiadis et al. 1977). As shown in Figure 5, positive clones hybridized only to plasmids containing β-globin sequences. The congruity of hybridization (the match between the globin mRNA sequence and the cloned gene sequence) was further tested using resistance to deoxyribonuclease S_1 of a globin [32P]-labeled cDNA annealed to the cloned sequence (Table 1). The sequence contained in the clone was able to protect radioactively labeled globin cDNA to the same extent as the plasmid cloned sequence representing the reverse transcript of the β-globin messenger RNA.

A Surprising Structural Feature of the Cloned Globin Gene Sequence

An initial restriction map of the cloned sequence demonstrated that the approximately 7000-base-

long fragment of mouse DNA contained globin hybridizing sequences approximately 1.5 and 4.5 kilobases (kb) from either end of the fragment (Fig. 6). A restriction-site analysis of the cloned reverse transcript of the β-globin mRNA revealed a single *Bam*HI site in both mouse and rabbit globin sequences (Tilghman et al. 1977; Efstratiadis et al. 1977) corresponding to amino acids 98–100 of the β-globin protein. A single *Bam*HI site was also found to cleave the hybridizing β-globin sequence in this cloned segment of the genome. However, two additional enzymes, *Hind*III and *Sst*I, which *do not* cleave the cloned mRNA sequence, *do* split the globin hybridizing sequence, which implies that they occur within the globin coding region. There are several possible explanations for this puzzling observation. First, it is possible that the sequence cloned is different from that of the adult β-globin sequence,

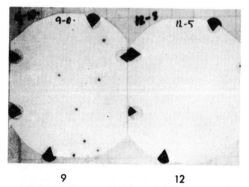

Figure 4. Identification of hybrid phage containing globin sequences. Plates containing 2000–3000 hybrid phage plaques were blotted and hybridized to [32P]-labeled globin cDNA essentially as described by Benton and Davis (1977). Plate 9 has eight positive phage; plate 12 has none.

PROBE

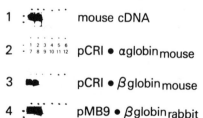

1   mouse cDNA

2   pCRI • αglobin$_{mouse}$

3   pCRI • βglobin$_{mouse}$

4   pMB9 • βglobin$_{rabbit}$

Figure 5. In situ hybridization of λgt*WES*·βG1 and λgt*WES*·βG2 to α- and β-containing chimeric plasmids. Replica filters were prepared from phage suspensions according to Kramer et al. (1976) and hybridized to the [32P]DNA probes as follows: (1) mouse α and β cDNA; (2) pcR1·mouse α globin; (3) pcR1·mouse β globin; (4) PMB9·rabbit β globin. The individual grids, numbered in filter 2, contained: (1) λgt*WES*·βG2; (2) λgt*WES*·βG1; (3–8) λgt*WES*·mouse DNA hybrids which were negative in the original screening; (9) λgt*WES*·mouse rDNA (Tiemeier et al. 1977); (10–12) λgt*WES*·mouse DNA hybrids which contain reiterated sequences.

Table 1. Hybridization of ^{32}P-labeled Globin cDNA to Hybrid Phage DNA

Hybrid DNA	Percentage hybridization (S_1-nuclease-resistant)
λgt*WES*·MβG1.0	45.3
λgt*WES*·MβG2.0	42.5
λgt*WES*·MC20 B4	4.2
λcI857	5.5
pCR1·α globin$_{mouse}$	43.1
pCR1·β globin$_{mouse}$	51.6

Hybridization of 0.1–0.2 µg of phage or plasmid DNA was performed in the presence of 1200 cpm ^{32}P-labeled globin cDNA as described by Tilghman et al. (1977).

for example, an embryonic sequence. Although this remains a possibility, another interpretation seems more likely in view of the electron microscopic data indicated below. Together, these data suggest that the structural globin gene sequence is *divided* by a segment of DNA that *does not* appear in the globin mRNA itself.

That this is indeed the case is indicated by an electron microscopic R-loop analysis (Thomas et al. 1976) of the type illustrated in Figure 7. Here we have taken mouse globin mRNA and annealed it to the cloned fragment. As shown in the figure, two R-loop structures may be visualized virtually adjacent to each other but separated by a looped-out, double-stranded region of DNA for which there is no displaced single-stranded DNA. This structure is seen in three independent isolates (nonsibling) of this 7-kb-long segment of the mouse genome, as well as in the separately cloned 15-kb fragment which also contains a mouse β-globin sequence (D. C. Tiemeier et al., in prep.). The correlation of

detailed electron microscopic analyses of these data together with that of the restriction map above suggests that the globin gene is divided into two coding segments which are separated by an intervening sequence of DNA approximately 550 bases long (Fig. 7) (S. Tilghman et al., in prep.).

Possible Interpretations of the New Finding

It is clear that the existence of the interrupted β-globin gene sequence can be interpreted in several ways. The possibility that this structure represents an artifact of the cloning procedure appears to be ruled out by the fact that three independently isolated, nonsibling clones contain the same R-loop structure. The possibility that this structure is a consequence of chromosomal DNA having been somehow transposed in the plasmacytoma from which the DNA was originally isolated appears less likely in view of our finding the same type of interrupted structure in a second β-globin gene encoded in the 15-kb segment of DNA which we have cloned (D. C. Tiemeier et al., in prep.).

None of the above possibilities allow us to conclude firmly that the sequence we have isolated is the same as that transcribed in globin-producing tissue. On the other hand, the agreed-upon length of a β-globin mRNA precursor, approximately 1600 bases (Curtis and Weissmann 1976; Ross 1976; Kwan et al. 1977; Bastos and Aviv 1977), is certainly long enough to accommodate the globin structural and intervening sequences. Although it is entirely possible that this intervening sequence plays some role in inactivating gene structures not destined to be expressed in a given cell line, this explanation requires more as-

Figure 6. Location of restriction endonuclease sites and globin sequences in λgt*WES*·MβG2. The top line represents the DNA genome of λgtWES·MβG2. The numbers refer to kilobase pairs from the left end of the hybrid. The 7000-base-long insert is illustrated as an open rectangle set off by vertical lines. The symbols in the left and right arms represent restriction enzyme sites: (◇) *Bam*HI; (↓) *Bgl*II; (↑) *Hind*III; (▽) *Hpa*I; (▼) *Sst*I. Note that there are additional *Hpa*I sites in the arms; only those directly adjacent to the insert are presented here. The insert is presented on an expanded scale in the second through the sixth lines. In this case, the numbers refer to the sizes of fragments, in kilobase pairs, generated by digestion with the restriction enzyme shown at the left. Fragments which hybridize to β-globin nucleic acid probes are represented as open rectangles (Tilghman et al. 1977). Fragments which do not hybridize are simply represented as horizontal lines. The hatched area within the insert displayed on the first and second lines represents an estimate of the region containing β-globin sequences based on these hybridization data and R-loop mapping in the electron microscope.

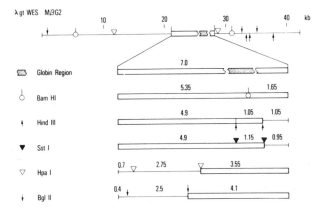

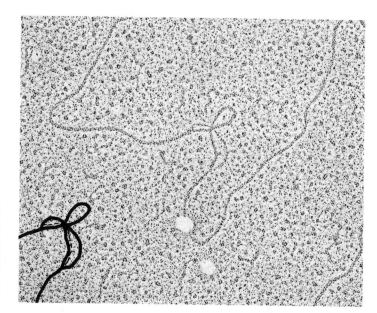

Figure 7. Electron microscopic representation of R loop formed between the cloned mouse globin sequence and mouse globin mRNA. The diagram represents double-stranded DNA (heavy line), displaced single-stranded DNA (solid fine line), and DNA-mRNA hybrid (heavy line and dashed fine line).

sumptions than appear warranted at this point. The greater likelihood seems that this sequence is transcribed into a precursor. If so, it raises the possibility that mRNA production occurs via a bighting-ligating mechanism in which the intervening sequence is clipped out of the RNA after transcription. Such a mechanism immediately raises new possibilities in interpreting physiologic and genetic data. First of all, it raises the possibility that a single polypeptide chain can be coded for by two (or more) distant structural gene sequences. It raises the further possibility that the processing reactions may play a role in the coherent assembly of genetic information. It also raises the possibility that such permutations may greatly amplify the genetic repertoire of a somatic cell. And finally, it offers a new way to think about the special problem of increasing the diversity of immunoglobulin genes and joining immunoglobulin constant- and variable-region gene sequences. Fortunately, the technology seems advanced to a state which should soon permit us to test these interesting possibilities.

Acknowledgments

We are most grateful to Ms. Catherine Kunkle who assisted us and is in large part responsible for the preparation of the manuscript. We are also grateful to Drs. Thomas Maniatis and Bernard Mach for providing plasmid clones used in these studies. We are also most grateful to Dr. Dolph Hatfield who provided us with the original sample of RPC-5 and for his patient advice.

REFERENCES

BASTOS, R. N. and H. AVIV. 1977. Globin RNA precursor molecules: Biosynthesis and processing in erythroid cells. *Cell* **11**: 641.

BENTON, W. D. and R. W. DAVIS. 1977. Screening λgt recombinant clones by hybridization to single plaques *in situ*. *Science* **196**: 180.

CURTIS, P. J. and C. WEISSMANN. 1976. Purification of globin messenger RNA from dimethylsulfoxide-induced Friend cells and detection of a putative globin messenger RNA precursor. *J. Mol. Biol.* **106**: 1061.

EFSTRATIADIS, A., F.C. KAFATOS, and T. MANIATIS. 1977. The primary structure of rabbit β-globin mRNA as determined from cloned DNA. *Cell* **10**: 571.

ENQUIST, L., D. TIEMEIER, P. LEDER, R. WEISBERG, and N. STERNBERG. 1976. Safer derivatives of bacteriophage λgt·λC for use in cloning of recombinant DNA molecules. *Nature* **259**: 596.

HARDIES, S. C. and R. D. WELLS. 1976. Preparative fractionation of DNA restriction fragments by reversed phase column chromatography. *Proc. Natl. Acad. Sci.* **73**: 3117.

KRAMER, R. A., J. R. CAMERON, and R. W. DAVIS. 1976. Isolation of bacteriophage λ containing yeast ribosomal RNA genes: Screening by *in situ* RNA hybridization to plaques. *Cell* **8**: 227.

KWAN, S.-P., T. G. WOOD, and J. B. LINGREL. 1977. Purification of a putative precursor of globin messenger RNA from mouse nucleated erythroid cell. *Proc. Natl. Acad. Sci.* **74**: 178.

LANDY, A., C. FOELLER, R. RESZELBACH, and B. DUDOCK. 1976. Preparative fractionation of DNA restriction fragments by high pressure column chromatography on RPC-5. *Nucleic Acids Res.* **3**: 2575.

LEDER, P., D. TIEMEIER, and L. ENQUIST. 1977. EK2 derivative of bacteriophage lambda useful in the cloning of DNA from higher organisms: The λgt*WES* system. *Science* **196**: 175.

PEARSON, R. L., J. F. WEISS, and A. D. KELMERS. 1971.

Improved separation of transfer RNA's on polychloro-triflouorethylene-supported reversed-phase chromatography columns. *Biochim. Biophys. Acta* **228**: 770.

POLSKY, F., M. H. EDGELL, J. G. SEIDMAN, and P. LEDER. 1977. High capacity gel electrophoresis for purification of fragments of genomic DNA. *Anal. Biochem.* (in press).

ROSS, J. 1976. A precursor of globin messenger RNA. *J. Mol. Biol.* **106**: 403.

ROUGEON, F. and B. MACH. 1977. Cloning and amplification of α and β mouse globin gene sequences synthesized *in vitro. Gene* **1**: 229.

THOMAS, M., J. R. CAMERON, and R. W. DAVIS. 1974. Viable molecular hybrids of bacteriophage lambda and eukaryotic DNA. *Proc. Natl. Acad. Sci.* **71**: 4579.

THOMAS, M., R. L. WHITE, and R. W. DAVIS. 1976. Hybridization of RNA to double-stranded DNA: Formation of R-loops. *Proc. Natl. Acad. Sci.* **73**: 2294.

TIEMEIER, D., L. ENQUIST, and P. LEDER. 1976. An improved derivative of a bacteriophage λ EK2 vector useful in the cloning of recombinant DNA molecules: λgt*WES* ·λB. *Nature* **263**: 526.

TIEMEIER, D. C., S. M. TILGHMAN, and P. LEDER. 1977. Purification and cloning of a mouse ribosomal gene fragment in phage λ. *Gene* (in press).

TILGHMAN, S., D. TIEMEIER, F. POLSKY, M. EDGELL, J. SEIDMAN, P. LEDER, L. ENQUIST, B. NORMAN, and A. LEDER. 1977. Cloning specific segments of the mammalian genome: Bacteriophage λ containing mouse globin and surrounding gene sequences. *Proc. Natl. Acad. Sci.* **74**: 4406.

31

Reprinted from *Natl. Acad. Sci. (USA) Proc.* **76**:4464–4468 (1979)

Cloning of integrated Moloney sarcoma proviral DNA sequences in bacteriophage λ

(recombinant DNA/retrovirus integration)

G. F. Vande Woude*, M. Oskarsson*, L. W. Enquist*, S. Nomura[†], M. Sullivan[‡], and P. J. Fischinger[†]

Laboratories of *Molecular Virology and [†]Viral Carcinogenesis, National Cancer Institute; and [‡]Laboratory of Molecular Genetics, National Institute of Child Health and Human Development, Bethesda, Maryland 20205

ABSTRACT　　We have identified integrated proviral DNA sequences of m1 and HT-1 isolates of Moloney sarcoma virus (MuSV) in *Eco*RI digests of transformed mink cell genomic DNA and have cloned these fragments in bacteriophage λ. Both the λ-HT1 phage recombinant, containing a 12.3-kilobase MuSV pair (kb) fragment, and the λ-m1 phage recombinant, containing a 7.0-kb fragment, possess full copies of the sarcoma viruses along with 5′ and 3′ host flanking sequences. The MuSV proviral DNA sequences, 6.7 kb for HT-1 and 5.2 kb for m1, are colinear by heteroduplex microscopy with the 1.5-kb difference in size accounted for by two ≈0.8-kb deleted regions in m1. Both integrated viral genomes are terminally redundant and have integrated at the same site in the provirus but at different sites on the host chromosome. The host sequences flanking integrated HT-1 MuSV have been identified as a single *Eco*RI restriction fragment of 5.6 kb in normal mink cells.

We have combined restriction endonuclease digestion, nucleic acid hybridization, and recombinant DNA methods to identify and isolate cellular DNA fragments carrying integrated retroviral DNA sequences. Specifically, we have studied two variants of the defective transforming Moloney murine sarcoma virus (MuSV) called m1 (1–3) and HT-1 (4). Although these variants were originally derived from a tumor-inducing uncloned MuSV stock (5), they have been maintained separately for more than 10 years in different hosts and cells. Both will transform cells, but they differ in their biochemical properties (6–8). About three-fourths of the MuSV genome is homologous to the genome of Moloney murine leukemia virus (M-MuLV); the remaining sequences are single-copy normal mouse cellular DNA commonly called sarcoma (*src*) sequences (6, 9, 10). In MuSV-transformed cells, the chromosomal arrangement of MuSV integrated sequences as well as the nature of the DNA surrounding the viral sequences is not well understood. Because the mouse genome contains multiple endogenous integrated proviruses with homology to MuLV (11), we used MuSV-transformed mink lung cells to simplify the identification of proviral sequences in transformed cells. Because there are no *Eco*RI restriction sites in DNA or provirus derived from parental M-MuLV and at least one MuSV isolate (12–14), we expected to obtain *Eco*RI fragments carrying the entire integrated provirus of each MuSV variant as well as flanking cellular DNA sequences. In this report, we describe the identification and cloning in the EK-2 vector λgtWES·λB of *Eco*RI fragments carrying the entire integrated MuSV genome derived from cellular DNA of m1- or HT-1-transformed mink lung cells.

MATERIALS AND METHODS

Cells and Viruses. Mink lung cells (American Type Culture Collection Mv 1 HC1 CCL64) were infected with feline xenotropic virus pseudotype of m1 MuSV with a 10-fold biological MuSV excess of focus-forming units (FFU) over helper virus (1, 8) and immediately cloned in microwells. Cells from a single nonproducer clone (rescuable with mink cell tropic helper viruses) were positive for the m1 MuSV-specific pP60[gag] (8). A subclone (MIMS-103), maintained for more than 1 year, was used for DNA isolation.

HT-1 MuSV-transformed mink cells were obtained by co-cultivating nonproducer HT-1 MuSV-transformed hamster tumor cells (4) with feline embryo fibroblasts (FEF) and infecting the mixture with feline leukemia virus (FeLV). Supernatant HT-1 MuSV(FeLV) pseudotype was used to infect FEF cells three times successively and then finally to infect mink cells at a multiplicity of infection of 0.5. The resulting mass transformed culture (HTMF) became nonproducing within several passages (≤10 FFU of FeLV per 10[7] cells) but was rescuable with FeLV (≥10[5] FFU). HTMF cells were negative for M-MuLV p30 and gp70 expression.

P323 virus, the FeLV pseudotype of the m3MuSV isolate (1, 3, 8), is related to m1 MuSV by hybridization (9). M-MuLV 1869, from J. Hartley, is free of polytropic recombinants (unpublished data).

M-MuLV and MuSV cDNA Probe Preparation and Conditions for Hybridization. cDNA was prepared in endogenous reverse transcription reactions with either M-MuLV (1869) or m3MuSV(FeLV) from P323 cells (8) by using actinomycin D (15). Specific activities were ≈1.3 × 10[7] cpm/μg for [³H]cDNA and ≈3 × 10[8] cpm/μg for [³²P]cDNA. Hybridization with viral RNA and protection of labeled viral RNA from RNase T1 indicated complete and stoichiometric transcription with $C_r t_{1/2}$ values of 0.06 mol sec/liter and 75–95% protection at cDNA/RNA molar ratios of 1 and 4, respectively (16).

Purification and Restriction Endonuclease Cleavage of Cellular DNA. Normal or MuSV-transformed mink cells were lysed in buffer A (0.5% Nonidet P40/3.6 mM calcium chloride/30 mM Tris·HCl, pH 7.5/5 mM magnesium acetate/125 mM potassium chloride/0.5 mM EDTA/0.25% sodium desoxycholate/1 mM *p*-chloromercuricbenzoate) at 60°C, and nuclei were collected by low-speed centrifugation. After two 60°C washes with buffer A, purified nuclei were incubated first

Abbreviations: MuSV, Moloney murine sarcoma virus; M-MuLV, Moloney murine leukemia virus; FFU, focus-forming units; FEF, feline embryo fibroblasts; FeLV, feline leukemia virus; HTMF, mink lung cells transformed by HT-1 MSV; kb, kilobase(s); RPC-5 chromatography, high-pressure liquid ion-exchange chromatography on RPC-5 column bed.

at 37°C for 30 min in 2 vol of buffer A containing RNase (50 μg/ml) and *Eco*RI (500 units/ml) and then at 55°C (1–3 hr) with 3 vol of K buffer (0.02 M Tris, pH 7.8/0.05 M EDTA/1% sodium dodecyl sulfate/100 μg of proteinase K per ml). This digest was extracted sequentially with equal volumes of phenol, phenol/chloroform (1:1, vol/vol), and chloroform and was dialyzed against 20 mM Tris, pH 7.8/0.2 mM EDTA. The final yield from 20 roller cultures was ≈ 20–30 mg of DNA.

Enrichment of DNA Fragments Containing MuSV Sequences. MuSV-specific sequences were enriched from total mink cell DNA by using procedures described by Tilghman *et al.* (17) and Tiemeier *et al.* (18). Purified mink cell DNA was digested to completion with *Eco*RI and sequentially subjected to chromatography on an RPC-5 column bed with high-pressure liquid ion-exchange conditions (RPC-5 chromatography), gel electrophoresis, Southern transfer (19), and hybridization with [^{32}P]cDNA$_{M-MuLV}$ probe prepared as described above. The final enrichment of DNA fragments containing MuSV ranged from 500- to 1000-fold.

Construction of Hybrid Phage and Isolation of λ MuSV Hybrids. Mink cell *Eco*RI fragments enriched for MuSV were ligated to the large *Eco*RI fragments of λgt WES·λB (17) and packaged into phage particles *in vitro* (20) (yield: 10^6 plaque-forming units/μg vector DNA). Of ≈5000 plaques screened (18) with cDNA$_{M-MuLV}$ probe, 5 were positive from HTMF DNA and one was positive from MIMS DNA. Initial isolates were purified twice on LE392 (17); high-titer stocks were prepared in *Escherichia coli* DP50*sup*F (21). All five independent isolates from HTMF DNA were identical by restriction enzyme analysis.

Purification of Hybrid λ Phage DNA. λ hybrid phage propagated in DP50*sup*F was concentrated by 10% polyethylene glycol precipitation, resuspended in buffer A without the chloromercuribenzoate but containing 6 mM 2-mercaptoethanol, DNase I (20 μg/ml), and RNase A (20 μg/ml), and incubated at 30°C for 30 min. This digest was extracted with 0.1 vol of Freon and sedimented at 100,000 × g for 1 hr through a step gradient of glycerol (2 vol 5% over 1 vol 45%) in buffer B (0.5% Nonidet P40/30 mM Tris·HCl, pH 7.5/125 mM potassium chloride/0.5 mM EDTA/6 mM 2-mercaptoethanol). The phage pellet was dissolved in buffer B and digested in K buffer, and the DNA was extracted as described above.

Physical and Biological Containment. This work was initiated with P4-EK2 containments as dictated by the 1976 guidelines; conditions were reduced to P2-EK2 as required in the revised guidelines.

RESULTS

Identification and Cloning of Cell DNA Fragments Containing Integrated MuSV DNA. DNA from mink lung cells transformed by m1 (MIMS) or HT-1 (HTMF) variants of MuSV was cleaved with *Eco*RI. We first fractionated the fragments by RPC-5 chromatography and then used size fractionation in agarose gels to facilitate identification of *Eco*RI fragments that hybridized to cDNA$_{M-MuLV}$ probe. Such a two-dimensional display of the MuSV-transformed mink genome is shown in Fig. 1. A single region of strong cDNA$_{M-MuLV}$ hybridization was observed in HTMF [12.3 kilobases (kb) pairs] and MIMS (7.0 kb) cell DNA. In both cases the hybridizing fragments eluted early from the column, well before the bulk of cellular *Eco*RI fragments. The additional faint bands observed varied with different probe preparations as in Fig. 1 *B* and *C*; this was a consequence of contamination of our probe with other mouse RNA species present in the M-MuLV preparations. We have never found strong hybridization between viral cDNA and normal mink lung cell DNA in a similar analysis. After preparative gel electrophoresis, MuSV-enriched fragments were

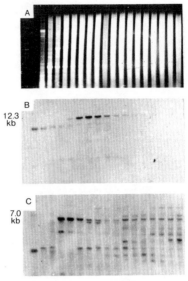

FIG. 1. Detection of MuSV proviral DNA sequences. MuSV-transformed mink cell DNA was digested with *Eco*RI and fractionated by RPC-5 chromatography; fractions were electrophoretically separated on 0.7% agarose gels (17, 18). Only the first half of the column fractions are presented. (*A*) Ethidium bromide-stained gel of HTMF cell DNA fractions. (*B*) Autoradiogram of HTMF DNA of *A* transferred to nitrocellulose paper (19) and hybridized with ^{32}P-labeled cDNA$_{M-MuLV}$. (*C*) Autoradiogram of MIMS cell DNA fractionated as in *A*, transferred, and hybridized with cDNA$_{M-MuLV}$ as in *B*.

inserted in the EK2 vector λgt WES·λB. The hybrid phages so isolated are referred to here as λ-m1 and λ-HT-1 and carry the 7.0-kb and 12.3-kb MSV transformed mink cell *Eco*RI fragments, respectively.

Characterization of the *Eco*RI Fragments Containing MuSV Sequences. The initial characterization of the hybrid phages included hybridization with M-MuLV and MuSV cDNA probes, heteroduplex formation, and restriction endonuclease digestions. By hybridization, both the m1 and HT-1 MuSV variants have been shown to contain sequences of M-MuLV (6, 7, 9). The hybridization of M-MuLV and MuSV cDNA probe to λ-m1 or λ-HT-1 DNA is summarized in Table 1. About 35 and 74% of the cDNA$_{M-MuLV}$ probe hybridized to λ-m1 and λ-HT-1, respectively. The latter hybridization indicates that λ-HT-1 contains more M-MuLV-related sequences (see below). As an additional control, we included a cDNA probe derived from another MuSV isolate, m3 (9). This probe contains the single-copy murine DNA *src* sequences in addition to the M-MuLV sequences present in MuSV. In this case, both λ-m1 and λ-HT-1 hybridized to about the same high extent. We conclude that the λ hybrids contain DNA fragments with MuSV-specific sequences.

We next compared the two MuSV clones by heteroduplex analysis: purified cloned *Eco*RI fragments from λ-m1 (7.0 kb) and λ-HT-1 (12.3 kb) were mixed, denatured, and reannealed and the resulting heteroduplexes were analyzed in the electron microscope. The hybridization experiments in Table 1 predicted homoduplex formation between m1 and HT-1 MuSV sequences but we were unsure as to the similarity, if any, of the cell DNA sequences flanking the integrated MuSV genomes. A typical heteroduplex between cell DNA fragments carrying m1 and HT-1 integrated sequences is shown in Fig. 2. The extensive homoduplex formation interrupted by two single-

Table 1.   Hybridization of MuSV or M-MuLV [³H]cDNA to λ-m1 and λ-HT-1 DNA

Nucleic acid	% hybridization with cDNA probe	
	[³H]MuSV	[³H]M-MuLV
λ-m1	75*	35
λ-HT-1	70	74
HMW virion RNA (MuSV)	100†	39
HMW virion RNA (M-MuLV)	81	100
Calf thymus DNA	0	0

Each hybridization mixture contained in 0.05 ml: 0.01 M Tris·HCl, pH 7.4; 0.75 M NaCl; 2 mM EDTA; 0.05% sodium dodecyl sulfate; 2.5 μg of calf thymus DNA; 1100 or 1400 trichloroacetic acid-insoluble cpm of cDNA$_{M\text{-}MuSV}$ or cDNA$_{M\text{-}MuLV}$, respectively; and 0.2 μg of sonicated hybrid phage DNA. The reaction mixtures were heated at 95°C for 5 min, quenched at 4°C, incubated at 66°C, and digested with S1 nuclease (15). Computations include background correction of 4.8% cDNA$_{MuSV}$ and 5.0% cDNA$_{M\text{-}MuLV}$. Sheared calf thymus DNA and high molecular weight m3-MuSV (P323) or M-MuLV (Moloney 1869) RNA served as negative and positive controls.
* Values were normalized to the homologous nucleic acid.
† [³H]cDNA$_{MuSV}$ was prepared from m3MuSV(FeLV) and contained cDNA homologous to the helper virus RNA. Also, the m3MuSV may contain additional sequences in common with MuLV not present in either m1 or HT-1MuSV (1, 8, 9).

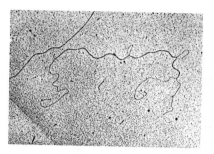

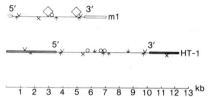

FIG. 2.   Heteroduplex and restriction map analysis of λ-m1 and λ-HT-1. (*Top*) Heteroduplex between purified 7.0-kb λ-m1 and 12.3 kb λ-HT-1 *Eco*RI fragments. Fragments were mixed, denatured, neutralized, reannealed, and spread as described by Tiemeier *et al.* (18). (*Center*) Diagram of heteroduplex. HT-1 MuSV regions deleted in m1 are indicated by the dashed line loops. The nonhomologous single-stranded DNA ends represent the 5′ and 3′ host flanking sequences. Length is shown in *Bottom*. (*Bottom*) Physical maps of integrated m1 and HT-1 MuSV sequences cloned in λ. The diamond symbol shows approximate locations of HT-1 sequences deleted in m1. The adjacent nonhomologous host flanking sequences of the cloned *Eco*RI fragments are represented by open bars (m1) and hatched bars (HT-1). Restriction sites: ○, *Bgl* II; ×, *Xho* I; ↑, *Xba* I; ∨, *Sac* I; ↓, *Hind*III.

stranded loops should localize the MuSV-specific sequences. We suggest that the single-stranded tails protruding from each end of the MuSV sequences define nonhomologous flanking cellular sequences. This analysis and restriction endonuclease digestion (see below) indicates that both m1 and HT-1 are integrated at the same site on the MuSV provirus but at different sites in the host chromosome. By measuring the lengths of the single- and double-stranded regions and by knowing the sizes of the m1 and HT-1 cloned *Eco*RI fragments, we interpreted the heteroduplex as depicted in Fig. 2 *bottom*. The m1 fragment contains 5.2 kb of MuSV sequences flanked by about 0.3 and 1.5 kb of cellular DNA. The HT-1 fragment carries 6.7 kb of MuSV sequences including 0.7- and 0.8-kb segments not present in λ-m1. The 6.7-kb HT-1 MuSV sequences are flanked by 3.3 and 2.3 kb of cellular sequences.

In experiments to be presented elsewhere, we hybridized M-MuLV virion RNA to λ-m1 and λ-HT-1 DNA and were able to orient the MuSV DNA sequences with respect to the 3′ and 5′ ends of virion RNA. These studies showed that the cloned fragments do indeed carry the entire genome of m1 and HT-1, that the integrated sequences are collinear with the virion RNA, and that the presumptive flanking cellular sequences do not hybridize to virion RNA. In addition, both cloned fragments transform cells after transfection and are rescuable with helper leukemia virus (D. Blair, personal communication).

Results of analysis of λ-m1 and λ-HT-1 by cleavage with various restriction enzymes support our conclusion that both MuSV fragments have integrated at the same region in the viral DNA but into different cellular DNA sequences. We constructed a physical map of the cloned MuSV fragments, by using heteroduplex and restriction enzyme data, indicating the flanking cell DNA sequences and the regions of homology of m1 and HT-1 (Fig. 2). The pairs of *Sac* I and *Xba* I sites found in both cloned fragments were of particular interest. These sites are near what we define from heteroduplex analysis as the cell DNA–proviral DNA junction sequences and are suggestive of a direct repeated sequence at or near each end of the integrated viruses. Such direct repeats have been demonstrated in unintegrated as well as integrated avian retroviruses (22–24) and in M-MuLV DNA synthesized *in vitro* (25). In the latter study,

the M-MuLV terminal repeated sequences were direct, about 0.6 kb long, and contained an *Sac* I and an *Xba* I site about 100 base pairs apart. We find the *Xba* I/*Sac* I fragments from both λ-m1 and λ-HT-1 to be about 120 base pairs long. The fine-structure restriction map of these clones will be documented elsewhere. There is significant correlation between the map deduced by Gilboa *et al.* (25) and the M-MuLV portions of both λ-MuSV clones.

Direct repeated sequences should be substrates for the *E. coli* general recombination system (26). Homologous intramolecular recombination would delete the MuSV DNA between the repeated sequences, leaving one repeat between the flanking cellular sequences. Indeed, this process occurs during growth of λ-m1 and λ-HT-1 in the *rec*⁺ *E. coli* host. We noted, in addition to the major *Eco*RI fragment cloned in each λ, minor *Eco*RI fragments at 2 and 6 kb for λ-m1 and λ-HT-1, respectively. These minor bands are the sizes expected if the MuSV genome had been excised from the major fragments. The minor bands also hybridized with cDNA$_{M\text{-}MuLV}$, indicating the presence of some viral sequences. Phages carrying these deleted fragments were isolated by CsCl gradient centrifugation or EDTA treatment of lysates (27). Subsequent analysis by restriction endonuclease digestion and heteroduplex formation indicated that they contained a single MuSV terminal repeat between both cellular flanking sequences (unpublished data). Although we cannot rule out other mechanisms for such site-specific recombination from these experiments, we suggest that both λ-m1 and λ-HT1 carry, at each end of the complete MSV sequences, short, direct repeats of MuSV DNA that may be substrates for general recombination.

Identification of Cellular *Eco*RI Fragments Containing the MuSV Integration Site. To begin analysis of the MuSV integration site, we looked for an *Eco*RI fragment in normal mink cells that would hybridize with the mink host portions of a nick-translated λ-HT-1 probe (Fig. 3A). Strong hybridization was observed only with a 5.6-kb fragment eluting early from the column. This is consistent with our estimate that 5.6 kb of the 12.3-kb λ HT-1 represented host sequences. With fractionated HTMF cell DNA, the 12.3-kb fragment, representing the integrated MuSV genome, hybridized intensely (Fig. 3B), but we again observed hybridization to a 5.6-kb fragment similar to that observed in normal mink DNA. This fragment was not detected with the cDNA$_{M-MuLV}$ probe (Fig. 1B). If we assume that the 5.6-kb fragments of both the normal and the transformed cells are the same, this fragment could contain the

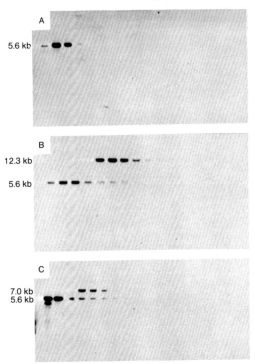

FIG. 3. Identification of the HT1 MuSV integration site. Cell DNA was fractionated and hybridized as in Fig. 1 except that the hybridization probe was prepared by nick-translation (28) of λ-HT-1 DNA (5×10^7 cpm/μg). Only the first half of RPC-5 chromatography column fractions are presented. No hybridization was observed in the later fractions. (A) Normal mink lung cell DNA, showing hybridization with a 5.6-kb *Eco*RI fragment. (B) HTMF DNA showing hybridization with 12.3-kb fragment containing the HT-1 provirus (cf. Fig. 1B) and 5.6-kb sequence as in A above. (C) MIMS DNA showing hybridization with the 7.0-kb fragment containing the m1 provirus (cf. Fig. 1C) and the 5.6-kb sequences as in A. The additional 5.4-kb fragment hybridizing in the MIMS DNA is unexplained.

HT-1 integration site. It is not clear why this fragment is still present in HTMF cells. One trivial explanation could be that the uncloned HTMF cells contain a population of normal cells. However, its appearance could also represent sites where viral DNA had been lost through homologous recombination similar to what we observed in *E. coli*. Alternatively, if these cells are diploid for the chromosome carrying the normal fragment, MuSV could be integrated into one homolog and not the other. The presence of mink related cell *src* sequences in the 5.6-kb fragment can be excluded because it did not hybridize with the λ-m1 probe (not shown).

A strong indication that the flanking cellular sequences for integrated m1 and HT-1 are different comes from a reciprocal hybridization experiment (Fig. 3C). Here, nick-translated λ-HT-1 was hybridized with MIMS DNA fractionated by RPC-5 chromatography. The 7-kb band carrying the m1 MuSV genome was readily apparent, but no detectable hybridization was discernible in the 12.3-kb region where integrated HT-1 MuSV was found in HTMF DNA. Predictably, the 5.6-kb fragment defining the HT-1 integration sequences was again detected in the MIMS cell DNA. The hybridization with the 5.4-kb fragment in this DNA is not currently understood. When the experiments described in Fig. 3 were performed with λ-m1 probe, we observed no hybridization with the 5.6-kb fragment (not shown). Thus, as indicated in the heteroduplex analysis (Fig. 2), the flanking host sequences, and therefore the integration sites for m1 and HT-1 MuSV, are nonhomologous.

The virtual absence of any other regions of hybridization when the cloned HT-1 fragment is used as probe underscores the unique nature of the fragment in the transformed genome. The many faint bands seen with the cDNA$_{M-MuLV}$ probe (see Fig. 1) were absent when nick-translated cloned DNA was used as probe. The 12.3-kb (HTMF) and 7.0-kb (MIMS) fragments seem to contain the only detectable MuSV sequences in the transformed mink lung cells.

We conclude from these data that HT-1 and m1 MuSV integrated into different regions of the mink genome and, for HT-1, integration may have been a simple reciprocal recombination event.

DISCUSSION

We have isolated the integrated proviral DNA forms of two defective transforming murine retrovirus genomes by cloning fragments of transformed cell DNA in phage λ. These methods present powerful tools for further analysis of retrovirus molecular biology. For example, we show here how one can compare viral genomes in cases in which it could not be done easily before by conventional methods. In addition, we demonstrate that one can determine unique sites of integration in the virus and in the cellular DNA.

Because they are defective, MuSV virions can be produced in quantity only under certain conditions. For example, only MuSV 124, which is produced in large excess over the helper virus, has previously been physically mapped in any detail (12, 29, 30). Our heteroduplex analysis (Fig. 2) represents the direct comparison of two MuSV isolates. It is of interest that, even though these defective genomes have been propagated separately for years in the presence of unrelated helper viruses and in heterologous hosts, they have maintained substantial homology both in DNA sequence and in genome organization. We might have expected more divergence because the only consistent selective pressures were for cellular transformation and ability to grow with helper virus. It has been demonstrated previously that, although m1 can express the pP60^{gag} polyprotein (8), the HT-1 variant can not. We were surprised to find the HT-1 genome to be about 1.5 kb larger than m1 and to show homology in the region predicted to encode the P60 product.

Further work is needed to understand why no immunologically reactive product seems to be made from this virus.

Several features of the integrated MuSV variants are noteworthy. First, in the whole complex array of *Eco*RI fragments derived from the transformed cell genome, only one *Eco*RI fragment contained MuSV sequences in either m1- or HT-1 transformed mink lung cells. Second, each viral DNA had been inserted by using what appears to be the same region of the proviral DNA. Although not documented here, both integrated MuSV sequences are collinear with virion RNA (unpublished data). Third, the flanking cellular sequences of each integrated MuSV were quite dissimilar as determined by heteroduplex analysis and by hybridization. Finally, each integrated genome was bracketed by a short direct repeat of MuSV sequences.

Convincing evidence exists for both multiple and preferred sites for avian and mammalian retrovirus integration (24, 31–35). For the MuSV variants used here, at least two distinct sites in the mink genome can be identified, and only direct sequence analysis will reveal local similarity in the actual integration sites. It is important to note that the m1-transformed mink cells were isolated from a single focus, and progeny cells were repeatedly cloned before we extracted DNA. The presence of a single m1 MuSV copy might not be surprising. On the other hand, the HTMF cells were obtained by mass infection, with no cell cloning. After about 20 weekly cell passages with selection for transformed cells, the MuSV specific sequences were found only in the 12.3-kb fragment. It is likely that other MuSV integration sites were available but were used infrequently, caused slower growth, or rendered the cell inviable. Also, the provirus may rapidly excise from other integration sites, possibly through selective pressure of virus or host.

The two forms of MuSV studied here have integrated by using the same regions of the viral genome. A significant feature of this region is the direct repeat found bracketing the integrated proviral DNA. The presence of direct repeats implies that excision could be expected to occur by homologous recombination. One might expect MuSV-transformed cells to yield a class of nontransformed revertants with only the repeat sequence remaining. The direct repeats of integrated proviral MuSV DNA could be generated during integration or during synthesis. One example of integration could occur by homologous recombination of a circular MuSV DNA genome bearing a single copy of the repeat with an identical repeat on the host genome. We can exclude this model because we have cloned in phage λ the 5.6-kb normal mink fragment (Fig. 3) representing the HT-1 MuSV host flanking sequences and it does not contain the 600 bp repeat (unpublished data). Another possibility would involve the presence of *both* direct repeats on the integrating MuSV DNA molecule so that integration occurs at the joint between the repeats. The host DNA site then need not carry the repeated sequences and the recombination process need not involve homologous recombination. Both forms of recombinantion are site specific with respect to the viral genome. The mechanisms to promote such site specific recombination is not known and could involve both virus- and host-specified elements. Circular forms of avian retrovirus DNA carrying one or two repeating sequences have been described (22, 23) and it is likely that the 600-base-pair direct repeat observed by Gilboa *et al.* (25) for M-MuLV DNA synthesized *in vitro* is present in the 9-kb circular proviral DNA molecules found *in vivo* (36).

Note Added in Proof. Using a specific "src" probe derived from λ-HT-1, we have isolated from normal mouse cell DNA a 140-kb *Eco*R1 fragment containing a 1.0-kb region colinear by heteroduplex mapping and restriction endonuclease analyses to MuSV "src." The fragment bears no homology to the M-MuLV portion of MuSV and does not transform cells in culture.

We thank D. C. Tiemeier, W. L. McClements, and J. V. Maizel for helpful discussions and suggestions during the course of work.

1. Fischinger, P. J., Nomura, S., Peebles, P. T., Haapala, D. K. & Bassin, R. H. (1972) *Science* **176**, 1033–1035.
2. Fischinger, P. J., Nomura, S., Tuttle-Fuller, N. & Dunn, J. (1974) *Virology* **59**, 217–229.
3. Oskarsson, M. K., Elder, J. H., Gautsch, J. W., Lerner, R. A. & Vande Woude, G. F. (1978) *Proc. Natl. Acad. Sci. USA* **75**, 4694–4698.
4. Huebner, R. J., Hartley, J. & Rowe, W. (1966) *Proc. Natl. Acad. Sci. USA* **56**, 1164–1169.
5. Moloney, J. B. (1966) *Natl. Cancer Inst. Monogr.* **22**, 139–142.
6. Scolnick, E. M., Howk, R. S., Anisowicz, A., Peebles, P. T., Scher, C. D. & Parks, W. P. (1975) *Proc. Natl. Acad. Sci. USA* **72**, 4650–4654.
7. Parks, W. P., Howk, R. S., Anisowicz, A. & Scolnick, E. M. (1976) *J. Virol.* **18**, 491–503.
8. Robey, W. G., Oskarsson, M. K., Vande Woude, G. F., Naso, R. B., Arlinghaus, R. B., Haapala, D. K. & Fischinger, P. J. (1977) *Cell* **10**, 79–89.
9. Frankel, A. E. & Fischinger, P. J. (1976) *Proc. Natl. Acad. Sci. USA* **73**, 3705–3709.
10. Frankel, A. E. & Fischinger, P. J. (1977) *J. Virol.* **21**, 153–160.
11. Chattopadhyay, S. K., Lowy, D. R., Teich, N. M., Levine, A. S. & Rowe, W. P. (1974) *Cold Spring Harbor Symp. Quant. Biol.* **39**, 1085–1101.
12. Canaani, E., Duesberg, P. & Dina, D. (1977) *Proc. Natl. Acad. Sci. USA* **74**, 29–33.
13. Verma, I. M. & McKennett, M. A. (1978) *J. Virol.* **26**, 630–645.
14. Smotkin, D., Gianni, A. M., Rosenblatt, S. & Weinberg, R. A. (1975) *Proc. Natl. Acad. Sci. USA* **72**, 4910–4913.
15. Benveniste, R. E. & Scolnick, E. M. (1973) *Virology* **51**, 370–382.
16. Nomura, S. (1978) *Virology* **91**, 444–452.
17. Tilghman, S. M., Tiemeier, D. C., Polsky, F., Edgell, M. H., Seidman, J. G., Leder, A., Enquist, L. W., Norman, B. & Leder, P. (1977) *Proc. Natl. Acad. Sci. USA* **74**, 4406–4410.
18. Tiemeier, D. C., Tilghman, S. M., Polsky, F., Seidman, J. G., Leder, A., Edgell, M. H. & Leder, P. (1978) *Cell* **14**, 237–245.
19. Southern, E. M. (1975) *J. Mol. Biol.* **98**, 503–517.
20. Enquist, L. W. & Sternberg, N. (1979) *Methods Enzymol.*, in press.
21. Leder, P., Tiemeier, D. & Enquist, L. (1977) *Science* **196**, 175–177.
22. Taylor, J. M., Hsu, T. W. & Lai, M. M. C. (1978) *J. Virol.* **26**, 479–484.
23. Shank, P. R., Hughes, S., Kung, H. J., Majors, J. E., Quintell, N., Guntaka, R. V., Bishop, J. M. & Varmus, H. E. (1978) *Cell* **15**, 1383–1395.
24. Hughes, S. H., Shank, P. R., Spector, D. H., Kung, H. J., Bishop, J. M., Varmus, H. E., Vogt, P. K. & Breitman, M. L. (1978) *Cell* **15**, 1397–1410.
25. Gilboa, E., Goff, S., Shields, A., Yoshimura, F., Mitra, S. & Baltimore, D. (1979) *Cell* **16**, 863–874.
26. Bellet, A. J. D., Busse, H. G. & Baldwin, R. L. (1971) in *The Bacteriophage Lambda*, ed. Hershey, A. D. (Cold Spring Harbor Laboratory, Cold Spring Harbor, NY), pp. 501–513.
27. Nash, H. A. (1974) *Virology* **57**, 207–216.
28. Maniatis, T., Jeffrey, A. & Kleid, D. G. (1975) *Proc. Natl. Acad. Sci. USA* **72**, 1184–1188.
29. Andersson, P., Goldfarb, M. P. & Weinberg, R. A. (1979) *Cell* **16**, 63–75.
30. Hu, S., Davidson, N. & Verma, I. M. (1977) *Cell* **10**, 469–477.
31. Keshet, E. & Temin, H. M. (1978) *Proc. Natl. Acad. Sci. USA* **74**, 3372–3376.
32. Collins, C. J. & Parsons, J. T. (1977) *Proc. Natl. Acad. Sci. USA* **74**, 4301–4305.
33. Battula, N. & Temin, H. M. (1978) *Cell* **13**, 387–396.
34. Lemons, R. S., Nash, W. G., O'Brien, S. J., Benveniste, R. E. & Sherr, C. J. (1978) *Cell* **14**, 995–1005.
35. Steffen, D. & Weinberg, R. A. (1978) *Cell* **15**, 1003–1010.
36. Yoshimura, F. K. & Weinberg, R. A. (1979) *Cell* **16**, 323–332.

32

The Construction and Use of Hybrid Plasmid Gene Banks in *Escherichia coli*

John Carbon, Louise Clarke, Christine Ilgen, and Barry Ratzkin

Department of Biological Sciences, University of California, Santa Barbara, California 93106

INTRODUCTION

The recent development of techniques for the cloning and amplification of DNA segments linked by biochemical methods into bacterial plasmid or phage vectors has now made it possible to isolate particular gene systems from any source for further study [see Nathans and Smith (19) for a review of restriction endonucleases and their use in restructuring and cloning of DNA]. Basically, DNA cloning is accomplished by the joining of a DNA segment (or a group of segments) *in vitro* to a DNA vector (plasmid or virus) capable of replication in a suitable cell, infecting the host organism with the hybrid DNA, and then isolating individual clones of cells (or virus) that are each the progeny of a single infected cell.

Although several different eukaryotic gene systems have been established on hybrid plasmids in *E. coli* cells, it is still uncertain whether any functionally active protein product can be made from eukaryotic genes in the prokaryotic environment (2,13,18). A definitive test for meaningful expression of the cloned DNA segment would be complementation of an auxotrophic mutation in the bacterial host cell, such that the foreign DNA is producing an enzymatically active protein capable of relieving the metabolic defect in the host. Recently, Struhl et al. (23) have reported that a segment of yeast DNA cloned onto a bacteriophage λ vector is capable of complementing the *his*B463 mutation in *E. coli*. Although the exact nature of this complementation has not yet been established, it seems possible that certain segments of eukaryotic DNA may yield functionally active protein products in the *E. coli* cell.

Several experimental requirements must be met in order to set up a really definitive test for the ability of any given eukaryotic gene system to be expressed and to complement an auxotrophic mutation in *E. coli*. Thus, the efficiency of the cloning procedure used must be high enough to ensure that sufficient transformant clones containing hybrid DNA plasmids are obtained to be representative of the entire genome of the organism under study. Secondly, it is preferable to use DNA segments produced by *random scission* of the parent DNA by hydrodynamic shear rather than by restriction endonuclease action, in order to ensure that the desired gene system remains intact on at least a portion of the cleaved DNA segments. In addition, a reasonably rapid and convenient *screening procedure* should be available to test recombinant DNA plasmids for the ability to complement bacterial mutations. The experimental system used must at least permit the isolation of a large number of *E. coli* gene systems on hybrid plasmids, in order to show in a convincing way that the cloning and complementation procedures are adequate for the task in hand.

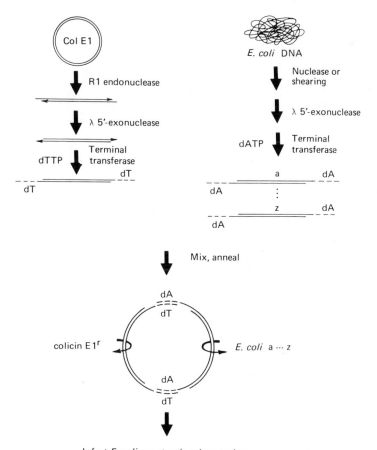

FIG. 1. Method for construction and selection of hybrid Co1E1-*E. coli* DNA plasmids. (From ref. 3.)

For the above reasons, we have used the poly(dA · dT) "connector" method for joining DNA (Fig. 1), a procedure that can give a high yield of recombinant DNA circles *in vitro,* and that can readily be used with sheared DNA samples (3,4,12,15,24). With this method, we have prepared hybrid circular DNA *in vitro* from poly(dT)-tailed plasmid ColE1 DNA (*Eco*RI-generated linear molecules) and randomly sheared segments of *E. coli,* yeast (*S. cerevisiae*), or *Drosophila* DNAs tailed with poly(dA). Transformation efficiencies using hybrid DNAs prepared in this manner are extremely high (about 10^3 unique transformants per microgram DNA), with essentially all of the transformants containing hybrid plasmids. Thus it has been possible to establish transformant colony banks containing a large number of clones that carry different hybrid plasmids representative of most of the genome of the parent organism.

E. coli strains harboring the fertility plasmid, F, along with the non-transmissible plasmid ColE1 transfer both plasmids with high efficiency to an F^- recipient (5,8,10). We find that hybrid ColE1 plasmids are also transferred efficiently from F^+ donors to auxotrophic F^- recipients, and this F-mediated transfer provides a convenient screening technique for the identification of specific hybrid plasmid clones in the transformant banks. This procedure has permitted the identification of over a hundred different hybrid plasmid clones carrying about 60 known *E. coli* gene systems on hybrid ColE1 plasmids.

Using preparations of mixed hybrid plasmid DNA purified from the total colony banks, we have transformed suitable *E. coli* auxotrophs and selected for and isolated a number of specific segments of *E. coli* DNA; more importantly, we have been able to isolate and characterize segments of yeast DNA that complement the *hisB* and *leu-6* mutations in *E. coli.* In principle, the methods described should permit the ready cloning of any segment of DNA for which a selection or mutation usable in complementation analysis is available.

MATERIALS AND METHODS

Bacterial Strains

The following strains, all derivatives of *E. coli* K-12, were used as recipients for transformations or in bacterial mating experiments: JA198 (*ΔtrpE5 recA thr-1 leu-6 lacY str*ʳ), JA199 (F^+ *ΔtrpE5 leu-6 hsr*$_K^-$ *hsm*$_K^+$), JA200 (F^+ *ΔtrpE5 recA thr-1 leu-6 lacY*), JA208 (*ΔaraC766 ΔlacZ-Y514 recA*), MV10 (*ΔtrpE5 thr-1 leu-6 lacY*)(11), MV12 (*ΔtrpE5 recA thr-1 leu-6 lacY*)(11), DM493 (W3110 *ΔtrpE5*), and CH754 (*argH metE xyl trpA36 recA56*). The F-plasmid was derived from *E. coli* strain Ymel.

Construction of Hybrid ColE1-DNA (*E. coli*) Annealed Circles

Linear poly(dT)-tailed plasmid CoIE1 DNA $[(L_{RI}exo)-(dT)_{150}]$ was prepared as previously described (3). High molecular weight *E. coli* DNA was purified from strain CS520 (HfrC *trpA58 metB glyVsu58*) as described (3), resuspended at 100 μg/ml in 0.01 M Tris-HCl (pH 7.5), 0.01 M NaCl, 0.001 M Na$_2$EDTA (STE), and fragmented by hydrodynamic shearing. DNA was sheared in a stainless steel cup (capacity = 1 ml) at 0°C for 45 min using a setting of 4.5 (approximately 5,400 rpm) on a Tri-R Stir-R motor (model S63C) fitted with a Virtis shaft and micro homogenizer blades. The fragmented CS520 DNA (average MW = 8.4 × 10^6 daltons) was treated with λ 5'-exonuclease, and poly(dA)$_{150}$ extensions were added to the 3'-ends of the DNA with the calf thymus deoxynucleotidyl terminal transferase (3,12,15). An equimolar mixture of CoIE1 DNA $[(L_{RI}exo)-(dT)_{150}]$ and CS520 DNA $[(L_{sh}exo)-(dA)_{150}]$ was annealed, and the DNA was concentrated by precipitation in 67% ethanol, dissolved in STE, and examined under the electron microscope (3). The annealed mixture contained approximately 25% hybrid DNA circular molecules.

Establishment of an *E. coli* Hybrid Plasmid Colony Bank

Twelve micrograms of annealed hybrid DNA were used to transform strain JA200 (C600 *recA*/F$^+$) according to a modification of the method of Mandel and Higa (16) described by Wensink et al. (24), except that after exposure to DNA, cells were diluted 10-fold into L-broth and grown for 30 min. The cells were plated in the presence of colicin E1 as previously described (3,4).

A separate control culture of JA200 was simultaneously treated in an identical manner to the transformed culture described above, except that cells were not exposed to DNA. From the number of colicin E1-resistant cells in this mock culture, it was estimated that about 67% of the colicin-resistant cells in the transformed culture contained CoIE1 plasmids, and the remaining clones were colicin tolerant but contained no plasmid (6). Twelve out of fifteen clones from the transformant culture were resistant to colicin E1 and sensitive to colicin E2 (11), and the remainder were resistant to both colicins. Use of the Triton lysis procedure (4) on 10 clones from a pilot experiment identical to the one described here and electrophoretic analysis of supercoiled plasmid DNAs in 1.2% agarose gels revealed that about 90% of the clones in the transformant culture (which were resistant to colicin E1 and sensitive to colicin E2) contained plasmids appreciably larger than CoIE1, whereas the remaining 10% contained no plasmid. From these data, we conclude that about 70% of the clones in the colony bank contain hybrid plasmids.

The colonies in the collection were arranged in a grid pattern of 48 per plate such that they could easily be transferred via a wooden block of 48 needles to standard 96-well MicroTest II dishes (Falcon Plastics), with each well containing 0.2 ml L-broth, colicin E1, and 8% dimethyl sulfoxide (20). The colonies in the collection could, therefore, be individually maintained, permanently stored at −80°C, and used repeatedly by inoculating fresh plates.

Isolation of hybrid plasmid DNA, and subsequent transformations were carried out as previously described (3,4). Standard techniques for bacterial selections, use of indicator plates, and liquid and replica matings were employed (17).

Establishment of the Yeast and *Drosophila* Hybrid Plasmid Banks

Yeast DNA was isolated from *S. cerevisiae,* strain X2180-1Aa SUC 2 *mal gal2* CUP1. Preparations of DNA from *D. melanogaster* were generously supplied by Dr. Norman Davidson (California Institute of Technology) and by Dr. David Finnegan and Dr. David Hogness (Stanford University). DNA was sheared to an average size of 8 to 9×10^6 daltons and tailed with poly(dA)$_{100}$ as described above. After annealing with ColE1 DNA[L_{RI}exo-(dT)$_{100}$], transformations of strain JA199 and establishment of colony banks were as described above.

Electron Microscopy

DNA was spread by the aqueous method of Davis, Simon, and Davidson (7), and all contour lengths were measured relative to reference ColE1 relaxed circles or *Eco*RI-cleaved ColE1 DNA linears on the same grid.

RESULTS

Establishment of a Collection of *E. coli* Clones Containing Hybrid ColE1-DNA (*E. coli*) Plasmids

We have determined the transformant colony bank size needed to obtain a plasmid collection which represents 90 to 99% of the *E. coli* genome as follows: given a preparation of cell DNA that has been fragmented to a size such that each fragment represents a fraction (f) of the total genome, then the probability (p) that a given unique DNA sequence is present in a collection of n transformant colonies is given by the expression:

$$p = 1 - (1 - f)^n$$

or

$$n = \frac{\ln (1 - p)}{\ln (1 - f)}$$

Thus, if a preparation of *E. coli* DNA were randomly sheared to an average size of 8.5×10^6 daltons for the construction of annealed hybrid circular DNA, a colony bank of only 720 transformants would be adequate to give a probability of 90% that any *E. coli* gene would be on a hybrid plasmid in one of the clones (Table 1; it is assumed that the desired gene is small in comparison with the size of the cloned fragments). At a probability level of 99%, the colony bank size (n) is only about 1,400 colonies for *E. coli*. As the genetic complexity of the organism increases, n at high probability levels increases dramatically (Table 1); at $p = 99\%$, the bank size for yeast is 4,600 colonies, whereas for *Drosophila* it is 46,000 colonies.

The construction of hybrid ColE1-DNA (*E. coli*) annealed circles using poly(dA-dT) connectors was carried out as previously described (3), except that the *E. coli* DNA was fragmented by hydrodynamic shearing (instead of restriction endonuclease cleavage) to an average size of $8.4 \pm 3.0 \times 10^6$ daltons, determined by measuring 47 molecules in the electron microscope. The preparation of annealed hybrid ColE1-DNA (*E. coli*) contained approximately 25% circles, 12% branched circles, 53% linears, 2% branched linears, and 8% unscorable tangles (100 molecules scored by electron microscopy).

The annealed DNA preparations can be used to transform various *E. coli* auxotrophs, selecting directly for the desired plasmid by complementation. For example, in Table 2 are shown transformation data obtained in this type of experiment (3). ColE1 circular plasmid DNA alone, when used to transform strain SB2(C600) to colicin E1 immunity, routinely yielded approximately 6×10^4 transformants per microgram DNA on L-broth-colicin E1 plates. The transformants were resistant to colicin E1 but sensitive to colicin E2 (11). Annealed hybrid DNA, when used in similar trans-

TABLE 1. *Colony "bank" sizes (n) needed to contain a particular hybrid plasmid transformant at various probability levels*

DNA source	Average size of DNA fragment cloned (daltons)	"Bank" size (n), no. of colonies		
		$p = 0.90$	$p = 0.95$	$p = 0.99$
E. coli	8.5×10^6	720	940	1,440
Yeast	1×10^7	2,300	3,000	4,600
Drosophila	1×10^7	23,000	30,000	46,000

The above calculations are based on the formula, $p = 1 - (1 - f)^n$, and assume that each transformant colony in the "bank" arises from an independent transformation event and that each hybrid molecule transforms with the same efficiency. It is also assumed that the length (x) of the desired DNA segment is small in comparison with the length (L) of the DNA fragment actually cloned, in order to minimize the effect of random breaks occurring within the desired length. More accurately, a corrected f value (f^*) could be obtained from the expression, $f^* = \left(1 - \frac{x}{L}\right)f$, and substituted for f in the above probability equation. (From ref. 4.)

TABLE 2. *Transformation efficiency of annealed circular hybrid DNA*[a]

DNA	Recipient	Selection	Transformants/μg DNA
ColE1 (I)	SB2(C600)	Colicin E1^r	6×10^4
ColE1 (L_{RI} exo)-(dT)$_{150}$	SB2(C600)	Colicin E1$_r$	<1
ColE1 (L_{RI} exo)-(dT)$_{150}$ + *E. coli* (L_{RI} exo)-(dA)$_{150}$	SB2(C600)	Colicin E1^r	$\sim 10^3$
"	NL20-127 (K12 ΔaraC766 recA)	Ara$^+$	0.2 (2/2 colicin E1^r)
"	MV10 (C600 ΔtrpE5 rec$^+$)	Trp$^+$	4 (12/45 colicin E1^s)

[a] In these experiments, the *E. coli* DNA was fragmented by digestion with endonuclease *Eco*RI, rather than by hydrodynamic shear. (From ref. 3.)

formations, gave approximately 10^3 transformants per microgram DNA. Again, these transformants were resistant to colicin E1 but sensitive to colicin E2. The ColE1[(L_{RI}exo)-(dT)$_{150}$] DNA gave fewer than one colicin E1-resistant transformant per microgram when used alone. Hybrid DNA was then used to transform a ΔtrpE5 strain (the *trpE* region is deleted) to Trp$^+$, or a ΔaraC strain to Ara$^+$. In both cases, transformants were readily obtained which could be shown to contain hybrid plasmids carrying the *trp* operon (pLC5; 15×10^6 daltons) or the *ara-leu* region (pLC3; 18×10^6 daltons).

We used the same amount (12 μg) of annealed hybrid ColE1-DNA (*E. coli*) that was used to directly select for a specific hybrid plasmid-containing strain (3) to transform *E. coli* strain JA200 (C600 ΔtrpE5 recA/F$^+$). In this experiment we selected instead for the vector determinant, colicin E1 resistance. Thus, in a single transformation and selection, a collection or "bank" of clones was obtained which carry different hybrid ColE1-DNA (*E. coli*) plasmids representative of a large portion of the bacterial genome. After exposure to DNA, the transformed cell culture was not incubated long enough to permit significant cell division before plating. Thus, each transformed clone was the consequence of a distinct and separate transformation event. In addition, the recipient for transformation was a strain harboring the sex factor, F, in order to permit the identification of a clone carrying a particular hybrid plasmid by F-mediated transfer of the hybrid through replica mating of the colony collection to a particular *E. coli* F$^-$ auxotroph (*see below*). A *recA* transformation recipient was chosen to avoid recombination of hybrid plasmids with host chromosomal DNA.

Approximately 2,100 colicin E1-resistant transformants were picked and transferred to plates in a grid array of 48 colonies per plate. The colonies were also individually maintained and stored as 8% DMSO cultures in MicroTest dishes at −80°C. Use of a Triton lysis procedure (4; H. Boyer, *personal communication*) on 10 clones from a pilot experiment and analysis of supercoiled plasmid DNAs in 1.2% agarose gels revealed that about 70% of the clones in the collection (1,400 transformant colonies) contained plasmids appreciably larger than plasmid ColE1. The remaining 30% of the colonies were tolerant to colicin E1 but probably contained no plasmid (6). Many of the latter colonies were also resistant to colicin E2. In more recent experiments, we have been able to reduce the background of colicin E1-tolerant cells by pregrowing the transformation recipient in 0.05% sodium deoxycholate (C. Ilgen and J. Carbon, *unpublished observations*).

Use of F-Mediated Transfer for the Identification of Specific Hybrid Plasmid-Bearing Clones in the Colony Bank

Once a transformant colony bank is established, it is essential to have a simple, rapid way to identify a desired hybrid plasmid-bearing clone within the collection. It has been shown that strains which harbor the fertility plasmid F and which also contain the nontransmissible plasmid ColE1 will transfer both plasmids to an F⁻ recipient with high efficiency (5,8,10). Preliminary reconstruction experiments using strain MV12/pLC19 (ColE1-*trp* plasmid, ref. 3) to which an F factor had been transferred indicated that this ColE1-DNA (*E. coli*) hybrid transferred readily to an F⁻/*trp* recipient (see Table 5). Furthermore, the transfer was of high enough efficiency to be easily detected by replica mating on plates, using appropriate selections and counterselections.

A number of replica mating experiments were therefore performed using the entire colony collection and various F⁻ auxotrophs. Some of these experiments were carried out in our laboratory and will be discussed in detail below. The remainder were done by other laboratories using the same colony collection. Table 3 lists the hybrid plasmids tentatively identified to date and some of the markers carried by these plasmids (Fig. 2). On the average, three hybrid plasmid containing clones were identified in the total collection for each marker sought in a replica mating experiment. The list includes about 40 known *E. coli* genes, some of which were assigned after clones harboring hybrid plasmids which carried neighboring genes were identified. For example, the four clones in the collection which were found by replica mating to contain ColE1-*xyl* plasmids were tested for overproduction of glycine-tRNA synthetase, the product of the closely neighboring *glyS* gene (Fig. 2), and two of the four clones did produce elevated levels of this enzyme (G. Nagel, *personal communication*).

The transformant colony collection appears to contain hybrid plasmids representative of nearly the entire *E. coli* genome, in that the probability of finding any cloned gene system chosen at random appears to be high

TABLE 3. *Hybrid plasmids identified in the colony bank*

Approximate map location (min)	*E. coli* markers complemented	pLC code
1	*araC*	24–41
4	*dnaE*[a]	26–43
7	*proA*[b]	28–33
		44–11
9	*lacZY*	20–30
12	*dnaZ*[a]	5–1
		5–2
		6–2
		10–24
		10–26
		30–3
		30–4
22–26	*flaKLM*[c]	24–46
		35–44
		36–11
27	*trpE*	4–6
		5–23
		29–41
		32–12
		32–27
		41–15
35–38	*flaGH, cheB*[c]	21–2
		24–15
	flaGH, cheB, mot[c]	1–28
		1–29
	mot, cheA, flaI[c]	27–20
		38–14
		38–36
35–38	*flaD*[c]	7–18
		13–12
	flaD, hag, flaN[c]	24–16
		26–7
	flaN, flaBCOE, flaAPQR[c]	41–7
38	*his*[b]	14–29
		26–21
43	*glpT*[d]	3–46
		8–12
		8–24
		8–29
		14–12
		19–24
		42–17
51	*recA, srl*[b]	17–43
		18–42
		21–33
		22–40
		24–32

(*Table 3 continues* →)

318

TABLE 3 (*Continued*)

Approximate map location (min)	E. coli markers complemented	pLC code
	recA[b]	17–38
		24–27
		30–20
70	glyS, xyl[e]	1–3
		44–22
	xyl	10–15
		32–9
75	ilv[b]	21–35
		22–3
		22–31
		26–3
		27–15
		30–15
		30–17
		44–7
	cya[f]	23–3
		29–5
		36–14
		41–4
		43–44
79	argH	20–10
		41–13
81	dnaB[a]	11–9
		44–14
89	dnaC[a]	4–39
		8–9
		25–8
		30–24
		31–39
90	serB, trpR[f]	32–33
	trpR, thr[f]	35–1
		35–21

[a] R. McMacken (*personal communication*).
[b] L. Margossiane and A. J. Clark (*personal communication*).
[c] M. Silverman and M. Simon (*personal communication*).
[d] J. Weiner (*personal communication*).
[e] G. Nagel (*personal communication*).
[f] J. Schrenk and D. Morse (*personal communication*).

Genes on hybrid plasmids identified as complementing specific *E. coli* markers in this table are not necessarily the only *E. coli* genes carried by the plasmids. In addition, plasmids listed here as complementing a specific marker do not always represent all the plasmids in the total collection which carry that particular *E. coli* gene. For example, the Co1E1-*lacZY* hybrid plasmid (pLC20–30) was identified by screening as the only plasmid in the collection capable of complementing a *lacY* mutation (see Table 4). pLC20–30 was later found to also complement *lacZ*, but other Co1E1-*lacZ* plasmids may occur in the colony bank. (From ref. 4.)

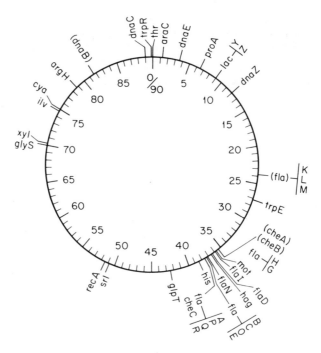

FIG. 2. Relative positions on the *E. coli* genetic map of gene systems complemented by hybrid Co1E1-DNA (*E. coli*) plasmids. For references, see Table 3. (From ref. 4.)

(about 80%). For example, in our laboratories the collection was screened for hybrid plasmids capable of complementing any of six mutations chosen at random (*araC*, *lacY*, *trpE*, *argH*, *xyl*, and *metE*). Of these, only the *metE* mutation could not be complemented by any hybrid plasmid from the bank (see Table 3). All of the known *E. coli* motility gene systems were found on hybrid plasmids in this same collection (M. Silverman and M. Simon, *personal communication*).

Characterization of Several ColE1-DNA (*E. coli*) Hybrid Plasmids from the Colony Bank

A semiconfluent to confluent patch on selective plates in the mating experiments described above was scored as representing a clone carrying the desired hybrid plasmid. Phenotypic reversion of a recipient marker could, however, have been a consequence of suppression if the recipient marker was a point mutation and the hybrid plasmid carried a suppressor transfer RNA gene, or a consequence of suppression possibly resulting from overproduction of an RNA or protein whose gene was carried by a hybrid plasmid. Complementation of a recipient marker in the F-mediated replica matings could also have been the result of Hfr formation (if the recipient was *recA⁺* or if the hybrid plasmid carried a *recA⁺* region) or F′ formation and subsequent mobilization of donor chromosomal genes. However, chromosomal mobilization generally occurs at a much lower frequency than F-mediated hybrid plasmid transfer (see Table 5) and could usually be distinguished from it.

To help establish that clones tentatively identified in the colony bank did carry the desired hybrid plasmids, we further characterized a number of these strains and their plasmids. Table 4 lists the hybrid plasmid strains identified by four replica mating experiments selecting for hybrid plasmids carrying the *ara, trp, arg,* and *xyl* regions. From one to six candidates were obtained from each screening of the bank. Since the parent bank strain was *lacY,* the entire colony collection was also screened for ColE1-*lacY* bearing clones by replica plating the bank onto indicator plates. One Lac⁺ candidate was found by this method.

Covalently closed, supercoiled hybrid plasmid DNA was purified from at least one representative of each marker class in Table 4. These DNAs were then used to transform either the original recipients or a recipient with a mutation in the same gene as the original. In all cases tested, a high frequency of transformation to the expected phenotype was obtained (Table 4), and all transformants screened were both resistant to colicin E1 and resistant to fr male-specific phage. These results indicated that the genes responsible for reversion of recipient markers in the original mating experiments were carried by ColE1-DNA (*E. coli*) hybrid plasmids. It is improbable in the case of the *ara, trp,* and *lac* plasmids that reversion was due to suppressor tRNA genes carried by the hybrids, since the mutations complemented by these plasmids were deletions. Suppression by over-production of a gene product, such as a tRNA, a protein with weak catalytic activity or a ribosomal component, is difficult to rule out in any case. However, two of the ColE1-*xyl* plasmids were shown to carry the closely neighboring *glyS* gene (G. Nagel, *personal communication*), and the ColE1-*lac* plasmid, originally detected in a *lacY* strain, transformed to Lac⁺ a strain which carries a deletion in at least *lac Y* and *Z.* Thus, these three hybrid plasmids certainly carry the desired portion of *E. coli* DNA.

The contour lengths of a number of hybrid plasmids were measured by electron microscopy of relaxed circles using ColE1 DNA (4.2×10^6 daltons; ref. 1) as a standard on the same grid. Data for four of these measurements are found in Table 4. Other plasmids measured were pLC18-42 ($11.9 \pm 0.1 \times 10^6$ daltons), pLC21-33 ($10.3 \pm 0.1 \times 10^6$ daltons), pLC32-33 ($18.1 \pm 0.1 \times 10^6$ daltons), pLC36-14 ($14.9 \pm 0.2 \times 10^6$ daltons), and pLC50 (ColE1-*lac,* $13.6 \pm 0.1 \times 10^6$ daltons; isolated with the same annealed hybrid DNA but in a separate transformation). The average size of the *E. coli* inserts based on this relatively small sample is $9.9 \pm 3.4 \times 10^6$ daltons, as com-

TABLE 4. *Characterization of several hybrid plasmids*

Mating recipient	Selection	pLC code	Plasmid size (×10⁻⁶ daltons)	Transformation recipient	Selection	Transformants/ μg DNA
JA208 ($\Delta araC766$)	Ara⁺	24–41	—ᵃ	JA208 ($\Delta araC766$)	Ara⁺	3 × 10⁴
DM493 ($\Delta trpE5$)	Trp⁺	4–6	—	MV12 ($\Delta trpE5$)	Trp⁺	—
		5–23	—			—
		29–41	—			4 × 10⁴
		32–12	—			4 × 10⁴
		32–27	—			—
		41–15	—			—
CH754 ($argH$)	Arg⁺	20–10	11.8 ± 0.2	CH754 ($argH$)	Arg⁺	>2 × 10⁵
		41–13	10.0 ± 0.1			>3 × 10⁵
CH754 (xyl)	Xyl⁺	1–3	17.0 ± 0.1	CH754 (xyl)	Xyl⁺	>2 × 10⁵
		10–15	—			>10⁴
		32–9	—			
		44–22	—			
MV (*lacY*)/ F⁺/ColE1-DNA (*E. coli*) (screened on MacConkey- lactose plates)	Lac⁺	20–30	19.4 ± 0.1	JA208 ($\Delta lacZ$-Y 514)	Lac⁺	>3 × 10³

ᵃ A dash indicates "not measured." (From ref. 4.)

pared with the average size of the sheared CS520 DNA ($8.4 \pm 3.0 \times 10^6$ daltons) used originally to construct the plasmids.

Efficiency of the F-Mediated Transfer of ColE1 Hybrid Plasmids

The F-mediated transfer of three ColE1-*trp* plasmids was studied (Table 5). Strain MV12/pLC19(F⁻) has been described (3) and harbors a ColE1-*trp* plasmid which carries all or most of the *trp* operon genes. This strain was made F⁺ by mating with strain Ymel and designated JA200/pLC19. Strains JA200/pLC29-41 and JA200/pLC32-12 are F⁺ clones from the colony bank which were identified as carrying ColE1-*trpE* plasmids (see Tables 3 and 4). The F⁻ strains, MV12/pLC29-41 and MV12/pLC32-12, were constructed by transformation of strain MV12(F⁻) with purified pLC29-41 and pLC32-12 DNAs isolated from their respective strains in the colony bank (see Table 4). These six strains were used as donors in mating experiments described in the legend to Table 5, with strain DM493 (*ΔtrpE5/F⁻*) serving as the recipient in all cases. All of the ColE1-*trp* plasmids were donated with high efficiency from strains that also contained F. No transfer was detected from strains that contained only the hybrid plasmids. Of the recipients in the matings receiving hybrid plasmids, 88 to 98% of them also received F and were able to further transfer ColE1-*trp* plasmids. A small fraction of the original recipients, however, picked up only the hybrid plasmids, since they could no longer transfer these plasmids and were resistant to phage fr. Several of the ColE1-*arg* and ColE1-*xyl* plasmids were also very efficiently transferred from F⁺ strains (D. Richardson and J. Carbon, *unpublished observations*). These data indicate that

TABLE 5. *Efficiency of transfer of ColE1-*trp *plasmids from F⁺ and F⁻ strains*

Donors	Recipient	% Donors transferring hybrid plasmid in 1 hr	% Trp⁺ recipients that were F⁺
MV12/pLC19 (F⁻)	DM493 (*ΔtrpE5/F⁻*)	<0.004	—
MV12/pLC29-41 (F⁻)	"	<0.002	—
MV12/pLC32-12 (F⁻)	"	<0.002	—
JA200/pLC19 (F⁺)	"	64	88
JA200/pLC29-41 (F⁺)	"	46	98
JA200/pLC32-12 (F⁺)	"	40	96

Liquid matings were performed as described by Miller (17) at a 10:1 ratio of recipients to donors. A 1-hr incubation with gentle shaking at 37°C was followed by Vortex interruption, dilution, and plating onto selective plates. The fraction of Trp⁺ recipients that were also F⁺ was determined by picking 50 recipients from each mating, replica mating these as donors, using JA198 (*ΔtrpE5str^r*) as a recipient in each case, and selecting for Trp⁺. Those donors which did not transfer the Trp⁺ phenotype were confirmed as being F⁻ by their resistance to phage fr. (From ref. 4.)

ColE1-DNA (*E. coli*) plasmids behave in a manner similar to plasmid ColE1 in an F⁺ background and that the hybrid plasmids and the F factor act as physically independent units.

Mixed Hybrid Plasmid DNA Preparations Representing a Large Portion of the *E. coli* Genome

An alternative way to maintain the genetic information of a complete genome as a collection of hybrid plasmids is in the form of a covalently closed, hybrid plasmid DNA preparation. Such a preparation was made by scraping a set of master replica plates containing all the clones in the colony bank, allowing the cells to grow nonselectively for approximately two generations, amplifying plasmid DNA in the presence of chloramphenicol, and purifying supercoiled total plasmid DNA. To identify individual plasmids, we then used this DNA to transform the five recipients listed in Table 6 to the Ara⁺, Trp⁺, Arg⁺, Xyl⁺, and Lac⁺ phenotypes. The observed and calculated transformation frequencies are given in Table 5. At the DNA levels used, no Ara⁺ or Lac⁺ transformants were found, but the experiment yielded a disproportionately large number of Arg⁺ transformants. Thus, this purified DNA preparation was not as representative of the bacterial genome as the colony bank and appeared to contain a relatively large amount of ColEl-*arg* DNA but less ColEl-*ara* and ColEl-*lac* DNAs.

Generally, we have observed that strains harboring various hybrid plasmids may grow at different growth rates and that certain hybrid plasmids segregate at high frequency and others not at all. These observations have indicated to us that the most suitable way to maintain a hybrid plasmid

TABLE 6. *Efficiency of transformation by mixed plasmid DNA prepared from the entire colony bank*

Transformation recipient	Selection	Transformants/μg DNA	
		Observed	Expected
JA208($\Delta araC766$)	Ara⁺	<0.5	20
MV12($\Delta trpE5$)	Trp⁺	6.3×10^2	2×10^2
CH754(*argH*)	Arg⁺	3.3×10^4	4×10^2
CH754(*xyl*)	Xyl⁺	5.9×10^3	8×10^2
JA208($\Delta lacZ$-*Y514*)	Lac⁺	<0.5	>2

Transformations and selections were carried out as described (3,4) using 2 μg DNA for each experiment. The expected number of transformants per microgram DNA was calculated from known efficiencies of transformation of purified individual plasmid DNAs using these same recipients (Table 4) and from the number of representatives of each plasmid in the colony bank, assuming all plasmids were approximately the same size. (From ref. 4.)

collection without loss of specific plasmids is as a set of individual clones, each harboring a unique plasmid.

Establishment of Hybrid Plasmid Colony Banks Derived from Yeast (*S. cerevisiae*) and *Drosophila* DNAs

The methods described above have also been applied to DNA prepared from either yeast (*S. cerevisiae*) or *Drosophila*. The DNA samples were sheared by high-speed stirring as described in Materials and Methods before reaction with the λ exonuclease and the terminal transferase. Using 25 μg of annealed hybrid DNA (ColE1 DNA as vector) to transform *E. coli* strain JA199 (hsr$_k^-$ hsm$_k^+$), we have obtained about 30,000 *unique* transformants to colicin E1 resistance in each case. Of these, at least 80% could be shown to contain hybrid plasmid DNA, although in both cases many of the hybrids contained a relatively small insert of about 1 kilobase (kb) or less in length. Only 20% of the clones contained relatively large hybrid plasmids with inserts ranging from 2.5 to 20 kb. The reason for the large number of relatively small inserts in these banks is still unclear.

These collections are of sufficient number to contain essentially all of the yeast genome, but because of the small size of most of the hybrid ColE1-Dm DNA plasmids, it is likely that only 10 to 20% of the *Drosophila* genome is represented in this initial collection. The collections have been stored in three forms: (a) as individual colonies (4,300 for yeast; 10,000 for *Drosophila*) in 8% DMSO suspension at −80°C; (b) as a mixture of transformed cells; and (c) as a mixture of hybrid plasmid DNA extracted from all of the transformants.

Complementation of *E. coli* Mutations by Hybrid ColE1-Yeast DNA Plasmids

THE leu-6 MUTATION

The *E. coli* strain that was transformed with joined ColE1-yeast DNA to form the hybrid plasmid colony bank carried two auxotrophic mutations, Δ*trpE5* and *leu-6* (strain JA199). As a first step toward the screening of the hybrid plasmids for ability to complement *E. coli* mutations, we plated a mixture of 15,000 colonies from the original transformation selection plates onto supplemented minimal media, selecting for growth in the absence of either tryptophan or leucine.

On minimal salts-glucose media supplemented with tryptophan but without leucine, healthy Leu$^+$ colonies were observed occurring at a frequency of one Leu$^+$ colony per 10^6 cells plated. Sixteen of the slower growing Leu$^+$ colonies were purified and subjected to further investigation. All 16 of the Leu$^+$ transformants were resistant to colicin E1 and sensitive to colicin E2, a property of cells carrying hybrid Co1E1 plasmids (11). In addition, all transferred the Leu$^+$ phenotype to suitable F$^-$ *leu-6* recipients at high frequency (about 10^{-3}). Plasmid DNA was isolated and purified by CsC1-ethidium bromide banding from eight of the Leu$^+$ transformants. Seven of these plasmid DNAs were capable of transforming strain JA199 to Leu$^+$. One plasmid, pY*eleu*10, which produced the fastest growing transformants, was investigated further. Circular pY*eleu*10 DNA has a molecular weight of 13.4 × 10^6 daltons, as determined by direct length measurements on electron photomicrographs using Co1E1 DNA (4.2 × 10^6 daltons) as a standard.

The data presented in Table 7 indicate that pY*eleu*10 DNA is capable of transforming strain JA199 (*leu-6*) to both colicin E1 resistance and Leu$^+$ with high frequency (>10^5 transformants per microgram DNA). Thus, it is clear that the hybrid plasmid pY*eleu*10 carries genetic information capable of complementing the *leu-6* mutation in *E. coli*.

The segment of cloned DNA in pY*eleu*10 was shown to be yeast DNA by measuring the rates of reassociation of labeled single-stranded plasmid DNA fragments in the presence of various unlabeled single-stranded DNAs. In Fig. 3 are shown the rates of reassociation of single-stranded pY*eleu*10 DNA in the presence of a large excess of single-stranded salmon sperm DNA, *E. coli* DNA, and yeast DNA. Although salmon sperm and *E. coli* DNAs have little or no effect on the rate of reassociation of the pY*eleu*10 sequences, single strands of yeast DNA greatly increase the rate, as would be expected if a portion of pY*eleu*10 DNA is derived from yeast DNA. Reassociation kinetics data of this type indicate that about 70% of the pY*eleu*10 DNA is derived from yeast DNA. This is in excellent agreement with the length measurements of pY*eleu*10 DNA, from which a molecular weight of 9.2 × 10^6 daltons was derived for the cloned yeast DNA segment, or 69% of the total plasmid DNA (assuming a complete 4.2 × 10^6 daltons Co1E1 segment is present in pY*eleu*10). Assuming that a single copy of the DNA segment cloned in pY*eleu*10 is present in the yeast genome, the reassocia-

TABLE 7. *Transformation efficiencies of Co1E1-yeast DNA hybrid plasmids*

Plasmid DNA	Transformants/μg DNA		
	Colicin E1^R	Leu$^+$	His$^+$
pY*eleu*10	2.7 × 10^5	3.1 × 10^5	
pY*ehis*1	1.5 × 10^4		2.2 × 10^3
pY*ehis*2	2.6 × 10^4		2.5 × 10^3
pY*ehis*3	2.4 × 10^4		2.5 × 10^3

Hybrid plasmid DNAs were isolated and purified as previously described (3,4). The transformation recipients were strain JA199 (for pY*eleu*10) and K12 *hisB463* (for the pY*ehis* plasmid DNAs).

326

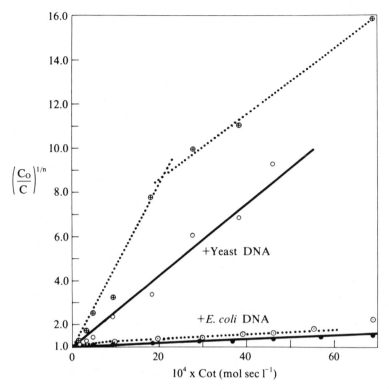

FIG. 3. Reassociation kinetics of labeled single-stranded pY*eleu*10 DNA fragments (0.085 μg/ml) in the presence of: salmon sperm DNA, 1,000 μg/ml (●———●); *E. coli* DNA, 196 μg/ml (⊙----⊙); yeast (*S. cerevisiae*) DNA, 176 μg/ml (○———○); and yeast DNA, 526 μg/ml (⊕----⊕). The pY*eleu*10 DNA was [3H]-labeled to a specific activity of 2.4 × 10⁶ cpm/μg by nick translation in the presence of [3H]dATP (8.33 c/mmole) and [3H]dTTP (8.36 c/mmole) by the method of M. Dieckmann and P. Berg (*personal communication*), as described by Schachat and Hogness (21). The reassociation rate experiments were carried out as previously described (21,24).

tion rates give a haploid genome size of 4.7×10^9 daltons, in excellent agreement with recent estimates (14).

THE hisb463 *MUTATION*

Mixed hybrid plasmid DNA isolated from a mixture of all transformants in the yeast hybrid plasmid bank can be used to transform *E. coli* auxotrophs, selecting for the desired complementation. Since Struhl, Cameron, and Davis (23) have recently reported complementation of the *E. coli* *hisB463* mutation by a yeast DNA segment cloned on a bacteriophage λ vector, we have attempted the isolation of a hybrid ColE1-yeast DNA plasmid capable of complementing this same mutation. Transformation of a K-12 *hisB463* strain with 2 μg of mixed hybrid plasmid DNAs derived from the total collection gave His$^+$ transformants at a frequency of 3 His$^+$ colonies per 10^5 transformants to colicin E1 resistance. No His$^+$ colonies were obtained in the absence of hybrid plasmid DNA, in line with the previous observation that the *hisB463* mutation is nonrevertable and is most probably a deletion that affects IGP dehydratase activity (23).

All of the His$^+$ transformants appeared to contain hybrid ColE1 plasmids and were colicin E1 resistant but sensitive to colicin E2. Plasmid DNA was isolated and purified from three of these clones (designated pYe*his*1, pYe*his*2, and pYe*his*3). These three plasmids appeared to be identical in length when measured on electron photomicrographs using ColE1 DNA as a standard. The total molecular weight of the pYe*his* plasmid is 10.7 × 10^6 daltons, indicating that a segment of DNA of molecular weight 6.5 × 10^6 daltons had been cloned.

Purified pYe*his* plasmid DNAs were capable of transforming a *hisB463* *E. coli* strain to colicin E1 resistance and to His$^+$ with high frequency (see Table 7). Direct selection for His$^+$ consistently yielded 10-fold fewer transformants than when colicin E1 resistance was selected (Table 7). This observation is similar to that reported by Struhl et al. (23) for the λgt-Sc*his* phage containing a *hisB* complementing segment of yeast DNA. The purified phage yields His$^+$ transformants at only 1% of the expected value, possibly due to inefficient expression or functioning of the yeast gene product.

Reassociation rate experiments have shown that the segment of yeast DNA in pYe*his*2 is, in part, identical to that cloned by Struhl et al. (23) in λgt-Sc*his* (see Fig. 4). Labeled single-stranded fragments of pYe*his*2

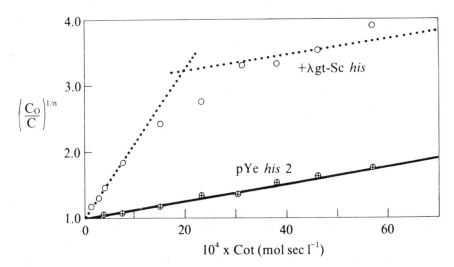

FIG. 4. Reassociation kinetics of labeled single-stranded pYe*his*2 DNA fragments (0.07 μg/ml) in the presence of: salmon sperm DNA, 1,000 μg/ml (⊕——⊕); and λgt-Sc*his* DNA, 125 μg/ml (○-----○). The pYe*his*2 DNA was nick-translated to a specific activity of 2.2 × 10^6 cpm/μg as described in the legend to Fig. 3.

DNA reassociate appreciably faster in the presence of unlabeled single-stranded λgt-S*chis* DNA. The experiment indicates that about 40% of the pY*ehis*2 DNA or 4.3 × 10⁶ daltons of DNA is identical to the yeast DNA segment cloned in λgt-S*chis*. The remaining 2.2 × 10⁶ dalton segment of cloned DNA in pY*ehis*2 is not present on λgt-S*chis*, although the phage apparently contains other sequences not present on the plasmid. Since the yeast DNA segment in λgt-S*chis* is an *Eco*RI-generated fragment whereas the segment in pY*ehis*2 was produced by random scission, it is not surprising that there is only a partial overlap between the two segments. pY*ehis*2 DNA contains a single *Eco*RI cleavage site, which is apparently identical to one of the two sites defining the yeast DNA segment in λgt-S*chis*.

DISCUSSION

Hybrid Plasmid Colony Banks

Many of the advantages of having particular gene systems isolated and cloned on plasmids present in multiple copies per bacterial cell have been described (3,4,11). This chapter reports procedures by which most of the genetic information contained in a bacterial or simple eukaryote genome can be established as collections of hybrid ColE1 plasmids, each of which contains a unique segment of DNA of approximate molecular weight 8 to 10 × 10⁶ daltons. We have maintained these plasmid collections in two ways: first, as colony banks of separate transformant clones, each harboring a unique hybrid plasmid, and, secondly, as covalently closed, supercoiled DNA preparations isolated from the total colony banks.

There are a number of ways to identify specific plasmids of interest within these collections. In this chapter we have used complementation of various *E. coli* mutations to isolate plasmids carrying specific genes by F-mediated transfer of hybrid plasmids from the F⁺ clones in the colony bank to F⁻ auxotrophs on selective plates, or by transformation of auxotrophs and direct selection using the mixed plasmid DNA preparation. In several cases, neighboring genes have been found to occur on the same hybrid plasmid.

Maintaining a plasmid collection representative of most of the *E. coli* genome in the form of a colony bank permits the use of additional methods to identify specific plasmids. The presence of hybrid ColE1-DNA (*E. coli*) plasmids in multiple copies per cell frequently leads to overproduction of the products of genes carried by these plasmids (11). Overproduction of a protein—a repressor, for example—might result in pleiotrophy or auxotrophy in cells harboring specific plasmids because the gene system(s) under the control of this repressor might not be inducible. Thus, screening the clones in the colony bank for auxotrophy may result in further assignment of regulatory genes to plasmids or reveal hitherto unknown regulatory phenomena.

Another method for identification of specific genes on plasmids in colony banks is the hybridization technique described by Grunstein and Hogness (9). Many gene systems cannot be identified by F-mediated transfer and selection, either because a suitable auxotrophic recipient is unavailable (e.g., most tRNA and ribosomal RNA genes) or because the transformant would not be viable. For example, strains carrying a tRNA-derived suppressor mutation on an amplifying plasmid may not survive because of overproduction of a deleterious su^+ gene product. We are presently using hybridization techniques to detect plasmids in the collections that carry *E. coli* and eukaryotic tRNA and ribosomal RNA genes. Recently, the spacing of tRNA genes on cloned segments of *Drosophila* DNA has been described (P. Yen, C. Ilgen, J. Carbon, and N. Davidson, *unpublished observations*).

Several difficulties arise in attempting to maintain colony banks of the type described in this chapter. In our experience, hybrid plasmids derived from *E. coli* DNA seem to be readily maintained in *recA* strains if selective pressure, either colicin E1 resistance or complementation of a host marker, is kept on the strains. In the absence of selective pressure many of these plasmids segregate, however. In *recA*$^+$ strains some plasmids appear to be lost, presumably through recombination with the genome or through segregation, and others become smaller, possibly by looping out segments of the *E. coli* DNA. For these reasons, we have maintained the colony collections as sets of individual clones, have attempted to keep selective pressure on plasmid-bearing strains, and have stored the original transformant collections in the form of separate cultures in 8% DMSO at $-80°C$. We have gone back to these original cultures to inoculate fresh plates for the experiments described here and have avoided continued growth of clones and replication of plasmids.

Expression of Eukaryotic DNA in *E. coli*

Although several laboratories have established eukaryotic DNA segments on suitable cloning vectors in *E. coli,* little evidence has been presented regarding the meaningful expression of the cloned DNA to form enzymatically active protein products. Recently, Struhl et al. (23) have reported that *E. coli hisB* mutations are complemented by an *Eco*RI fragment of yeast DNA cloned on a phage λ vector; however, the mechanism of the complementation is still unclear.

When a large fraction of the yeast genome is represented in a collection of hybrid plasmids derived from randomly sheared DNA, it is surprisingly easy to find hybrid plasmids capable of complementing *E. coli* auxotrophic mutations. Thus, the yeast-ColE1 hybrid plasmids complementing the *hisB* and *leu-6* mutations were found without the need to screen the collection for complementation of hundreds of different *E. coli* mutations. In fact, the "success rate," in terms of finding *E. coli* mutations that are complemented by hybrid plasmids from the yeast DNA-ColE1 hybrid collection, was about 20% (two successful, *leu-6* and *hisB,* out of only nine different auxotrophic mutations attempted). It is possible that the "success rate" is even higher than 20% since apparent complementation of mutations other than *hisB* and *leu-6* has been observed in preliminary experiments but not verified as yet by isolation of pure plasmid DNAs for transformation tests.

The mechanism of the expression of yeast DNA segments in *E. coli* is still unclear. Thus, we do not know that the normal yeast transcription and translation initiation signals are being used in the bacterial system. One possibility might be that random transcription products are made from fortuitous promoter sequences on the yeast DNA, and that the *E. coli* ribosomes are capable of initiating at or near the normal ribosome-binding sequences on the resulting RNA. Transcription and read-through from vector promoter sequences do not seem likely since Struhl et al. (23) have shown that the yeast DNA segment in λgt-Sc*his* can be inserted in either orientation without affecting expression.

The exact nature of the complementation phenomena we are observing remains to be explained. In the case of the *hisB463* complementation, the mutation is probably a deletion, and thus some form of suppression seems unlikely (23). The *leu-6* mutation does revert with measurable frequency, however, so that in this case suppression effects have not been ruled out. In addition, we have not yet determined which of the four genes in the *leu* operon (*leuA, B, C,* or *D*) is defective in the *leu-6* strain. Additional studies with a set of well-defined and mapped *leu* mutants (22) should clarify this point.

Finally, the relative ease with which we have found yeast DNA segments capable of complementing *E. coli* mutations suggests that it should now be possible to isolate many different yeast gene systems on hybrid plasmids for studies on the genetic organization and expression of the yeast genome. It is possible that yeast DNA, being derived from a relatively primitive eukaryote, can be expressed in the prokaryotic host cell, but that DNA from higher eukaryotes would be mostly inert. Our current studies with *Drosophila melanogaster* DNA are designed to answer this important question.

Biohazard Considerations

F-mediated transfer of ColE1-like plasmids provides a valuable tool for the identification of specific hybrid plasmids constructed of ColE1 DNA and the DNAs of prokaryotic and simple eukaryotic organisms. However, the procedure[1] converts normally nontransmissible hybrid plasmids, such

---

[1] The current NIH guidelines on recombinant DNA research specifically prohibit the cloning of eukaryotic DNA on transmissible plasmid vectors. Thus, we are no longer performing cloning experiments in *E. coli* cells that contain the fertility plasmid F.

as those derived from Co1E1, into a state which permits their ready transfer to other bacterial hosts. Because of biohazard considerations, we do not recommend the establishment in an F^+ background of strains or banks of clones containing hybrid plasmids constructed of Co1E1 DNA and random fragments of DNA from higher eukaryotic organisms until the biological properties of such plasmids have been further characterized.

ACKNOWLEDGMENTS

The authors would like to thank Denise Richardson for excellent technical assistance. We also are indebted to R. Ratliff for a generous gift of deoxynucleotidyl terminal transferase, and to K. Struhl and R. Davis for a sample of λgt-S*chis* DNA. This work was supported by NIH Grants CA-11034 and CA-15941. One of us (B.R.) received support by a grant from Abbott Laboratories. Portions of this work have been published elsewhere (3,4).

REFERENCES

1. Bazaral, M., and Helinski, D. R. (1968): Circular DNA forms of colicinogenic factors E1, E2 and E3 from Escherichia coli. *J. Mol. Biol.,* 36:185–194.
2. Chang, A. C. Y., Lansman, R. A., Clayton, D. A., and Cohen, S. N. (1975): Studies of mouse mitochondrial DNA in Escherichia coli: structure and function of the eukaryotic-prokaryotic chimeric plasmids. *Cell,* 6:231–244.
3. Clarke, L., and Carbon, J. (1975): Biochemical construction and selection of hybrid plasmids containing specific segments of the Escherichia coli genome. *Proc. Natl. Acad. Sci. U.S.A.,* 72:4361–4365.
4. Clarke, L., and Carbon, J. (1976): A colony bank containing synthetic Co1E1 hybrid plasmids representative of the entire Escherichia coli genome. *Cell,* 9:91–99.
5. Clowes, R. C. (1964): Transfert génétique des facteurs colicinogénes. (Symp. on Bacteriocinogy, Paris, May, 1975.) *Ann. Inst. Pasteur,* 107 (Suppl. 5):74–92.
6. Davis, J. K., and Reeves, P. (1975): Genetics of resistance to colicins in Escherichia coli K-12: cross-resistance among colicins of group A. *J. Bacteriol.,* 123:102–117.
7. Davis, R., Simon, J., and Davidson, N. (1971): Electron microscope heteroduplex methods for mapping regions of base sequence homology in nucleic acids. *Methods Enzymol.,* 21:413–428.
8. Fredericq, P., and Betz-Bareau, M. (1953): Transfert génétique de la propriété colicinogéne chez Escherichia coli. *C. V. Seanc. Soc. Biol. Paris,* 147:2043.
9. Grunstein, M., and Hogness, D. S. (1975): Colony hybridization: a method for the isolation of cloned DNAs that contain a specific gene. *Proc. Natl. Acad. Sci. U.S.A.,* 72:3961–3965.
10. Hardy, K. B. (1975): Colicinogeny and related phenomena. *Bacteriol. Rev.,* 39:464–515.
11. Hershfield, V., Boyer, H. W., Yanofsky, C., Lovett, M. A., and Helinski, D. R. (1974): Plasmid Co1E1 as a molecular vehicle for cloning and amplification of DNA. *Proc. Natl. Acad. Sci. U.S.A.,* 71:3455–3459.
12. Jackson, D. A., Symons, R. H., and Berg, P. (1972): Biochemical method for inserting new genetic information into DNA of simian virus 40: circular DNA molecules containing lambda phage genes and the galactose operon of Escherichia coli. *Proc. Natl. Acad. Sci. U.S.A.,* 72:2904–2909.
13. Kedes, L. H., Chang, A. C. Y., Houseman, D., and Cohen, S. N. (1975): Isolation of histone genes from unfractionated sea urchin DNA by subculture cloning in E. coli. *Nature,* 255:533–538.

14. Lauer, G. D., and Klotz, L. C. (1975): Determination of the molecular weight of Saccharomyces cerevisiae nuclear DNA. *J. Mol. Biol.,* 95:309–326.
15. Lobban, P. E., and Kaiser, A. D. (1973): Enzymatic end-to-end joining of DNA molecules. *J. Mol. Biol.,* 78:453–471.
16. Mandel, M., and Higa, A. (1970): Calcium-dependent bacteriophage DNA infection. *J. Mol. Biol.,* 53:159–162.
17. Miller, J. (1972): *Experiments in Molecular Genetics.* Cold Spring Harbor Laboratory, Cold Spring Harbor, New York.
18. Morrow, J. F., Cohen, S. N., Chang, A. C. Y., Boyer, H. W., Goodman, H. M., and Helling, R. B. (1974): Replication and transcription of eukaryotic DNA in Escherichia coli. *Proc. Natl. Acad. Sci. U.S.A.,* 71:1743–1747.
19. Nathans, D., and Smith, H. O. (1975): Restriction endonucleases in the analysis and restructuring of DNA molecules. *Annu. Rev. Biochem.,* 44:273–293.
20. Roth, J. (1970): Genetic techniques in studies of bacterial metabolism. *Methods Enzymol.,* 17:3–35.
21. Schachat, F. H., and Hogness, D. S. (1973): Repetitive sequences in isolated Thomas circles from Drosophila melanogaster. *Cold Spring Harbor Symp. Quant. Biol.,* 38:371–381.
22. Somers, J. M., Amzallag, A., and Middleton, R. B. (1973): Genetic fine structure of the leucine operon of Escherichia coli. *J. Bacteriol.,* 113:1268–1272.
23. Struhl, K., Cameron, J. R., and Davis, R. W. (1976): Functional genetic expression of eukaryotic DNA in Escherichia coli. *Proc. Natl. Acad. Sci. U.S.A.,* 73:1471–1475.
24. Wensink, P. C., Finnegan, D. J., Donelson, J. E., and Hogness, D. S. (1974): A system for mapping DNA sequences in the chromosomes of Drosophila melanogaster. *Cell,* 3:315–325.

33

Reprinted from *Science* **202**:1279–1284 (1978)

CLONING HUMAN FETAL γ GLOBIN AND MOUSE α-TYPE GLOBIN DNA: PREPARATION AND SCREENING OF SHOTGUN COLLECTIONS

F. R. Blattner, A. E. Blechl, K. Denniston-Thompson, H. E. Faber, J. E. Richards, J. L. Slightom, P. W. Tucker, and O. Smithies

The development of recombinant DNA techniques permitting clonal replication of eukaryotic DNA segments in bacteria has brought a revolutionary change in approach to genetic research. This set of techniques allows a DNA fragment containing a gene of interest to be replicated as a clone in a bacterium and opens the way to isolation of mammalian genes.

The first mammalian clones to be isolated in this way were complementary DNA (cDNA) copies of messenger RNA's (mRNA's) (*1*). Cloning of cDNA's takes advantage of the fact that comparatively pure mRNA's for specific genes can be isolated from cells specialized to produce large amounts of specific protein. However, this approach is limited to genes that produce an RNA product. While cDNA clones made from mRNA are useful, they lack intervening and regulatory sequences present in the genome.

A second major advance was the cloning of specific DNA segments partially purified from genomic DNA by column chromatography, R-loop centrifugation, or agarose gel electrophoresis (*2*). But physical DNA fractionation techniques are only capable of modest purification, and the effort required can be considerable.

The full power of the recombinant DNA technique lies in its application to unfractionated DNA. This approach (termed a "shotgun experiment") calls for the construction of a collection of recombinant DNA molecules which contains a large enough sample of cloned DNA fragments from a target genome to ensure that the desired single copy gene (or genes) is represented. In this approach, the entire purification of a DNA segment is accomplished by clonal replication. However, the method depends on the ability to recognize the DNA fragment of interest after it has been cloned.

The shotgun technique has been applied to organisms of low or moderate complexity such as *Escherichia coli* or yeast (*3*), as well as to the isolation of multiple copy genes from more complex organisms (*4*). In this and the following report (*5*), we describe the initial use of the shotgun strategy with mammalian genomes. We have used the procedure to isolate genomic clones for mouse α-type globin and human fetal γ globin DNA.

Both regulatory and technical difficulties have had to be overcome before the shotgun method could be used to isolate single copy mammalian genes. The U.S. regulations pertaining to the cloning of mammalian DNA have required that the cloning vectors and all techniques that might affect the biological containment of these vectors must be certified by the National Institutes of Health (NIH) (*6*). For these studies we have developed the cloning vectors Charon 3A, Charon 4A, Charon 16A, and Charon 21A (*7*) and techniques for in vitro encapsidation of the recombinant DNA molecules (*8*). These vectors and techniques have been certified by NIH as meeting the EK2 level of biological containment.

The second problem was the generation of very large collections of independent recombinant DNA molecules. Depending on the fragment size into which the genome is divided, from 10^6 to 10^7 individual clones are needed to overcome the statistical problem of sampling the mammalian genome. In principle, this could be done by scaling up each of the steps used in recombinant DNA experiments. However, a more desirable approach was to improve the efficiency of the techniques, particularly of the very inefficient transfection step in which phage DNA containing inserts of foreign DNA is introduced into *E. coli* cells. We therefore replaced this step with a modification of the technique for in vitro encapsidation of λ phage DNA (*9–11*) which led to an increase of about 100-fold in efficiency.

The strategy that was adopted for our initial experiments was to produce a shotgun collection by the simplest possible means. Our primary interest is the study of the human and mouse genomes, especially genes such as the globins, which show differential expression at different times of development and are involved in well-characterized genetic diseases (*12*).

Complete Eco RI digests of mouse and human DNA were the source of "random" genomic target molecules. These could be efficiently inserted into the cloning vector Charon 3A by ligation of their cohesive Eco RI termini. The use of complete Eco RI digests has the disadvantage that fragments exceeding the capacity of the vector will be difficult or impossible to clone. However, Charon 3A will accommodate a large range of Eco RI fragments [0 to 11 kilobase pairs (kbp) in theory]. Furthermore, some of the globin gene fragments produced by Eco RI digestion are known to be in this size range (*13*).

We expected that the most difficult step would be the screening of such a large population of phages to identify one carrying the gene of interest. Surprisingly, no new techniques were needed. We found that the radioactive plaque screening technique (*14*) could be scaled up by the use of cafeteria trays in place of petri dishes, so that a full mammalian genome could, in principle, be screened on two such plates.

Since probes for an increasing repertoire of genes are becoming available through cDNA cloning, one of the most attractive features of the plaque hybridization technique is the ability to screen for several genes at once. We initially decided to screen with a mixture of radioactive probes for four different genes. Although our efforts were frustrated by technical problems and we only obtained clones for globin genes in the first experiments, the use of multiple probes remains an important option for future work.

Target DNA's were purified by a method similar to that of Blin and Stafford (*15*). Mouse DNA was isolated from livers of random bred CD1 adults (Charles River Mouse Farms). In compliance with NIH regulations, human

Table 1. Comparison of three methods for introducing λ DNA into bacteria (9–11, 18, 19). The efficiency of maturation is defined as the number of plaque-forming units formed per input DNA molecule. (1 μg of λ DNA corresponds to 2×10^{10} molecules.) The efficiency of ligation is the ratio of plaque-forming units produced by digested and ligated vector and target DNA's to the number of plaque-forming units produced by undigested vector DNA. The fraction of output phages containing inserts varies, depending on the vector used. The amounts of target and vector Ch3A DNA needed to produce 10^7 clones were calculated on the basis of the efficiency of maturation. The volume of cultures needed to produce 10^7 clones was estimated on the assumption it would be possible to scale up the procedures directly.

Method	Calcium shock	Sphero-plasts	In vitro packaging
Efficiency of DNA maturation	10^{-6}	10^{-5}	10^{-3}
Efficiency of ligation	5×10^{-3}	5×10^{-3}	5×10^{-3}
Fraction of output phages containing foreign DNA	1/3	1/3	1/3
Number of λ DNA molecules needed to produce 10^7 clones	6×10^{15}	6×10^{14}	6×10^{12}
Mass of vector DNA needed to produce 10^7 clones	240 mg	24 mg	240 μg
Mass of target DNA needed to produce 10^7 clones	24 mg	2.4 mg	24 μg
Volume of cultures needed	400 liters	30 liters	2 liters

DNA was isolated from a primary cell culture of embryonic fibroblasts (*16*).

Before propagation of the Charon 3A vector, the Eco RI fragment carrying the *lac* Z gene, which had been inserted for safety testing, was removed by cutting with Eco RI, rejoining the cohesive ends, and transfecting spheroplasts. The resulting phage was designated Ch3AΔ1ac. Retention of all safety mutations was verified (data not shown). The target DNA and vector DNA with the λ cohesive ends annealed were digested separately with Eco RI and mixed in a 1:1 molar ratio of ends (vector: target, 7:1, by weight). The overall DNA concentration was 750 μg/ml. Ligation of DNA molecules at such a high concentration (*17*) favors the production of concatenated molecules whose lengths are many times that of λ. Most of the ligated DNA failed to enter a 0.7 percent agarose gel (data not shown).

The statistical distribution of fragments in such concatenates ought to follow a simple distribution with zero, one,

and more than one target fragment occurring between vector fragments about equally often. In half these cases, vector fragments bracketing a target fragment would be expected to be oriented in the direction required to obtain infective particles. We would expect that clones containing two or more Eco RI fragments from unrelated portions of the target genome would not be uncommon, although clones containing large multiple inserts would be eliminated when the size of DNA exceeded the limit that could be fitted into a λ capsid.

Three techniques were considered for producing viable phage from naked recombinant DNA molecules: calcium shock transfection (*18*), spheroplast transformation (*19*), and in vitro packaging (*9*–*11*). Table 1 gives a comparison of the efficiencies of these techniques. Although the first two methods may be suitable for production of a few thousand phages, the volume of competent cells as well as the amount of vector and target DNA becomes prohibitive when a minimum of 10^6 or 10^7 individual phages must be produced from a ligation mixture. Therefore, we used in vitro packaging to introduce the recombinant molecules into bacteria. In this technique, cell-free extracts made from λ lysogens provide proheads, tails, and all other compo-

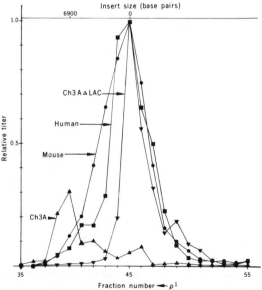

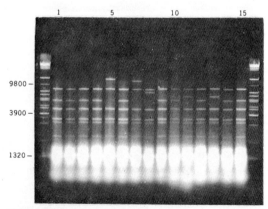

Fig. 1 (left). Isopycnic centrifugation of Charon 3A phage shotgun collections of mouse and human genomes. The human shotgun (L3) (■), the mouse shotgun (●), and a control tube containing Charon 3A (▲) and Charon 3AΔ1ac (▼) were centrifuged to equilibrium in a CsCl gradient of average density $\rho = 1.5$ g/ml in a Beckman SW 50.1 rotor at 30,000 rev/min. The tubes were punctured, and fractions were collected and titered. To distinguish Ch3A from Ch3AΔ1ac in the control tube, plates containing 5-bromo-4-chloro-3-indolyl-β-D-galactoside were used (*7*). The peak fraction of each gradient was normalized to 1.0 before plotting. The results were virtually identical for the other human shotgun L2. Fig. 2 (right). DNA inserts in random phages from a human shotgun collection. Fifteen plaques were picked at random from the human shotgun L3. Lysates were grown, DNA was extracted from 2-ml portions (*29*) digested with Hpa I, and subjected to electrophoresis through 0.7 percent agarose gels that were stained with ethidium bromide and photographed under ultraviolet illumination. DNA fragments of known size were run in the outside channels. The sizes are 1320, 1990, 3110, 3600, 3900, 4590, 4925, 5690, 5950, 6760, 7560, 9570, 12400, and 19800 base pairs, a few of which are indicated in the figure. Samples 5, 7, 8, 11, and 13 show inserts manifested by changes in the migration of the band at 5.8 kbp. The sizes of the inserts measured from this gel ranged from 1 to 6.5 kbp. Comparable results were obtained for the other human shotgun L2 and the mouse shotgun.

nents needed to assemble phage in the test tube. Encapsidated DNA is then introduced into cells by infection.

For preparing a shotgun collection of phages containing random-length fragments, the packaging method must be efficient and its efficiency must be reasonably constant for molecules of different length. In addition, to preserve the safety features of the Charon vector system, there must be no packaging of endogenous λ DNA from the lysogens used to produce extracts, and no recombination between the vector DNA and this endogenous λ DNA.

Our technique (20) incorporates the basic procedure of Becker and Gold (9) which we modified by the use of extracts from strain NS428 described by Sternberg *et al.* (11) and by the addition of putrescine to the buffer (25) suggested by the work of Hohn and Murray (10). Use of the lysogenic strain NS428 to produce all extracts contributes to biological safety in two ways. (i) Since the cells and

Table 2. Construction of shotgun collections for mouse and human DNA. Charon 3AΔlac DNA and human or mouse target DNA were digested and ligated (17). The ligated DNA was encapsidated by in vitro packaging in the volumes shown (20). The titer of the packaged phages was then determined to ascertain the proportion of input phage DNA molecules encapsidated. The Ch3AΔlac control shows the efficiency of packaging achieved for DNA which had not been digested and ligated.

DNA	DNA (μg)		Total packaging volume (ml)	λ DNA equivalents in	Viable phage out	Packaging efficiency
	Ch3AΔlac	Genomic				
Human L2	20.4	2.52	1.57	4.97×10^{11}	4.55×10^6	9.1×10^{-6}
Human L3	61.2	7.56	4.71	1.49×10^{12}	3.76×10^6	2.5×10^{-6}
Mouse	40.8	5.04	3.14	9.88×10^{11}	2.1×10^7	2.1×10^{-5}
Ch3AΔlac control	0.7	0.0	0.208	1.7×10^{10}	7.2×10^7	4.2×10^{-3}

the prophage are recombination-deficient, recombination in the extracts between the Charon vector and the endogenous λ DNA is eliminated. (ii) Since the cells cannot produce A protein, prior packaging of endogenous λ DNA during induction of lysogens is blocked at the first step. There is no need to physically inactivate (by ultraviolet irradiation) prophage DNA from the extracts. Use of the spermidine-putrescine buffer along with isolated A protein allows a high efficiency of packaging for all size classes of recombinant molecules. Current NIH regulations require that these safety features should be verified for each set of encapsidation extracts (21).

Three shotgun collections, two of human and one of mouse DNA fragments, were constructed (Table 2). In each case, more than 10^6 phages were produced. The efficiency of packaging of ligated DNA was between 10^{-2} and 10^{-3} of that obtained with untreated vector DNA.

Two methods were used to determine the distribution of inserts in these populations (Figs. 1 and 2). The first method, which can also be used preparatively to enrich for phages with inserts, was to subject samples of the shotguns to isopycnic centrifugation in CsCl gradients. Both mouse and human shotgun preparations exhibit broad density profiles with shoulders of density greater than that of the parent phage Ch3AΔlac (Fig. 1). From the area under the shoulder, we estimate that up to 40 percent of the phages in each shotgun contain inserts. However, very few appear to contain inserts larger than 6.7 kbp, the size of the inserts in the Ch3A marker (7).

The second method was to examine Hpa I digests of the DNA from plaques selected at random from the shotguns (Fig. 2). Inserts can be detected by changes in the mobility of the 5.8-kbp Hpa I fragment which carries the Eco RI cloning site. If the insert has Hpa I sites, more than one band will be produced. For example, Fig. 2 shows that 5 of the

15 candidates from human shotgun L3 had inserts and, in agreement with the density studies, the insert sizes ranged from 0.96 to 6.7 kbp. Thus the insert sizes observed covered a smaller range than the theoretical maximum capacity of Ch3A.

The Benton-Davis screening procedure consists of hybridization of radioactive nucleic acid probes to nitrocellulose replicas of lawns containing

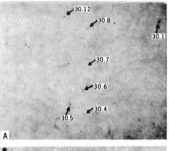

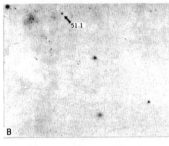

Fig. 3. Initial screening of mouse and human shotgun collections; 300,000 plaques each were screened by hybridization to nitrocellulose replicas (22, 23). (A) In the case of mouse, ^{32}P-labeled probes specific for rat insulin, rat growth hormone, and mouse α and β globin were used (24); poly(rA) was not included in the hybridization mixture. (B) For the human collection, a ^{32}P-labeled probe specific for human α, β, and γ globin was used; poly(rA) was included in the hybridization mixture (23). The figure presents approximately 25 percent of one of two autoradiograms taken from each filter. Arrows point to the signals that appeared on both films. Plaques were picked from beneath these signals and numbered as illustrated.

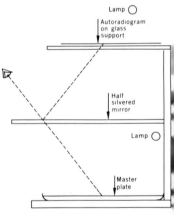

Fig. 4. The "Tucker" box. This device, based on a technique widely used for crystallographic model building, proved indispensable for picking candidates from megaplates requiring P3 containment. It allowed accurate picking of plaques under a laminar flow hood without contamination of the autoradiogram or the plate. Since frequent alignment marks were used (22), it also permitted compensation for the approximately 1 percent shrinkage of the nitrocellulose replica. A half-silvered mirror is supported midway between the top surface of the megaplate and the autoradiogram, which rests on a sheet of glass. The sides of the box, which provide support, are omitted from the drawing. The box was designed to fit easily within a laminar flow hood. Variable transformers were used on the lamps to allow light intensity adjustment for best illumination. The observer sees images of the autoradiographic spots and alignment marks in the same plane as the top agar of the megaplate and can pick the corresponding plaques.

Fig. 5. Characterization of mouse genomic clones by competition hybridization. (A) Three phages, Mm30.5, Mm117Hbα, and Mm28.2, labeled 1, 2, and 3, respectively, of each were plated and two replicas of each were hybridized with a ^{32}P-labeled mouse globin cDNA probe (24). The top group was hybridized in the absence of poly(rA); the bottom group was hybridized with poly(rA). (B) Hpa I restriction digests of DNA from the same phages were prepared and subjected to agarose gel electrophoresis in two duplicate runs on a single gel (29). The ethidium bromide staining pattern of the gels is shown. Southern transfers (27) were made of the replicate gel portions, and each was hybridized with the ^{32}P-labeled mouse globin probe with (bottom) and without (top) poly-(rA).

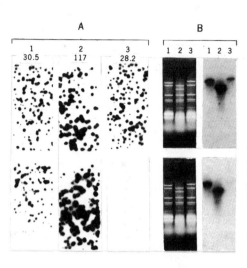

A

B

| 1 | 2 | 3 |
| 30.5 | 117 | 28.2 |

bacteriophage plaques in order to identify particular phages that contain homologous DNA sequences (14). Since plaques must be at least marginally separated to allow detection, it is necessary to allow about 2 to 4 cm² of lawn per 1000 phages screened. Our screening was scaled to the film (33 by 43 cm) used for medical x-rays.

Phages were plated on megaplates that were poured in aluminum foil–lined cafeteria trays (35 by 45 cm) (22). Large sheets of nitrocellulose were used to lift replicas. The DNA–DNA hybridization was done at 68°C in photography trays (23).

The first experiment to yield a genomic clone for a single copy gene also revealed that the mouse genome contains numerous clonable DNA segments that give very strong positive hybridization to radioactive polyadenylate [poly(A)] regions in the probes. In this experiment four megaplates (Nos. 28 to 31) from the mouse shotgun were screened. The autoradiogram from plate 30 is shown in Fig. 3A. A mixture of radioactive probes was used, including cDNA made from mouse α- and β-globin mRNA as well as nick-translated plasmid DNA's containing cDNA copies of rat insulin and rat growth hormone mRNA's (24). Forty-two positive hybridizing spots, including faint ones (Fig. 3A), were picked by the use of an alignment system (Fig. 4). All candidates were replated on standard petri dishes and rescreened by hybridization. Twenty-one candidates did not hybridize upon rescreening; these candidates correlated well with the faintest and most questionable spots on the original autoradiogram.

One of the candidates still hybridized specifically to the globin probe on retesting. This clone has been shown to contain α-type globin DNA (5) and has been designated Charon 3AMm30.5α, which can be abbreviated Mm30.5. Depending on the emphasis, this designation can be abbreviated by omitting any elements from the name but preserving their order.

The remaining 20 candidates on retesting hybridized both to globin probe and to a mixture of the three other probes. A few of these were tested further and shown to hybridize to all four probes individually. Since the only characteristic shared by the probes was their content of poly(A) regions, we suspected that A·T-rich (A, adenine; T, thymine) regions of the genome were being cloned. Mol *et al.* (25) have shown that the mammalian genome contains A·T-rich regions that can be isolated on oligo dT columns. Jeffreys and Flavell (26) have also included poly(rA) (riboadenylate) in their recent Southern hybridization (27) experiments to reduce the background of nonspecific hybridization. Therefore, to test the A·T hypothesis, competition hybridization experiments with poly(rA) were done by both plaque and Southern gel hybridization techniques. Figure 5 shows the results for one of the candidates, Mm28.2. Hybridization to the band containing the insert in Mm28.2, as well as hybridization to the plaques of this phage, could both be inhibited fully with poly(rA). All 20 of the questionable candidates were then tested for inhibition with poly(rA) by the plaque hybridization technique. All were inhibited, although hybridization to Mm30.5 and the control Mm117Hbα was

unaffected by poly(rA). We consequently conclude that the hypothesis that these are clones of A·T-rich mouse DNA segments is supported.

Gel electrophoresis of Hpa I digests of eight of the clones of A·T-rich DNA was carried out. All had different-sized inserts. This shows that the A·T sequences occur in many contexts in the genome. We found that they could be avoided by including poly(rA) in the hybridization mixture.

In the search for human globin genes, we used nine master plates to screen a total of 3×10^6 phage plaques, of which approximately 1×10^6 contained fragments of genomic DNA. Five master plates from human shotgun L3, screened with mouse globin probe without poly(rA), yielded 107 globin candidates. None of these were globin-positive on retesting in the presence of poly(rA). Four other master plates, three from L3 and one from L2, were screened with human globin probe in the presence of poly(rA). They yielded ten candidates and, from these, one human clone, Hs51.1, was obtained (from shotgunL2). The autoradiogram from master plate 51 permitted the isolation of this human globin clone (Fig. 3B). It is evident (Fig. 3, compare A and B) that the addition of poly(rA) to the hybridization mixture improved the efficiency of the technique. It is possible that the use of probe from the same species also increases the probability of success, but we doubt it because both Hs51.1 and Mm30.5 hybridize well to both mouse and human probe.

To determine whether Mm30.5 and Hs51.1 are in fact clones of fragments of DNA that are present in the genome, Southern hybridization experiments were done. Eco RI was used to excise the hybridization positive DNA fragments from each of these clones. The fragments were partially purified on agarose gels, labeled with ^{32}P by nick translation (24), and used as probes for Southern hybridization to Eco RI–digested human or mouse DNA. The autoradiogram of the mouse experiment shows a prominent band of 4.7 kbp, the same size as the insert in Mm30.5 (5). In addition, when the autoradiogram was exposed for longer times, several weaker bands of 2.34, 5.25, 10.5, 12.0, 14.7, and 18.5 kbp were seen. Thus Mm30.5 is clearly a genomic clone of a 4.7-kbp fragment, but some sequences present in the cloned fragment are also present elsewhere in the genome. The experiment with Hs51.1 probe using human DNA also shows a band at the same position as the 2.7-kbp insert in Hs51.1 (5). In addition, a band of 7.4 kbp is seen. This band is rougly equal in intensity to the 2.7-kbp band. No additional bands were seen with increased exposure of this autoradiogram. Therefore, we conclude that both Mm30.5 and Hs51.1 are genomic clones.

We have demonstrated the feasibility of cloning single copy genes from shotgun collections of phages constructed from unfractionated complete Eco RI digests of human and mouse DNA. Very little pure DNA is required for such experiments, and it need be manipulated only to the extent of digestion and ligation into the cloning vehicle. The key to obtaining this high efficiency is the use of in vitro packaging as the method for introducing recombinant DNA molecules into bacteria (9–11).

Two clones specific for globin sequences were isolated, one from each of the target genomes. Mouse globin cDNA is perfectly satisfactory for detecting plaques of the human globin clone Hs51.1, and human globin cDNA will detect the mouse globin clone Mm30.5. This principle will likely be useful in other cases for cloning the same gene from different species.

Our first approach was to screen each shotgun megaplate with several probes in the hybridization mixture. In this way we had hoped to isolate several interesting genes at once. Although this method is sound in principle, it is necessary that each probe used be free of anything that could hybridize nonspecifically. In many cases it may be simpler to screen the plates with each probe individually, since up to five nitrocellulose replicas can be taken from a single master plate. In this way, if any of the probes proves unsatisfactory, the whole experiment is not lost.

The major difficulty encountered has been the relatively common occurrence of simple sequence DNA's in the mammalian genome. Clones of A·T-rich DNA fragments yield hybridization-positive plaques when using probes containing stretches of poly(A) or poly(T) unless the hybridization is inhibited by poly(rA).

A second problem, of which we were aware at the outset, concerns the use of shotguns made from complete Eco RI digests. Such digests of mammalian DNA yield about 10^6 independent fragments, many of which contain only parts of the genes of interest. This causes problems in detecting clones corresponding to all the parts of desired genes, since some may not have sufficient DNA homologous to the probe to give a strong signal.

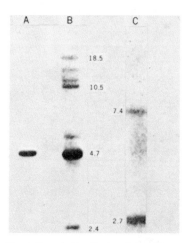

Fig. 6. Genomic identification of cloned fragments. Mouse and human DNA samples (25 μg) were completely digested with Eco RI and subjected to electrophoresis on 0.7 percent agarose gels. DNA was transferred to nitrocellulose filters by the Southern method (27) and hybridized. Radioactive probes were prepared by nick translation of partially purified cloned fragments that were isolated from the genomic clones by electrophoresis through 0.7 percent agarose gels. (A) ^{32}P-labeled 4.7-kbp fragment from Mm30.5 hybridized to mouse genomic DNA; the exposure time was 12 hours. (B) Same as (A), except that the exposure time was 40 hours. (C) ^{32}P-labeled 2.7-kbp fragment from Hs51.1 hybridized to human genomic DNA; the exposure time was 24 hours. Panel (C) has been photographically enlarged twofold to compensate for differences in the lengths of the gels.

The present type of shotgun collections are very simple to construct and use and are fine when the target is known to be on an Eco RI fragment smaller than 7 kbp. However, we are constructing new shotguns from DNA that contains partial as well as complete Eco RI digests, using our EK2 vector Charon 4A. This vector accepts inserts up to 22 kbp and will not grow with inserts less than 8 kbp. We expect that fewer clones will need to be screened with this strategy and that large (and small) gene fragments will be less likely to be missed.

The ability which we have demonstrated to isolate single copy mammalian genes from random shotgun collections of phages opens the way to the study of almost any gene from organisms of any complexity, provided that a reasonable probe can be obtained (28).

References and Notes

1. R. Higuchi, G. V. Paddock, R. Wall, W. Salser, *Proc. Natl. Acad. Sci. U.S.A.* 73, 3146 (1976); T. Maniatis, S. G. Kee, A. Efstratiadis, F. C. Kafatos, *Cell* 8, 163 (1976); F. Rougeon, P. Kourilsky, B. Mach, *Nucleic Acids Res.* 2, 2365 (1975); T. H. Rabbitts, *Nature (London)* 260, 221 (1976).
2. S. M. Tilghman, D. C. Tiemeier, F. Polsky, M. H. Edgell, J. G. Seidman, A. Leder, L. W. Enquist, B. Norman, P. Leder, *Proc. Natl. Acad. Sci. U.S.A.* 74, 4406 (1977); S. Tonegawa, E. Brack, N. Hozumi, R. Schuller, *ibid.*, p. 3518.
3. L. Clarke and J. Carbon, *Cell* 9, 91 (1976); R. A. Kramer, J. R. Cameron, R. W. Davis, *ibid.* 8, 227 (1976); K. Struhl, J. R. Cameron, R. W. Davis, *Proc. Natl. Acad. Sci. U.S.A.* 73, 1471 (1976).
4. L. H. Kedes, A. C. Y. Chang, D. Houseman, S. N. Cohen, *Nature (London)* 255, 533 (1975).
5. O. Smithies, A. E. Blechl, K. Denniston-Thompson, N. Newell, J. E. Richards, J. L. Slightom, P. W. Tucker, F. R. Blattner, *Science* 202, 1284 (1978).
6. *Fed. Regist.* 41, 27902 (1975).
7. F. R. Blattner, *et al.*, *Science* 196, 161 (1977); F. R. Blattner, D. O. Kiefer, D. D. Moore, J. deWet, B. G. Williams, Application for EK2 Certification of a Host Vector System for DNA Cloning, Supplement IX: "Data on Charon 21A."
8. H. E. Faber, D. Kiefer, F. R. Blattner, Application for EK2 Certification of a Host Vector System for DNA cloning, Supplement X: "Data on In Vitro Packaging Method."
9. A. Becker and M. Gold, *Proc. Natl. Acad. Sci. U.S.A.* 72, 581 (1975).
10. B. Hohn and K. Murray, *ibid.* 74, 3259 (1977).
11. N. Sternberg, D. Tiemeier, L. Enquist, *Gene* 1, 255 (1977).
12. H. F. Bunn, B. G. Forget, H. M. Ranney, *Human Hemoglobins* (Saunders, Philadelphia, 1977); D. J. Weatherall and J. B. Clegg, *The Thalassemia Syndromes* (Blackwell, Oxford, 1972).
13. S. M. Tilghman *et al.*, *Proc. Natl. Acad. Sci. U.S.A.* 74, 4406 (1977); J. G. Mears, F. Ramirez, D. Leibowitz, A. Bank, *Cell* 15, 15 (1978); J. G. Mears *et al.*, *Proc. Natl. Acad. Sci. U.S.A.* 75, 1222 (1978); Y. W. Kan, J. P. Holland, A. M. Dozy, S. Charache, H. H. Kazazian, *Nature (London)* 258, 162 (1975).
14. W. D. Benton and R. W. Davis, *Science* 196, 180 (1977).
15. N. Blin and D. W. Stafford, *Nucleic Acid Res.* 3, 2303 (1976). Confluent monolayers of human fibroblasts (16) in four 500-ml roller bottles were rinsed with phosphate-buffered saline. Cells were lysed by 5 minutes of incubation in 50 ml of 1 percent Sarkosyl NL30, 0.1*M* EDTA, *p*H8, and proteinase K (100 μg/ml) at 56°C. Boiled ribonuclease (100 μg/ml) was added and the mixture was incubated at 56°C for 2 hours. The DNA fraction was extracted three times with a 1:1 mixture of phenol and chloroform-isoamyl alcohol (24:1, by volume). Solid CsCl was added to a final density of 1.70 g/cm³, and the preparation was centrifuged for 60 hours at 30,000 rev/ min (SW50.1 rotor). The DNA band was collected and dialyzed against 0.01*M* tris-Hcl, *p*H 7.0, 0.01*M* NaCl, 0.001*M* EDTA. Approximately 5 g of mouse liver, frozen in liquid nitrogen, was powdered by grinding under liquid nitrogen; then 50 to 100 ml of 0.1*M* EDTA, *p*H 8, with 1 percent Sarkosyl NL30, was heated to 50° to 55°C with magnetic stirring, and proteinase K (100 μg/ml) was added. The frozen liver powder was slowly added over a 1/2-hour period. Incubation at 56°C continued for 2 hours, and the DNA isolation was completed as above.
16. A human embryonic fibroblast culture was prepared by Dr. Robert DeMars in 1971 from a first trimester female abortus carefully dissected to free it from maternal cells, and it was divided in portions after one subculture in vitro and stored in liquid nitrogen. DeMars showed that the cells had a normal female karyotype, that they senesced after prolonged cultivation in vitro, and that contamination with maternal cells was below the limit of detection (10^{-3}) as measured by the fraction of azaguanine-resistant cells. (The mother was heterozygous for the Lesch-Nyhan mutation; the embryo was homozygous for normal alleles [J. S. Felix and R. DeMars, *J. Lab Clin. Med.* 77, 596 (1971)].
17. H. Jacobsen and W. H. Stockmeyer, *J. Chem. Phys.* 18, 1600 (1950); J. C. Wang and N. Davidson, *J. Mol. Biol.* 19, 469 (1969); B. G. Williams, F. R. Blattner, R. K. Jaskunas, M. Nomura, *J. Biol. Chem.* 252, 7344 (1977).
18. W. Mandel and A. Higa, *J. Mol. Biol* 53, 159 (1970).

19. W. D. Henner, I. Kleber, R. Benzinger, *J. Virol.* **12**, 741 (1973).

20. Strain NS428, N100 (λAam 1 1*b*2*red*3*c*I-857*S*am7), was a gift from N. Sternberg and L. Enquist. Strain dg805, W3350 (λdgal805*c*I-857*S*am7) was the gift from A. Becker. Strain NS428 was grown with shaking at 30° to 33°C in 600 ml NZY broth cultures (7) to $A_{575} = 0.2$, as measured on a Bausch and Lomb spectrophotometer (Spec 20). An additional 600 ml of 64°C NZY broth was added to the cultures to bring the temperature tc 45°C, and the flask was incubated at 39°C with shaking for 1 hour (New Brunswick model G25). To test for successful induction a few milliliters of culture were shaken with chloroform and checked for the expected clearing. Freeze thaw lysate (FTL) was prepared by resuspending each pellet in 0.4 ml of cold 10 percent sucrose in 0.05*M* tris-HCl, *p*H 7.4, and transferring it to a 10-ml Oak Ridge type polycarbonate centrifuge tube. Egg white lysozyme (20 μl) (2 mg/ml, freshly dissolved in 0.25*M* tris-HCl, *p*H 7.4) was added with mixing, and the tubes were frozen in liquid nitrogen. Sonic extract (SE) was prepared by resuspending the NS428 pellet in 1.8 ml of buffer A (20 m*M* tris-HCl, *p*H 8, 3 m*M* MgCl$_2$, 0.05 percent 2-mercaptoethanol, 1 m*M* EDTA-KOH, *p*H 7), and immediately sonicated with cooling in an ice-salt bath until an opalescent solution free of intact cells was obtained. The preparation was centrifuged for 6 minutes at 4500*g*, and the supernatant was either used immediately or stored in liquid nitrogen. Protein A (pA) was prepared (A. Becker, personal communication) from six induced cultures of λdg805 prepared as described above for NS428, except that the heating was initiated at an absorbancy of 0.8. The induced cells were centrifuged and resuspended in 100 ml of buffer A and sonicated with cooling; the debris was removed by centrifugation. Polyethyleneimine (Miles, code 23-444) was dissolved to a final concentration of 10 percent in buffer A, and the solution was adjusted to *p*H 8 with HCl. This polyethyleneimine solution was added with stirring in the cold until no further precipitate formed (approximately 10 ml). The precipitate was collected by centrifugation and resuspended with an electric tissue homogenizer in 500 ml of 0.05*M* ammonium succinate (*p*H 6) containing 0.035 percent (by volume) 2-mercaptoethanol and centrifuged again; the pellet was resuspended in 500 ml of 0.1*M* ammonium succinate (as above) and centrifuged again. The pellet was resuspended in 0.2*M* ammonium succinate. The solution was again centrifuged, and the supernatant was saved. Neutralized ammonium sulfate (NAS) was prepared by adjusting saturated ammonium sulfate containing 0.1 percent 2-mercaptoethanol to *p*H 6.5; then 220 ml of NAS was added to the supernatant with stirring in the cold for 20 minutes. The precipitate was removed by centrifugation and 290 ml of NAS plus 15 g of solid ammonium sulfate was added to the supernatant. The precipitate was stirred for 30 minutes in the cold, collected, and resuspended in 5.5 ml of buffer A without MgCl$_2$ and dialyzed against the same buffer. One volume of glycerol was added, and the preparation was subdivided and stored in liquid nitrogen. Immediately before packaging, the FTL was thawed (<5°C) and held on ice for 45 minutes. Next, 0.05 ml of buffer M1 [6 m*M* tris-HCl (*p*H 7.4), 30 m*M* spermidine, 60 m*M* putrescine, 18 m*M* MgCl$_2$, 15 m*M* ATP (from a 0.1*M* stock at *p*H 7.0), 0.2 percent (by volume) 2-mercaptoethanol] was added, mixed, and centrifuged at 35,000 rev/min at 4°C for 25 minutes in a precooled type 65 rotor. Buffer A (30 μl) was mixed with 4 μl DNA (<1 μg), 4 μl buffer Ml, 20 μl thawed sonic extract, and 1 to 2 μl pA and incubated 15 minutes at room temperature. FTL (150 μl) was added and the mixture was incubated for 60 minutes at room temperature. The resulting phages were then plated.

21. To verify that the safety features of the Charon 3A vector were not compromised by the use of in vitro packaging, two tests were done on each batch of extracts as required by the NIH. (i) To determine whether endogenous DNA from the extracts was packaged, sham experiments were done in which no exogenous DNA was added to a 10× scale packaging mixture (2 ml final volume). The entire output was plated on 20 plates with lawns of bacteria permissive for the prophage in NS428. In five such experiments no phages were seen (<1.5 × 10⁻⁷ to <7.5 × 10⁻⁹ of the level that would be produced if vector DNA had been added). (ii) To determine whether amber mutations of the vector, 10× scale packaging experiments were done with vector DNA alone. In this case, the entire output was plated on a *su*° bacterial lawn to determine levels of amber⁺ revertants. In five such experiments the highest level of reversion seen was 2 × 10⁻⁷.

22. Cafeteria trays (35 by 45 cm) were lined with aluminum foil, sterilized with ethanol and ultraviolet and filled with 1.5 liters of bottom agar (7). After hardening in a still air hood, they were covered with sterile glass and transferred to P3 containment. Approximately 5 × 10⁵ phages were mixed with 8 ml of stationary phase bacteria plus 8 ml of 0.01*M* MgCl$_2$ and 0.01*M* CaCl$_2$ and incubated at 37°C for 10 minutes. The mixture was then added to 150 ml of melted top agar (7), mixed, spread evenly over the warm, level surface of the megaplate and allowed to harden for 30 minutes before incubation at 37°C overnight. After development of the lawn the plate was chilled at least 12 hours in a refrigerator. The glass tops were lined with sheets of paper to reduce condensation. Nitrocellulose sheets type BA85 (33 by 45 cm) or rolls (33 cm by 3 m) (Schleicher and Schuell) were cut to size, rinsed in distilled water and 6× SSC (1× SSC is 0.15*M* NaCl, 0.015*M* sodium citrate, *p*H 7.2) and blotted on 3MM paper. These were placed on the surface of the chilled plate for 4 minutes. To aid in subsequent alignment, a series of marks spaced every 3 inches were made by penetration with a syringe needle dipped in India ink. The filter was then transferred to a bed of 3-MM paper saturated with 1.5*M* NaCl, 0.2*N* NaOH for 4 minutes and then neutralized for 4 minutes on a bed of the same paper saturated with 0.5*M* tris-HCl, *p*H 7.2, blotted and allowed to dry. Up to five filters can be lifted from a single plate with chilling between transfers. The nitrocellulose was then incubated between 3MM sheets for 2 hours at 80°C, and washed for 1 hour with gentle shaking at 68°C in 6× SSD (6× SSC, 0.02 percent Ficoll, 0.02 percent polyvinylpyrrolidone, 0.02 percent bovine serum albumin, 0.5 percent sodium dodecyl sulfate) [see D. T. Denhardt, *Biochem. Biophys. Res. Commun.* **23**, 641 (1966)].

23. The hybridization mixture contained poly(rA) (30 μg/ml; Sigma code P-8878), denatured sonicated *E. coli* DNA (10 μg/ml) in 6× SSD. For each megaplate filter, 1 × 10⁸ count/min of probe (24) was added to 300 ml of the mixture in a flat-bottomed photography tray. The tray was covered with a glass sheet, sealed within a plastic garbage bag, and incubated for 24 hours, with gentle shaking at 68°C. The filter was washed two or three times at 68°C (1 hour each) with shaking in 1 liter of 3 × SSC containing 0.5 percent sodium dodecyl sulfate, blotted, dried in air, and mounted for autoradiography beneath Kodak XR-1 film and a DuPont Quanta III intensifier screen. Exposure was for 2 days at −90°C. For Southern transfers of mini lysate gels (27), the hybridization procedure was the same, except that the concentration of probe was about 3 × 10⁴ count/min per milliliter. For Southern transfers of genomic DNA gels (26), about 5 × 10⁵ count/min per milliliter probe was used and hybridization was for 40 hours.

24. Human globin mRNA was provided by Drs. Arthur Bank and Francesco Ramirez and mouse globin mRNA was purified from mouse reticulocytes and provided by Dr. Jeffrey Ross. Mouse or human unfractionated mRNA was converted to [α-³²P]-labeled cDNA essentially by the method described [D. L. Kacian and J. C. Myers, *Proc. Natl. Acad. Sci. U.S.A.* **73**, 2191 (1976)]. The plasmids pRGH-1 [P. H. Seeburg, J. Shine, J. A. Martial, J. D. Baxter, H. M. Goodman, *Nature (London)* **270**, 486 (1977)] and pAU-1 [A. Ullrich, J. Shine, J. Chirgwin, R. Q. Pictet, E. Tischer, W. J. Rutter, H. M. Goodman, *Science* **196**, 1313 (1977)], containing rat growth hormone and rat insulin sequences, respectively, were a gift of Dr. Howard Goodman. The DNA's were made radioactive essentially according to the procedure described by T. Maniatis, A. Jeffrey, and D. G. Kleid [*Proc. Natl. Acad. Sci., U.S.A.* **72**, 1184 (1975)]. Double-stranded probes were denatured with NaOH.

25. J. N. Mol, R. A. Flavell, P. Borst, *Nucleic Acids Res.* **3**, 2367 (1967).

26. A. J. Jeffreys and R. A. Flavell, *Cell* **12**, 429 (1977).

27. E. M. Southern, *J. Mol. Biol.* **98**, 503 (1975).

28. *Note added in proof*: Similar work has just been published by T. Maniatis, R. C. Hardison, E. Lacey, C. O'Connell, D. Quon, G. K. Sim, A. Efstratiadis, *Cell*, **15**, 703 (1978).

29. For rapid screening of clones by agarose gel electrophoresis of restriction digests, DNA was prepared directly from lysates. 0.4 ml SDS mix (0.25*M* EDTA, 0.5*M* tris-HCl, *p*H 9, 2.5 percent recrystallized sodium dodecyl sulfate) was added to 2 ml of clarified phage lysate and heated for 30 minutes at 70°C. At this point the samples contained no live phage and could be removed from P3 containment. Then 0.5 ml of 8*M* potassium acetate was added and the mixture was placed on ice for 15 minutes and centrifuged for 20 minutes at 12,000*g*. The supernatant was precipitated with two volumes of ethanol and centrifuged for 30 minutes at 27,000*g*. After the traces of ethanol were removed with a cotton-tipped swab, the pellet was dissolved in 0.4 ml of 0.3*M* sodium acetate, and precipitated again with ethanol. The pellet was dissolved in 50 μl of 1m*M* EDTA, *p*H 8; 1 to 5 μl of the preparation was digested with 2 units of restriction enzyme before agarose gel electrophoresis [T. M. Shinnick, E. Lund, O. Smithies, F. R. Blattner, *Nucleic Acids Res.* **2**, 1911 (1975)].

30. This is paper 2281 from the Laboratory of Genetics at the University of Wisconsin and paper 8 in the series, "Charon Phages for DNA Cloning." Supported by NIH grants GM21812 (F.R.B.), AM20120 and GM20069 (O.S.), CA09075 (K.D.T.), GM06526 and GM07131 (P.W.T.), and GM07133 (J.E.R. and A.E.B.); and an NIH research career development award (to F.R.B.). We thank Howard Goodman for his collaboration with us in developing these procedures while visiting in our laboratory; Andrew Becker for hospitality in his laboratory; and D. O. Kiefer, N. Borenstein, T. C. Szeto, E. Kopetsky, and J. Kucera for technical assistance. This work was done under the NIH guidelines which require EK2, P3 containment.

339

Part V

EXPRESSION OF EUKARYOTIC GENETIC INFORMATION IN *ESCHERICHIA COLI*

Editors' Comments
on Papers 34 Through 38

Up to this point, we have presented some of the discoveries that resulted in recombinant DNA technology, the currently used vectors and techniques, and the use of these procedures for isolation of single copy genes from complex genomes. In Part V, we have collected several papers describing the expression of eukaryotic genes in prokaryotic cells. Of the many potential uses of recombinant DNA technology, few have attracted as much attention (both for economic and humanitarian reasons) as the production of medically important substances. For example, interferon, the rare hormones, the expensive vaccines, and the diagnostic reagents such as specific antibodies, all have been suggested as candidates for application of recombinant DNA technology. The papers in this section demonstrate that such applications are possible and may well be economically feasible.

The initial propagation and expression of eukaryotic DNA in bacterial cells was described by Morrow and colleagues (Paper 34).

This group isolated ribosomal genes from the *Xenopus laevis* genome and inserted them into the plasmid vector pSC101. They observed that when this plasmid segregated into minicells, RNA was recovered that hybridized specifically with *X. laevis* ribosomal DNA. The authors noted at the time, "since the *X. laevis* rDNA... does not code for protein, we do not yet know whether synthesis of proteins can be carried out by eukaryotic DNA in *E. coli*." Although RNA complementary to eukaryotic DNA was being made in *E. coli*, it was not clear if the DNA was being transcribed faithfully, obeying the correct stop and start signals.

These questions were answered two years later when Struhl, Cameron, and Davis (1976) discovered that a specific fragment of the yeast chromosome carried by a λ vector could complement a histidine negative mutation in *E. coli*. During this same time, Ratzkin and Carbon (Paper 35) discovered that several yeast fragments encoded products that could complement some *E. coli* amino acid auxotrophs. Included were leucine and histidine deficiencies in *E. coli* as well as a leucine deletion mutation in *Salmonella typhimurium*. These exciting discoveries suggested that there was no significant barrier to eukaryotic gene expression in *E. coli*.

But it was not destined to be that simple. In 1977 several lines of evidence indicated that the primary DNA sequence of many higher eukaryotic genes contained intervening sequences that must be removed ("spliced out") after transcription to obtain the synthesis of the correct peptide (see Abelson 1979). *E. coli* was thought to be incapable of removing these intervening sequences. The successful expression of yeast functions in *E. coli* reflected the absence of intervening sequences in most of the structural genes. However, expression of products from higher eukaryotes in *E. coli* seemed unlikely. Another problem encountered was that post-translational cleavage and modification of some peptides was required to produce the active protein. Again it seemed unlikely that *E. coli* could accomplish these events. However, there were ways to get around these problems.

Villa-Kamaroff et al. (Paper 36) present a solution to the problem of intervening sequences blocking expression in *E. coli*. These authors constructed recombinant plasmids that contained and expressed in *E. coli*, the genetic information for rat proinsulin by using complementary DNA (cDNA) copies of the specific hormone messenger RNA. Since these were mature mRNA species, the intervening sequences (if any were present) had been removed. The cDNA was inserted in the proper frame for translation using the terminal transferase tailing procedure and the Pst I cleavage site within the β-lactamase gene. (β-lactamase inactivates ampicillin and is the product of the ampicillin resistance gene in the pBR322 plasmid. It is an unusual protein in that it is transported through the *E. coli* cell membrane and can be released by osmotic shock into the culture medium.) The recombinant plasmids constructed were able to produce fusion peptides of the hormone sequences and the β-lactamase protein. A significant advance was the sensitive antigen-antibody "sandwich" technique that detected bacterial colonies producing these peptides (see Broome and Gilbert 1978; Sanzey et al. 1976; Skalka and Shapiro 1977, for methods of immunological detection of colonies and phage plaques producing new products from recombinant DNA products).

Using a similar approach, Chang et al. (1978) reported the construction of plasmids that expressed a protein with enzymatic properties, immunological reactivity, and size characteristics of mouse dihydrofolate reductase. Moreover, *E. coli* cells that contained such recombinant plasmids were resistant to trimethoprim, indicating that the mouse enzyme functioned in *E. coli*. A selectable marker, such as mouse dihydrofolate reductase, that functions in both mammalian and *E. coli* cells may well be used to construct a mammalian version of the yeast–*E. coli* "shuttle" plasmids (Part II, Papers 17 and 18) as well as to study the signals that control transcription and translation in many organisms.

Paper 37 by Martial and coworkers represents another approach to not only producing mammalian gene products in *E. coli* but also for synthesizing them in quantity and controlling the production using well-characterized regulatory systems. This group synthesized DNA complementary to human growth hormone messenger RNA and linked it to a fragment of the *E. coli* *trp*D gene in the plasmid pBR322. A stable fusion protein was made that constituted about 3 percent of the bacterial protein. This fusion protein reacted specifically with antibodies to authentic human growth hormone. The synthesis of this novel peptide was regulated by the control region of the *E. coli* trp operon carried by the plasmid. Another well-characterized regulatory system that has been incorporated into plasmid vectors is that of the lactose operon. The promoter, operator, and a portion of the *lacz* gene have been inserted into colE1 derivative plasmids and used for controlled expression of phage lambda repressor (Backman et al. 1976) and *Xenopus laevis* ribosomal RNA (Polisky et al. 1976).

The final paper in this volume by Burrell and colleagues (Paper 38) describes the insertion of DNA sequences from the Hepatitis

B virus (HBV) into the plasmid vector pBR322. By fusing HBV sequences to the β-lactamase gene using the Pst I cleavage site and terminal transferase tailing methods, these authors were able to obtain expression of a hybrid protein with characteristics of the HBV core antigen. Either HBV core antigen coding DNA did not contain intervening sequences or their presence did not matter for expression of antigenicity. The importance of this discovery was stated by the authors as follows: "Such clones would be...useful for...large scale production of HBV DNA and viral antigens for diagnostic purposes and, possibly, vaccine production."

The findings in these last three papers emphasize the promise that recombinant DNA research holds for future medical applications as well as basic research into gene structure and function. Other studies highlighted in the papers presented in this volume have already yielded information about the basic organization of eukaryotic genomes and some of the mechanisms of control active during transcription and expression of those genomes. Such information might have taken decades to discover with conventional techniques. Uses and advances of recombinant DNA technology continue to be reported at a rapid rate. As an example, two issues of *Science* have been devoted entirely to the dramatic discoveries that have resulted from this technology (*Science* 1977 and 1980). We hope that the basic background covered by these 38 papers will enable others to keep pace with this fascinating revolution in biology.

REFERENCES

Abelson, J. 1979. RNA Processing and the Intervening Sequence Problem. *Ann. Rev. Biochem.* **48**:1035–1069.

Backman, K., M. Ptashne, and W. Gilbert. 1976. Construction of Plasmids Carrying the cl Gene of Bacteriophage λ. *Natl. Acad. Sci. (USA) Proc.* **73**:4174–4178.

Broome, S., and W. Gilbert. 1978. Immunological Screening Method to Detect Specific Translation Products. *Natl. Acad. Sci. (USA) Proc.* **75**:2746–2749.

Chang, A. C. Y., J. H. Nunberg, R. J. Kaufman, H. A. Erlich, R. T. Schimke, and S. N. Cohen. 1978. Phenotypic Expression in *E. coli* of a DNA Sequence Coding for Mouse Dihydrofolate Reductase. *Nature* **275**: 617–624.

Polisky, B., R. J. Bishop, and D. H. Gelfand. 1976. A Plasmid Cloning Vehicle Allowing Regulated Expression of Eukaryotic DNA in Bacteria. *Natl. Acad. Sci. (USA) Proc.* **73**:3900–3904.

Sanzey, B., O. Mercereau, T. Ternynzk, and P. Kourilsky. 1976. Methods for Identification of Recombinants of Phage λ. *Natl. Acad. Sci. (USA) Proc.* **73**:3394–3397.

Science. 1977. **196.**

Science. 1980. **209.**

Skalka, A., and L. Shapiro. 1977. *In situ* Immunoassays for Gene Translation Products in Phage Plaques and Bacterial Colonies. *Gene* **1**:65–79.

Struhl, K., J. R. Cameron, and R. W. Davis. 1976. Functional Genetic Expression of Eukaryotic DNA in *Escherichia coli. Natl. Acad. Sci. (USA) Proc.* **73**:1471–1475.

34

Reprinted from *Natl. Acad. Sci. (USA) Proc.* **71**:1743–1747 (1974)

Replication and Transcription of Eukaryotic DNA in *Escherichia coli*

(restriction/plasmid/transformation/recombination/ribosomal DNA)

JOHN F. MORROW*††, STANLEY N. COHEN†, ANNIE C. Y. CHANG†, HERBERT W. BOYER§,
HOWARD M. GOODMAN¶, AND ROBERT B. HELLING§||

Departments of *Biochemistry and †Medicine, Stanford University School of Medicine, Stanford, California 94305; and
Departments of §Microbiology and ¶Biochemistry and Biophysics, University of California, San Francisco, Calif. 94143

ABSTRACT Fragments of amplified *Xenopus laevis* DNA, coding for 18S and 28S ribosomal RNA and generated by *Eco*RI restriction endonuclease, have been linked *in vitro* to the bacterial plasmid pSC101; and the recombinant molecular species have been introduced into *E. coli* by transformation. These recombinant plasmids, containing both eukaryotic and prokaryotic DNA, replicate stably in *E. coli*. RNA isolated from *E. coli* minicells harboring the plasmids hybridizes to amplified *X. laevis* rDNA.

Recombinant DNA molecules constructed *in vitro* from separate plasmids (1, 2) by the joining of DNA fragments having cohesive termini (3, 4) generated by the *Eco*RI restriction endonuclease (5, 6) can form biologically functional replicons when introduced into *Escherichia coli* by transformation (7). The *E. coli* tetracycline resistance plasmid, pSC101 (1, 8) (molecular weight 5.8 × 10⁶), is useful for selection of recombinant plasmids in *E. coli* transformants, since insertion of a DNA segment at its single *Eco*RI cleavage site does not interfere with expression of its tetracycline resistance gene(s) or with the replication functions of the plasmid (1, 2).

This report describes the *in vitro* linkage of pSC101 and eukaryotic DNA cleaved by *Eco*RI endonuclease, and subsequent recovery of recombinant DNA molecules from transformed *E. coli* in the absence of selection for genetic properties expressed by the eukaryotic DNA. The amplified rDNA (coding for 18S and 28S ribosomal RNA) of *Xenopus laevis* was used as a source of eukaryotic DNA, since it has been well characterized and can be isolated in quantity (9, 10). Recombinant plasmids containing both *X. laevis* and pSC101 DNA replicate stably in *E. coli*, where they are capable of synthesizing RNA complementary to *X. laevis* rDNA.

MATERIALS AND METHODS

DNA coding for ribosomal RNA of *X. laevis*, isolated by CsCl-gradient centrifugation, and ³²P-labeled 18S and 28S *X. laevis* ribosomal RNA were the generous gifts of Dr. D. D. Brown. Bacterial strains and the tetracycline resistance plasmid pSC101 have been described (1, 2, 8). Covalently-closed circular plasmid DNA was isolated as described (8, 11), or

Abbreviations: rRNA, ribosomal RNA; rDNA, amplified DNA containing the genes for 18S and 28S rRNA; *Eco*RI, the RI restriction and modification host specificity of *E. coli* controlled by the fi⁺ plasmid, pHB1.

‡ Present address: Carnegie Institution of Washington, Department of Embryology, 115 W. University Parkway, Baltimore, Maryland 21210.

|| Present address: Department of Botany, University of Michigan, Ann Arbor, Mich. 48104.

by an adaptation of a NaCl-sodium dodecyl sulfate cleared-lysate procedure (12, 13). Transformation of *E. coli* by plasmid DNA (7), isolation of *E. coli* minicells (14), heteroduplex analysis by electron microscopy (15), DNA·RNA hybridization (16, 17), and analysis of fragments generated by *Eco*RI endonuclease by agarose gel electrophoresis (refs. 1, 6, and 18; Helling, Goodman and Boyer, in preparation) have been described elsewhere. Molecular weights of fragments were calculated from their mobility in gels relative to the mobility of fragments of λ DNA cleaved by *Eco*RI endonuclease. Radioactive labeling of RNA in *E. coli* minicells was according to Roozen *et al.* (19); ³H-Labeled RNA was isolated from minicells by a modification of a procedure described (17).

Purification of *Eco*RI restriction endonuclease (20) and *E. coli* ligase (the generous gift of Drs. P. Modrich and I. R. Lehman) (21) have been described. *E. coli–X. laevis* recombinant plasmids were constructed *in vitro* as follows: the reaction mixture (60 µl) contained 100 mM Tris·HCl (pH 7.5), 50 mM NaCl, 5 mM MgCl₂, 1.0 µg of pSC101 plasmid DNA, 2.5 µg of *X. laevis* rDNA, and excess *Eco*RI restriction endonuclease (1 µl, 2 units). After a 15-min incubation at 37°, the reaction mixture was placed at 63° for 5 min to inactivate the *Eco*RI endonuclease. A 3-µl sample was examined by electron microscopy to assess digestion. The remainder was refrigerated at 0.5° for 24 hr to allow association of the short cohesive termini; melting temperature (Tm) was 5–6° (3).

The reaction mixture for ligation of phosphodiester bonds was adjusted to a total volume of 100 µl and contained, in addition to the components of the endonuclease reaction, 30 mM Tris·HCl (pH 8.1), 1 mM sodium EDTA, 5 mM MgCl₂, 3.2 nM NAD, 10 mM (NH₄)₂SO₄, 5 µg of bovine-serum albumin, and 9 units of *E. coli* DNA ligase (21). All components were chilled to 0.5° before their addition to the reaction mixture. Ligase reactions were incubated at 14° for 45 min, and returned to 0.5° for 48 hr. Additional NAD and ligase were added, and the mixture was incubated at 15° for 30 min and then for 15 min at 37°. A 3-µl sample of the mixture was examined by electron microscopy for reassociation of fragments. Ligated DNA was used directly in the plasmid transformation procedure (7).

RESULTS

Cleavage of rDNA of X. laevis. Linear molecules of *X. laevis* rDNA (molecular weight about 50 × 10⁶, as determined by electron microscopy) were treated with excess *Eco*RI endonuclease. After complete digestion, about 44% of the molecules had a molecular weight of 3.1 × 10⁶ (Fig. 1) and a second major class (25%) of fragments had a molecular weight of 4.3

Proc. Nat. Acad. Sci. USA 71 (1974)

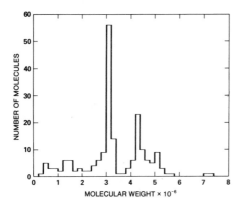

FIG. 1. Histogram of molecular weights of fragments of *X. laevis* rDNA generated by *Eco*RI endonuclease. *X. laevis* rDNA was completely digested by *Eco*RI endonuclease and mounted for electron microscopy. 195 molecular lengths were measured with a Philips EM-300 electron microscope. Molecular weights, which have been multiplied by 10^6, were calculated from lengths of fragments relative to a simian virus 40 (SV40) DNA standard (molecular weight, 3.4×10^6) included on the same grid.

$\times 10^6$ (Fig. 1). In addition, some fragments (9%) having molecular weights of 5.1×10^6 were observed. The occurrence of molecules smaller than 3.1×10^6 may result from length heterogeneity of the untreated rDNA preparation.

Separation of *X. laevis* rDNA cleaved by *Eco*RI endonuclease by agarose gel electrophoresis (Fig. 2) confirmed the electron microscope data and identified other size classes of fragments. Fragments equivalent to the two major DNA

lengths seen by electron microscopy were observed in the gels (molecular weight estimates from gel, 3.0×10^6 and 4.2×10^6), and two minor bands having estimated molecular weights of 3.9×10^6 and 4.8×10^6 were seen. Additional bands, which are not readily apparent in Fig. 2, were observed in the original gel. Cleavage of the pSC101 plasmid DNA by the *Eco*RI endonuclease occurred at one site, resulting in formation of a single linear fragment having a molecular weight of about 5.8×10^6 (Fig. 2 and refs. 1 and 2).

Analysis of Recombinant Plasmids. A mixture of pSC101 DNA and *X. laevis* rDNA, both treated with *Eco*RI endonuclease, was ligated after random association of the short cohesive termini had occurred. This DNA was used to transform *E. coli* strain C600 $r_K^- m_K^-$, and tetracycline-resistant transformants ($3.3 \times 10^3/\mu g$ of pSC101 DNA) were selected and numbered consecutively CD1, CD2, etc. Plasmid DNA isolated from three of the transformants yielded three DNA fragments after digestion with the *Eco*RI endonuclease, and ten yielded two fragments (Table 1, Fig. 2). Each plasmid contained a fragment corresponding to linear pSC101 DNA (5.8×10^6, molecular weight). The other fragments had molecular weights (estimated from gels) of 4.2×10^6, 3.9×10^6, or 3.0×10^6, which correspond to the molecular weight estimates for certain of the *X. laevis* rDNA fragments generated by the endonuclease. An example of each type of recombinant plasmid was selected for further study.

Buoyant densities of *X. laevis* rDNA and of *Eco*RI digests of these four representative recombinant plasmids were compared (Fig. 3). Each of the plasmid DNA species contains a fragment generated by *Eco*RI endonuclease having a buoyant density about equal to that of amplified *X. laevis* rDNA ($\rho = 1.729$ g/cm^3) (29), in addition to a fragment having the buoyant density of pSC101 DNA ($\rho = 1.710$ g/cm^3). We infer from the similar buoyant densities of the three *X. laevis* rDNA

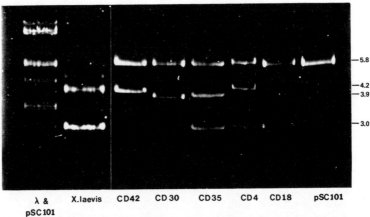

FIG. 2. Agarose gel electrophoresis of fragments generated by *Eco*RI endonuclease. DNA was isolated and digested to completion with excess *Eco*RI endonuclease. Electrophoresis was in a 0.7% agarose slab gel (17 cm × 13 cm × 3 mm) at 3 V/cm in 0.09 M Tris base–2.8 mM EDTA–0.09 M boric acid (pH 8.3) buffer at 25° for 15 hr. After completion of electrophoresis, the DNA was stained with 4 μg/ml of ethidium bromide and photographed under a long-wave UV lamp (refs. 1, 6, and 18; Helling, Goodman and Boyer, in preparation). Electrophoresis was from top to bottom in the figure (i.e., the anode is at the *bottom*) and the source of DNA is indicated. The figure shown is a composite of two separate gels. An *Eco*RI endonuclease digest of a mixture of bacteriophage λ DNA and pSC101 DNA was used for molecular weight standardization, and is shown in the *left-hand column*. The bands seen in this column had estimated molecular weights of (counting from the *bottom*): 2.09, 3.03, 3.56, 3.7, 4.7, 5.8 (pSC101), 13.7, and 15.7, all × 10^{-6} (Helling, Goodman and Boyer, in preparation). The 15.7×10^6 fragment results from joining of the terminal cohesive ends of λDNA, which are located on the 2.09 and 13.7 × 10^6 fragments. The calculated molecular weights of the major fragments present in *X. laevis* rDNA treated with *Eco*RI endonuclease and in the plasmids shown are indicated in the figure. All values have been multiplied by 10^{-6}.

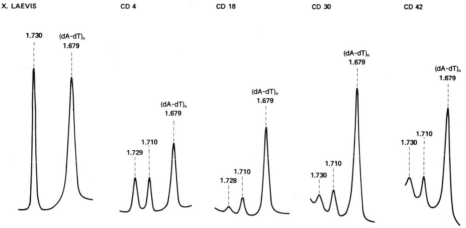

FIG. 3.   Analytical ultracentrifugation of DNA, cleaved by *Eco*RI endonuclease, isolated from *E. coli* transformants carrying various recombinant plasmids and *X. laevis* rDNA. Covalently-closed circular plasmid DNA was cleaved as indicated in *Methods* and Fig. 2. Centrifugation in CsCl ($\rho = 1.710$ g/cm^3) was for 28–36 hr at 44,000 rpm in the presence of (dA-dT)$_n$($\rho = 1.679$ g/cm^3) density marker. Densitometer tracings of photographs taken during centrifugation are shown.

fragments present in the separate plasmids, that *Eco*RI endonuclease cleavage of *X. laevis* rDNA produces fragments that are similar to each other in (G+C) composition. However, we did not have a sufficient amount of rDNA to carry out this analysis directly.

Electron microscope analysis of a heteroduplex formed between *X. laevis* rDNA and the plasmid CD42 (Fig. 4) shows that the plasmid contains DNA nucleotide sequences present

TABLE 1.   *X. laevis–E. coli recombinant plasmids*

Plasmid DNA	Molecular weight of *Eco*RI plasmid fragments estimated by gel electrophoresis ($\times 10^{-6}$)	Molecular weight from contour length ($\times 10^{-6}$)	Buoyant density in CsCl (g/cm^3)
Tetracycline-resistant clone			
CD4	5.8, 4.2, 3.0	13.6	1.721
CD7	5.8, 4.2	—	—
CD12, CD20, CD45, CD47, CD51	5.8, 3.0	—	—
CD14	5.8, 4.2, 3.0	—	—
CD18	5.8, 3.0	9.2	1.720
CD30	5.8, 3.9	10.0	1.719
CD35	5.8, 3.9, 3.0	—	—
CD42	5.8, 4.2	10.6	1.720
pSC101	5.8	6.0	1.710

The procedures used for isolation of DNA, agarose gel electrophoresis, CsCl gradient centrifugation, and calculation of a molecular weight and buoyant density are indicated in *Methods*. The reproducibility of molecular weight estimates from the gel mobility of DNA fragments is ±5%. The standard deviation of contour lengths used for molecular weight estimates was 2–4%.

in *X. laevis* rDNA. Moreover, in this and other heteroduplexes, two separate plasmid DNA molecules were seen to form duplex regions with a single strand of *X. laevis* rDNA, consistent with the observation (22, 23) that the rDNA sequences of *X. laevis* are tandemly repeated.

[32]P-Labeled 18S and 28S *X. laevis* rRNA were hybridized to DNA obtained from the plasmids CD4, CD18, CD30, and CD42 (Fig. 5). CD4 DNA, which contains both the 3.1 $\times$ 10^6 and 4.3 $\times$ 10^6 fragments of *X. laevis* rDNA, anneals almost equally with both the 18S and 28S rRNA species. CD18 plasmid DNA hybridizes principally with 28S *X. laevis* rRNA, while the DNA of plasmids CD30 and CD42 anneals primarily with 18S ribosomal RNA. These data suggest that the latter two plasmids contain DNA fragments having a similar nucleotide sequence and suggest that the *X. laevis* rDNA fragment of CD30 may lack a short sequence contained in the 4.3 $\times$ 10^6 fragment carried by CD42. Contamination of the 18S rRNA preparation with 28S rRNA may account for the small amount of hybridization of 18S ^{32}P]RNA observed with the DNA of the CD18 plasmid, which carries only the 3.0 $\times$ 10^6 *Eco*RI fragment of *X. laevis* rDNA.

Transcription of X. laevis DNA in E. coli Minicells. It has been reported that various plasmids can segregate into *E. coli* minicells (14, 19, 24, 25) at the time when these spheres are budded off from the parent *E. coli* (26). The lack of chromosomal DNA in minicells has made them particularly suitable for studies of plasmid DNA, RNA, and protein synthesis (19, 27, 28) in the absence of a background of chromosomal macromolecular synthesis. The minicell-producing *E. coli* strain P678-54 was transformed to tetracycline resistance with plasmid DNA isolated from C600 $r_K^-m_K^-$ containing CD4, CD18, or CD42. Transformation as well as transfer of recombinant plasmids to this and to other $r_K^+m_K^+$ strains occurred at a frequency reduced by 50- to 1000-fold, compared with the $r_K^-m_K^-$ strain. Minicells containing the plasmids were isolated and incubated with [3H]uridine; RNA purified from such minicells was hybridized with *X. laevis* rDNA immobilized on nitrocellulose membranes, in order to determine whether the *X. laevis* rDNA linked to the pSC101 replicon

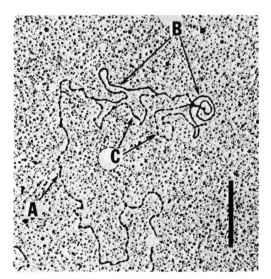

Fig. 4. Electron photomicrograph of heteroduplex of *X. laevis* rDNA and two separate CD42 plasmid DNA molecules. (*A*) Single strand of *X. laevis* rDNA. (*B*) Double-stranded regions of homology between plasmid CD42 and *X. laevis* rDNA. (*C*) Single-stranded regions corresponding in length to the DNA segment of CD42 derived from pSC101. CD42 DNA was nicked by x-irradiation (0.8 break/strand) to permit strand separation. The molecular weight of the double-stranded heteroduplex region is 4.5×10^6, relative to CD42 DNA standard on the same grid (12 measurements, standard deviation 0.25×10^6). The contour length of the DNA segment separating the two duplex regions is consistent with a single-strand molecular weight of 1.5×10^6 for this segment (2 measurements). The *bar* indicates 1 μm. About one-third of *X. laevis* rDNA single strands hybridized with one or more molecules of CD42 DNA.

is transcribed in *E. coli*. These results (Table 2) show that RNA species capable of annealing with purified *X. laevis* rDNA are synthesized in *E. coli* minicells carrying the recombinant plasmids CD4, CD18, and CD42, but not by minicells carrying the pSC101 plasmid alone.

DISCUSSION

The presence of *X. laevis* rDNA replicating as a part of a bacterial plasmid in *E. coli* has been demonstrated by five separate criteria: (*1*) Treatment of recombinant plasmid DNA molecules with the *Eco*RI restriction endonuclease generates fragments that have molecular weights indistinguishable from those of the cleaved *X. laevis* ribosomal DNA and linear pSC101 DNA. (*2*) Uncleaved recombinant plasmids have buoyant densities in CsCl intermediate to the buoyant densities of pSC101 and *X. laevis* rDNA. The plasmid fragments generated by the *Eco*RI restriction enzyme are similar in buoyant density to *X. laevis* rDNA, and reflect the (G+C) content of amplified ribosomal RNA genes from that organism (9, 10). (*3*) Electron microscope heteroduplex analysis of recombinant plasmid DNA molecules isolated from *E. coli* indicates sequence homology with *X. laevis* rDNA. In some instances more than one plasmid molecule was observed to anneal with a single strand of rDNA, which is known to have tandemly repeated sequences. (*4*) 18S and 28S ribosomal [32P]RNA obtained from *X. laevis* anneals with recombinant

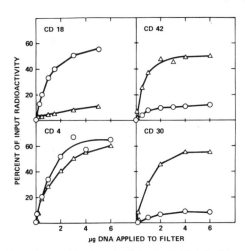

Fig. 5. Hybridization of *X. laevis* 32P-labeled ribosomal RNA (specific activity 9×10^6 cpm/μg) with DNA of recombinant *X. laevis–E. coli* plasmids. (△) 18S ribosomal RNA; (○) 28S ribosomal RNA. Hybridization on nitrocellulose membranes was carried out at DNA excess. Filters containing 2 μg of pSC101 DNA bound less than 0.1% of input RNA under the experimental conditions used. 900 cpm (0.01 μg) input [32P]RNA was used for each filter.

plasmid DNA molecules isolated from *E. coli*. (*5*) RNA synthesized by *E. coli* minicells carrying recombinant plasmids hybridizes to rDNA from *X. laevis*, providing a functional demonstration of the presence of *X. laevis* rDNA in *E. coli*.

It has been reported (1) that fragments, generated by *Eco*RI endonuclease, of the large antibiotic resistance plasmid R6-5 can randomly associate to form new plasmid DNA species which in turn can be linked to pSC101 *in vitro* and introduced

TABLE 2. [*3H*]*RNA synthesized by E. coli minicells*

Plasmid carried by minicells	Input cpm	[3H]RNA counts hybridized to		
		X. laevis rDNA		pSC101 DNA
		0.2 μg	0.4 μg	1.8 μg
CD42	4810	905 (19%)	1436 (30%)	961 (20%)
CD18	3780	389 (10%)	—	1277 (34%)
CD4	5220	789 (15%)	—	1015 (19%)
pSC101	4170	0 (0%)	—	1500 (36%)

Minicells containing plasmids were isolated as described by Cohen *et al.* (14) and were incubated with [3H]uridine (50 μCi/ml, 30 Ci/mol) as described by Roozen *et al.* (19) for 10 min at 37°. Minicells collected by centrifugation were resuspended in Tris· HCl (20 mM, pH 7.5)–5 mM MgCl₂–1 mM EDTA, pH 8.0, and rapidly frozen and thawed three times. RNA was extracted as described (17). Hybridization assays were carried out on nitrocellulose membranes as described (16, 17) at saturating levels of pSC101 DNA. Hybridizations involving *X. laevis* DNA were not performed at DNA excess. Counts bound to blank filters (5–10 cpm) were subtracted from experimentally determined values. There is no cross-hybridization between bacterial ribosomal RNA and *X. laevis* rDNA (28). 3H counts eluted from filters containing *X. laevis* DNA were rendered acid-soluble by ribonuclease A (20 μg/ml, 0.30 M NaCl–0.030 M Na citrate, 1 hr, 37°) but were resistant to pancreatic deoxyribonuclease [20 μg/ml, 10 mM Tris·HCl (pH 7.4), 10 mM MgCl₂, 1 hr, 37°].

into *E. coli* by transformation. These earlier investigations used markers carried by *Eco*RI endonuclease-generated fragments incorporated into the recombinant plasmid to select for transformants. In the present experiments, tetracycline-resistant transformants consisting of pSC101 linked to *X. laevis* were recovered by random association (ratio: 4 *X. laevis* rDNA termini per pSC101 terminus) of DNA fragments in the absence of selection for any *X. laevis* genetic marker. Since the majority (75%) of the tetracycline-resistant clones contain only the pSC101 DNA molecule, the conditions used may favor cyclization of this molecule; tandem pSC101 plasmid DNA molecules were not observed in this recombination-competent host.

Although transcription of *X. laevis* rDNA was detected in *E. coli* C600 transformants (unpublished data), the amount of *X. laevis*-hybridizable material isolated from such cells represented too small a fraction of the total ³H-labeled RNA to be studied easily. *E. coli* minicells thus provide a powerful tool for the investigation of macromolecular synthesis coded by *X. laevis* rDNA and by other eukaryotic genes joined to bacterial plasmids, since minicells are devoid of bacterial chromosomal DNA and plasmid species segregating into minicells are capable of plasmid-specific synthesis of functional gene products (27, 28). Moreover, study of control of eukaryotic gene transcription may be possible in *E. coli* minicells, since mechanisms regulating the repression and derepression of anthranilate synthetase (28) and tetracycline resistance (27) have been shown to be operative in these cells.

Since the *X. laevis* rDNA inserted into the pSC101 plasmid does not code for protein, we do not yet know whether synthesis of proteins can be carried out by eukaryotic DNA in *E. coli*. In addition, it is not clear whether the RNA synthesized on *X. laevis* rDNA template in *E. coli* minicells is an accurate transcript of the eukaryotic genes. However, these experiments demonstrate that transcription of cloned *X. laevis* rDNA does occur in a prokaryotic organism. In addition, the location of an *Eco*RI cleavage site near the boundary between the 18S and 28S rRNA genes (as shown by the data in Fig. 5) suggests that investigation of the transcription of specific *X. laevis* rRNA cistrons in *E. coli* may be practical.

X. laevis rDNA replicates stably in *E. coli* for at least 100 generations (unpublished data) as part of the pSC101 plasmid replicon, and can be recovered from transformed *E. coli* in amounts required for *in vitro* analysis by procedures commonly used for the isolation of bacterial plasmids. Cloned *X. laevis* DNA can then be introduced into other *E. coli* strains by transformation. The ability to clone specific fragments of DNA from a complex genome provides a potentially valuable tool for the study of organization and function of eukaryotic genomes. Our results using cloned *X. laevis* rDNA suggest that the amplified ribosomal RNA genes of this organism may not be contained in homogeneous units of repeated sequences, as previously supposed (9, 10), but that some degree of heterogeneity with respect to the size of the repeat unit may occur in the rDNA sequence. The gel electrophoresis and DNA·RNA hybridization data reported here indicate that at least two *Eco*RI sites are present per *X. laevis* repeat unit and suggest that the 3.0 × 10⁶-dalton fragment generated by *Eco*RI endonuclease may be paired with either the 3.9 × 10⁶ or the 4.2 × 10⁶-dalton fragment in two kinds of commonly occurring repeat units existing in the preparation of *X. laevis* rDNA we have studied. The molecular weight estimated for the intact repeat unit (7.2 to 7.5 × 10⁶) is in

general agreement with data obtained by Hourcade *et al.* (23) and by A. Forsheit, N. Davidson and D. D. Brown (personal communication).

The procedure reported here offers a general approach utilizing bacterial plasmids for the cloning of DNA molecules from various sources, provided that both molecular species have cohesive termini made by a restriction endonuclease, and that insertion of a DNA segment at the cleavage site of the plasmid does not interfere with expression of genes essential for its replication and selection.

These studies were supported by Grants AI08619, GM 14378, and CA14026 from the National Institutes of Health, by Grant NSF-GB30581 from the National Science Foundation, by Grant NP-112H from the American Cancer Society and by a USPHS Career Development Award to S.N.C. J.M. thanks P. Berg for stipend support during this work. R.B.H. was a USPHS Special Fellow. We wish to thank Mary Betlach (UCSF) for her dedicated assistance.

1. Cohen, S. N., Chang, A. C. Y., Boyer, H. W. & Helling, R. W. (1973) *Proc. Nat. Acad. Sci. USA* 70, 3240–3244.
2. Chang, A. C. Y. & Cohen, S. N. (1974) *Proc. Nat. Acad. Sci. USA* 71, 1030–1034.
3. Mertz, J. E. & Davis, R. W. (1972) *Proc. Nat. Acad. Sci. USA* 69, 3370–3374.
4. Sgaramella, V. (1972) *Proc. Nat. Acad. Sci. USA* 69, 3389–3393.
5. Hedgepeth, J., Goodman, H. M. & Boyer, H. W. (1972) *Proc. Nat. Acad. Sci. USA* 69, 3448–3452.
6. Dugaiczyk, A., Hedgepeth, J., Boyer, H. W. & Goodman, H. M. (1974) *Biochemistry*, in press.
7. Cohen, S. N., Chang, A. C. Y. & Hsu, L. (1972) *Proc. Nat. Acad. Sci. USA* 69, 2110–2114.
8. Cohen, S. N. & Chang, A. C. Y. (1973) *Proc. Nat. Acad. Sci. USA* 70, 1293–1297.
9. Dawid, I. B., Brown, D. D. & Reeder, R. H. (1970) *J. Mol. Biol.* 51, 341–360.
10. Birnstiel, M. L., Chipchase, M. & Speirs, J. (1971) *Prog. Nucl. Acid Res. Mol. Biol.* 11, 351–389.
11. Cohen, S. N. & Miller, C. A. (1970) *J. Mol. Biol.* 50, 671–687.
12. Hirt, B. (1967) *J. Mol. Biol.* 26, 365–369.
13. Guerry, P., LeBlanc, D. J. & Falkow, S. (1973) *J. Bacteriol.* 116, 1064–1066.
14. Cohen, S. N., Silver, R. P., McCoubrey, A. E. & Sharp, P. A. (1971) *Nature New Biol.* 231, 249–251.
15. Davis, R. W., Simon, M. N. & Davidson, N. (1971) in *Methods in Enzymology*, eds. Grossman, L. & Moldave, K. (Academic Press. New York), Vol. 21, pp. 413–428.
16. Brown, D. D. & Weber, C. S. (1968) *J. Mol. Biol.* 34, 661–680.
17. Cohen, S. N. & Hurwitz, J. (1968) *J. Mol. Biol.* 37, 387–406.
18. Sharp, P. A., Sugden, B. & Sambrook, J. (1973) *Biochemistry* 12, 3055–3063.
19. Roozen, K. J., Fenwick, R. G., Jr. & Curtiss, R., III (1971) *J. Bacteriol.* 107, 21–23.
20. Greene, P. J., Betlach, M. C., Goodman, H. M. & Boyer, H. W. (1973) "DNA replication and biosynthesis," in *Methods in Molecular Biology*, ed. Wickner, R. B. (Marcel Dekker, Inc. New York), in press.
21. Modrich, P., Anraku, Y. & Lehman, I. R. (1973) *J. Biol. Chem.* 248, 7495–7511.
22. Wensink, P. C. & Brown, D. D. (1971) *J. Mol. Biol.* 60, 235–247.
23. Hourcade, D., Dressler, D. & Wolfson, J. (1973) *Proc. Nat. Acad. Sci. USA* 70, 2926–2930.
24. Inselburg, J. (1970) *J. Bacteriol.* 102, 642–647.
25. Levy, S. B. & Norman, P. (1970) *Nature* 227, 606–607.
26. Adler, H. I., Fisher, W. D., Cohen, A. & Hardigree, A. A. (1967) *Proc. Nat. Acad. Sci. USA* 57, 321–326.
27. van Embden, J. & Cohen, S. N. (1973) *J. Bacteriol.* 116, 699–709.
28. Frazer, A. C. & Curtiss, R., III (1973) *J. Bacteriol.* 115 615–622.
29. Sinclair, J. H. & Brown, D. D. (1971) *Biochemistry* 10, 2761–2769.

35

Reprinted from *Natl. Acad. Sci. (USA) Proc.* **74**:487–491 (1977)

Functional expression of cloned yeast DNA in *Escherichia coli*

(plasmid ColE1/leucine operon/histidine operon/eukaryote DNA/complementation)

BARRY RATZKIN AND JOHN CARBON

Section of Biochemistry and Molecular Biology, Department of Biological Sciences, University of California, Santa Barbara, Calif. 93106

ABSTRACT A collection of hybrid circular DNAs was constructed *in vitro* using the poly(dA·dT) "connector" method; each hybrid circle contained one molecule of poly(dT)-tailed DNA of plasmid ColE1 (made linear by digestion with *Eco*RI endonuclease) annealed to a poly(dA)-tailed fragment of yeast (*Saccharomyces cerevisiae*) DNA, produced originally by shearing total yeast DNA to an average size of 8×10^6 daltons. This DNA preparation was used to transform *E. coli* cells, selecting colicin-E1-resistant clones that contain hybrid ColE1–yeast DNA plasmids. Sufficient numbers of transformant clones were obtained to ensure that the hybrid plasmid population was representative of the entire yeast genome. Various hybrid ColE1–yeast DNA plasmids capable of complementing *E. coli* auxotrophic mutations were selected from this population. Plasmid pYe*leu*10 complements several different point or deletion mutations in the *E. coli* or *S. typhimurium leuB* gene (β-isopropylmalate dehydrogenase); plasmids pYe*leu*11, pYe*leu*12, and pYe*leu*17 are specific suppressors of the *leuB6* mutation in *E. coli* C600. Plasmid pYe*his*2 complements a deletion in the *E. coli hisB* gene (imidazole glycerol phosphate dehydratase). Complementation of bacterial mutations by yeast DNA segments does not appear to be a rare phenomenon.

A key question invoked by the recent discovery of methods to biochemically construct recombinant DNAs concerns the ability of fragments of eukaryotic DNA inserted into plasmid or phage vectors to be expressed meaningfully in a bacterium such as *Escherichia coli*. In order for expression to occur, components of the bacterial cell must transcribe the sense strand of the foreign DNA into mRNA and translate this mRNA into functional protein. In addition, further post-translational modifications may also be required to obtain an active product. Major differences apparently exist between eukaryotes and prokaryotes in several of these vital components and processes, such as the RNA polymerases, ribosomal subunits, translational initiation requirements, and post-transcriptional and post-translational modifications.

Several laboratories have shown that transcription of cloned eukaryotic DNAs in *E. coli* can occur in a bacterial cell; however, there is no evidence that normal transcription start and stop signals are recognized in the bacterial system (1–3). However, Struhl *et al.* (4) have recently shown that a segment of yeast DNA cloned on a phage vector can complement *hisB* mutations in *E. coli*, a strong indication that a functional gene product is being formed in this heterologous system.

These considerations have also led us to use a sensitive assay for the expression of cloned eukaryotic DNA, the ability to suppress or complement bacterial auxotrophic mutations *in vivo*. Thus, it should be possible to detect a relatively low level of expression of a cloned eukaryotic gene by the relief of a metabolite requirement in a mutant bacterial strain. In addition, a wide variety of *E. coli* strains containing mutations in known genes involved in amino acid and nucleotide biosynthesis are readily available, so that the generality of any complementation phenomena could be established. We also have selected DNA from an organism (*Saccharomyces cerevisiae*) that has many biosynthetic pathways in common with *E. coli*. Because the size of the yeast genome is only 2–3 times that of *E. coli* (5), the number of different recombinant plasmids that we would need to screen in order to cover the entire genome would not be excessively large (6).

Using the poly(dA·dT) "connector" method (7, 8) to join randomly sheared yeast DNA segments to plasmid ColE1 DNA (L$_{RI}$), we have isolated and characterized ColE1–yeast DNA recombinant plasmids that can complement *leuB* and *hisB* mutations in *E. coli*. The frequency with which the interspecies complementation is observed suggests that it is not a rare phenomenon.

METHODS

Bacterial Strains. The recipient for transformations with poly(dT)-tailed ColE1 DNA (L$_{RI}$, indicating linear, generated by digestion with endonuclease *Eco*RI) annealed to a collection of poly(dA)-tailed sheared yeast DNA was strain JA199 (*hsdM⁺ hsdR⁻ lacY⁻ leuB6 ΔtrpE5/F⁺*). It was constructed by transducing strain MV10 (*hsdM⁺ hsdR⁻ lacY⁻ leuB6 thr-1 ΔtrpE5*) (9) to threonine independence with P1kc phage grown on strain HB94 (*hsdM⁺ hsdR⁻ thr⁺*) as described by Wood (10). Strain KL380 (*alaS5 ara⁻ argA⁻ lacZ⁻ leuB6 metE⁻ recA⁻ strA⁻*) was obtained from K. B. Low; strain K-12 *hisB463* from K. Struhl and R. Davis; *E. coli* strains CV512 (*leuA371/F⁺*), CV514 (*leuB401/F⁺*), CV516 (*leuB61/F⁺*), CV522 (*leuC222/F⁺*), CV524 (*leuD211/F⁺*), and *Salmonella typhimurium* strains *leuA124*, *leuB698*, *leuC5076*, and *leuD657* were obtained from J. Calvo (11, 12).

Media. LB broth, Vogel's, and Bonner's E media were used (13). Amino acids were added at 50 μg/ml. YPD medium contained 2% peptone, 1% yeast extract, and 2% glucose.

DNA Preparations. Covalently closed, supercoiled ColE1 DNA was prepared as described (6, 14). Yeast DNA was isolated from strain X2180-1A a (*SUC2 mal gal2 CUP1*) (obtained from the Yeast Genetics Stock Center, University of California, Berkeley). Yeast cells were grown in YPD medium to an OD$_{600}$ of 7 and harvested by centrifugation. Twenty grams of cells was resuspended in 10 ml of buffer containing 0.5 M Na$_4$EDTA, 0.01 M Tris·HCl (pH 7), and 10 ml of proteinase K (2 mg/ml) in the same buffer at pH 9.5) and 18 ml of sodium *N*-lauroylsarcosinate (1% in the same buffer, pH 9.5) was added. The cells were broken at 20,000 pounds/inch² (140 MPa) in a French pressure cell (American Instrument Co.). The lysate was then incubated at 50° for 12 hr. These buffers and the incubation procedure are described by Lauer and Klotz (5). The viscous lysate was extracted with redistilled phenol, and the DNA was precipitated by adding 2 volumes of ethanol and spooled out on a hooked glass rod. The DNA was further purified by isopropanol fractionation (15), digestion with pre-heated pancreatic RNase, phenol extraction, and a final isopropanol

Abbreviations: Leu⁺, leucine-independent; His⁺, histidine-independent.

fractionation. This DNA was used for the poly(dA)-tailing reaction.

DNAs from *E. coli* strain JA199 and its F⁻ derivative JA194 were prepared by standard procedures (14, 15). Cellular DNAs for reassociation experiments were banded in CsCl/ethidium bromide density gradients prior to use.

Preparation of Recombinant ColE1–Yeast DNAs and Transformation. Poly(dT)-tailed ColE1 DNA (L_{RI}) and poly(dA)-tailed sheared yeast DNA (average molecular weight = 8 to 10 × 10⁶) were prepared as described by Clarke and Carbon (6, 14). Annealing of the two DNAs in an equimolar ratio produced a preparation with 10% circles, 16% tailed circles, 61% linears, and 14% tangles (visualized by standard electron microscopy).

Twenty-five micrograms of this annealed DNA preparation was used to transform strain JA199 to colicin E1 resistance as described by Clarke and Carbon (6), except that after the cells were diluted 10-fold with LB broth, they were grown for 90 min at 37° before plating in the presence of colicin E1.

Isolation of Recombinant Plasmid DNAs. A modification of the Hirt extraction procedure (16) was used. Cultures were incubated in the presence of chloramphenicol (200 μg/ml) to amplify plasmid for 16 hr, harvested, and treated with lysozyme/EDTA in the cold and lysed at room temperature with 1% sodium dodecyl sulfate for 15 min. The lysate was then kept at 0° for 30 min or until it was thoroughly chilled. The chromosomal DNA was pelleted by centrifugation of the lysate at 19,000 rpm in the Sorvall SS-34 rotor for 30 min. The supernatant liquid was slowly decanted, diluted 2-fold with STE buffer (10 mM Tris·HCl, pH 8, 10 mM NaCl, 1 mM EDTA), and made 1 M in NaCl. Phenol extraction was carried out for 20 min at 0° The aqueous layer was removed, mixed with 2 volumes of cold absolute ethanol, and kept at −20° for 2 hr. The precipitate was collected by centrifugation, and purified by ethidium bromide/CsCl banding.

Labeling of Plasmid DNA and Reassociation Kinetics. DNA was labeled by nick translation with 30 μM deoxyribonucleoside triphosphates, [³H]dTTP (8.4 Ci/mmol), and [³H]dATP (8.3 Ci/mmol) by the method of M. Dieckmann and P. Berg (personal communication) as described by Schachat and Hogness (17). pYe*leu*10 DNA was labeled to a specific activity of 2.4 × 10⁶ cpm/μg and pYe*his*2 DNA to 2.2 × 10⁶ cpm/μg.

The reassociation kinetics of plasmid DNAs at 65° in the presence and absence of various driver DNAs was measured by an S1 nuclease assay as previously described (17, 18).

Enzymes. *Eco*RI endonuclease and λ 5′-exonuclease were isolated as described (14); DNA polymerase I was a gift from A. Kornberg and deoxynucleotidyl terminal transferase was from R. Ratliff. S1 nuclease (200,000 units/mg) was purchased from Miles Laboratories, Inc.

Biohazard Considerations. Although F-mediated transfer of ColE1-like plasmids provides a valuable tool for the identification of complementing hybrid plasmids, the procedure converts normally non-transmissible plasmids into a state that permits ready transfer to other bacterial hosts. A recent revision (September, 1976) of the National Institutes of Health guidelines specifically prohibits the cloning of foreign DNA in K-12 hosts that contain wild-type conjugative plasmids. We have therefore discontinued the use of F-mediated transfer. Other procedures to detect complementation, such as isolation of hybrid plasmid DNA from F⁻ cells followed by transformation of suitable auxotrophic recipients, would appear to offer less risk of escape of the hybrids into the environment, and are in line with the current guidelines.

This work was carried out under P2 laboratory conditions. All bacteria and DNA preparations were destroyed by autoclaving or exposure to Clorox solution before disposal.

RESULTS

In order to set up a definitive test for the ability of any given eukaryotic gene system to be expressed and to complement an auxotrophic mutation in *E. coli*, the efficiency of the cloning procedure used must be high enough to insure that transformant clones containing hybrid DNA plasmids are obtained in numbers sufficient to be representative of the entire genome of the organism under study. In addition, it is preferable to use DNA segments produced by random scission (hydrodynamic shear), rather than by restriction endonuclease action, to ensure that the desired gene system remains intact on at least a portion of the cleaved DNA segments. For example, previous studies have shown that the use of the poly(dA·dT) "connector" method (7, 8) to join randomly sheared *E. coli* DNA with linear ColE1 DNA yields a preparation that will transform *E. coli* cells to colicin E1 resistance with high efficiency, thereby establishing a collection of transformants containing hybrid plasmids representative of the entire *E. coli* genome using only 10–15 μg of annealed DNA (6, 14).

A transformation of strain JA199 with 25 μg of poly(dT)-tailed plasmid ColE1 DNA (L_{RI}) annealed to poly(dA)-tailed sheared yeast DNA produced 190,000 colonies resistant to colicin E1. Since there was a doubling of cell numbers from the time of transformation to the plating of the transformed cells, this should represent 90,000 unique transformants. Plates from a control experiment in which cells were not exposed to DNA contained only a fifth as many colonies. Assuming that the haploid yeast genome contains 10¹⁰ daltons of DNA and that the sheared pieces of yeast DNA are 10⁷ daltons in size, then only 4600 transformants would be necessary for a 99% probability that the pool of recombinant plasmids would contain any particular yeast DNA segment (6). Thus, the large pool of transformants ensures that we have cloned essentially all of the yeast genome and that a random selection of a few thousand transformants would be representative of the genome. Ten clones that were colicin-E1-resistant and E2-sensitive were picked at random and plasmid DNA was isolated. Eight of these yielded plasmid DNA that was larger than ColE1 DNA.

The colicin-resistant colonies were treated in three ways. A permanent bank of 4300 colonies was stored in 7% dimethylsulfoxide/LB broth cultures in microtiter dishes as described by Clarke and Carbon (6). Forty thousand colonies were also scraped off the plates and either plated out selecting for leucine-independent (Leu⁺) cells on selective media, or used to isolate mixed plasmid DNA for subsequent transformations.

Hybrid ColE1–Yeast DNA Plasmids That Complement a *leuB* Mutation. The *leu-6* mutation present in *E. coli* strain JA199 and in all K-12 C600 strains is a lesion in the *leuB* gene, which encodes for β-isopropylmalate dehydrogenase (J. Calvo, personal communication; ref. 19). This enzyme catalyzes the oxidative decarboxylation of β-isopropylmalate to form α-keto-isocaproic acid, on the pathway of leucine biosynthesis in both *E. coli* and yeast.

Portions of a suspension of 40,000 JA199 transformant colonies containing hybrid ColE1–yeast DNA plasmids were plated on minimal medium with tryptophan and without leucine and incubated at 30°. After 3–4 days, large and small Leu⁺ colonies appeared at a frequency of 10⁻⁶. Several of the small colonies were selected and tested for resistance to colicins E1 and E2 (9). The plates were then further incubated at room

Biochemistry: Ratzkin and Carbon

Proc. Natl. Acad. Sci. USA 74 (1977) 489

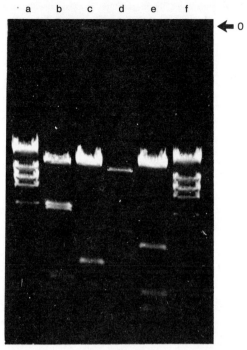

FIG. 1. Fractionation by 1.2% agarose gel electrophoresis of complete endonuclease *Eco*RI digests of various hybrid ColE1–yeast DNA plasmids; lanes a and f, phage λ DNA; lane b, pY*eleu*10 DNA; lane c, pY*eleu*11 DNA; lane d, pY*eleu*12 DNA; lane e, pY*eleu*17 DNA. Each well contained 1 μg of DNA digested with excess endonuclease *Eco*RI as previously described (1, 8). Electrophoresis was carried out at 8 V/cm for 2 hr and the DNA was visualized as described (1). The sizes of the λ fragments are (from the top, in daltons): 13.7×10^6, 4.7×10^6, 3.7 to 3.5×10^6, 3.0×10^6, and 2.1×10^6.

Table 1. Transformation of *E. coli* auxotrophs by ColE1–yeast hybrid plasmids

Plasmid DNA	Molecular weight $\times 10^{-6}$	Allele transformed	Transformants/μg DNA	
			Colicin-E1 resistant	Leu$^+$ or His$^+$*
pY*eleu*10	13.5	*leuB6*	270,000	310,000
		leuB401	21,000	24,000
		leuB61	26,000	16,000
pY*eleu*11	8.9	*leuB6*	70,000	3,900
		leuB401	140,000	0
		leuB61	42,000	0
pY*eleu*12	10.3	*leuB6*	70,000	7,600
		leuB401	17,000	0
		leuB61	8,800	0
pY*eleu*17	13.2	*leuB6*	70,000	7,500
		leuB401	73,000	0
		leuB61	18,000	0
pY*ehis*1	10.7	*hisB463*	15,000	1,300*
pY*ehis*2	10.7	*hisB463*	26,000	1,700*
pY*ehis*3	10.7	*hisB463*	26,000	1,900*

These data have been corrected for the presence of both spontaneous Leu$^+$ revertants and colicin-E1-resistant mutants. Molecular weights were determined by electron microscopy using ColE1 DNA (4.2×10^6) as standard.
* Histidine-independent transformant.

temperature for 2 weeks, after which time very small (1–2 mm) colonies appeared. Two of the intermediate-size Leu$^+$ colonies and 15 of the very small ones that were colicin-E1-resistant and E2-sensitive were also tested for the ability to donate their Leu$^+$ character to strain KL380 on minimal medium containing arginine, methionine, and streptomycin. All except one transferred the Leu$^+$ character at a high frequency (10^{-3}–10^{-4} per donor cell). The Leu$^+$ recipients were also colicin-E1-resistant and E2-sensitive, an indication that hybrid ColE1 plasmids had been transferred (9).

Covalently closed, supercoiled plasmid DNA was isolated from the 16 Leu$^+$ colicin-E1-resistant transformants described above. The purified plasmid DNAs were treated with excess endonuclease *Eco*RI and the digests were fractionated by electrophoresis on 1.2% agarose gels. Four different patterns of *Eco*RI restriction fragments were observed among this group (Fig. 1). Plasmid DNA (pY*eleu*10) from the fastest-growing Leu$^+$ transformant gave a unique fragment pattern, quite different from that displayed by any of the slower-growing transformants. The majority of the latter group gave the restriction pattern shown by pY*eleu*12 DNA, although two other plasmid types were obtained, as typified by pY*eleu*11 and pY*eleu*17 (Fig. 1). These four plasmids are of different sizes, with molecular weights ranging from 8.9×10^6 to 13.5×10^6 (see Table 1). None of the *Eco*RI fragments obtained from this

plasmid group appear to be identical, although in each case the largest fragment is presumed to contain the 4.2×10^6 dalton ColE1 segment.

These plasmid DNAs (pY*eleu*10, pY*eleu*11, pY*eleu*12, and pY*eleu*17) were used to transform three different *E. coli leuB* mutants, selecting for either Leu$^+$ or colicin E1 resistance. As shown in Table 1, plasmid pY*eleu*10 DNA is capable of transforming all of the *leuB* mutants (*leuB6*, *leuB61*, and *leuB401*) (11) to both Leu$^+$ and colicin E1 resistance with high frequency ($>10^5$ transformants per μg). Although the other three plasmid DNAs would readily transform all of the *leuB* strains to colicin E1 resistance, these plasmids would complement only the mutation in *leuB6* (Table 1). pY*eleu*10 DNA is not capable of transforming to Leu$^+$ strains bearing mutations in other genes of the *leu* operon, such as *leuA371*, *leuC222*, or *leuD211* (11).

If pY*eleu*10 carries a segment of yeast DNA that specifies the synthesis of a functional β-isopropylmalate dehydrogenase, it should complement deletions of the *leuB* region in the bacterial host cell. Although well-characterized *leuB* deletions in *E. coli* K-12 were not available, several *leu* deletion mutants in *Salmonella typhimurium* have been mapped by Calvo and Worden (12). The pY*eleu*10 plasmid was transferred from strain JA199 into the *Salmonella* deletion mutants *leuA124*, *leuB698*, *leuC5076*, and *leuD657* (obtained from Joseph Calvo), by F-mediated transfer (6). Leu$^+$ *Salmonella* colonies were obtained only from the JA199/pY*eleu*10 × *leuB698* cross, with a frequency of 10^{-3} Leu$^+$ recipients per donor cell. Plasmid DNA isolated from four of the *Salmonella* Leu$^+$ strains gave an *Eco*RI restriction fragment pattern identical to that from authentic pY*eleu*10 DNA. Thus, the presence of pY*eleu*10 DNA correlates with the transfer of a Leu$^+$ phenotype to a strain harboring a deletion in *leuB*.

Isolation of Hybrid Plasmids That Complement a *hisB* Deletion. Mixed plasmid DNA isolated from the 40,000

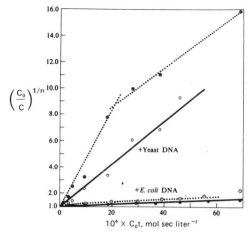

FIG. 2. Reassociation kinetics of labeled single-stranded pY$eleu$10 DNA fragments (0.085 μg/ml) in the presence of single-stranded fragments of: (●——●) salmon sperm DNA, 1000 μg/ml; (⊙··⊙) *E. coli* DNA, 196 μg/ml; (○——○) yeast (*S. cerevisiae*) DNA, 176 μg/ml; and (⊕··⊕) yeast DNA, 526 μg/ml. The experiment was carried out as described in *Methods*. C, DNA concentration in moles of nucleotide per liter; C_0, initial DNA concentration; t, time in seconds; n, 0.45 (18).

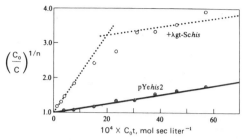

FIG. 3. Reassociation kinetics of labeled single-stranded pY$ehis$2 DNA fragments (0.07 μg/ml) in the presence of single-stranded fragments of: (⊕——⊕) salmon sperm DNA, 1000 μg/ml; and (○··○) λgt-Sc*his* DNA, 1.25 μg/ml. See *Methods* for details.

transformant colonies was used in attempts to transform several *E. coli* auxotrophic strains to prototrophy. Among these strains was *hisB463*, which lacks an active imidazole glycerol phosphate dehydratase. Struhl *et al.* (4) have previously isolated a λ phage containing a segment of yeast DNA (λgt-Sc*his*) that suppresses this mutation. Two micrograms of our mixed hybrid plasmid DNA was used to transform strain *hisB463*, selecting for His⁺. Three His⁺ colonies were detected after 5 days of incubation at 30°, as compared to 10⁵ colicin-E1-resistant colonies. These colonies were colicin-E1-resistant, but sensitive to colicin E2. Plasmid DNAs were isolated from these strains and used to transform strain *hisB463* again. As shown in Table 1, these purified plasmid DNAs (pY*ehis*1, pY*ehis*2, and pY*ehis*3) transformed the *hisB463* strain to colicin E1 resistance and to His⁺ with high frequency. Transformation to His⁺ occurred at a lower frequency than to colicin E1 resistance, although prolonged incubation produced more His⁺ colonies.

All three pY*ehis* plasmids are of similar size, with a molecular weight of 10.7 × 10⁶. They each contain a single *Eco*RI endonuclease restriction site within the segment of cloned yeast DNA, and are probably identical.

The pY*eleu*10 and pY*ehis*2 Plasmids Contain Yeast DNA. We have shown that purified pY*eleu*10 and pY*ehis* DNAs can suppress the appropriate *leu* or *his* mutation in *E. coli*; it is also necessary to establish the source of the cloned DNA segments. In order to prove that these plasmids contain segments of yeast DNA, we labeled pY*eleu*10 and pY*ehis*2 DNAs by nick translation with DNA polymerase I and ³H-labeled deoxynucleoside triphosphates and determined if single-stranded plasmid DNA would reassociate with authentic yeast DNA.

The reassociation of single-stranded pY*eleu*10 DNA was driven well by yeast DNA single strands, but not appreciably by single-stranded *E. coli* DNA (Fig. 2). If one assumes that 70% of pY*eleu*10 is yeast DNA (from the total pY*eleu*10 molecular weight, 13.5 × 10⁶, minus the ColE1 segment, 4.2 × 10⁶), then we can calculate from the initial slopes of the reassociation rate curves that the yeast segment in pY*eleu*10 is 0.11–0.12% of the

yeast genome. If we further assume one copy of the pY*eleu*10 segment in the yeast genome, then the total yeast genome would be 7.8 to 8.5 × 10⁹ daltons, well within the range of values in the literature (5 to 10 × 10⁹ daltons) (5). Apparently this segment of DNA is a unique fragment of the yeast genome.

Fig. 3 shows the reassociation kinetics of single-stranded pY*ehis*2 DNA in the presence or absence of single-stranded λgt-Sc*his* DNA, from the recombinant λ phage containing a segment of yeast DNA that suppresses the *hisB463* mutation (4). If the pY*ehis*2 plasmid and λgt-Sc*his* contain similar segments of yeast DNA, then the addition of λgt-Sc*his* single-stranded DNA should increase the rate of reassociation of single-stranded pY*ehis*2 DNA. As shown in Fig. 3, λgt-Sc*his* DNA markedly increases the rate of reassociation of pY*ehis*2 DNA. The fraction of λgt-Sc*his* DNA that is homologous to the pY*ehis*2 DNA can be calculated as above to be equal to 22% of the λgt-Sc*his* DNA (30 × 10⁶ daltons), equivalent to 6.6 × 10⁶ daltons of DNA. From the molecular weights of pY*ehis*2 and ColE1 DNAs and assuming a full-length ColE1 segment is present in pY*ehis*2, the calculated mass of the cloned yeast DNA segment would be 6.7 × 10⁶ daltons. pY*ehis*2 DNA contains a single *Eco*RI site, which is apparently identical to one of the two sites defining the yeast segment in λgt-Sc*his* DNA.

Efficiency of the Complementation. One measure of the efficiency of complementation or suppression of an *E. coli* mutation by a cloned yeast DNA segment would be the relative growth rates of the transformant and wild-type cells in minimal unsupplemented media. For example, *hisB463* strains containing the λgt-Sc*his* prophage are reported to grow with a 2.7 hr generation time in minimal medium without histidine, versus a 2.1 hr generation time for a *his*⁺ wild-type under the same conditions (4). Colonies of *leuB*/pY*eleu*10 strains growing on minimal solid media without leucine appear to grow at about 1/3–1/2 the rate of wild-type colonies. In liquid minimal media without leucine, a *leuB*⁺ transductant of a *leuB6* strain grew twice as fast (1 hr doubling time) as did strain *leuB6*/pY*eleu*10 (2 hr doubling time). However, in media supplemented with leucine, both strains grow with a 1 hr doubling time. Apparently, the presence of the yeast DNA segment in pY*eleu*10 partially compensates for the *leuB* mutation, but has no deleterious effects on the bacterial cell.

leuB6 strains bearing the pY*eleu*11, pY*eleu*12, and pY*eleu*17 plasmids grow very slowly when plated on solid media without leucine, requiring 3–4 days at 30° to form tiny colonies just visible to the naked eye. Apparently the suppression effect associated with the presence of these plasmids in *leuB6* strains is quite inefficient.

Although the selections for the pY*eleu* and pY*ehis* containing strains were carried out at 30° (the optimum temperature for yeast growth), strains bearing these plasmids grew equally well at 37° on selective media.

DISCUSSION

It now seems clear that segments of yeast DNA capable of complementing a variety of *E. coli* mutations can be readily isolated. The actual mechanism of complementation by the cloned DNA is not clear. The *hisB463* mutation that is complemented by the yeast DNA cloned in pYe*his*2 and λgt-Sc*his* is thought to be a deletion (4). The *leuB6* mutation in strain JA199 is a point mutation, but we have observed high frequency transfer of the Leu⁺ phenotype to a *Salmonella typhimurium* *leuB* deletion by F-mediated transfer from a pYe*leu*10 strain. However, even if a deletion mutation is complemented, we are not certain that the yeast DNA in pYe*leu*10 contains the homologous gene from yeast, *leu2* (20). The different milieu of the *E. coli* cell may allow a product from some other yeast gene to function as a β-isopropylmalate dehydrogenase.

Certainly the mechanism of this type of complementation or suppression can differ in various cases. For example, in addition to pYe*leu*10, three other hybrid ColE1–yeast DNA plasmids have been found that suppress the *leuB6* mutation, although with a much lower efficiency than does pYe*leu*10. These plasmids all display different patterns of cutting by endonuclease *Eco*RI. Although pYe*leu*10 will strongly complement any of the *leuB* mutations we have tried (Table 1), these other pYe*leu* plasmids appear to suppress only the *leuB6* mutation in C600 strains. The possibility exists of missense suppression mediated by yeast tRNA or aminoacyl-tRNA synthetases produced from yeast genes on the plasmid, since it is well known that tRNA misaminoacylation occurs readily in *E. coli*–yeast heterologous systems (21).

We believe, then, that many yeast DNA segments in *E. coli* will be readily expressed; the barriers to translation and transcription are somehow overcome. However, not all genes will be expressed similarly. For example, the pYe*his*2 plasmid and λgt-Sc*his* phage are expressed somewhat more efficiently than is pYe*leu*10, as measured by growth rates of transformants on unsupplemented minimal media. pYe*his*2 transformants to His⁺ do not occur at the same frequency as do transformants to colicin E1 resistance; few His⁺ transformants appear at first but increasing numbers appear with longer incubation times. pYe*leu*10 transformants to Leu⁺ and to colicin E1 resistance do occur at the same frequency, and they appear more synchronously. These differences could be due to differing rates of expression as alluded to by Struhl *et al.* (4) or to the expression of other genes on the plasmids that hinder cell growth in the case of pYe*his*2.

The level of success with which we have found hybrid ColE1–yeast DNA plasmids that suppress *E. coli* mutations is fairly good. Out of 11 different mutations tried with our re-

combinant collection, two were successfully suppressed and another two to three show apparent complementation, but have not yet been completely characterized. In one case, a *trpAB* deletion has been shown to be suppressed by a ColE1–yeast recombinant plasmid, and the cloned DNA segment has been shown to be yeast DNA. Thus, it seems possible that productive expression of yeast (and perhaps other eukaryotic) DNA in *E. coli* can occur more readily than has previously been anticipated.

We are grateful to Denise Richardson and Gary Tschumper for valuable technical assistance; to K. Struhl and R. Davis for a gift of λgt-Sc*his* DNA, and to J. Calvo for supplying several key *leu* mutant strains. Strain JA199 was constructed by Margaret Nesbitt. This work was supported by Research Grants CA-11034 and CA-15941 from the National Cancer Institute, U.S. Public Health Service, and by a research grant from Abbott Laboratories.

1. Morrow, J. F., Cohen, S. N., Chang, A. C. Y., Boyer, H. W., Goodman, H. M. & Helling, R. B. (1974) *Proc. Natl. Acad. Sci. USA* **71**, 1743–1747.
2. Chang, A. C. Y., Lansman, R. A., Clayton, D. A. & Cohen, S. N. (1975) *Cell* **6**, 231–244.
3. Kedes, L. H., Chang, A. C. Y., Houseman, D. & Cohen, S. N. (1975) *Nature* **255**, 533–538.
4. Struhl, K., Cameron, J. R. & Davis, R. W. (1976) *Proc. Natl. Acad. Sci. USA* **73**, 1471–1475.
5. Lauer, G. D. & Klotz, L. C. (1975) *J. Mol Biol.* **95**, 309–326.
6. Clarke, L. & Carbon, J. (1976) *Cell* **9**, 91–99.
7. Lobban, P. E. & Kaiser, A. D. (1973) *J. Mol. Biol.* **78**, 453–471.
8. Jackson, D. A., Symons, R. H. & Berg, P. (1972) *Proc. Natl. Acad. Sci. USA* **69**, 2904–2909.
9. Hershfield, V., Boyer, H. W., Yanofsky, C., Lovett, M. A. & Helinski, D. R. (1974) *Proc. Natl. Acad. Sci. USA* **71**, 3455–3459.
10. Wood, W. B. (1966) *J. Mol. Biol.* **16**, 118–133.
11. Somers, J. M., Amzallag, A. & Middleton, R. B. (1973) *J. Bacteriol.* **113**, 1268–1272.
12. Calvo, J. M. & Worden, H. E. (1970) *Genetics* **64**, 199–214.
13. Miller, J. (1972) *Experiments in Molecular Genetics* (Cold Spring Harbor Laboratory, Cold Spring Harbor, N.Y.).
14. Clarke, L. & Carbon, J. (1975) *Proc. Natl. Acad. Sci. USA* **72**, 4361–4365.
15. Marmur, J. (1961) *J. Mol. Biol.* **3**, 208–218.
16. Hirt, B. (1967) *J. Mol. Biol.* **26**, 365–369.
17. Schachat, F. H. & Hogness, D. S. (1973) *Cold Spring Harbor Symp. Quant. Biol.* **38**, 371–381.
18. Wensink, P. C., Finnegan, D. J., Donelson, J. E. & Hogness, D. S. (1974) *Cell* **3**, 315–325.
19. Kessler, D. P. & Englesberg, E. (1969) *J. Bacteriol.* **98**, 1159–1169.
20. Satyanarayana, T., Umbarger, H. E. & Lindegren, G. (1968) *J. Bacteriol.* **96**, 2012–2017.
21. Jacobson, K. B. (1971) in *Progress in Nucleic Acid Research and Molecular Biology*, eds. Davidson, J. N. & Cohn, W. E. (Academic Press, New York), Vol. 11, pp. 461–488.

36

Reprinted from *Natl. Acad. Sci. (USA) Proc.* **75**:3727–3731 (1978)

A bacterial clone synthesizing proinsulin

(rat preproinsulin/cDNA cloning/solid-phase radioimmunoassay/DNA sequence/fused proteins)

LYDIA VILLA-KOMAROFF*, ARGIRIS EFSTRATIADIS*, STEPHANIE BROOME*, PETER LOMEDICO*,
RICHARD TIZARD*, STEPHEN P. NABER†, WILLIAM L. CHICK†, AND WALTER GILBERT*

* Biological Laboratories, Harvard University, Cambridge, Massachusetts 02138; and † Elliot P. Joslin Research Laboratory, Harvard Medical School, and the Peter Bent Brigham Hospital, Boston, Massachusetts 02215

ABSTRACT We have cloned double-stranded cDNA copies of a rat preproinsulin messenger RNA in *Escherichia coli* χ1776, using the unique *Pst* endonuclease site of plasmid pBR322 that lies in the region encoding amino acids 181–182 of penicillinase. This site was reconstructed by inserting the cDNA with an oligo(dG)·oligo(dC) joining procedure. One of the clones expresses a fused protein bearing both insulin and penicillinase antigenic determinants. The DNA sequence of this plasmid shows that the insulin region is read in phase; a stretch of six glycine residues connects the alanine at position 182 of penicillinase to the fourth amino acid, glutamine, of rat proinsulin.

Can the structural information for the production of a higher cell protein be inserted into a plasmid in such a way as to be expressed in a transformed bacterium? To attack this problem, we used as a model rat insulin, an interesting protein that can be identified by immunological and biological means.

Although mature insulin contains two chains, A and B, it is the product of a single longer polypeptide chain. The hormone is initially synthesized as a preproinsulin structure (1, 2). A hydrophobic leader sequence of 23 amino acids at the amino terminus of the nascent chain is cleaved off, presumably as the polypeptide chain moves through the endoplasmic reticulum (2–4), producing a proinsulin molecule. The proinsulin chain folds up and then the C peptide is cleaved from its middle (5). Thus each of the two (nonallelic) insulin genes in the rat (6–8) encodes a polypeptide 109 amino acids long, whose initial structure is NH$_2$—leader sequence—B chain—C peptide—A chain.

Ullrich *et al.* (9) have cloned double-stranded cDNA copies of rat preproinsulin mRNA isolated from pancreatic islets and determined sequences covering much of those two genes. We have made double-stranded cDNA copies of mRNA from a rat insulinoma (10) and cloned these in the *Pst* (*Providencia stuartii* endonuclease) site of pBR322 (11), which lies within the penicillinase gene.

The *Escherichia coli* penicillinase is a periplasmic protein, the gene for which was recently sequenced (12). Penicillinase is synthesized as a preprotein with a 23 amino acid leader sequence (12, 13), which presumably serves as a signal to direct the secretion of the protein to the periplasmic space, and is removed as the protein traverses the membrane. Insertion of the structural information for insulin into the penicillinase gene should cause expression of the insulin sequence as a fusion product transported outside the cell.

MATERIALS AND METHODS

Bacterial Strains. *E. coli* K-12, strain HB101 [*hsm⁻*, *hrs⁻*, *recA⁻*, *gal⁻*, *pro⁻*, *str*ʳ (14)] was initially obtained from H. Boyer. *E. coli* K-12 strain χ1776 (15) (F⁻, *tonA53*, *dapD8*,

minA1, supE42, Δ40[gal–uvrB], λ⁻, minB2, rfb-2, nalA25, oms-2, thyA57, metC65, oms-1, Δ29[bioH–asd], cycB2, cycA1, hsdR2) was provided by R. Curtiss.

DNA and Enzymes. pBR322 DNA, a gift from A. Poteete, was used to transform *E. coli* HB101. Plasmid DNA was purified according to the procedure of Clewell (16). Avian myeloblastosis virus reverse transcriptase (RNA-dependent DNA polymerase), *E. coli* DNA polymerase I, and terminal transferase were gifts from T. Papas, M. Goldberg, and J. Wilson, respectively. Restriction enzymes were purchased from Bethesda Research Labs and New England BioLabs.

RNA Purification. An x-ray-induced, transplantable rat beta cell tumor (10) was used as source of preproinsulin mRNA. Tumor slices (20 g per preparation) were homogenized, and a cytoplasmic RNA (about 2 mg/g of tissue) was purified from a postnuclear supernatant by Mg^{2+} precipitation (17), followed by extraction with phenol and chloroform, and enriched for poly(A)-containing RNA by oligo(dT)-cellulose chromatography (18). About 4% of the material binds to the column (data from eight preparations). Further purification of the oligo(dT)-cellulose-bound material by sucrose gradient centrifugation and/or polyacrylamide gel electrophoresis showed that the preproinsulin mRNA was a minor component of the preparation.

Double-Stranded cDNA Synthesis. Oligo(dT)-cellulose-bound RNA was used directly as template for double-stranded cDNA synthesis (19), except that a specific p(dT)$_8$dG-dC primer (Collaborative Research) was utilized for reverse transcription. The concentrations of RNA and primer were 7 mg/ml and 1 mg/ml, respectively. All four [α-^{32}P]dNTPs were at 1.25 mM (final specific activity 0.85 Ci/mmol). The reverse transcript was 2% of the input RNA, and 25% of it was finally recovered in the double-stranded cDNA product.

Construction of Hybrid DNA Molecules. pBR322 DNA (5.0 μg) was linearized with *Pst*, and approximately 15 dG residues were added per 3′ end by terminal transferase at 15° in the presence of 1 mM Co^{2+} (20) and autoclaved gelatin at 100 μg/ml. Similarly, dC residues were added to 2.0 μg of double-stranded cDNA, which was then electrophoresed in a 6% polyacrylamide gel. Following autoradiography, molecules in the size range of 300 to 600 base pairs (0.5 μg) were eluted from the gel (21). Size selection was done after tailing rather than before because previous experience had indicated that occasionally impurities contaminating DNA extracted from gels inhibits terminal transferase. The eluted double-stranded cDNA was concentrated by ethanol precipitation, redissolved in 10 mM Tris·HCl at pH 8, mixed with 4 μg of dG-tailed pBR322, and dialyzed versus 0.1 M NaCl/10 mM EDTA/10 mM Tris, pH 8. The mixture (4 ml) was then heated at 56° for 2 min, and annealing was performed at 42° for 2 hr. The hybrid DNA was used to transform *E. coli* χ1776.

Transformation and Identification of Clones. Transformation of *E. coli* χ1776 (an EK2 host) with pBR322 (an EK2 vector) was performed in a biological safety cabinet in a P3 physical containment facility in compliance with NIH guidelines for recombinant DNA research published in the *Federal Register* [(1976) **41**, 27902–27943].

χ1776 was transformed by a transfection procedure (22) adapted to χ1776 by A. Bothwell (personal communication) and slightly modified as follows: χ1776 was grown in L broth (23) supplemented with diaminopimelic acid at 10 μg/ml and thymidine (Sigma) at 40 μg/ml to OD_{590} of 0.5. Cells (200 ml) were sedimented at 500 × g and resuspended by swirling in 1/10th vol of cold buffer containing 70 mM $MnCl_2$, 40 mM Na acetate at pH 5.6, 30 mM $CaCl_2$, and kept on ice for 20 min. The cells were repelleted and resuspended in 1/30th of the original volume in the same buffer. Two milliliters of the annealed DNA preparation was added to the cells. Aliquots of this mixture (0.3 ml) were placed in sterile tubes and incubated on ice for 60 min. The cells were then placed at 37° for 2 min. Broth was added to each tube (0.7 ml) and the tubes were incubated at 37° for 15 min; 200 μl of the cells was spread on sterile nitrocellulose filters (Millipore) overlaying agar plates containing tetracycline at 15 μg/ml. (The filters were boiled to remove detergents before use.) The plates were incubated at 37° for 48 hr. Replicas of the filters were made by a procedure developed by D. Hanahan (personal communication): The nitrocellulose filters containing the transformants were removed from the agar and placed on a layer of sterile Whatman filter paper. A new sterile filter was placed on top of the filter containing the colonies and pressure was applied with a sterile velvet cloth and a replica block. A sterile needle was used to key the filters. The second filter was placed on a new agar plate and incubated at 37° for 48 hr. The colonies on the first filter were screened by the Grunstein–Hogness technique (24), using as probe an 80-nucleotide-long fragment of cDNA produced by *Hae* III digestion of high specific activity cDNA (9). Positive colonies were rescreened by hybrid-arrested translation (25) as described in the legend of Table 1.

Radioimmunoassays. Two-site solid-phase radioimmunoassays were performed (28). Cells from colonies to be tested were transferred with an applicator stick onto 1.5% agarose containing 30 mM Tris·HCl, pH 8, lysozyme at 0.5 mg/ml, and 10 mM EDTA; released antigen was adsorbed to an IgG-coated polyvinyl disk during a 1-hr incubation at 4°. The wash buffer contained streptomycin sulfate at 300 μg/ml and normal guinea pig serum (Grand Island Biological Co.) instead of normal rabbit serum. Guinea pig antiserum to bovine insulin was purchased from Miles Laboratories.

Standard (liquid) radioimmunoassays were performed using the back titration procedure employing alcohol precipitation of insulin–antibody complexes (29).

DNA Sequencing. DNA sequencing was performed as described by Maxam and Gilbert (30).

RESULTS

Construction and Identification of cDNA Clones. We isolated poly(A)-containing RNA from a transplantable rat insulinoma. This preparation contained preproinsulin mRNA, because it directed the synthesis in a cell-free system of a product precipitable with anti-insulin antibody (data not shown). However, the mRNA yield after further purification was not sufficient for cloning, and therefore we decided to clone cDNA synthesized from the total preparation. In an attempt to enrich the reverse transcript for insulin sequences, we utilized the DNA sequence reported by Ullrich *et al.* (9) to choose a specific primer, $(dT)_8dG$-dC. The product of double-stranded cDNA synthesis (19) was extended by a short oligo(dC) tail about 15 nucleotides in length, and sized on a polyacrylamide

Table 1. Hybrid-arrested translation and immunoprecipitation of the cell-free products

Source of arresting DNA	Radioactivity, cpm/20 μl			% Immuno-precipitable*
	Acid insoluble	Immuno-precipitable		
		− Insulin	+ Insulin	
Control I (−DNA, −RNA)[†]	2,570			
Control II (−DNA, +RNA)[‡]	35,700	12,300	310	36.2
pBR322	28,800	7,850	245	29.0
Clone 3	15,100	3,630	264	26.9
Clone 13	19,600	5,190	350	28.4
Clone 15	18,600	4,850	252	28.7
Clone 16	29,200	8,830	247	32.2
Clone 17	24,000	6,700	316	30.0
Clone 18	15,900	3,690	251	25.8
Clone 19	8,650	587	277	5.0
Clone 20	15,100	4,070	231	30.6
Clone 21	21,100	5,170	223	26.7

Plasmid DNA (about 3 μg) was digested with *Pst*, precipitated with ethanol, and dissolved directly in 20 μl of deionized formamide. After heating for one minute at 95° each sample was placed on ice. Following the addition of 1.5 μg of oligo(dT)-cellulose-bound RNA, piperazine-*N,N′*-bis(2-ethanesulfonic acid) (Pipes) at pH 6.4 to 10 mM, and NaCl to 0.4 M, the mixtures were incubated for 2 hr at 50°. They were then diluted by the addition of 75 μl of H_2O and ethanol precipitated in the presence of 10 μg of wheat germ tRNA, washed with 70% (vol/vol) ethanol, dissolved in H_2O, and added to a wheat germ cell-free translation mixture (26) containing 10 μCi of [³H]leucine (60 Ci/mmol). Fifty-microliter reaction mixtures were incubated at 23° for 3 hr and then duplicate 2-μl aliquots were removed for trichloroacetic acid precipitation. From the remainder two 20-μl aliquots were treated with ribonuclease, diluted with immunoassay buffer, and analyzed for the synthesis of immunoreactive preproinsulin by means of a double antibody immunoprecipitation (27) in the absence or presence of 10 μg of bovine insulin. The washed immunoprecipitates were dissolved in 1 ml of NCS (Amersham) and assayed in 10 μl of Omnifluor (New England Nuclear) by liquid scintillation counting.
* Calculated using the formula [(immunoprecipitable radioactivity in the absence of insulin) − (immunoprecipitable radioactivity in the presence of insulin)]/[(acid-insoluble radioactivity) − (acid-insoluble radioactivity of control I)].
† Reaction mixture incubated in the absence of added RNA.
‡ Cell-free translation by the direct addition of oligo(dT)-cellulose-bound RNA into the reaction mixture.

gel. A broad size cut averaging 500 base pairs was selected in order to enrich for full-length sequences. We inserted these molecules into the *Pst* site of pBR322 after elongating the 3′-terminal extension of the cleavage site with oligo(dG). We used this oligo(dG)·oligo(dC) joining procedure in order to reconstruct the *Pst* recognition sequence (ref. 31; W. Rowenkamp and R. Firtel, personal communication); approximately 40% of the inserts were excisable with *Pst* after cloning. From about 0.25 μg of tailed cDNA we obtained 2355 transformants in *E. coli* strain χ1776. To identify clones containing insulin sequences, we first screened one-third of the transformants, using as a probe an 80-nucleotide-long *Hae* III fragment of cDNA synthesized from oligo(dT)-bound RNA because the results of Ullrich *et al.* (9) suggested that such a fragment should be insulin specific. About 20% of the clones were positive, but restriction analysis of plasmid DNA from a few candidates showed that the inserts were not insulin sequences. We concluded that our probe was not pure and rescreened some of the positive clones, using hybrid-arrested translation (25). This method is based on the principle that mRNA in the form of an RNA·DNA hybrid does not direct cell-free protein synthesis. We incubated aliquots of oligo(dT)-bound RNA with linearized

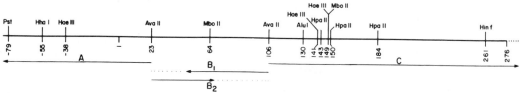

FIG. 1. Restriction map of the insertion in clone pI19. Each restriction site is identified by a number indicating the 5′-terminal nucleotide generated by cleavage at the message strand. Nucleotides are numbered beginning with the first base of the sequence encoding proinsulin. Nucleotides in the 5′ direction from position 1 in the message strand are identified by negative numbers, beginning with −1. Arrows indicate the sequenced fragments; those pointing to the left indicate sequences derived from the antimessage strand, and those pointing to the right indicate sequences derived from the message strand. The uniquely labeled restriction fragments were generated as follows: Following excision with *Pst*, DNA of the insertion was digested with *Ava* II and end labeled. Fragments A and C purified on a polyacrylamide gel were sequenced directly because the *Pst* ends do not label significantly. Fragment B was strand separated on a polyacrylamide gel and sequenced in both directions. The exact number of C·G pairs in the right-hand tail before the *Pst* site could not be counted.

DNA from nine clones under conditions favoring DNA·RNA hybridization (32), added them to cell-free translation systems, and assayed for a specific inhibition of insulin synthesis. Table 1 shows that one of the plasmids, pI19, inhibited the synthesis of immunoprecipitable material. Restriction endonuclease digestions of the *Pst*-excised insert of pI19 with several enzymes generated fragments whose sizes were consistent with the sequence of Ullrich *et al.* (9). We confirmed the presence of insulin DNA in pI19 by direct DNA sequence analysis and screened the rest of the clones with purified pI19 insert labeled by nick translation. About 2.5% (48/1745) of the clones hybridized strongly to this probe. There must have been enrichment for insulin sequence at some step of our procedure, because hybridization analysis using cloned insulin DNA as probe showed the presence of only 0.3% insulin mRNA in the original oligo(dT)-bound RNA.

Sequence Information. Fig. 1 shows the restriction map of the insertion in clone pI19 and Fig. 2 shows the sequence of the insert. It corresponds to rat insulin I (5, 33) and encodes the entire preproinsulin chain with the exception of the first two amino acid residues of the reported preregion (1). It therefore extends the sequence determined by Ullrich *et al.* (9) by twenty-five 5′-terminal nucleotides. It also verifies the reported amino acid residues for positions −14, −17, −18, and −20; it identifies the previously uncertain residue −15; and it identifies the unknown residue −19. However, the residues at positions −16 and −21 differ from those reported (1).

The sequence deviates from that determined by Ullrich *et al.* (9) at the region immediately after the UGA terminator, where a GAGTC sequence occurs, predicting a *Hinf* cleavage

site that we have experimentally verified. Furthermore, only moderate agreement exists between the two sequences for the next 15 nucleotides of the 3′ untranslated region.

Expression. Almost two-thirds of the clones carrying inserts were ampicillin resistant; thus the active site of penicillinase must lie between amino acid residues 23 and 182 (12). The degree of resistance was variable, suggesting the expression of different sequences from the inserts in the form of fused translation products, probably differing in length and stability.

We therefore screened colonies of the 48 clones containing insulin sequence for the presence of insulin antigenic determinants, using a solid-phase radioimmunoassay (28). Polyvinyl sheets coated with antibody molecules will bind specific antigens released from bacteria. The immobilized antigen can then be detected by autoradiography following exposure of the sheets to ^{125}I-labeled antibody. This method permits detection of as little as 10 pg of insulin in a colony. We coated plastic disks with anti-insulin antibody and used ^{125}I-labeled anti-insulin to detect solely insulin antigenic determinants. Disks coated with anti-penicillinase antibody and exposed to ^{125}I-anti-insulin detect the presence of a fused protein, as do disks coated with anti-insulin and exposed to radioiodinated anti-penicillinase.

One clone, pI47, gave positive responses with all of the combinations described above; this indicates the presence of a penicillinase–insulin hybrid polypeptide. Fig. 3 shows some of the results. To determine whether this fused protein is secreted, we grew clone pI47 in liquid culture and extracted the proteins in the periplasmic space by osmotic shock, a method that does not lyse bacteria (34). Fig. 4 shows that the insulin

FIG. 2. DNA sequence of the insertion in clone pI19. Nucleotides are numbered using the convention described in Fig. 1. Accordingly, amino acids are numbered beginning with the first amino acid of proinsulin, while the last amino acid of the leader sequence (pre region) is numbered as −1. Restriction endonuclease cleavage sites experimentally verified are underlined and identified. The arrows indicate, in order, the ends of the leader sequence and the peptides B, C, and A. Two nucleotides indicated by double underlining are uncertain.

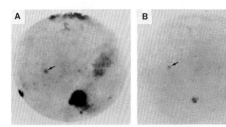

FIG. 3. Initial detection of penicillinase-insulin hybrid polypeptides in an insulin cDNA clone. Cells from colonies of the 48 insulin cDNA clones and from control colonies, χ1776 and χ1776-pBR322, were applied to an agarose/lysozyme/EDTA plate. Positive controls, 5 ng of insulin and 5 ng of penicillinase, each in 1 μl of wash buffer, also were spotted on plate. Antigen was adsorbed to an IgG-coated polyvinyl disk during a 1-hr incubation at 4°. Immobilized antigen was labeled by setting the plastic disk on a solution containing radioiodinated anti-insulin IgG. The autoradiographs are of disks precoated with anti-insulin IgG (A) or anti-penicillinase IgG (B), exposed on Kodak X-Omat R film using a Du Pont Cronex Lightning Plus intensifying screen for 12 hr at −70°. The arrows indicate the signal generated by clone pI47. The large exposed area in the lower right of (A) is the positive control for insulin detection.

antigen was recovered in the distilled water wash of the shock procedure. Table 2 shows that the insulin antigen in the wash is also detectable and quantifiable by a standard radioimmunoassay. The yield of antigen depended on the growth medium; antigen was released by cells grown in M9/glucose/amino acids medium but not by cells grown in brain/heart infusion. We estimate a recovery of about 100 molecules per cell.

Structure of the Fused Protein. We sequenced pI47 to determine the sequence around the junctions. Fig. 5 shows that a proinsulin I cDNA lies in the *Pst* site in the correct orientation and in phase, so that a fused protein can be synthesized. In pI19, the insert is in the correct orientation, but not in phase. In pI47 the oligo(dG)·oligo(dC) region encodes six glycines that connect the penicillinase sequence, ending at amino acid 182 (alanine), to the fourth amino acid (glutamine), of the proinsulin sequence. The cDNA sequence in pI47 extends 26 base pairs past the UGA terminator. Thus, we infer the structure of the fused protein to be penicillinase(24–182)-(Gly)$_6$-proinsulin(4–86).

DISCUSSION

The coding regions of eukaryotic structural genes are often interrupted by introns (35–38), whose transcripts are spliced out of the mature mRNA. Because prokaryotes do not appear to process their messengers, double-stranded cDNA made from a mature messenger is the material of choice to carry eukaryotic structural information into bacteria.

By using cDNA cloning technology and an extremely sensitive method to assay expression, we were able to construct a derivative of *E. coli* strain χ1776 carrying an insulin gene sequence and to detect the synthesis and secretion into the periplasmic space of a fused protein carrying antigenic determinants of both insulin and penicillinase. This was accomplished simply by inserting double-stranded cDNA carrying the structural information for insulin into a restriction site within the structural gene for penicillinase. Not only is the fused DNA sequence expressed as a chain of amino acids, but also the polypeptide folds so as to reveal insulin antigenic shapes. Thus we expect soon to be able to demonstrate biological function for this, or for a similar, fused protein.

We anticipate that the joining of cDNA sequences to nucleotides that lie ahead of the *Pst* site in the penicillinase gene

Table 2. Immunoreactive insulin concentration in distilled water wash of osmotic shock procedure

Exp.	Insulin, μunits/ml	Cells/ml
1	318	1.5×10^{10}
2	166	6.0×10^{9}
3	386	4.2×10^{10}

Duplicate 0.1-ml aliquots of each sample prepared as described in the legend to Fig. 4 were assayed (29) in a final volume of 0.4 ml using rat insulin standard, a gift from J. Schlichtkrull. One unit = 48 μg. The NaCl/Tris wash, the 20% sucrose wash, and the media of χ1776-pI47 as well as the water wash from osmotic shock of χ1776-pBR322 gave values below the sensitivity of the assay (25 μunits/ml).

will also produce fused and secreted molecules. Moreover, if the fusion replaces the preproinsulin leader with that of penicillinase it is likely that the new protein will also be secreted by the *E. coli* cell and may even be correctly matured by cleavage of the leader sequence.

Clearly, we have exploited a general method that should lead to the expression and secretion of any eukaryotic protein provided another protein, such as penicillinase, will serve as a carrier, by virtue of its leader sequence. Moreover, the secretion of the eukaryotic protein sequence to the periplasm or extracellular space will both permit its harvest in a purified form and probably eliminate intracellular sources of instability.

Often just an expression of antigens is the goal. In a "shotgun" screening, the existence of a fused protein antigen could be used to identify transformants carrying desired eukaryotic gene fragments. On the other hand, the insertion of a DNA fragment coding for surface antigenic determinants of a virus into a carrier protein should lead to the secretion of a fused protein that could serve as a vaccine, even though no entirely correct virus product is ever produced.

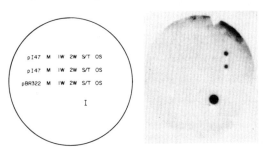

FIG. 4. Release of insulin antigen from χ1776-pI47 cells by osmotic shock. One liter of χ1776-pI47 cells growing at 37° in M9 medium supplemented with 1 g of tryptone, 0.5 g of yeast extract, and 0.5% glucose was harvested at a density of 5 × 10⁷ cells per ml and washed two times in 10 ml of cold 10 mM Tris·HCl, pH 8/30 mM NaCl. The cells were then osmotically shocked (34) in the following manner: The final wash pellet was resuspended in 10 ml of 20% sucrose per 30 mM Tris·HCl, pH 8, at room temperature, made 1 mM in EDTA, shaken at room temperature for 10 min, centrifuged out, resuspended in 10 ml of cold distilled water, shaken in an ice bath for 10 min, and again pelleted. The resulting supernatant was termed the "water wash." As a control, 1 liter of χ1776-pBR322 was grown and treated in a similar manner. Aliquots (1 μl) of each fraction to be assayed for the presence of insulin antigen were applied to the surface of a 1.5% agar plate. (A) Positions of each fraction on the plate. M, medium; 1W, first wash supernatant; 2W, second wash supernatant; S/T, sucrose/Tris supernatant; OS, distilled water wash; I, insulin. (B) Autoradiograph showing results of a two-site radioimmunoassay of these fractions. Antigen was adsorbed to a polyvinyl disk and labeled by using anti-insulin IgG. The labeled areas correspond to the water washes and the positive control (1 ng insulin). A spectrophotometric assay for β-galactosidase (23) indicated that no more than 4% of cells lyse during this procedure.

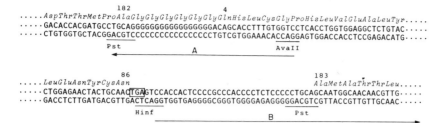

FIG. 5.   Partial DNA sequence of the insertion in clone pI47. Clone pI47 DNA was digested with *Hin*f and two fragments, H1 and H2 (ca. 1700 and 280 base pairs long, respectively) were isolated. H1 contains the amino-terminal portion of the penicillinase gene and the bulk of the cDNA insert. H1 was digested with *Ava* II, end labeled, and digested again with *Pst*. A fragment 39 nucleotides long (fragment A, arrow) was isolated and sequenced. Fragment H2 was end labeled and digested with *Alu* I (which cuts at the region corresponding to amino acid 200 of penicillinase). A fragment 88 base pairs long (fragment B, arrow) was isolated and sequenced. The termination sequence TGA is boxed.

We thank David Baltimore, Philip Sharp, and Salvador Luria for the use of the Massachusetts Institute of Technology P3 laboratory. We thank Macy Koehler for help with the figures; Fotis Kafatos for use of facilities; Philip Sharp, Al Bothwell, Shirley Tilghman, Doug Hanahan, and Richard Firtel for discussions. W.G. is an American Cancer Society Professor of Molecular Biology. W.L.C. is an Established Investigator of the American Diabetes Association. This work was supported by National Institutes of Health Grants AM 21240 and GM 09541-17 to W.G. and AM 15398 to W.L.C.

1. Chan, S. J., Keim, P. & Steiner, D. F. (1976) *Proc. Natl. Acad. Sci. USA* **73**, 1964–1968.
2. Chan, S. J. & Steiner, D. F. (1977) *Trends Biochem. Sci.* **2**, 254–256.
3. Blobel, G. & Dobberstein, B. (1975) *J. Cell Biol.* **67**, 835–851.
4. Blobel, G. & Dobberstein, B. (1975) *J. Cell Biol.* **67**, 852–862.
5. Steiner, D. F., Kemmler, W., Tager, H. S. & Peterson, J. D. (1974) *Fed. Proc. Fed. Am. Soc. Exp. Biol.* **33**, 2105–2115.
6. Smith, L. F. (1966) *Am. J. Med.* **40**, 662–666.
7. Clark, J. L. & Steiner, D. F. (1969) *Proc. Natl. Acad. Sci. USA* **62**, 278–285.
8. Markussen, J. & Sundby, F. (1972) *Eur. J. Biochem.* **25**, 153–162.
9. Ullrich, A., Shine, J., Chirgwin, J., Pictet, R., Tischer, E., Rutter, W. J. & Goodman, H. M. (1977) *Science* **196**, 1313–1319.
10. Chick, W. L., Warren, S., Chute, R. N., Like, A. A., Lauris, V. & Kitchen, K. C. (1977) *Proc. Natl. Acad. Sci. USA* **74**, 628–632.
11. Bolivar, F., Rodriguez, R. L., Greene, P. J., Betlach, M. C., Heyneker, H. L., Boyer, H. W., Crossa, J. H. & Falkow, S. (1977) *Gene* **2**, 95–113.
12. Sutcliffe, J. G. (1978) *Proc. Natl. Acad. Sci.* **75**, 3737–3741.
13. Ambler, R. P. & Scott, G. K. (1978) *Proc. Natl. Acad. Sci. USA* **75**, 3732–3736.
14. Boyer, H. W. & Rouland-Dussoix, D. (1969) *J. Mol. Biol.* **41**, 459–472.
15. Curtiss, R., III, Pereira, D. A., Hsu, J. C., Hull, S. C., Clarke, J. E., Maturin, L. J., Sr., Goldschmidt, R., Moody, R., Inoue, M. & Alexander, L. (1977) in *Recombinant Molecules: Impact on Science and Society. Proceedings of the 10th Miles International Symposium*, eds. Beers, R. F., Jr., & Bassett, E. G. (Raven, New York), pp. 45–56.
16. Clewell, D. B. (1972) *J. Bacteriol.* **110**, 667–676.
17. Palmiter, R. (1974) *Biochemistry* **13**, 3603–3615.
18. Aviv, H. & Leder, P. (1972) *Proc. Natl. Acad. Sci. USA* **69**, 1408–1412.
19. Efstratiadis, A., Kafatos, F. C., Maxam, A. M. & Maniatis, T. (1976) *Cell* **7**, 279–288.
20. Roychoudhury, R., Jay, E. & Wu, R. (1976) *Nucleic Acid Res.* **3**, 101–116.
21. Gilbert, W. & Maxam, A. M. (1973) *Proc. Natl. Acad. Sci. USA* **70**, 3581–3584.
22. Enea, V., Vovis, G. F. & Zinder, N. D. (1975) *J. Mol. Biol.* **96**, 495–509.
23. Miller, J. M. (1972) *Experiments in Molecular Genetics* (Cold Spring Harbor Laboratory, Cold Spring Harbor, New York), pp. 431–435.
24. Grunstein, M. & Hogness, D. S. (1975) *Proc. Natl. Acad. Sci. USA* **72**, 3961–3965.
25. Paterson, B. M., Roberts, B. E. & Kuff, E. L. (1977) *Proc. Natl. Acad. Sci. USA* **74**, 4370–4374.
26. Roberts, B. E. & Paterson, B. M. (1973) *Proc. Natl. Acad. Sci. USA* **70**, 2330–2334.
27. Lomedico, P. T. & Saunders, G. F. (1976) *Nucleic Acids Res.* **3**, 381–391.
28. Broome, S. & Gilbert, W. (1978) *Proc. Natl. Acad. Sci. USA* **75**, 2746–2749.
29. Makula, D. R., Vichnuk, D., Wright, P. H., Sussman, K. E. & Yu, P. L. (1969) *Diabetes* **18**, 660–689.
30. Maxam, A. M. & Gilbert, W. (1977) *Proc. Natl. Acad. Sci. USA* **74**, 560–564.
31. Boyer, H. W., Betlach, M., Bolivar, F., Rodriguez, R. L., Heyneker, H. L., Shine, J. & Goodman, H. M. (1977) in *Recombinant Molecules: Impact on Science and Society. Proceedings of the 10th Miles International Symposium*, eds. Beers, R. F., Jr. & Bassett, E. G. (Raven, New York), pp. 9–20.
32. Casey, J. & Davidson, N. (1977) *Nucleic Acids Res.* **4**, 1539–1552.
33. Humbel, R. E., Bosshard, H. R. & Zahn, H. (1972) in *Handbook of Physiology, Section 7* (*Endocrinology*), eds. Steiner, D. F. & Freinkel, N., (American Physiological Society, Washington, DC), Vol. 1, pp. 111–132.
34. Neu, H. C. & Heppel, L. A. (1965) *J. Biol. Chem.* **240**, 3685–3692.
35. Tilghman, S. M., Tiemeier, D. C., Seidman, J. G., Peterlin, B. M., Sullivan, M., Maizel, J. V. & Leder, P. (1978) *Proc. Natl. Acad. Sci. USA* **75**, 725–729.
36. Jeffreys, A. J. & Flavell, R. A. (1977) *Cell* **12**, 1097–1108.
37. Breathnach, R., Mandel, J. L. & Chambon, P. (1977) *Nature* **270**, 314–319.
38. Gilbert, W. (1978) *Nature* **271**, 501.

37

Reprinted from *Science* **205**:602–607 (1979)

Human Growth Hormone: Complementary DNA
Cloning and Expression in Bacteria

J. A. Martial, R. A. Hallewell, I. D. Baxter, and H. M. Goodman

Abstract. *The nucleotide sequence of a DNA complementary to human growth hormone messenger RNA was cloned; it contains 29 nucleotides in its 5' untranslated region, the 651 nucleotides coding for the prehormone, and the entire 3' untranslated region (108 nucleotides). The data reported predict the previously unknown sequence of the signal peptide of human growth hormone and, by comparison with the previously determined sequences of rat growth hormone and human chorionic somatomammotropin, strengthens the hypothesis that these genes evolved by gene duplication from a common ancestral sequence. The human growth hormone gene sequences have been linked in phase to a fragment of the* trp D *gene of* Escherichia coli *in a plasmid vehicle, and a fusion protein is synthesized at high level (approximately 3 percent of bacterial protein) under the control of the regulatory region of the* trp *operon. This fusion protein (70 percent of whose amino acids are coded for by the human growth hormone gene) reacts specifically with antibodies to human growth hormone and is stable in* E. coli.

Growth hormone, along with at least two other polypeptide hormones, chorionic somatomammotropin (placental lactogen) and prolactin, forms a set of proteins with amino acid sequence homology and to some extent overlapping biological activities (*1, 2*). Since the genes of this set of proteins probably have a common ancestral origin (*1*), they constitute an excellent model to study the evolution, structure, and differential regulation of related genes. In addition, since human growth hormone is of considerable medical importance and its supply is limited, the synthesis of growth hormone in bacteria might provide the required alternate source of this critical hormone.

We have previously isolated and analyzed bacterial clones containing copies of complementary DNA (cDNA) transcripts of messenger RNA's (mRNA's) for these hormones. The complete sequence of rat pregrowth hormone mRNA (*3*) has been reported; in addition, sequence data have been presented for fragments of about 550 bases complementary to part of the coding (amino acid residues 24 to 191) and 3' untranslated portions of human chorionic somatomammotropin (hCS) (*3, 4*) and human growth hormone (hGH) mRNA's (*5*). A partial sequence of rat prolactin has been determined by Gubbins *et al.* (*6*). These sequence data showed that, whereas the growth hormone genes of

the rat and man had significant homology, they also had diverged substantially, such that they differed more than the genes for the functionally distinct human hormones hCS and hGH.

We now report the synthesis, cloning, and sequence analysis of cDNA containing the entire coding and most of the noncoding portions of hGH mRNA. We also describe the insertion of these sequences into an "expression plasmid" containing part of the *Escherichia coli* tryptophan (*trp*) operon whose construction has been realized by Hallewell and Emtage (*7*). We describe the use of this plasmid to promote the inducible bacterial synthesis of high levels of a hybrid protein, 70 percent of which is composed of amino acids coded for by the hGH gene.

Human growth hormone mRNA isolation. Polyadenylated RNA was isolated (*8*) from human pituitary tumors removed by transphenoidal hypophysectomy. To obtain an indication of the integrity and the relative abundance of growth hormone mRNA in each sample, the individual mRNA preparations were translated in the wheat germ cell-free system, and the products were analyzed by electrophoresis on sodium dodecyl sulfate–polyacrylamide gels (Fig. 1). Among the translation products of the five acromegalic tumor RNA's (Fig. 1, lanes 1 to 5), the most prominent band corresponds to a protein of approximately 24,000 daltons. This protein is assumed to be human pregrowth hormone since it is similar in size to rat pregrowth hormone (Fig. 1, "rat") and is precipitated by antiserum to hGH (data not shown). This assumption is further justified by comparison with the translation products of polyadenylated RNA isolated from bovine pituitary (Fig. 1, "cow") and from a human prolactin-producing tumor (Fig. 1, lane 6). Both of these RNA's directed the synthesis of a protein similar in size to human and rat pregrowth hormone, but also directed the synthesis of a larger quantity of a protein of higher molecular weight, presumably preprolactin. The tumors show variation in the extent to which hGH mRNA is present, as measured by their translational activities. Nevertheless, hGH mRNA appears to be the most abundant mRNA species in the acromegalic tumors (Fig. 1, lanes 1 to 5). These results are consistent with and verify the clinical diagnoses made prior to surgery.

Molecular cloning of hGH cDNA. The polyadenylated RNA from the tumors that appeared to have the greatest abundance of hGH mRNA by the translational assay were pooled for synthesis of double-stranded cDNA (Fig. 2). Portions of the double-stranded cDNA were analyzed by restriction endonuclease digestion before and after treatment with S1 nuclease. A high proportion of the cDNA was about 1000 nucleotides long (Fig. 2, lanes a and b), the length expected for hGH mRNA, assuming analogy with rat pregrowth hormone mRNA (3). Endonuclease Hae III digestion of the DNA generated a 550-base pair (bp) fragment (Fig. 2, lanes h and j) previously reported to occur in hGH and hCS cDNA's (4, 5). The prominence of this band supports the idea that the cDNA is highly enriched in hGH gene sequences. This is further suggested by the finding of a fragment of about 400 bp generated by digestion with Hinf I and Sma I (Fig. 2, lane g). The fragment of about 500 bp generated by Pvu II (Fig. 2, lanes e and i) extends beyond the previously cloned 550 bp fragment, which contains only one Pvu II site. However, its presence is predictable from the sequence of rat growth hormone cDNA (3), and by the conservation between species of the amino acid sequence in this region (9). The fragments of about 350 bp and 150 bp generated by combined digestion with Pvu II and Bgl II (Fig. 2, lane f) would also be anticipated from the previously determined structure of the 550 bp hGH fragment and knowledge of the existence of the additional Pvu II site. Therefore, this cDNA preparation appears to be highly enriched in full-length copies of hGH mRNA.

The uncleaved cDNA was cloned in the plasmid pBR322 and *E. coli* χ1776 in a P3 physical containment facility (10) by methods similar to those previously described (3). Briefly, the cDNA was first treated with S1 nuclease and subsequently with DNA polymerase I in the presence of the four deoxynucleoside triphosphates to generate blunt-ended cDNA molecules. Synthetic DNA containing the site for the restriction endonuclease Hind III was then added to

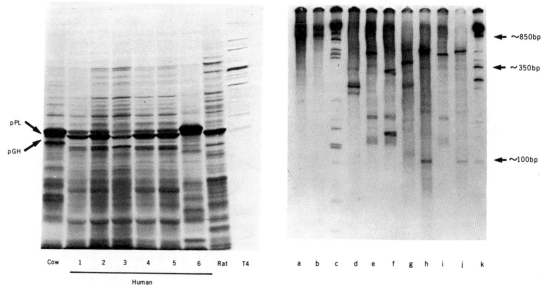

Fig. 1 (left). Translation products of mRNA isolated from growth hormone and prolactin-producing tumors and from bovine pituitary. These tissues were stored in liquid nitrogen shortly after their removal until preparation of the RNA. A portion of polyadenylated RNA isolated from each human tumor, the bovine pituitary, and the cultured rat pituitary tumor (GC) cells was used as a messenger in the wheat germ cell-free protein synthesis system (20). The ^{35}S-labeled proteins were analyzed by electrophoresis and autoradiography on sodium dodecyl sulfate–polyacrylamide gels (12.5 percent) (20). The translation products from five acromegalic and one prolactin-secreting tumor are shown in lanes 1 to 5 and lane 6, respectively. The lanes labeled "cow" and "rat" show the translation products from bovine pituitary and rat GC cell RNA. Lane 7 shows the bacteriophage T4 proteins (21) used as molecular weight markers. The arrows indicate the bands corresponding to pregrowth hormone (pGH) and preprolactin (pPL).   Fig. 2 (right). Analysis of cDNA synthesized from mRNA extracted from the growth hormone–producing pituitary tumors. The polyadenylated RNA from tumors two, four, and five were pooled (135 μg) and used as a template for synthesis of ^{32}P-labeled double-stranded cDNA as described (3). Samples of this cDNA were cleaved with various restriction endonucleases before and after S1 nuclease digestion. Samples of this cDNA were cleaved with various restriction endonucleases before and after electrophoresis on a 4.5 percent polyacrylamide gel (3). (Lane a) Uncleaved cDNA before S1 digestion; (lane b) uncleaved cDNA after S1 digestion; (lane c) bacteriophage fd DNA, Hpa II (molecular weight markers); (lane d) cDNA, Pst I + Bgl II; (lane e) cDNA, Pvu II; (lane f) cDNA, Bgl II + Pvu II; (lane g) cDNA, Hinf I + Sma I; (lane h) cDNA, Hae III; (lane i) cDNA, Pvu II; (lane j) cDNA, Hae III; (lane k) bacteriophage fd DNA, Hae III (molecular weight markers).

each end of the cDNA, and cohesive ends were generated by digestion with endonuclease Hind III. The resulting cDNA was purified on gel and ligated to Hind III–cut and bacterial alkaline phosphatase–treated pBR322 plasmid DNA

(8). Bacteria were then transformed with this recombinant plasmid. Colonies with recombinant DNA containing plasmids, selected by antibiotic resistance (tet^{s}amp^R) were grown, and the plasmid DNA's were isolated. The DNA was digested

with Hind III, treated with various other restriction endonucleases, and analyzed by gel electrophoresis. One clone contained an insert of about 800 bp whose digestion by Hae III, Pvu II, and Bgl II generated fragments similar in size

```
                                  -26                              -20                                       -10
                                  met ala thr gly ser arg thr ser leu leu leu ala phe gly leu leu cys leu pro trp
       GG AUC CUG UGG ACA GCU CAC CUA GCU CCA AUG CCU ACA CGC UCC CGG ACG UCC CUG CUC CUG GCU UUU GGC CUG CUC UGC CUG CCC UGC

                         1                              10                                   20
     leu gln glu gly ser ala phe pro thr ile pro leu ser arg leu phe asp asn ala met leu arg ala his arg leu his gln leu ala
     CUU CAA GAG GGC AGU GCC UUC CCA ACC AUU CCC UUA UCC AGG CUU UUU CAC AAC GCU AUG CUC CGC GCC CAU CGU CUC CAC CAG CUG GCC

                        30                             phe glu glu ala tyr ile pro lys glu gln lys tyr ser phe leu gln asn pro gln thr ser leu cys phe
     phe asp thr tyr gln glu      40                                   50
     UUU GAC ACC UAC CAG GAG UUU GAA GAA GCC UAU AUC CCA AAG GAA CAG AAG UAU UCA UUC CUG CAG AAC CCC CAG ACC UCC CUC UGU UUC

                        60                             70                                   80
     ser glu ser ile pro thr pro ser asn arg glu glu thr gln gln lys ser asn leu glu leu leu arg ile ser leu leu leu ile gln
     UCA GAG UCU AUU CCG ACA CCC UCC AAC AGG GAG CAA ACA CAA CAG AAA UCC AAC CUA GAG CUG CUC CGC AUC UCC CUG CUC CUC AUC CAG

                        90                             100                                  110
     ser trp leu glu pro val gln phe leu arg ser val phe ala asn ser leu val tyr gly ala ser asp ser asn val tyr asp leu leu
     UCG UGG CUG GAG CCC GUG CAG UUC CUC AGG AGU GUC UUC GCC AAC AGC CUG GUC UAC GGC GCC UCU GAC AGC AAC GUC UAU GAC CUC CUA

                        120                            130                                  140
     lys asp leu glu glu gly ile gln thr leu met gly arg leu glu asp gly ser pro arg thr gly gln ile phe lys gln thr tyr ser
     AAC GAC CUA GAG GAA GGC AUC CAA ACG CUG AUG CGG AGG CUG GAA GAU GGC ACC CCC CGG ACU GGG CAG AUC UUC AAG CAG ACC UAC AGC

                        150                            160                                  170
     lys phe asp thr asn ser his asn asp asp ala leu leu lys asn tyr gly leu leu tyr cys phe arg lys asp met asp lys val glu
     AAG UUC GAC ACA AAC UCA CAC AAC GAU GAC CCA CUA CUC AAG AAC UAC GGG CUG CUC UAC UGC UUC AGG AAG GAC AUG GAC AAG GUC GAG

                        180                            190 191
     thr phe leu arg ile val gln cys arg ser val glu gly ser cys gly phe AM
     ACA UUC CUG CGC AUC GUG CAG UGC CGC UCU GUG GAG GGC AGC UGU CGC UUC UAG CUG CCC GGG UGG CAU CCU GUG ACC CCU CCC CAG UGC

 CUC UCC UGG CCC UGG AAC UUG CCA CUC CAG UGC CCA CCA GCC UUG UCC UAA UAA AAU UAA GUU GCA UCA AAA AAA AAA
```

Fig. 3. Nucleotide sequence of hGH mRNA and the amino acid sequence of human pregrowth hormone. The sequence was determined according to the procedure of Maxam and Gilbert (11) from 5'- or 3'-end-labeled restriction fragments of chGH800/pBR322. The sequence between the two internal Hae III sites was taken from our previous work (5). Most of it, and all of the sequence outside these two internal Hae III sites was resequenced by the chain-termination technique (12) as described in detail elsewhere (22) using as single-stranded templates the chGH800 Hind III fragment recloned in the vector M13 mp5 and as primers restriction fragments of chGH800/pBR322. The RNA sequence has been taken from the DNA sequence. The amino acid sequence has been deduced from the RNA sequence using the genetic code. The termination codon, UAG, is designated by the symbol AM for "amber."

4/UUU/phe	3/UCU/ser	3/UAU/tyr	2/UGU/cys
10/UUC/phe	7/UCC/ser	5/UAC/tyr	3/UGC/cys
1/UUA/leu	3/UCA/ser	0/UAA/OC	0/UGA/OP
0/UUG/leu	1/UCG/ser	1/UAG/AM	2/UGG/trp
2/CUU/leu	0/CCU/pro	1/CAU/his	1/CGU/arg
10/CUC/leu	6/CCC/pro	2/CAC/his	4/CGC/arg
4/CUA/leu	2/CCA/pro	2/CAA/gln	0/CGA/arg
16/CUG/leu	1/CCG/pro	11/CAG/gln	2/CGG/arg
2/AUU/ile	1/ACU/thr	0/AAU/asn	2/AGU/ser
6/AUC/ile	4/ACC/thr	9/AAC/asn	5/AGC/ser
0/AUA/ile	5/ACA/thr	1/AAA/lys	0/AGA/arg
4/AUG/met	2/ACG/thr	8/AAG/lys	5/AGG/arg
0/GUU/val	2/GCU/ala	2/GAU/asp	0/GGU/gly
3/GUC/val	6/GCC/ala	9/GAC/asp	8/GGC/gly
0/GUA/val	1/GCA/ala	6/GAA/glu	0/GGA/gly
4/GUG/val	0/GCG/ala	9/GAG/glu	3/GGG/gly

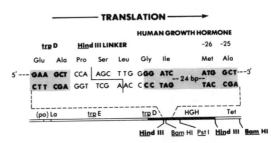

Fig. 4. Codon usage in hGH mRNA. The numbers indicate how many times the codons are used in the region of hGH mRNA coding for the prehormone; OC, OP, and AM designate the stop codons ochre, opal, and amber, respectively.   Fig. 5. Postulated nucleotide sequence around the Hind III site in the hybrid gene of expression plasmid ptrpED50-chGH800. Plasmid ptrpED50 was constructed from ptrpED5-1 (7) by linearizing the plasmid with Hind III, filling in the protruding 5' ends with the use of DNA polymerase I (Klenow fragment from Boehringer) and ligating synthetic decamers containing a Hind III site (collaborative research) to the blunt-ended material (3, 8). After digestion with Hind III the plasmid was separated from residual linker molecules by chromatography on Sephadex G-200, recircularized with T4 DNA ligase, and used to transform E. coli W3110 trpoE∇1(23) by a standard procedure (24). Plasmid DNA isolated from one of these colonies was digested with Hind III and treated with alkaline phosphatase (4, 8). A portion (5 μg) of this DNA was end-labeled with [γ-³²P]-ATP with the use of T4 DNA kinase (Boehringer) and cut by Hae III. The DNA sequence of the labeled fragments was determined by chemical cleavage (11), after they were isolated by polyacrylamide electrophoresis. The cloned chGH800 DNA was cleaved from pBR322 with Hind III and isolated by polyacrylamide gel electrophoresis. This DNA was ligated to similarly cleaved and alkaline phosphatase-treated ptrpED50. The ligation mixture was used to transform E. coli strains W3110 trpoE∇1 and RR1 (25) in a P3 facility, and transformants resistant to ampicillin were selected. Resistant colonies were examined for the presence of inserted chGH800 sequences by gel analysis of such plasmids after digestion with restriction endonucleases Bam HI and Pst I. Plasmids with the growth hormone initiator codon proximal to the trpD gene sequence showed bands of 250 and 900 bp, whereas plasmids with the inserted cDNA in the opposite orientation showed bands of 250 and 350 bp.

to those from the 550 bp hGH cDNA clone (data not shown) and to the digested uncloned cDNA (Fig. 2). This suggested that this clone did contain cDNA complementary to full-length or nearly full-length hGH mRNA. (This clone is designated chGH800/pBR322.)

Sequence analysis of cloned DNA. The nucleotide sequence of the cloned DNA was determined by the chemical cleavage method of Maxam and Gilbert (*11*) and the chain-termination technique of Sanger, Nicklen, and Coulson (*12*). The hGH mRNA sequence and the corresponding amino acid sequence of human pregrowth hormone can be derived from the DNA sequence (Fig. 3). The amino acid sequence determined from the DNA sequence is consistent with the known amino acid sequence of hGH (*13*) with the following exceptions: the DNA sequence predicts glutamine, asparagine, glutamine, glutamic acid, glutamine, aspartic acid, asparagine, and glutamine at amino acid positions 29, 47, 49, 74, 91, 107, 109, and 122, respectively, while the protein sequence indicates glutamic acid, aspartic acid, glutamic acid, glutamine, glutamic acid, asparagine, aspartic acid, and glutamic acid. It is likely that the DNA sequence is correct in this regard since it is sometimes difficult in protein sequence analysis to differentiate aspartic acid from asparagine and glutamic acid from glutamine. The amino acid sequence of the signal peptide portion of human pregrowth hormone had not been previously determined and is deduced from the mRNA sequence. If translation begins with the methionine codon "in phase," 26 codons proximal to the first amino acid of growth hormone (Fig. 3), then the primary translation product of hGH mRNA would be a protein of 24,851 daltons, a value in agreement with the cell-free translation data shown in Fig. 1.

A comparison of the amino acid and nucleic acid sequence homologies between rat growth hormone, hGH, and hCS and their respective mRNA's is shown in Table 1. In the coding regions, there is higher homology between the nucleic acid sequences than between the amino acid sequences. This difference is consistent with the already mentioned view that the genes of these related hormones evolved from a common evolutionary precursor gene, and is further supported by the marked homology in the 5'-noncoding portions of the mRNA's for rat and human growth hormone. (Data for the 5'-noncoding region of hCS are not yet available.) Human growth hormone has more homology with hCS than with rat growth hormone,

Table 1. Amino acid and nucleic acid sequence homology of growth hormone, chorionic somatomammotropin, and their mRNA's. Data for rat growth hormone (rGH) and human chorionic somatomammotropin (hCS) and their mRNA's are from (*3, 4*). For hCS, only data for amino acids residues 24 to 191 (and the corresponding portion of the mRNA) and a portion of the noncoding 3'-region corresponding to the cloned 550 bp fragment are used for comparisons, since data for the other portions are not available. The amino acid sequence of the prepeptide portion of hCS was determined by Sherwood *et al.* (*27*).

Source	Homology (percent)	
	hGH versus rGH	hGH versus hCS
Nucleic acid:		
5'-Noncoding	73	
Presequence	76	
Coding	76	92
3'-Noncoding	38*	94
Amino acid:		
Presequence	58	84
Coding	67	86

*The homology in this region can be increased to 55 percent by adding appropriate gaps in the sequences. This procedure also reveals that there is a homology of 27 out of 30 bases in the region of the AAUAAA (*28*). Similar conservation between species but with different sequences are found when human and rabbit β-globin (*29*) and human and rat insulin (*30*) are compared.

especially for the 3'-noncoding portions. This finding supports the hypothesis developed earlier that the chorionic somatomammotropin and growth hormone genes probably evolved by a gene duplication mechanism (*1*) at some time after the separation of the human and rat species. In addition, the fact that both hormones exist in both species implies that the same hormones may have evolved independently more than once.

Figure 4 shows the codons used for hGH mRNA. As is the case with rat growth hormone (*3*) and hCS (*6*) mRNA's, there is a nonrandom selection of codons. This appears to be mostly due to the preference for G (guanine) or C (cytosine) over A (adenine) or U (uracil) for the third position of the triplet codon. This is also the case with most (*3, 14*) but not all (*14*) eukaryotic mRNA's whose structures are known.

Construction of a plasmid for growth hormone expression. To see whether hGH gene sequences can be expressed in bacteria, we used the plasmid p*trp*ED5-1 (*7*), which contains the regulatory region [(po)La], the first gene (*trp*E), and 15 percent of the second gene (*trp*D) of the *E. coli trp* operon. Cells containing p*trp*ED5-1 normally synthesize small amounts of *trp* gene products. However, if *trp* operon transcription is derepressed

by addition of 3β-indolylacrylic acid, synthesis of *trp* gene products increases, so that within 3 hours *trp* proteins account for about 30 percent of the total cellular proteins (*7*). We hoped that by placing the hGH gene sequence under control of the *trp* operon, not only would it be expressed, but a higher level of hGH production could be obtained than was previously achieved with rat growth hormone gene sequences under control of the β-lactamase gene (*15*).

The hGH sequences from chGH800 were inserted at the Hind III site of the *trp*D gene sequence as described in the legend to Fig. 5. In order to insert the hGH codons in phase with those of the *trp*D sequence, the Hind III site in the *trp*D gene was manipulated in such a way as to shift the reading frame of any DNA inserted through the Hind III site of the plasmid by one base. To do this, p*trp*ED5-1 was cleaved with Hind III, the protruding 5'-ends "filled in" with the use of DNA polymerase I (Klenow fragment) and synthetic DNA decamers containing the Hind III site ligated to the blunt-ended material. This DNA was then digested with Hind III to produce new cohesive Hind III ends, and plasmid molecules were recircularized with the use of DNA ligase after the residual Hind III linker molecules were removed on a Sephadex (G-200) column. This material was used to transform *E. coli* and, after selection for ampicillin-resistant colonies, the plasmid DNA was isolated from one of the transformants. The DNA sequence at the Hind III site of the newly constructed plasmid (designated p*trp*ED50) was determined and showed that the enzymatic reactions had altered the reading frame as predicted. In this way, when chGH800/pBR322 was cleaved with Hind III and the hGH sequences ligated to the Hind III site of the newly constructed plasmid, the codons of hGH would be in phase with those of the *trp*D gene, provided that the hGH gene sequences were inserted in the proper orientation. This was achieved by obtaining several clones, isolating plasmid DNA, and determining the orientation of the cloned segment by restriction endonuclease analysis (Fig. 5, legend).

As is indicated in Fig. 5, the construction of the hGH "expression" plasmid was such that the anticipated product would be a fusion protein containing the NH₂-terminal region of the *trp*D protein, amino acids coded by the 5'-untranslated portion of hGH mRNA, the 26 amino acids of the signal peptide, and all of the amino acids of hGH.

Synthesis of growth hormone in bacteria. To determine whether the newly constructed gene can direct the synthesis of large amounts of a new fused polypeptide, and whether its expression is regulated by the *trp* promoter, cells containing the expression plasmid were derepressed for *trp* transcription, and proteins were labeled for 5 minutes with ¹⁴C-labeled amino acids. Figure 6 shows an autoradiogram of sodium dodecyl sulfate–polyacrylamide gel of such proteins labeled at various times from 0 to 4 hours after induction of the *trp* operon. Two proteins (53,000 and 32,000 daltons) seem to be specifically derepressed by the inducer. The higher molecular weight protein is the *trp*E gene product (7). The 32,000-dalton protein has approximately the anticipated size for the *trp*D–hGH fusion protein (34,000 daltons). It is immunoprecipitated by antiserum to hGH (Fig. 6, lane a) but not by the control antiserum (Fig. 6, lane b), and precipitation can be blocked by a large excess of hGH (Fig. 6, lanes c to h). Some of the *trp*E protein is immunoprecipitated by antiserum to hGH, but the amount is less when an excess of competitor hGH is added or control antiserum is used. This

result might be expected for two reasons. Precipitation of *trp*E protein by control antiserum may be due to the high abundance of this protein. More interestingly, specific precipitation of *trp*E by hGH antiserum (Fig. 6, lane c) and blockage of precipitation by an excess of competitor hGH (Fig. 6, lane e) may be the result of association of the *trp*E protein and the *trp*D–hGH fused polypeptide. The *trp*E and *trp*D proteins are normally associated in *E. coli* as a tetramer containing two subunits of each protein (16); the resulting enzymatic activity (anthranilate synthetase) requires the *trp*E protein and the NH₂-terminal 30 percent of the *trp*D protein (17). Thus, the fused *trp*D–hGH protein may contain those *trp*D residues required for binding *trp*E. All of these lines of evidence suggest that the 32,000-dalton protein is a fused *trp*D–hGH polypeptide.

On the basis of the relative quantity of radioactivity incorporated into the *trp*D–hGH gene product, the fusion product appears to be a major protein made by the bacteria, constituting 3 percent of the total bacterial protein synthesis. Thus,

the natural hGH gene sequence can be expressed at a high level in bacteria.

The *trp*E and *trp*D protein molecules ordinarily accumulate at a similar rate (molar ratio 1:1) when the *trp* operon is induced (7). However, the molar ratio of *trp*E to *trp*D–hGH is about 6:1, indicating that synthesis of *trp*D–hGH is only 17 percent of the expected level. The reduced level of synthesis is not due to instability of the fused polypeptide since a "pulse-chase" experiment has shown that no significant degradation of the fused polypeptide occurs during the 60-minute period after incorporation of the label (data not shown). There is some evidence that the chick ovalbumin protein is also synthesized in *E. coli* at lower levels than expected (18).

Hypopituitary dwarfism is a fairly common disease treatable only by replacement with hGH (19). Growth hormone may also be useful in the treatment of other disorders. However, the potential uses of this hormone have not been adequately investigated because its only source is pituitaries from human cadavers. In order to have an adequate supply of this hormone, it is necessary to find alternative means of producing it; the synthesis of hGH in bacteria may provide such a means.

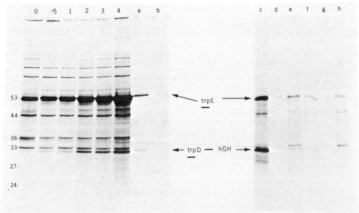

References and Notes

1. T. A. Bewley, J. S. Dixon, C. H. Li, *Int. J. Pept. Protein Res.* **4**, 281 (1972); H. D. Niall, M. L. Hogan, R. Sayer, I. Y. Rosenblum, F. C. Greenwood, *Proc. Natl. Acad. Sci. U.S.A.* **68**, 866 (1971).
2. A. G. Frantz, in *Peptide Hormone*, J. A. Parson, Ed. (University Park Press, Baltimore, 1976), p. 199.
3. P. H. Seeburg, J. Shine, J. A. Martial, J. D. Baxter, H. M. Goodman, *Nature (London)* **270**, 5637 (1977).
4. P. H. Seeburg, J. Shine, J. A. Martial, A. Ullrich, J. D. Baxter, H. M. Goodman, *Cell* **12**, 157 (1977).
5. P. H. Seeburg, J. Shine, J. A. Martial, J. D. Baxter, H. M. Goodman, in preparation; H. M. Goodman, P. H. Seeburg, J. Shine, J. A. Martial, J. D. Baxter, *Alfred Benson Symposium 1978* (Munsgaard, Copenhagen, in press).
6. E. J. Gubbins, R. A. Maurer, J. L. Hartley, J. E. Donelson, *Nucleic Acids Res.* **6**, 915 (1979); N. E. Cooke and J. A. Martial, unpublished results.
7. R. A. Hallewell and S. Emtage, in preparation.
8. A. Ullrich, J. Shine, J. Chirgwin, R. Pictet, E. Tischer, W. J. Rutter, H. M. Goodman, *Science* **196**, 1313 (1977).
9. M. O. Dayhoff, *Atlas of Protein Sequence and Structure* (National Biomedical Research Foundation, Washington, D.C., 1972), vol. 5, pp. D-201–D-204.
10. "Recombinant DNA research, revised guidelines," *Fed. Reg.* **43**, 60080 (1978); "Recombinant DNA research: actions under guidelines," *ibid.* **44**, 21730 (1979).
11. A. Maxam and W. Gilbert, *Proc. Natl. Acad. Sci. U.S.A.* **74**, 560 (1977).
12. F. Sanger, S. Nicklen, A. R. Coulson, *ibid.*, p. 5463.
13. Reference (9), p. D-202, was used as the "standard" hGH sequence, except that an additional Gln residue was placed at position 69 as indicated in (1) and (2). References (1) and (2)

Fig. 6. Autoradiograms of sodium dodecyl sulfate–polyacrylamide gels (10 percent) of ¹⁴C- and ³⁵S-labeled proteins from bacteria harboring the p*trp*ED50-chGH800 expression plasmid. Cultures of strains W3110 *trp*oEVI and RRI harboring this plasmid were induced with 3β-indolylacrylic acid and 3-ml samples labeled for 5 minutes with 2 μCi of ¹⁴C-labeled amino acids (W3110 *trp*oEVI) or 10 μCi of ³⁵S-labeled methionine (RRI) as described (7). Samples labeled at zero, 0.5, 1, 2, 3, and 4 hours were centrifuged and resuspended (50 μl) by sonication (26) prior to loading (5 μl per gel slot) in sodium dodecyl sulfate sample buffer (20). Samples were immunoprecipitated by means of the SAC technique (20) in order to collect antigen-antibody complexes. Immunoprecipitations contained 10 μl of sonicated cells, 390 μl of 0.5 percent NP40 (Particle Data Laboratory, Elmhurst, Ill.) in phosphate saline buffer (0.025M potassium phosphate, pH 7.4, 0.1M NaCl), 20 μl of rabbit antiserum to hGH (Antibodies Inc., 1000 unit/ml) or 20 μl of nonimmune rabbit antiserum each diluted 50-fold in phosphate saline buffer containing bovine serum albumin (2 mg/ml). Competitor hGH was added at 40 μg per reaction mixture. The *E. coli* proteins in lane 0 were used as molecular weight markers (8). (a and b) Immunoprecipitates of ¹⁴C-labeled proteins from the 4-hour time point with (a) antiserum against hGH and with (b) nonimmune serum. (c to e) Immunoprecipitates of ³⁵S-labeled proteins from the 4-hour time point, with (c) antiserum against hGH, (d) nonimmune serum, and (e) antiserum against hGH together with an excess of competitor hGH. (f to h) Immunoprecipitates of ³⁵S-labeled proteins from the zero time point with the use of (f) antiserum against hGH, (g) nonimmune serum, and (h) antiserum against hGH together with an excess of competitor hGH.

present slightly different versions of the hGH sequence, which match the sequence predicted from the cDNA in several residues, which do not match the standard sequence.

14. L. McReynolds, B. W. O'Malley, A. D. Nisbet, J. E. Fothergill, D. Givol, S. Fields, M. Robertson, G. G. Brownlee, *Nature (London)* **273**, 723 (1978).
15. P. H. Seeburg, J. Shine, J. A. Martial, R. D. Ivarie, J. A. Morris, A. Ullrich, J. D. Baxter, H. M. Goodman, *ibid.* **276**, 5690 (1978).
16. J. Ito and C. Yanofsky, *J. Bacteriol.* **97**, 734 (1969).
17. C. Yanofsky, V. Horn, M. Bonner, S. Stasiowski, *Genetics* **69**, 409 (1971).
18. T. H. Fraser and B. J. Bruce, *Proc. Natl. Acad. Sci. U.S.A.* **75**, 5936 (1978).
19. J. M. Tanner, *Nature (London)* **237**, 433 (1972).
20. J. A. Martial, J. D. Baxter, H. M. Goodman, P. H. Seeburg, *Proc. Natl. Acad. Sci. U.S.A.* **74**, 1816 (1977).
21. P. A. O'Farrell and L. H. Gold, *J. Biol. Chem.* **248**, 7066 (1973).
22. B. Cordell, G. Bell, E. Tischer, F. M. DeNoto, A. Ullrich, R. Pictet, W. J. Rutter, H. M. Goodman, in preparation.
23. N. E. Murray and W. J. Brammar, *J. Mol. Biol.* **77**, 615 (1973).
24. D. M. Glover, in *New Techniques in Biophysics and Cell Biology*, R. H. Pain and B. J. Smith, Eds. (Wiley, New York, 1976), vol. 8, pp. 125–145.
25. F. Bolivar, R. L. Rodriguez, M. C. Betlach, H. W. Boyer, *Gene* **2**, 75 (1977).
26. P. H. O'Farrell, *J. Biol. Chem.* **250**, 4007 (1975).
27. L. M. Sherwood, Y. Burstein, I. Schechter, *Conference on Precursor Processing in The Biosynthesis of Proteins* (New York Academy of Sciences, 2 to 4 May 1979), Abstr.; S. Birken, D. L. Smith, R. E. Canfield, I. Boime, *Biochem. Biophys. Res. Commun.* **74**, 106 (1977).
28. N. J. Proudfoot and G. G. Brownlee, *Nature (London)* **263**, 211 (1976).
29. A. Efstratiadis, F. C. Kafatos, T. Maniatis, *Cell* **10**, 571 (1977).
30. G. I. Bell, W. F. Swain, R. Pictet, B. Cordell, H. M. Goodman, W. J. Rutter, *Nature (London)*, in press.
31. Supported by a grant from Eli Lilly Co. and a postdoctoral fellowship from the British Science Research Council (to R.A.H.). We thank Dr. C. Yanofsky for communicating information prior to publication and for helpful suggestions on the manuscript; J. Messing for M13mp5 and advice on its use; Drs. W. Swain and P. O'Farrell for their respective gifts of plasmid pBR322 and radioactive T4 protein; Drs. P. Seeburg and J. Shine for helpful discussion and participation in some experiments; and D. Coit and E. Tischer for technical assistance. Avian myeloblastosis virus reverse transcriptase was provided by the Office of Program Resources and Logistics, NCI. J.D.B. and H.M.G. are investigators of the Howard Hughes Medical Institute.

38

Copyright © 1979 by Macmillan Journals Ltd.

Reprinted from *Nature* **279**:43–47 (1979)

Expression in *Escherichia coli* of hepatitis B virus DNA sequences cloned in plasmid pBR322

C. J. Burrell*, Patricia Mackay
Department of Bacteriology, University of Edinburgh Medical School, Edinburgh UK

P. J. Greenaway, P. H. Hofschneider†
Microbiological Research Establishment, Porton Down, Salisbury, Wiltshire, UK

K. Murray
Department of Molecular Biology, University of Edinburgh, King's Buildings, Edinburgh UK

INFECTION with hepatitis B virus (HBV) is widespread in man. Between 3 and 15% of healthy blood donors in Western Europe and the US show serological evidence of past infection and about 0.1% are chronic carriers of the virus. In many African and Asian countries the prevalence is much higher and the majority of the adult population have been infected, while 5–10% of the population are chronically infected[1]. Most infections are subclinical and are followed by apparently complete recovery with the development of virus-specific antibody. However, a significant proportion of infections (probably 1–5%) may produce chronic sequelae including persistent infection, chronic hepatitis of various types, cirrhosis and possibly primary liver cancer.

Plasma from some blood donors and patients infected with HBV contains 42-nm spherical particles (Dane particles)[2] which have serological and biochemical properties, suggesting that they are the infective virions of hepatitis B. These have an outer envelope containing the hepatitis B surface antigen (HBsAg) and an inner core (diameter 27 nm) bearing a second unrelated antigen, the hepatitis B core antigen (HBcAg). Within the core is a double-stranded circular DNA molecule of molecular weight ~2×10^6 which has a large variable gap in one strand, and an endogenous DNA-dependent DNA polymerase activity that can fill in this gap in *in vitro* reactions[3]. There is some evidence that the total virus genome length may be around one-third greater than the 2×10^6 daltons found in single molecules in which case productive infection of a cell may require simultaneous infection by at least two genetically different particles[3]. A third antigen, the hepatitis B e antigen (HBeAg), which is probably also virus-coded, is found free in the plasma of some infected individuals and possibly also in association with Dane particles. Passively or actively acquired antibody to HBsAg (anti-HBs) confers some immunity to subsequent HBV challenge, and prototype vaccines composed of purified inactivated HBsAg prepared from the plasma of infected carriers are being evaluated[4]. However, the virus cannot be grown in tissue culture and normally infects only man and apes. This means that molecular studies of the virus and its genome have been based on the limited amounts of material obtainable from the plasma of infected individuals. Such studies could be advanced considerably by insertion of HBV DNA into a bacterial plasmid or phage to allow its production in quantity from cloned, purified single molecules. Such clones would also be useful for studies of the expression of HBV gene products in bacterial cells, and for large-scale production of HBV DNA and viral antigens for diagnostic purposes and, possibly, vaccine production.

Restriction of HBV DNA

The published information on digestion products of HBV DNA (Dane particle DNA) with restriction endonucleases[5,6] is limited and so additional analyses of this type were carried out before attempting cloning experiments. The amount of HBV DNA available was limited (about 90 ng from 5 ml of plasma), so the DNA was first labelled with ^{32}P in the endogenous DNA polymerase reaction[3]. DNA from bacteriophage λ was then added as a carrier to titrate the restriction enzyme and to provide reference fragments of known size. The digests were analysed by electrophoresis in agarose gels and the results of some of these experiments are shown in Fig. 1.

The heterogeneity of the undigested labelled HBV DNA (tracks *b* and *e*) precluded a detailed analysis of the restriction digests. The multiple bands of rather similar size could be due in part to the presence of linear and circular forms of otherwise similar molecules, in part to differing degrees of repair synthesis, although repeated DNA preparations from the same plasma sample gave reproducible patterns; the heterogeneity could also represent a true molecular dispersity from a mixed population of virions. HBV DNA preparations from several individual donors were all heterogeneous and differed slightly from each other (an example is included in Fig. 1; tracks *a*, *b* and *c* compared with *d*, *e* and *f*) both before and after digestion with various restriction enzymes. Heterogeneity has also been observed by Landers *et al.*[6] in the products of the endogenous polymerase reaction which comprised two principal radioactive components representing linear and circular molecules with additional minor components due to incomplete repair. Examination of the HBV DNA (Fig. 1, tracks *b* and *e*) by electron microscopy showed that the population contained circular molecules of MW ~ 2×10^6 and linear molecules ranging from 0.5 to 10×10^6. The linear molecules represented about three times the concentration of circular molecules and only a few of them had a MW around 2×10^6, the majority being between 0.5 and 1×10^6. The heterogeneity observed on electrophoresis was, therefore, not due to circular and linear forms of similar length and few, if any, of the linear molecules were labelled in the polymerase reactions.

* Present address: Division of Virology, I.M.V.S., Box 14, Rundle St. Post Office, Adelaide, S. Australia.
† Permanent address: Max Planck Institute for Biochemistry, Am Klopferspitz, Martinsried bei Munich, FRG.

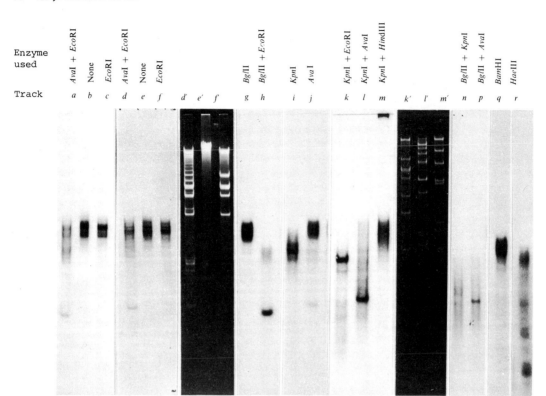

Fig. 1 Autoradiographs of restriction enzyme digests of HBV DNA after electrophoresis in agarose gels. Examples of HBV DNA isolated from the plasma of two blood donors are shown, the samples in tracks *a*, *b* and *c* from one donor being directly comparable with those in tracks *d*, *e* and *f* from the other; the sample used for the digests in tracks *d*, *e* and *f* was also used for all the other digests shown, and also for the preparative experiments for cloning. In addition to the autoradiographs, the figure includes examples, indicated by primed letters, of the gels stained with ethidium bromide to reveal fragments of the λ⁺ DNA included with the HBV DNA in the restriction reactions. HBV DNA was prepared from Dane particles isolated by two cycles of ultracentrifugation from clarified plasma obtained from individual HBsAg-positive blood donors essentially as described by Landers *et al.*[6]. The single-strand gaps in the DNA molecules were repaired and the DNA radioactively labelled in reactions with the endogenous DNA polymerase in which ³H-dCTP and ³H-dTTP (22 and 30 mCi μmol⁻¹, respectively; Radiochemical Centre Amersham) or ³²P-dGTP (300 mCi μmol⁻¹) were included[6]. The released core particles were further purified by sedimentation through 30% w/v sucrose solution and DNA was isolated by treatment with proteinase K (Boehringer–Mannheim; 2 mg ml⁻¹ in 0.6% SDS) followed by phenol extraction and dialysis. Restriction enzyme digests (37 °C. 1.5 h) were carried out in 10 mM Tris-HCl, pH 7.5, 10 mM MgCl₂, 10 mM 2-mercaptoethanol, 40 mM NaCl after addition of phage λ DNA (0.5–1 μg per reaction) and the reactions were stopped by heating at 70 °C for 5 min. The samples were applied to 1% w/v agarose gels[23] in 0.04 M Tris-acetate, pH 8.2 for electrophoresis (35 mA, 8 h). Gels were stained with ethidium bromide and photographed under UV light[24] and then either placed in alkali to denature DNA fragments for transfer to cellulose nitrate membrane filters[13] or dried on Whatman 3MM paper for autoradiography. The MW of the fragments of HBV DNA was estimated from their electrophoretic mobility[24] with reference to fragments of λ⁺ DNA in digests with R.*Eco*RI and R.*Hin*dIII included in the same gel[25]. These results are included in Fig. 2 which gives a provisional map of some of the restriction targets relative to each other.

Digestion of HBV DNA with R.*Eco*RI or R.*Bgl*II changed the pattern of the major bands only slightly (Fig. 1 tracks *c*, *f* and *g*) whereas digestion of the DNA with these two enzymes together (track *h*) gave a radioactive fragment, MW 0.75 × 10⁶ as the principal component, with little of the original DNA remaining. This is consistent with the introduction of a single break in circular molecules by either enzyme alone to give intact linear molecules; when the two enzymes acted together two fragments were formed, the smaller containing the entire labelled region. On digestion with R.*Kpn*I (track *i*) almost the whole group of bands was displaced down the gel corresponding with the loss from each of a fragment of MW ∼ 0.4 × 10⁶ but no corresponding radioactive fragment was found. This suggests that HBV DNA may contain at least two targets for R.*Kpn*I located outside the region repaired in the reaction with DNA polymerase and spanning a sequence common to all the molecules. The alternative explanation requiring cleavage of circular molecules at a single target to give linear molecules is less likely in view of the behaviour observed in the R.*Eco*RI and R.*Bgl*II digests. R.*Hae*III furnished a spectrum of fragments with a range of sizes (track *r*) in a pattern broadly similar to that published[5,6]. Digestion of HBV DNA with R.*Ava*I and with

R.*Bam*HI gave several radioactive fragments (tracks *j* and *g*); a major product of the R.*Ava*I digest had a MW of 0.88 × 10⁶, while radioactive fragments of 1.2 × 10⁶ and 1.8 × 10⁶ occurred in the R.*Bam*HI digests; in other digests with R.*Bam*HI smaller fragments were also observed. These results, together with the principal radioactive products found in various digests with pairs of restriction enzymes (for example, R.*Eco*RI and R.*Ava*I, track *a* or *d*) are summarised in Fig. 2. Within the constraints imposed by the dispersity of the HBV DNA preparations used, they provided the approximation of relative positions of restriction targets as shown.

The results imply that, in most molecules, the single-stranded gap repaired by the endogenous DNA polymerase lies within a relatively constant region of the DNA sequence; Landers *et al.*[6] similarly concluded, from an analysis of R.*Hae*III digests after different polymerase reaction times, that DNA repair took place largely within the same one-third to one-half of the total DNA sequence, although initiation of DNA repair could occur at variable sites within this region.

Digests of HBV DNA with R.*Eco*RI or R.*Bgl*II offer the possibility of cloning the entire HBV genome, while major fragments might be cloned from digests with R.*Bam*HI, R.*Kpn*I or R.*Ava*I, or from various double digests. For structural studies of the HBV genome it is desirable to clone the entire DNA molecule, but clones covering a range of fragments could well be more useful for attempts to demonstrate the expression of HBV sequences in *E. coli*. Micromethods based on radioimmunoassay can be applied to individual bacterial colonies or phage plaques for the detection of specific polypeptides[7]. As many eukaryotic genes will not be expressed in prokaryotic cells it appeared desirable to insert HBV DNA fragments within a prokaryotic gene so as to produce a fused polypeptide; for this purpose the R.*Pst* target in the *E. coli* plasmid pBR322 has been used successfully[8].

Cloning of HBV DNA fragments in pBR322

HBV DNA isolated from Dane particles from a single HBsAg positive, HBeAg positive donor (serotype *adyw*) was labelled to a low specific radioactivity with ^{3}H in a repair reaction with the endogenous polymerase in order to facilitate its handling. This preparation was then variously digested with R.*Ecor*RI, R.*Bam*HI, R.*Bgl*II, R.*Kpn*I and R.*Ava*I. Portions of the R.*Eco*RI and R.*Bam*HI digests were used for insertion at the respective sites of appropriately restricted pBR322 DNA by annealing and ligation with T4 DNA ligase. The fragments from the other restriction enzyme digests and the remainder of the R.*Eco*RI and R.*Bam* digests were treated with polynucleotide terminal transferase for addition of 3' oligo (dC) sequences[9]. These fragments were annealed to pBR322 DNA to which oligo (dG) sequences had been attached after cleavage with R.*Pst*. The DNA preparations were then used to transform competent cultures of *E. coli* HB101 and transformants were screened for the acquisition of HBV sequences on the following basis. Cells transformed with recombinants made using the R.*Pst* site in pBR322 retained their resistance to tetracycline, but became sensitive to ampicillin. Cells transformed with recombinants made using the R.*Bam*HI site in pBR322 became sensitive to tetracycline, but retained their resistance to ampicillin, while cells transformed with DNA cloned using the R.*Eco*RI site remained resistant to both antibiotics[10] and were detected by colony hybridisation[11] with ^{32}P-labelled DNA from Dane particles. The presence of HBV DNA sequences in colonies where antibiotic resistance and sensitivities indicated the insertion of additional DNA into the plasmid was confirmed by colony hybridisation.

Characterisation of pBR322–HBV hybrids

Plasmid DNA isolated from cell cultures that had been treated with chloramphenicol to amplify plasmid production[12] was analysed by gel electrophoresis before and after digestion with restriction endonucleases.

DNA fragments from several of the gels were transferred to cellulose nitrate filters for hybridisation[13] with HBV DNA labelled with ^{32}P by the endogenous polymerase reaction. Examples of these results are shown in Fig. 3 and some of the characteristics of the cloned segments are given in Table 1. The hybridisation observed with appropriate fragments does not establish conclusively that the cloned sequences were HBV-specific, for the cloned DNA and ^{32}P-labelled probe had been prepared from the same plasma sample. Thus contaminating non-viral DNA fragments that had been cloned inadvertently would also have been detected if the same non-viral DNA sequence was labelled by the endogenous polymerase; however, in other experiments such ^{32}P-labelled probes hybridised with DNA from HBV-infected liver tisue, but not with DNA from normal human liver. Furthermore, preparations of ^{32}P-labelled DNA made from Dane particles purified from four different blood donors gave the same patterns when hybridised against restricted DNA fragments from the recombinant plasmids, thus strengthening the view that the cloned sequences were in fact HBV DNA.

The results presented in Fig. 3 are examples taken from the large number of recombinant colonies obtained from the transformation experiments and relate to colonies described below.

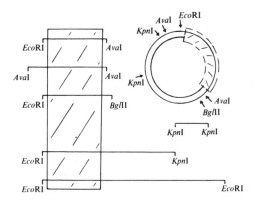

Fig. 2   The size and approximate relative order in the HBV genome of some of the major fragments in various restriction enzyme digests (Fig. 1) of HBV DNA. The site for R.*Eco*RI is taken arbitrarily as a reference point for the circular map. The shaded area indicates the region labelled with ^{32}P in the endogenous repair reaction.

Expression of HBV sequences

A population of the colonies carrying putative recombinant plasmids, selected on the basis of drug resistance and sensitivity characteristics, was screened for the production of HBV antigens. Three test systems were used in the solid phase method of radioimmunoassay described by Broome and Gilbert[7] which uses polyvinyl disks coated with IgG from specific antisera. These disks are placed in contact with cells producing antigen which binds to the IgG surface. When subsequently incubated with ^{125}I-labelled IgG from the same antiserum, the bound antigen retains the label and can be detected readily by autoradiography. The sera used were anti-HBs (human and hyperimmune animal sera), anti-HBc (human sera containing HBsAg and HBeAg and selected on the basis of a high anti-HBc titre) and anti-HBc and anti-HBe together (human sera containing HBsAg and high levels of anti-HBc and anti-HBe).

Clear positive results were obtained in the two test systems containing anti-HBc. Of some 350 colonies tested, 13 gave intense spots on autoradiography with the anti-HBc + anti-HBe antibodies (Fig. 4*a*). Most of these colonies remained strongly positive when re-tested in both the anti-HBc and anti-HBc + anti-HBe assay systems (Fig. 4*b*). None of the clones giving a negative response in the initial screening was subsequently positive, but some clones that were positive initially gave a negative result after subculturing, which probably reflects instability of some of the hybrid plasmids. All of the clones that were positive in the anti-HBc + anti-HBe system were positive when tested with anti-HBc alone, implying that none was producing detectable levels of HBeAg. If lysis of the bacterial colonies with phage λ before radioimmunoassay was omitted, only a very faint outline of the positive colonies was discernible (Fig. 4*c*). This is consistent with the presence of the antigen as a periplasmic polypeptide fused to the major part of β-lactamase (penicillinase) as anticipated. The sensitivity of this assay was shown to be comparable with that attainable with the same reagents in a solid phase microtitre well assay[14] by titration of 10 μl samples (diluted progressively from 1 in 400) of semi-purified HBcAg from human liver[15].

The immunological specificity of the reactions was confirmed by the following observations. Replicate assays with different anti-HBc sera identified the same positively reacting clones. The positive reactions in the HBcAg assay were abolished if the polyvinyl disks were coated with normal human IgG instead of specific anti-HBc IgG, or if the normal human serum used as diluent for the ^{125}I-labelled antibody was replaced by anti-HBc-positive serum from a different donor (by competition with excess unlabelled anti-HBc), or if ^{125}I-anti-HBS replaced ^{125}I-anti-HBc in the assay with the anti-HBc coated disks. Finally,

Fig. 3 Electrophoretic separation in agarose gels of restriction enzyme digests of recombinant plasmids comprising pBR322 and HBV DNA sequences. The examples shown are of plasmids (Table 1) that elicit production of antigenic material that reacts specifically with antibodies to HBcAg (Fig. 4). The left-hand panels show DNA fragments revealed by staining with ethidium bromide and photography under UV light. After photography, gels were soaked in alkali for denaturation of the DNA fragments which were then transferred to cellulose nitrate membrane filters[13] for hybridisation with ^{32}P-labelled HBV DNA and autoradiography; the right hand panels show the corresponding radioautographs. HBV DNA (~0.5 μg) labelled with ^{3}H was prepared from 40 ml clarified blood plasma from a donor with a high titre of HBsAg, serotype *adw* as described in the legend to Fig. 1 and ref. 6. In one experiment 0.5 μg *E. coli* DNA was added as a carrier, but in a second the carrier was omitted. The DNA was divided into six portions for various restriction

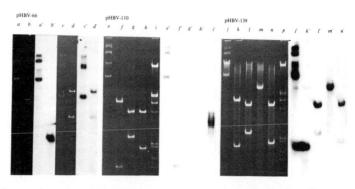

enzyme digests. Portions of R.*Eco*RI and R.*Bam*HI digests were incubated with pBR322 digested with the same enzyme in reactions with T4DNA ligase[26] (1 U ml^{-1}, 10 °C for 3 h followed by storage at 0 °C) in 66 mM Tris-HCl pH 7.2, 10 mM MgCl$_2$, 40 mM NaCl, 0.2 mM EDTA, 0.1 mM ATP, 10 mM 2-mercaptoethanol. The remainder of the R.*Kpn*I digest as well as R.*Kpn*I and R.*Bgl*II digests were used in terminal transferase reactions, but with the exception of the R.*Kpn*I digest the samples were first incubated with phage λ exonuclease (1.5 h at 0 °C in 50 mM Na glycinate, pH 9.5, 5 mM MgCl$_2$, 50 μg ml^{-1} bovine serum albumin (BSA) followed by phenol extraction and recovery of the DNA by ethanol precipitation) to remove the 5′ single-stranded projections left by the restriction enzymes. Poly(dC)sequences were attached to the 3′ termini by incubation with polynucleotide terminal transferase[9] (250 U ml^{-1} for 10–20 min at 27 °C) in 15 μl 100 mM potassium cacodylate, pH 7.0, 1 mM CoCl$_2$, 1 mM dCTP, 50 μg ml^{-1} BSA, and the reactions stopped by addition of excess EDTA. An approximately molar equivalent of pBR322 DNA digested with R.*Pst* and incubated in similar reactions with dGTP instead of dCTP (from Dr J. van den Berg) was then added and the samples diluted to 50 μl for annealing in 50 mM Tris-HCl, pH 7.5, 100 mM NaCl, 5 mM EDTA, heated to 60 °C and gradually cooled to room temperature over a period of about 5 h. Aliquots of the solutions (10 μl) were then incubated with competent cultures (0.1 μl) of *E. coli* HB101 prepared as described by Lederberg and Cohen[27] and incubated overnight at 37 °C on L-agar plates containing tetracycline (20 μg ml^{-1}) or ampicillin (50 μg ml^{-1}). Colonies were subcultured on to Millipore filters supported on appropriate agar plates to test their resistance or sensitivity to the two antibiotics and for colony hybridisation[11] with ^{32}P-labelled HBV DNA from Dane particles. Colonies giving positive hybridisation reactions were grown in liquid culture and then shaken overnight at 37 °C after addition of chloramphenicol (170 μg ml^{-1}) to amplify the number of plasmids within the cells which were then collected by centrifugation and treated with lysozyme and EDTA[12]. The resulting spheroplasts were lysed with Triton X-100 and the plasmid recovered by equilibrium centrifugation in CsCl solution (0.95 g per ml lysate) containing ethidium bromide (200 μg ml^{-1}). The plasmid bands were collected, extracted with propan-1-ol saturated with aqueous CsCl, dialysed, and samples digested with restriction endonucleases as described in the legend to Fig. 1. The results shown are from the following hybrid plasmids: pHBV-66 in tracks *a*, *b*, *c* and *d* in which *a* is the undigested plasmid and *b*, *c* and *d* are digests with R.*Pst*, R.*Kpn*I and R.*Bam*HI, respectively; pHBV-110 in tracks *e*, *f*, *g* and *h*, in which *e* is the undigested plasmid and *f*, *g* and *h* are digests with R.*Pst*, R.*Bam*HI and R.*Bam*HI + R.*Eco*RI, respectively; pHBV-139 in tracks *j*, *k*, *l*, *m* and *n*, in which *j* is the undigested plasmid and *k*, *l*, *m* and *n* are digests with R.*Pst*, R.*Bam*I, R.*Eco*RI, and R.*Bam*HI + R.*Eco*RI, respectively. Tracks *i* and *p* show reference digests of λ° DNA with R.*Eco*RI + R.*Hind*III[25] which also contained ^{32}P-labelled undigested DNA from Dane particles. The primed letters identify the samples on the autoradiograph, made on Kodak X-omat H X-ray film with an intensifying screen, of the cellulose nitrate filters after hybridisation with HBV DNA.

absorption of the radioactive anti-HBc with semipurified HBcAg[15], by overnight incubation at 4 °C followed by centrifugation to remove the excess antigen, abolished the positive result in the radioimmunoassay. Thus the detection of an as yet unidentified HBV-specific antigen by interaction with its specific antibody present in the sera of HBV-infected individuals remains a formal possibility, but it is unlikely, and the results are wholly consistent with the observed activity being that of HBcAg. The HBcAg polypeptide detected in the colonies, however, is unlikely to be identical with that occurring naturally for the cloning experiment was such as to produce polypeptides linked to β-lactamase[8]. However, it has not been established that this is the case and it remains possible that translation of β-lactamase sequences could be terminated and followed by reinitiation, but against this is the occurrence of clones that do not exhibit HBcAg activity, but which contain HBV DNA fragments similar to those that do. The point will be clarified by DNA sequence determination, which is in progress. All of the 13 HBcAg-positive clones had been made using the R.*Pst* site in pBR322, five with fragments from R.*Bam*HI digests of HBV DNA and eight from R.*Kpn*I digests. Not all of the plasmids have been isolated for analysis, but the three illustrated in Fig. 3 were all from cells giving positive reactions for HBcAg. The smallest fragment of HBV DNA in these plasmids (Table 1) was about 0.95 × 10⁶ daltons, but others with an HBV fragment

about half this size also gave positive reactions for HBcAg; values reported[3] for the MW of the naturally occurring HBcAg range from 17,000 to 80,000.

In an equivalent assay for HBsAg, similar, but faint positive reactions were obtained with four clones which are being analysed further. One might expect detection of expression of serological activity to be more difficult with HBsAg than with HBcAg as its protein moiety is markedly hydrophobic and hence

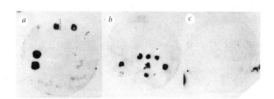

Fig. 4 Autoradiographs showing the detection by radioimmunoassay[7] of bacterial colonies expressing HBcAg. *a*, Four HBcAg-positive colonies amongst 52 colonies examined on one plate. *b*, The result obtained when initially positively reacting colonies, together with a random selection of colonies giving a negative reaction, were subcultured from stock plates and re-tested. *c*, A duplicate of *b* in which lysis of the colonies with phage λ before radioimmunoassay was omitted. Colonies of the bacteria were grown at 37 °C overnight on Millipore filters supported on nutrient plates containing tetracycline (to maintain selection for the plasmid). The cells were lysed by imprinting the filters for a few minutes on a lawn of bacteria confluently lysed with a virulent derivative of phage λ and then incubating further for several hours at 37 °C. Polyvinyl disks coated with anti-HBc + anti-HBe specific human IgG (60 μg ml^{-1}) were placed face down on the colonies, which had obviously lysed, and incubated at 4 °C for 3–4 h. The disks were then removed and washed thoroughly and vigorously to remove the considerable quantity of adherent viscous bacterial debris. Finally the disks were incubated overnight at 4 °C with homologous ^{125}I-labelled IgG (10⁵ c.p.s. per μg, 2 × 10⁴ c.p.s. per ml), washed thoroughly and autoradiographed on Kodak Blue X-ray film exposed for 2 d with an intensifying screen.

Table 1   Some properties of the plasmids shown in Fig. 3

| Hybrid plasmid | MW of fragment excised by R.*Pst* ×10⁻⁶ | Targets for restriction enzymes within the HBV sequences | | |
		R.*Eco*RI	R.*Bam*HI	R.*Ava*I
pHBV-66	1.2	–	+	+
pHBV-110	0.95	–	+	+
pHBV-139	1.16	–	+	+

In all the plasmids, the site for R.*Bam*HI within the HBV sequence is located about 0.7 × 10⁶ daltons from the R.*Pst* site near the R.*Eco*RI site.

tends to remain associated with lipid[16], and some experiments (but not others) suggest that carbohydrate residues on HBsAg glycoprotein may be required for full serological activity[17,18].

Conclusions and further implications

A large number of hybrid DNA molecules comprising pBR322 and various fragments of the HBV genome have been cloned and propagated in *E. coli*. It thus becomes possible to produce HBV DNA in the quantities required for detailed structural and sequence analysis and location of the various coding sequences in the viral genome. Such analysis with DNA from several independent isolates and clones will explain the basis of the heterogeneity found in DNA from Dane particles. The DNA will also be useful for further genetic manipulation related to studies of expression and as a source of a highly radioactively labelled probe for hybridisation experiments for both diagnostic purposes and fundamental studies.

When inserted within a normal coding sequence of the *E. coli* plasmid the DNA from hepatitis B virus, which normally infects only man and apes, can be expressed to give serologically active translation products. This suggests that at least the region of the HBV genome coding for the amino acid sequence necessary for serological activity probably does not contain inserted sequences, or 'introns', which have now been found in a number of eukaryotic and viral genes[19–22], for it is widely believed, although not proved, that *E. coli* cannot process transcripts of these sequences. None of the hybrid plasmids so far examined contains a target for *R.Eco*RI within the HBV sequences. It was thus possible to insert the entire plasmid into a derivative of bacteriophage λ using this restriction site and plaques of the recombinant phage gave positive reactions in the disk radio-immunoassay (results not shown). The HBV sequences may well prove more stable when propagated within the phage genome, especially as a lysogen, and appropriate exploitation of the phage regulatory systems should permit significantly increased yields of the antigens from *E. coli*, raising the possibility of large scale antigen production for diagnostic purposes and development of vaccines. The insertion of HBV DNA into a phage λ derivative has been described very recently by Fritsch *et al.*[28].

We thank Professor B. P. Marmion, Dr R. J. C. Harris and many of the staff at MRE, Porton, for their help, Glynis Leadbetter for radioimmunoassay reagents, Archie Barr and Anthony Keen for HBsAg-positive plasma, Johan van den Berg and Professor Charles Weissmann for reagents and help with terminal transferase reactions, Eric Gowans and Pamela Beattie for electron microscopy, Carol White and Eva Zehelein for technical assistance. This work was assisted in part by a grant from the Scottish Home and Health Department and was supported partly by Biogen S.A. Cloning experiments were performed in CIV conditions as recommended by GMAG; work on DNA from Dane particles was done in a CI laboratory and DNA from cells transformed with hybrid plasmids in a CII laboratory.

Received 22 February; accepted 16 March 1979.

1. Szmuness, W., Harley, E. J., Ikram, H. & Stevens, C. E. in *Viral Hepatitis* (eds Vyas, G. N., Cohen, S. N. and Schmid, R.) (Franklin Institute Press, Philadelphia, 1978).
2. Dane, D. S., Cameron, C. H. & Briggs, M. *Lancet* i, 695–698 (1970).
3. Robinson, W. S. *A. Rev. Microbiol.* 31, 357–377 (1977).
4. Purcell, R. H. & Gerin, J. L. *Am. J. clin. Path.* 70, 159–169 (1978).
5. Summers, J., O'Connell, A. & Millman, I. *Proc. natn. Acad. Sci. U.S.A.* 72, 4597–4601 (1975).
6. Landers, T. A., Greenberg, H. B. & Robinson, W. S. *J. Virol.* 23, 368–376 (1977).
7. Broome, S. & Gilbert, W. *Proc. natn. Acad. Sci. U.S.A.* 75, 2746–2749 (1978).
8. Villa-Komaroff, L. *et al. Proc. natn. Acad. Sci. U.S.A.* 75, 3727–3731 (1978).
9. Roychoudhury, R., Jay, E. & Wu, R. *Nucleic Acids Res.* 3, 101–116 (1976).
10. Bolivar, F. *et al. Gene*, 2, 95–133 (1977).
11. Grunstein, M. & Hogness, D. S. *Proc. natn. Acad. Sci.* 72, 3961–3965 (1975).
12. Clewell, D. B. *J. Bact.* 110, 667–676 (1972).
13. Southern, E. M. *J. molec. Biol.* 98, 503–517 (1975).
14. Purcell, R. H., Wong, D. C., Alter, H. J. & Holland, P. V. *Appl. Microbiol.* 26, 478–484 (1973).
15. Cohen, B. J. & Cossart, Y. E. *J. clin. Path.* 30, 709–713 (1977).
16. Howard, C. R. & Burrell, C. J. *Progr. med. Virol.* 22, 36–103 (1976).
17. Burrell, C. J., Proudfoot, E., Keen, G. A. & Marmion, B. P. *Nature new Biol.* 243, 260–262 (1973).
18. Shiraishi, H., Shirachi, R., Ishida, N. & Sekini, T. *J. gen. Virol.* 38, 363–368 (1978).
19. Jeffreys, A. J. & Flavell, R. A. *Cell* 12, 1097–1108 (1977).
20. Tilghman, S. M. *et al. Proc. natn. Acad. Sci.* 75, 725–729 (1978).
21. Breathnach, R., Mandel, J. L. & Chambon, P. *Nature* 270, 314–319 (1978).
22. Chow, L. T., Gelinas, R. E., Broker, T. R. & Roberts, R. J. *Cell* 12, 1–8 (1977).
23. Hayward, G. S. & Smith, M. G. *J. molec. Biol.* 63, 383–395 (1972).
24. Sharp, P. A., Sugden, W. & Sambrook, J. *Biochemistry* 12, 3055–3063 (1973).
25. Murray, K. & Murray, N. E. *J. molec. Biol.* 98, 551–564 (1975).
26. Weiss, B. *Meth. Enzym.* XXI D, 319 (1971).
27. Lederberg, E. M. & Cohen, S. N. *J. Bact.* 119, 1072–1074 (1974).
28. Fritsch, A., Pourcel, C., Charnay, P. & Tiollais, P., *C. r. hébd. Acad. Sci. Paris* 287, 1453–1456 (1978).

AUTHOR CITATION INDEX

381

SUBJECT INDEX

389

About the Editors

KATHERINE J. DENNISTON is a senior staff fellow in the Laboratory of Molecular Virology, National Cancer Institute, National Institutes of Health. She received the B.A. degree in biology from Mansfield State College and the Ph.D. degree in microbiology from The Pennsylvania State University.

Dr. Denniston was a postdoctoral fellow at The Pennsylvania State University from 1975 to 1976 and at the University of Wisconsin from 1976 to 1978. She is currently studying the structure of DNA obtained from defective interferring particles of herpes simplex virus type 1 using recombinant DNA technology.

LYNN W. ENQUIST is on the staff of the Laboratory of Molecular Virology, National Cancer Insitute, National Institutes of Health. He received the B.S. degree in bacteriology from South Dakota State University and the Ph.D. degree in microbiology from Virginia Commonwealth University.

Dr. Enquist was a postdoctoral fellow at the Roche Institute of Molecular Biology from 1971 to 1973. He was a staff fellow in the Laboratory of Molecular Genetics, National Institute of Child Health and Human Development, National Institutes of Health, from 1973 to 1977. Dr. Enquist is a commissioned officer in the Public Health Service. He is currently studying the mechanisms and control of recombination in DNA viruses.